Nanosensors for Futuristic Smart and Intelligent Healthcare Systems

I0031633

Edited by

Suresh Kaushik

Department of Chemistry
Indian Agricultural Research Institute
New Delhi, India

Vijay Soni

Department of Medicine
Weill Cornell Medicine
New York, NY, USA

Efstathia Skotti

Department of Food Science and Technology
Ionian University
Argostoli, Greece

CRC Press
Taylor & Francis Group
Boca Raton London New York

CRC Press is an imprint of the
Taylor & Francis Group, an **Informa** business

A SCIENCE PUBLISHERS BOOK

First edition published 2022
by CRC Press
6000 Broken Sound Parkway NW, Suite 300, Boca Raton, FL 33487-2742

and by CRC Press
4 Park Square, Milton Park, Abingdon, Oxon, OX14 4RN

Library of Congress Cataloging-in-Publication Data (applied for)

ISBN: 978-0-367-55434-7 (hbk)
ISBN: 978-0-367-55436-1 (pbk)
ISBN: 978-1-003-09353-4 (ebk)

DOI: 10.1201/9781003093534

Typeset in Times New Roman
by Radiant Productions

Preface

Healthcare sector is probably the most benefited from the application of nanotechnology. The concept of nanotechnology was proposed in 1965 by Richard Feynman, a physicist and Nobel Laureate. The main idea behind nanotechnology was to exploit the advantages of miniaturization of materials and explore the future of creating compelling and tinier devices. The standard working range of nanotechnology is 1 to 100 nanometers. The matter changes its behavior as its size reduced to nanoscale due to quantum size effects. One of the early applications of nanotechnology is in the field of nanosensors. A nanosensor is not necessarily a device merely reduced in size to few nanometers, but a device that makes use of the unique properties of nanomaterials and nanoparticles to detect and measure new types of events in the nanoscale. A typical sensor has three main modules: a receptor a transducer and a detector with a digital output. Hence, nanosensors are sensing devices with at least one of their sensing dimensions up to 100 nm. The nanostructure materials used in production of nanosensors include nanoscale wires, carbon nanotubes, thin films, nanoparticles and polymer nanomaterials.

In order to provide better-quality healthcare, it is very important that high standards of health care management are achieved by making timely decisions based on rapid diagnostics, smart data analysis and informatics analysis. Smart nanosensors are emerging as efficient and affordable analytical diagnostics tools for early-stage disease detection. Nanosensor can detect analytes or biomarkers in small quantity of samples such as blood, saliva, tears, sweat. A biological marker or biomarker is a characteristic that is objectively measured and evaluated as an indicator of normal biological processes, pathogenic processes or pharmacologic responses to a therapeutic intervention. Emerging nanomaterial science and flexible electronics have led to wearable biophysical nanosensors that are capable of monitoring human activities, body motions, and electrophysiological signals such as electrocephalogram and electrocardiogram. Wearable biochemical nanosensors are emerging for noninvasive detection of molecular-level indicators such as electrolytes and metabolites from biofluids. Nanosensors are widely used to detect antibodies, antigens or nucleic acids in crude samples such as saliva, sputum and blood based upon colorimetric, fluorescent or electrochemical detection approaches. Nanobiosensor offers many advantages such as being affordable, sensitive, specific, user-friendly, rapid and robust, equipment-free and deliverable to end user.

Wearable devices such as activity trackers and smart watches can provide unique insights into our health and well-being. During the coronavirus disease 2019 (COVID-19) pandemic, the potential of wearable health devices has become iv *Nanosensors for Futuristic Smart and Intelligent Healthcare Systems*

increasingly apparent. With the advances in point-of-care testing, chip-based and paper-based nanobiosensors have been developed for rapid diagnosis of infectious diseases, ensuring fast detection of analytes near to the patients facilitating a better disease diagnosis, monitoring and management. Hence, nanobiosensors can be a reliable and cost-effective way to detect specific pathogen in point-of-care settings. Wearable nanobiosensors have potential to provide continuous real-time physiological information via dynamic, noninvasive measurements of biochemical markers in biofluids, such as sweat, tears, saliva and interstitial fluids. Wearable sensors have received much attention since the arrival of smartphones and other mobile devices. Wearable monitoring platforms can lead insights into dynamic biochemical processes in biofluids by enabling continuous, real-time monitoring of biomarkers. Such real-time monitoring can provide information on wellness and health. As the disease can be diagnosed at an early stage, quick medical decision can be taken to start early treatment. Numerous potential point-of-care devices have been developed in recent years which are paving the way to next-generation point-of-care testing.

Significant advances in wireless communication and networking technologies have paved the way to envisage and design innovative healthcare services. Various wireless technologies have been used to transmit data within a wearable body area network. Wearable sensor nodes are deployed inside a wearable body area network to monitor physiological signals. The Internet of NanoThing, the interconnection of nanoscale devices to the existing communication networks, has the potential to bring a revolutionizing advancement in the field of real-time monitoring of healthcare services. A combination of multiplexed biosensing, microfluidic sampling and transport systems have been integrated, miniaturized and combined with flexible materials for improved wearability and ease of operation.

The overall theme of this book is to compile a comprehensive treatise on nanosensors for healthcare system. Specifically, we address the enthusiasm that nanosensors technology including wearable and wireless tools have provided to monitor health status in real-time and diagnosis of infectious disease particularly keeping the in view the current situation of pandemic COVID-19 disease worldwide, which might change the behavior of people in future. With this view, we have designed the book with two Sections-I and II, explaining the fundamentals and applied technologies in nanosensor based-medical devices used in healthcare systems. Under Section-I from Chapters 1 to 9, we have attempted to focus on the basic concept of nanosensor technologies applied in wearable and implantable medical devices used in disease diagnosis and monitoring health status, while sensing paradigms, wireless, array and microfluidics technologies applied in implantable and wearable devices for real-time monitoring health status are discussed under Section-II from Chapters 10 to 17. In Chapter 1, we provided the basics and recent advances of smart nanosensor technology in healthcare sector. Chapter 2 discusses about the development of nanosensor technology in biomarkers detection used for disease diagnosis, while Chapter 3 addresses infectious disease diagnosis including COVID-19 disease using innovative nanosensors. Chapters 4 to 9 are centered on recent advances in wearable and implantable devices providing a glimpse into the world of wearables. In Chapter 4, wearable devices for real-time disease monitoring are covered to provide insight into the point of care treatment (POCT), specially during pandemic coronavirus disease period. Nanocarbon-based sensor are discussed for wearable health monitoring parameters such as EEG, ECG, EMG in Chapter 5. Chapter 6 provides a basic background of electrochemical wearable sensors for applications in biomedical and healthcare system. Smart textile-based wearables nanosensors are addressed in Chapter 7, while emerging topic of electronic-skin (E-skin) is introduced in Chapter 8. Non-invasive and implantable wearable and dermal nanosystems applied for healthcare are covered in Chapter 9. Chapters 10 and 11 discuss the use of nanogenerator-based self-powered sensors in healthcare system and provide the recent development in this technology. Chapter 12 describes minimally invasive microneedle nanosensors focusing on COVID-19 disease. Wireless nanosensors used in healthcare system for monitoring health status in real-time are described in Chapter 13. Chapter 14 explores the potential role of nanosensors in Internet of Medical Things (IoMT). In Chapter 15, microfluidics chip technology for disease diagnosis is discussed using the dielectrophoresis technique. Chapter 16 provides the recent development in nanosensor array technology for multiplexed sensing in real-time monitoring of health status. The book will not be complete without a discussion on the use of artificial intelligence in healthcare systems (Chapter 17), as these are becoming fundamental to innovative approaches for smart and intelligence medical diagnosis and monitoring health status in real-time in futuristic healthcare sector.

We believe that this book will be very useful and valuable to researchers, scientists, engineers, technocrats working in development of nanosensors for smart healthcare systems. Fast, portable, new, and easy-to-use devices that involves nanosensors can be modified according to the information presented in this book. The completion of this book could not have been possible without help, inspiration, and encouragement from many people including our families. Finally, we would like to express our sincere gratitude to the leading authors, who accepted our invitation to join us and dedicated their valuable time and efforts to guarantee the success of the book.

Suresh Kaushik
Vijay Soni
Efstathia Skotti

Contents

List of Abbreviations

HRS	:	Hyper-Rayleigh scattering
QDNB	:	Quantum dot-nanobeads
HIV	:	Human Immunodeficiency Virus
BBB	:	blood-brain barrier
FRET	:	Fluorescence resonance energy transfer
CNT	:	Carbon nanotube
QD	:	Quantum dot
FITC	:	Fluorescein isothiocyanate
CNTs	:	Carbon nanotubes
SWCNTs	:	single-wall carbon nanotubes
MWCN	:	multi-wall carbon nanotubes
GO	:	graphene oxide
MNs	:	Magnetic nanoparticles
DMR	:	diagnostic magnetic resonance
CLIO	:	Cross-linked Iron Oxide Nanoparticles,
SQUID	:	Superconducting quantum interference devices
ELISA	:	Enzyme-linked immunosorbent assay
EIA	:	enzyme immunoassay
MRI	:	Magnetic resonance imaging
PET	:	Positron emission tomography
CT	:	Computed tomography
AgNPs	:	Silver nanoparticles
AuNPs	:	Gold nanoparticles
SPR	:	surface plasmon resonance
POCT	:	Point-of-care technology
FDA	:	Food and Drug Administration
t-TENG	:	textile tri-boelectric nanogenerators
PPG	:	Photoplethysmogram
e-skin	:	electronic skin
R2R	:	roll-to-roll
NPs	:	nanoparticles
CNT	:	carbon nanotubes
NW	:	Nanowires
PEDOT	:	polyethylenedioxythiophene
PPy	:	polypyrrole
PDES	:	polymerizable deep eutectic solvent

PANi	:	polyaniline
BTO	:	barium titanate
TENG	:	PDMS-based triboelectric nanogenerator
ECG	:	electrocardiogram
EEG	:	electroencephalogram
EMG	:	electromyogram
GO	:	graphene oxide
rGO	:	reduced graphene oxide
SNR	:	signal-to-noise ratio
E-textiles	:	electronic textiles
PTFE	:	polytetrafluoroethylene
SBS	:	poly(styrene-block-butadienstyrene)
PPTA	:	poly(p-phenylene terephthalamide)
PEDOT:PSS	:	polyurethane (PU)/poly(3,4-ethylenedioxythiophene) polystyrene sulfonate fibers,
ICPs	:	Intrinsically Conducting Polymers
PPy	:	polypyrrole
LCE	:	liquid crystal elastomer fiber
OSAHS	:	obstructive sleep apnea-hypopnea syndrome
MOF	:	metal-organic framework
IoT	:	Internet of Things
EMG	:	Electromyography
PDCA	:	2,6-pyridine dicarboxamide
PDMS	:	polydimethylsiloxane
PVA	:	poly(vinyl alcohol)
PLA	:	polylactic acid
PG	:	polyethylene glycol
PU	:	polyurethane
ISFET	:	ion-sensitive field-effect transistor
FISA	:	flexible integrated sensor array
FPCB	:	flexible printed circuit board
ISE	:	ion-selective electrodes
HMI	:	Human machine interface
CdSSe	:	Cadmium Sulphoselenide
MGA	:	Modified graphene aerogel
MWNTs/PDMS	:	multiwalled carbon nanotubes/poly(dimethyl siloxane)
DED	:	dry eye disease
PB	:	Prussian-Blue
PPD	:	poly-orthophenylene diamine
LOx	:	lactate-oxidase
CA	:	cellulose acetate
GOx	:	glucose oxidase
GRVs	:	glucose-responsive vesicles
HS-HA	:	hypoxia-sensitive hyaluronic acid

Section I

Sensing Paradigms of Wearables and Implantable Devices for Medical Diagnostics

CHAPTER 1

Smart Nanosensors in Healthcare
Recent Developments and Applications

Sneh Lata Gupta[1] and *Srijani Basu*[2,*]

1. Introduction

Nanotechnology is the study and application of matter that is nanometers in diameters (10^{-9} meter). As a result, nanostructures are quite alike biological molecules, which are nanostructures similar in size to biologic molecules such as proteins. Nanostructures can be made from a variety of materials including carbohydrates (sugars), polymers, or lipids, and have a variety of functional and physical characteristics. This structural flexibility provides room for development of nuanced miniaturized industrial, healthcare, and artificial assemblies such as DNA aggregates, nanorods or tubes, lipid vesicles, and dendritic polymers (Gnach et al. 2015, Bayda et al. 2019).

Nanotechnology has revolutionized the health care system. Some nanotechnology-based new innovations include the development of novel and sensitive diagnostic kits, sensors with improved signals, efficient drug delivery systems, implants, and much more. Nanosensors provide non-invasive, simple, and user-friendly early detection methods, which is the need of the hour, especially in a pandemic world. Nano based structure and devices are used for both early detection and treatment for further prevention of the disease. Nanomedicines are providing future hopes with nanorobots to better fight against human illnesses. Nanosensor based devices are used as sensing tools to administer analytes or diagnose metabolites in the body of human beings. Due to their small size, nanoparticles have the properties of bioavailability as they can easily enter into the capillary of the blood system and

[1] National Institute of Immunology, Aruna Asaf Ali Marg, New Delhi, India, 110067.
[2] Departments of Medicine, Weill Cornell Medicine, New York, NY, USA.
* Corresponding author: srb2003@med.cornell.edu

can also be taken up by cells of interest so the drug delivery can be directed and because of their small size, they are not recognized by the immune system and are biocompatible. Some of the nanoparticles can be used to track intracellular trafficking due to their specific electrical and/or optical characteristics (Maysinger 2007, Oyelere et al. 2007, Chen et al. 2008).

There are two types of nanosensors based on the different sensing mechanisms—chemical and mechanical nanosensors as explained in Table 1 (Saffioti et al. 2020).

Table 1. Description of the two types of nanosensors.

Chemical Nanosensors	Mechanical Nanosensors
Upon detection of analyte, there will be change in electrical conductivity that can be measured using nanomaterials.	Upon detection of analyte, nanomaterials are physically manipulated due to change in their electrical conductivity and this physical change provides a signal to be detected.
Example: Nanotubes and nanowires.	Example: Carbon nanotube (CNT) based fluidics shear stress mechanical sensors and cantilever-based sensors.

Saffioti et al. 2020

1.1 Nano-biosensors in Healthcare

For diagnostic devices that use the biological sample and for use in medical sensing apparatus, there is a preferential need for the development of rapid and sensitive detection. For detection, we utilize specific biomarkers such as antigen, cytokine, enzymes, or metabolites. Biological samples used for diagnosis include peripheral blood, tissue biopsies, urine, or saliva. Existing techniques of diagnosis are ELISA (Enzyme-linked immunosorbent assay), flow cytometry-based on multiple beads, electrochemical detection, and usage of semiconductor which measures conductance or electrical measurement based on detection of the change in resonant frequency. But all the above-mentioned techniques have some drawbacks such as some of these techniques are not high throughput and time consuming. Further electrochemical detection has poor sensitivity and in the case of electrical measurement, changes in resonant frequency are affected by changes in the viscosity of the medium. So, nano-sensors or nano biosensors are in utmost need as they can provide rapid and sensitive, high throughput sample analysis (Jackson and Mahmood 1994, Anderson et al. 2000).

A nano-biosensor includes four major components as shown in Figure 1 (Noah and Ndangili 2019):

1. Analyte in sample detection: DNA/RNA, ion, metabolite, protein, sugar, enzyme
2. Test Sample: Blood, saliva, tissue biopsies, urine
3. Signal transducers: Measurement of pH change by electrode, detection of electroactive substance by electrode used in electrochemical reaction, light by photon counter in optical devices, mass change by piezoelectric device, heat by thermometric devices
4. Detector: To measure signal such as digital or analog; for example, flow response or batch response curve in time

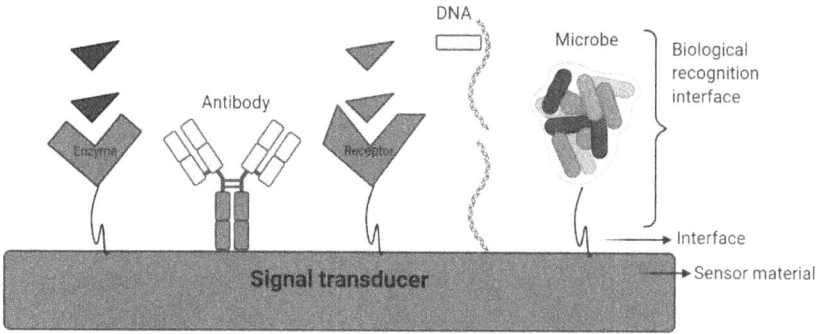

Fig. 1. A schematic representation of the various parts of a nano biosensor, which includes a bio-recognition element, transducer, analyte, and detector.

2. Nanosensors in Medical and Biological Applications

Nanosensors have been extensively used in the healthcare system, some of which include detection and diagnosis of infectious and non-infectious disease. In addition to that, it is tremendously used in drug delivery and gene delivery in a precise manner as target-oriented delivery is possible by various nanoparticles. Detection of metal or ion is useful in diagnostic purpose. There are many physical laws using nanoscale medicine and in analytical device that has been explored in recent years and have been summarized in Table 2.

Table 2. Advantage of having nanoscale size explored in health care.

Qualitative Parameters	Advantage in Medicine
Large surface to volume ratio	It allows nanoparticles to have higher interaction with biomolecules in a quick time. This helps in designing nano based sensor to be sensitive and faster in their response.
Nano scale (ultra-small size)	Miniaturization of analytical device and easily portable at point of care site. Due to its small size, nanosensors are easy to implant in body and less invasive in nature.
Sensitive detection	Useful for biological sample detection, e.g., using biopsies.
Viscosity nature	Useful in preparation of more complex nano fluidics.

This chapter describes the role of nanosensors in many areas of healthcare and medicine below:

2.1 Microbial Detection using Nanosensor

Single bacterium can be detected in 20 min time by *in situ* assay using fluorescence bio conjugated nanoparticles. For example, detection of Salmonella, an enteric pathogen, by using variation in silicon nitride cantilever surface variation (Kaittanis et al. 2010). Using nanowire device, a single virus can be detected such as influenza (Cheng et al. 2014).

2.2 Infectious Diseases

Infectious diseases are caused by bacteria, virus or fungi and are contagious in nature. Most microbes can spread quickly due to their fast multiplication and unpredictable mutation rate. This can lead to high morbidity and mortality and global spread due to better connectivity and higher global travelling. Rapid, accurate and inexpensive diagnostic devices and kits are required, especially for developing countries, which are user-friendly and robust in nature. The most popular and conventional technique that has been used so far include polymerase chain reaction (PCR), *in vitro* culture technique and differential staining followed by microscopy or serological test for antibodies for detection and diagnosis. But all these techniques are either time consuming or require skilled labor (Li et al. 2014, Wilson et al. 2014, Bates and Zumla 2015, Golding et al. 2016).

Nano diagnostic devices provide a novel and inexpensive method of efficient early detection of infectious disease pathogen load. One such example is the use of QD-nanobeads (QDNB) for amplifying fluorescent signal in dot blot assay to detect Hepatitis B Surface antigen at picogram amounts in one step leading to improved sensitivity and faster results (Zhang et al. 2014). Metallic nanoparticles used in diagnostic devices also help with rapid results due to their interactions with electromagnetic radiation. By interacting with radiations, they can absorb it. This property can be utilized in colorimetric assay. This principle was used to develop various assays as many molecules such as antibodies, pathogen specific antigen or enzymes are conjugated with gold nanoparticles to design electrochemical electrode, signal amplifiers and/or probes (Wang et al. 2017). Hyper-Rayleigh scattering (HRS) spectroscopy is based on scattering effect and in that gold nanorods are used to detect single-base-mismatch in HIV virus DNA or hepatitis C virus ssRNA in a fast and efficient manner (Darbha et al. 2008, Griffin et al. 2009). Usage of nanoprobes in magnetic barcoding to detect *Mycobacterium tuberculosis* (MTB) has also been reported. It is PCR based amplification of mycobacterial genes that were labelled using magnetic nanoprobes. It's a high throughput platform sensitive and low cost per sample technique to detect *Mycobacterium* and its drug resistant variants in patient sputum samples (Liong et al. 2013). Similarly, the same testing assay is used to detect other infectious pathogens such as *Staphylococcus aureus* and its resistant strain and also *Klebsiella pneumoniae*. To detect bacterial specific genes such as wcaG, mecA and fnbA, they used magnetic particles and quantum dots using fluorescent nanoparticle. These methods are quite sensitive and reliable to detect pathogens in a low-cost manner (Cihalova et al. 2017). Specific examples of each infectious disease are discussed below.

2.2.1 Covid-19

Coronavirus disease-2019 is caused by coronavirus, a single stranded RNA virus. In March 2019, World Health Organization declared Covid-19 as a pandemic. Nanotechnology can provide an excellent way in the area of spreading the infection, diagnostic as well as therapeutics. It has been reported that coronavirus does not survive on copper (Cu) surfaces and is non-functional at high temperature. In the preparation of mask, selection of fabrics along with nanosystems that are heat and/

Table 3. Usage of nanotechnology in the era of Covid-19 pandemic.

Purpose	Requirement	Nanotechnology Intervention
Prevention of virus spreads	Effective Mask	Combined usage of suitable fabrics + heat/light sensitive nanosystem.
Diagnostic	Portable kit at point of care site	Usage of artificial intelligence and Internet of things (IOT) to make smart device for large scale diagnosis.
Therapeutics	Nanomedicine	Requirement of biocompatible nanoparticles, which can be delivered in a target specific manner to combat virus.

Paliwal et al. 2020

or light sensitive is a good strategy to prevent the spread of virus infection (Paliwal et al. 2020). The use of nanosensor for diagnostic and therapeutic purposes has been shown in Table 3. A recent report on usage of microneedle array (MNA) to combat corona virus is shown, in which recombinant corona virus-based vaccine is shown to have antigen specific response. Antibody is generated in this case after two weeks of immunization (Kim et al. 2020).

2.2.2 Malaria

Usage of biomarkers or specific metabolites in specifically blood, urine, or saliva for malaria diagnosis is an important aspect to combat or eradicate malaria in endemic regions. Hemozoin is an important biomarker for early diagnosis (Hede et al. 2018, Obisesan et al. 2019). This biomarker is conjugated with paramagnetic nanoparticles present in parasites. Hemozoin, which is similar to β-hematin, is important in sensor devices for malarial diagnosis, which gives stable and reliable results. Another device uses metallic nanoparticles based on CuO, Al_2O_3, and Fe_2O_3 to develop an electrochemical sensor to detect β-hematin for malaria diagnosis. β-hematin present on solution gets reduced on an electrochemical rod made up of nanoparticle-based electrode (Obisesan et al. 2019).

2.2.3 Human Immunodeficiency Virus (HIV)

HIV is a viral disease and a major cause of health problems worldwide. Detection of the viral load remains challenging in the early phase of the disease. Nanosensors based on nanoplasmonic particles are immobilized by conjugation with antibodies to detect different viral subtypes that originated during evolution in a reliable manner. The signal is measured by capturing the virus on a nanoparticle surface coated with antibodies. In another report, modified sandwich assay was used to detect p24 HIV capsid antigen from human serum using gold nanoparticles and optoplasmonic transduction method for sensitive and specific detection. Using this method, they were able to detect as low as 10–17 g/ml, which is equivalent to one virion/10 ml of plasma (Inci et al. 2013, Kosaka et al. 2017).

Graphene is also used in the detection of HIV and for the development of sensors. In one of the reports, they combined antibodies on graphenes attached with the amine functional group to detect HIV and associated cardiovascular disease and rheumatoid arthritis (RA). In their assay, they used anti-p24, anti-cardiac troponin 1 (anti-cTn1), and anti-cyclic citrullinated peptide (anti-CCP) for detection of HIV, cardiovascular diseases, and RA, respectively. To visualize the signal in this assay, it can be combined with microscopy (Islam et al. 2019). Another report used a diagnostic chip consisting of a magnetic nanosensor and magnetic nanoparticles to detect HIV from body fluids such as blood plasma and saliva. This chip has 80 different analytes to detect HIV antigen in a highly sensitive and non-invasive way (Ng et al. 2019).

2.2.4 Bilharzia or Schistosomiasis

Mostly, cases of infection of this disease are found in the African population. This is a neglected tropical disease (Ajibola et al. 2018). The conventional method to detect this disease is to use antibodies against soluble egg antigen (specific bilharzia antigen). Such methods are less sensitive. Nanotechnology-based detection devices offer both simple and sensitive tools for diagnosis such as nano strips. Nano strips consist of gold nanoparticles conjugated with antibodies and have the potential to diagnose bilharzia antigen, which is the soluble egg protein. This type of nanosensor uses cyclic voltammetry and has been shown to positively correlate the concentration of soluble egg antigen to the current signal. These are more sensitive assays that can detect the range of ng/ml concentration of soluble egg protein antigen. There is another novel immunosensor screen printed device that is useful in *Schistosoma mansoni* antibodies detection. In this device, the soluble worm antigen is fixed onto the nanocarbon used in the screen-printed electrodes by glutaraldehyde chitosan-based cross-linking. Detection is based on cyclic as well as differential voltammetry. This method can also be used to detect nano level *Schistosoma mansoni* antibody at point of care diagnosis (Noah and Ndangili 2019).

2.2.5 Tuberculosis (TB)

TB is an infectious disease. A major problem with the treatment for this disease is the long duration of treatment and development of multiple drug-resistant strains. Advancement in drug delivery methods such as a nanoparticle-based encapsulated drug is important to reduce the duration and propagation of multiple drug-resistant strains (Brambilla et al. 2011).

2.3 Non-Infectious Diseases

Nanotechnology-based application is utilized in non-infectious diseases as well. Specific examples of each non-infectious disease are discussed below.

2.3.1 Neurodegenerative Disease

Given their small size, nanosensors have the ability to cross the blood-brain barrier (BBB) via endocytosis or transcytosis or both and so are useful in drug delivery avenue in neurodegenerative disorder therapeutics. Drug delivery carriers such as nano-based dendrimers, gels, emulsion, liposome, suspension, or polymeric particles can be used. Site-directed BBB delivery ensures that there will be lower neurotoxicity during treatment. For example, Alzheimer's disease is the most common cause of dementia in elderly people. Development of nanoparticles that can enter specifically into endothelial cells in the brain capillary help in early diagnosis and then treatment of this disease. Nanoparticles conjugated with circulating beta-amyloid have shown some promising results. Nanoparticle-based barcodes, sensors, and scanning tunnel microscopy play an important role in the detection of amyloid-beta 1–42 deposits (Brambilla et al. 2011).

2.3.2 Diabetes

Millions of people are suffering from diabetes worldwide and it has no cure so far. Therefore, monitoring the blood glucose levels is an important measure to reduce multi-organ complications. Current biosensors to measure blood glucose levels are based on electrochemical enzymatic reactions that measure glucose levels with accuracy and rapidity. However, periodic/discrete sampling and active patient participation, i.e., finger pricks to draw blood, is a prerequisite in this method (Pickup et al. 2008). To overcome this problem, advancement in detection using nanosensor based products is required. Such nanosensor based devices can monitor blood glucose level continuously. The benefits of detecting continuous blood glucose levels are a better measurement to understand the trend in blood glucose fluctuations over real-time. These devices can monitor blood glucose even during sleep. This is important as blood glucose levels can reach dangerously low levels during sleep. Using nanomaterial and nanosensors, we have improved detection of blood glucose levels by many folds (Pickup et al. 2008, Cash and Clark 2010).

Further, it provides several other advantages such as larger surface area, improved conductivity from enzyme to electrode, and space for additional catalytic steps. A modification in enzymatic electrode method-based detection is the use of carbon nanotube (CNT) nanomaterial, which gives better conductivity and larger surface area. Porous nanofiber can also improve detection as substrate glucose oxidase is loaded onto it (Zhu et al. 2010). Usage of graphene-based glucose detection composed of graphene and chitosan is quite sensitive with detection limits of 0.3603 mg/dl. Several other types of nanosensors used for measurement of blood glucose levels include FRET-based nanosensors and quantum dots based nanosensors.

Fluorescence resonance energy transfer (FRET) based nanosensor: It is a smart tattoo of skin as the fluorescent-based sensor is implanted in the skin for continuous measurement of blood glucose levels. Detection from these tattoos is temporary and measurements are made once a week or month and the tattoo degrades soon after. The sensor detects blood glucose by changing its fluorescence properties. An example of polymeric FRET nanosensor is boronic acid derivatives for blood glucose detection. They detect blood glucose via FRET mechanism. The principle behind detection via FRET is based on distance. In the absence of glucose binding, there will be no signal generation because the nanospheres are close together allowing for efficient energy transfer by FRET mechanism between fluorescence nanosensors. However, if glucose is bound to boronic acid derivatives, then it will cause swelling in boronic acid derivatives that lead to an increase in distance and no FRET condition. Boronic acid derivatives such as conjugation with hydrophobic group dye such as alizarine

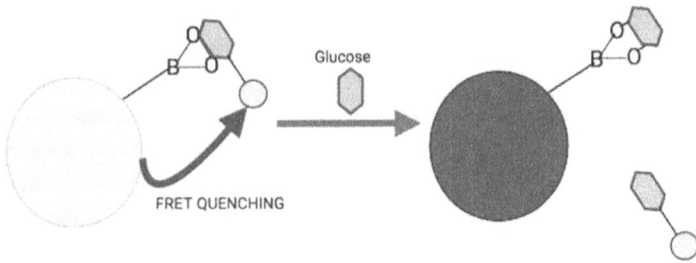

Fig. 2. Glucose sensors made by attaching a recognition element to the quantum dot and a quencher to the glucose molecule. In the absence of the glucose moiety, these molecules are bound together, quenching the QD fluorescence. Glucose displaces the quencher and increases fluorescence.

are made, which improves signal and detection measurement in glucose monitoring (Zenkl et al. 2008, Billingsley et al. 2010, Cash and Clark 2010).

Quantum dots based nanosensors: Quantum dot-based nanosensors have brilliant optical properties for use in glucose sensors such as minimal photo-bleaching and narrow fluorescence peaks. The QDs don't interact with glucose themselves and so have no inherent ability for recognition and must be attached to the recognition element for efficacious implementation (Cash and Clark 2010).

2.3.3 Cancer

Early diagnosis of cancer is important to treat the patient. Biosensors based on silicon nanochannel are currently being developed for the diagnosis of breast cancer and they offer multiple advantages including small size and high surface to volume ratio and excellent electrical properties. Similarly, in ovarian cancer single-walled carbon nanotube have an excellent property for optical detection and can detect biomarkers such as human epididymis protein 4 (Williams et al. 2018, Noah and Ndangili 2019). Imaging of tumor is also an important aspect for early detection. Nanoparticles consist of metal oxide coated with antibodies specific for receptors on cancer cells, which are expressed at higher levels compared to healthy cells and can be detected on MRI or CT scans due to high signal production. This is an example where nanotechnology helps in the identification of cancer for its type or stage of progression in a reliable manner (Rong et al. 2019). To diagnose prostate cancer, magnetic nanoparticles are used that are loaded with specific antibodies for prostate-specific antigen (PSA) (Gholami et al. 2020). Quantum dots coupled with breast cancer-related biomarker such as Her2 linked with the fluorescent marker are useful in high-resolution immunofluorescent labeling. There are various types of nano-based structure which are extensively used in medical purposes as described in Table 4 and Figure 3.

Nanosensor based analytical devices, which are in use to detect metabolites in various diseases, are listed in Table 5.

Table 4. Various types of nano based structures and their use in medicine/healthcare.

Structure	Size	Usage
Carbon nano tube	0.5–3 nm in diameter X 20–1000 nm length	Detection of protein biomarker and DNA.
Dendrimers	< 10 nm	Useful in contrast imaging and controlled drug delivery.
Nanocrystals	2–9.5 nm	Useful for poorly soluble drug formulation and biomarker labelling such as Her2 in breast cancer.
Nanoparticles	100–1000 nm	Useful in imaging such as MRI or ultrasound. Targeted drug delivery, permeation enhancer and reporters for different process such as angiogenesis, apoptosis, etc.
Quantum dots	2–9.5 nm	Cell, tissue or tumor visualization by optical detection.

Nahar et al. 2006

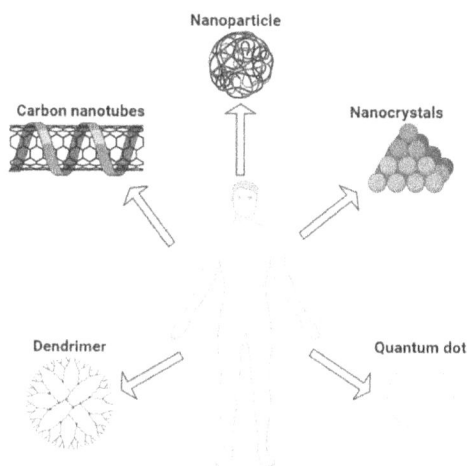

Fig. 3. Shows the various kinds of nanosensors and their use in the field of medicine and healthcare.

Table 5. Usage of nanosensor in healthcare diagnostics.

Disease	Detection
SARS (severe acute respiratory syndrome)	dsDNA and nucleocapsid protein
Zika virus detection	Zika specific antigen driven antibody detection
Clostridium Botulinum toxin	Botulinum toxin type A mediated antibody detection
Heavy metal ion toxicity such as Hg, Pb, Cu, Zn, etc.	Chelation based detection
Breast Cancer	Antibody mediated detection of breast cancer specific c-erbB-2 in saliva
Prostate cancer	Prostate specific antigen (PSA) mediated antibody detection

Jackson and Mahmood 1994, Anderson et al. 2000

2.4 Nanosensors in Biopsy

Taking a biopsy from brain tissue to test whether a tumor is benign or malignant in nature is important for further treatment. This is also fraught with many difficulties and nanotechnology provides an alternative that is non-invasive in nature. Nano-patterned based endoscopic pen is inserted inside brain tissue and it can collect proteins and cells from the specific site of the brain while brain tissue remains intact (Zottlel et al. 2019).

2.5 Nanoshells

To destroy cancerous cells locally instead of with chemotherapy, nanoshells can be used. Nanoshells absorb a different wavelength of light such as near-infrared light and convert it into heat (hyperthermic condition) inside the cancerous tissue and destroy it without interfering with the healthy organ/tissue and have fewer side effects. Similarly, magnetic nanoparticles can be used in Magnetic Resonance Imaging (MRI) as well as for tumor destruction by the generation of heat (Morton et al. 2010).

2.6 Implants

Nanomaterial based medical implants have been done successfully in various areas such as in the area of biocompatible orthopedic patients. This gives new hope for both soft tissue implant and hard tissue implant. Other applications include material used in bone substitutes or dental restoratives. Nanotechnologies provide a new tool such as artificial *in vitro* stimulated cells that can be used to replace or repair damaged tissue, and are therefore important in tissue engineering. This has the potential to revolutionize organ transplantation or implant organ artificially (Tasciotti et al. 2016, Janjic and Gorantla 2019, Yao and Martins 2020).

2.7 Drug Delivery

State-of-the-art methods of nanomedicine drug delivery have the potential to revolutionize the diagnosis, prevention, and have therapeutic applications for a variety of diseases. Replacing large size drug particles with nanomaterial-based drug delivery materials provides more directed drug delivery to the targeted cells. The obvious benefits of these are less toxicity to other non-targeted tissues and fewer side effects and more availability to targeted ones. For drug delivery, either nanomaterial or nano-sized liposomes are used. There are several advantages of using nano-based medicine such as enhanced permeability to epithelial and endothelial tissues. Drugs that are not water soluble can also be delivered using nanosensors and accumulation and long-term retention can be achieved in desired host tissues to provide more efficient drug delivery.

The advantage of nano-sensor based delivery can be utilized in tumor tissues. Due to the fast-growing nature of the tumor tissues, more nanomedicine gets accumulated inside these tissues. siRNA or miRNA using nanoparticle-based intracellular drug delivery is a new era of precise drug delivery. Dendrimers, liposomes and micelles

Table 6. Summary of the types of nanoparticles used in drug delivery.

Types of Nanoparticle	Usage in Drug Delivery
Charged nanoparticle	Help in targeted delivery and higher loading capacity.
Gold nanoparticle	Ease of synthesis and inert/nontoxic gold core.
Solid lipid nanoparticle (SLN)	Can incorporate both hydrophilic and hydrophobic type of drug in combination in higher amount and in stable manner.
Magnetic nanoparticle	Advantage in bioimaging and tracking.
Nanosuspensions	Oral drug delivery for poor soluble drug.
Micellar Nanoparticles	Help in hydrophobic drug delivery.
Mesoporous Silica nanoparticle	Can be endocytosed by mammalian cells and help in precise drug delivery.

nanostructures are used for targeted drug delivery, which contains drugs that are poorly soluble and have less absorption ability in targeted cells. Nanoparticles based drug delivery is shown in Table 6.

There are two ways of drug delivery—passive and active. In the passive way, drugs are encapsulated inside a cavity by utilizing its hydrophobic properties. At the targeted site, the drug is released. Chitosan is suitable for the material used for the encapsulation of nanoparticles which can be used for diagnosis and is also a therapeutic agent for various diseases such as cancer. Besides this, alginates, cellulose, xanthan gum, etc. are used as nanoparticles for drug delivery.

While in active mode, the drug is conjugated to the nanostructure carrier material. Active and passive targeting of the drug in cancer method is described in Table 7. The combination of therapy and diagnosis called theragnosis can be achieved with the nanosensors. These liposomes include nanoparticles amalgamate containing all elements of imaging, therapy, and a targeting material. Liposomes is the most extensively used for drug delivery and consist of phospholipid and steroid (50–450 nm size) and they tend to easily fuse with cell membrane because of analogy (Patra et al. 2018). Nowadays, the trend is to develop a multipurpose nanoparticle that can be simultaneously used for a variety of applications. For example, polymeric

Table 7. Usage of nanoparticle-based drug delivery in cancer treatment.

Passive Targeting	Active Targeting
Nanoparticles are encapsulated and delivery is based on enhanced permeability and retention (EPR).	It is based on receptor-ligand interaction and delivery is based on affinity between peptide/drug to tumor surface receptor.
In this targeting manner nanoparticle started depositing at tumor site gradually.	Nanocarrier based ligands tagged on surface for the tumor overexpressing receptor.
Limited use in solid tumor due to poor permeability.	Extensive use in solid tumor.
Example: polyethylene glycolate fused liposome-based delivery.	Example: transferrin ligand tagged on nanoparticles to target transferrin receptor, acetylcholine for acetylcholine receptor to treat brain tumor.

Attia et al. 2019

micelles for drug delivery in cancer and bio-imaging such as magnetic resonance imaging (MRI). Similarly, the dendrimer is developed with various conjugation such as Fluorescein isothiocyanate (FITC) for bio-imaging, paclitaxel for chemotherapeutic drugs, and folic acid to target folate receptors on cancer cells (Majoros et al. 2006, Nasongkla et al. 2006).

2.8 Gene Delivery

Gene delivery by using nanoparticles is an innovative approach in biomedical science. Nanoparticles and viruses are roughly the same sizes. Nanoparticles have certain advantages over virus-mediated delivery as this method is safer and so we can use this method for broad applications. The new methodology for utilization of nanoparticles based artificial viruses still need to be developed (Santhi et al. 2000).

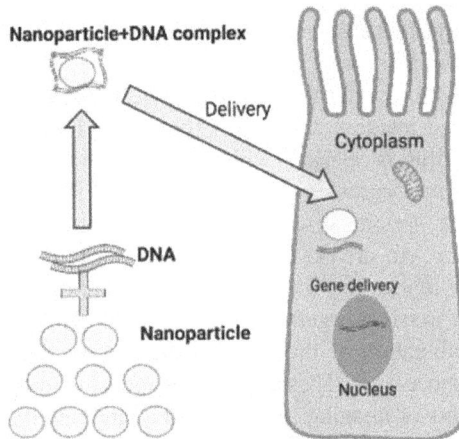

Fig. 4. A generalized method of gene delivery with the help of nanosensors.

2.9 Metabolite/Ion Detection

Biochemical nanosensor is used for the detection of ions or pH measurement and ion fluctuations (Patolsky et al. 2006). One such example is the measurement of Ca^{2+} fluctuation during stimulation in vascular smooth muscle cells. Here the nanosensor is inserted and fluctuation in Ca^{2+} levels is correlated with cellular stimulation (Vo-Dinh and Kasili 2005). Another example of biochemical nanosensor usage is in asthma patients where nitric oxide levels are monitored in breath to prevent the risk of asthmatic attack. This type of nanosensor consists of a nanotube-based field-effect transistor and a carbon nanotube network (Tallury et al. 2010).

3. Nanorobots

Nanorobots have a lot more potential in medical health care. With the recent advances in engineering and development of nanorobots, we have been able to achieve more success in drug delivery, *in vivo* imaging, and taking biopsies to name a few. Even with these advancements, more improvements and more human trials are needed to deal with the safety issues and complexity of the human body. Precise delivery of drugs such as magnetically driven nanorobots mediated drug delivery of fluorouracil can suppress tumor growth locally in mice. Another example of cargo delivery by biohybrid nanorobots is the use of nanoparticles coupled with genes and proteins that can be delivered via *Listeria monocytogenes* to monitor luminescence signal generated by organ-specific gene expression (Akin et al. 2007). Delivering stem cells to the damaged organs can have potential usage in regenerative medicine and/or cell transfer treatment therapy. Another example of magnetic nanorobotic payload delivery is the use of tissue plasminogen activator (TPA) intravenously. Once the nanorobots reach a blood clot site, they start rotating by application of externally applied magnetic field and induce local flow mixing of TPA in clotting site and help in thrombolysis (Cheng et al. 2014).

4. Future Perspectives

Development of nanotechnology-based devices that are handy, sensitive, and affordable will be in demand. It needs more innovation and high throughput with better resolution in measurement when used for the diagnostic point of care devices. It has tremendous potential in biomedical applications from early diagnosis to treatment of disease. Single-cell analysis or small metabolite detection is useful for monitoring various components of living cells in a real-time manner. Nanoscale size of receptor or pores or various functional component of cells gives a possibility to derive new nanoprobe for sensitive detection of biomolecules whether it be an enzyme or biomarker. Development of miniaturized and biocompatible devices that can detect in a non-invasive manner and fast manner shows the future hope for diagnosis and prevention of risk for life-threatening disease or better surveillance of infectious diseases. Another area of development is the integration of various drugs, fluorophore on nanoparticles, which can be used in both treatment and imaging. Sensitive detection of biomolecules such as DNA, protein, ions, and small molecules has potential in early diagnosis, drug screening, and biomarker development for a particular disease. It is an emerging trend to have nanosensor wearable devices and personalized medicine. Nanotechnology-based medicine and equipment will serve as a milestone in the health care system. There is a lot more scope to improve the nano-based platform for high throughput signal detection by using various nanoparticles. Nanoparticle-mediated drug delivery in a directed manner has tremendous potential as they have the advantage of fewer side effects. Molecular nanotechnology is a recent advancement of nanotechnology that will further our knowledge of emerging infectious diseases.

References

Ajibola, O., B. H. Gulumbe, A. A. Eze and E. Obishakin. (2018). Tools for detection of Schistosomiasis in resource limited settings. *Med Sci* (Basel) 6.
Akin, D., J. Sturgis, K. Ragheb, D. Sherman, K. Burkholder, J. P. Robinson, A. K. Bhunia, S. Mohammed and R. Bashir. (2007). Bacteria-mediated delivery of nanoparticles and cargo into cells. *Nat Nanotechnol* 2: 441–449.

Anderson, J. L., L. A. Coury, Jr. and J. Leddy. (2000). Dynamic electrochemistry: methodology and application. *Anal Chem* 72: 4497–4520.

Attia, M. F., N. Anton, J. Wallyn, Z. Omran and T. F. Vandamme. (2019). An overview of active and passive targeting strategies to improve the nanocarriers efficiency to tumour sites. *J Pharm Pharmacol* 71: 1185–1198.

Bates, M. and A. Zumla. (2015). Rapid infectious diseases diagnostics using smartphones. *Ann Transl Med* 3: 215.

Bayda, S., M. Adeel, T. Tuccinardi, M. Cordani and F. Rizzolio. (2019). The history of nanoscience and nanotechnology: from chemical-physical applications to nanomedicine. *Molecules* 25.

Billingsley, K., M. K. Balaconis, J. M. Dubach, N. Zhang, E. Lim, K. P. Francis and H. A. Clark. (2010). Fluorescent nano-optodes for glucose detection. *Anal Chem* 82: 3707–3713.

Brambilla, D., B. Le Droumaguet, J. Nicolas, S. H. Hashemi, L. P. Wu, S. M. Moghimi, P. Couvreur and K. Andrieux. (2011). Nanotechnologies for Alzheimer's disease: diagnosis, therapy, and safety issues. *Nanomedicine* 7: 521–540.

Cash, K. J. and H. A. Clark. (2010). Nanosensors and nanomaterials for monitoring glucose in diabetes. *Trends Mol Med* 16: 584–593.

Chen, P. C., S. C. Mwakwari and A. K. Oyelere. (2008). Gold nanoparticles: From nanomedicine to nanosensing. *Nanotechnol Sci Appl* 1: 45–65.

Cheng, R., W. Huang, L. Huang, B. Yang, L. Mao, K. Jin, Q. ZhuGe and Y. Zhao. (2014). Acceleration of tissue plasminogen activator-mediated thrombolysis by magnetically powered nanomotors. *ACS Nano* 8: 7746–7754.

Cihalova, K., D. Hegerova, A. M. Jimenez, V. Milosavljevic, J. Kudr, S. Skalickova, D. Hynek, P. Kopel, M. Vaculovicova and V. Adam. (2017). Antibody-free detection of infectious bacteria using quantum dots-based barcode assay. *J Pharm Biomed Anal* 134: 325–332.

Darbha, G. K., U. S. Rai, A. K. Singh and P. C. Ray. (2008). Gold-nanorod-based sensing of sequence specific HIV-1 virus DNA by using hyper-Rayleigh scattering spectroscopy. *Chemistry* 14: 3896–3903.

Gholami, A., S. M. Mousavi, S. A. Hashemi, Y. Ghasemi, W. H. Chiang and N. Parvin. (2020). Current trends in chemical modifications of magnetic nanoparticles for targeted drug delivery in cancer chemotherapy. *Drug Metab Rev* 52: 205–224.

Gnach, A., T. Lipinski, A. Bednarkiewicz, J. Rybka and J. A. Capobianco. (2015). Upconverting nanoparticles: assessing the toxicity. *Chemical Society Reviews* 44: 1561–1584.

Golding, C. G., L. L. Lamboo, D. R. Beniac and T. F. Booth. (2016). The scanning electron microscope in microbiology and diagnosis of infectious disease. *Sci Rep* 6: 26516.

Griffin, J., A. K. Singh, D. Senapati, E. Lee, K. Gaylor, J. Jones-Boone and P. C. Ray. (2009). Sequence-specific HCV RNA quantification using the size-dependent nonlinear optical properties of gold nanoparticles. *Small* 5: 839–845.

Hede, M. S., S. Fjelstrup, F. Lotsch, R. M. Zoleko, A. Klicpera, M. Groger, J. Mischlinger, L. Endame, L. Veletzky, R. Neher, A. K. W. Simonsen, E. Petersen, G. Mombo-Ngoma, M. Stougaard, Y. P. Ho, R. Labouriau, M. Ramharter and B. R. Knudsen. (2018). Detection of the malaria causing plasmodium parasite in saliva from infected patients using topoisomerase I activity as a biomarker. *Sci Rep* 8: 4122.

Inci, F., O. Tokel, S. Wang, U. A. Gurkan, S. Tasoglu, D. R. Kuritzkes and U. Demirci. (2013). Nanoplasmonic quantitative detection of intact viruses from unprocessed whole blood. *ACS Nano* 7: 4733–4745.

Islam, S., S. Shukla, V. K. Bajpai, Y. K. Han, Y. S. Huh, A. Kumar, A. Ghosh and S. Gandhi. (2019). A smart nanosensor for the detection of human immunodeficiency virus and associated cardiovascular and arthritis diseases using functionalized graphene-based transistors. *Biosens Bioelectron* 126: 792–799.

Jackson, K. W. and T. M. Mahmood. (1994). Atomic absorption, atomic emission, and flame emission spectrometry. *Anal Chem* 66: 252R–279R.

Janjic, J. M. and V. S. Gorantla. (2019). Nanomedicine: new hope for transplant paradigms lost? *Nanomedicine* (Lond) 14: 2645–2649.

Kaittanis, C., S. Santra and J. M. Perez. (2010). Emerging nanotechnology-based strategies for the identification of microbial pathogenesis. *Adv Drug Deliv Rev* 62: 408–423.

Kim, D., J. Y. Lee, J. S. Yang, J. W. Kim, V. N. Kim and H. Chang. (2020). The architecture of SARS-CoV-2 transcriptome. *Cell* 181: 914–921 e910.

Kosaka, P. M., V. Pini, M. Calleja and J. Tamayo. (2017). Ultrasensitive detection of HIV-1 p24 antigen by a hybrid nanomechanical-optoplasmonic platform with potential for detecting HIV-1 at first week after infection. *PLoS One* 12: e0171899.

Li, Z., D. Li, D. Zhang and Y. Yamaguchi. (2014). Determination and quantification of *Escherichia coli* by capillary electrophoresis. *Analyst* 139: 6113–6117.

Liong, M., A. N. Hoang, J. Chung, N. Gural, C. B. Ford, C. Min, R. R. Shah, R. Ahmad, M. Fernandez-Suarez, S. M. Fortune, M. Toner, H. Lee and R. Weissleder. (2013). Magnetic barcode assay for genetic detection of pathogens. *Nat Commun* 4: 1752.

Majoros, I. J., A. Myc, T. Thomas, C. B. Mehta and J. R. Baker, Jr. (2006). PAMAM dendrimer-based multifunctional conjugate for cancer therapy: synthesis, characterization, and functionality. *Biomacromolecules* 7: 572–579.

Maysinger, D. (2007). Nanoparticles and cells: good companions and doomed partnerships. *Org Biomol Chem* 5: 2335–2342.

Morton, J. G., E. S. Day, N. J. Halas and J. L. West. (2010). Nanoshells for photothermal cancer therapy. *Methods Mol Biol* 624: 101–117.

Nahar, M., T. Dutta, S. Murugesan, A. Asthana, D. Mishra, V. Rajkumar, M. Tare, S. Saraf and N. K. Jain. (2006). Functional polymeric nanoparticles: an efficient and promising tool for active delivery of bioactives. *Crit Rev Ther Drug Carrier Syst* 23: 259–318.

Nasongkla, N., E. Bey, J. Ren, H. Ai, C. Khemtong, J. S. Guthi, S. F. Chin, A. D. Sherry, D. A. Boothman and J. Gao. (2006). Multifunctional polymeric micelles as cancer-targeted, MRI-ultrasensitive drug delivery systems. *Nano Lett* 6: 2427–2430.

Ng, E., C. Yao, T. O. Shultz, S. Ross-Howe and S. X. Wang. (2019). Magneto-nanosensor smartphone platform for the detection of HIV and leukocytosis at point-of-care. *Nanomedicine* 16: 10–19.

Noah, N. M. and P. M. Ndangili. (2019). Current trends of nanobiosensors for point-of-care diagnostics. *J Anal Methods Chem* 2019: 2179718.

Obisesan, O. R., A. S. Adekunle, J. A. O. Oyekunle, T. Sabu, T. T. I. Nkambule and B. B. Mamba. (2019). Development of electrochemical nanosensor for the detection of malaria parasite in clinical samples. *Front Chem* 7: 89.

Oyelere, A. K., P. C. Chen, X. Huang, I. H. El-Sayed and M. A. El-Sayed. (2007). Peptide-conjugated gold nanorods for nuclear targeting. *Bioconjug Chem* 18: 1490–1497.

Paliwal, P., S. Sargolzaei, S. K. Bhardwaj, V. Bhardwaj, C. Dixit and A. Kaushik. (2020). Grand challenges in bio-nanotechnology to manage the COVID-19 pandemic. *Frontiers in Nanotechnology* 2.

Patolsky, F., G. Zheng and C. M. Lieber. (2006). Nanowire-based biosensors. *Anal Chem* 78: 4260–4269.

Patra, J. K., G. Das, L. F. Fraceto, E. V. R. Campos, M. D. P. Rodriguez-Torres, L. S. Acosta-Torres, L. A. Diaz-Torres, R. Grillo, M. K. Swamy, S. Sharma, S. Habtemariam and H. S. Shin. (2018). Nano based drug delivery systems: recent developments and future prospects. *J Nanobiotechnology* 16: 71.

Pickup, J. C., Z. L. Zhi, F. Khan, T. Saxl and D. J. Birch. (2008). Nanomedicine and its potential in diabetes research and practice. *Diabetes Metab Res Rev* 24: 604–610.

Rong, G., E. E. Tuttle, A. Neal Reilly and H. A. Clark. (2019). Recent developments in nanosensors for imaging applications in biological systems. *Annu Rev Anal Chem* (Palo Alto Calif) 12: 109–128.

Saffioti, N. A., E. A. Cavalcanti-Adam and D. Pallarola. (2020). Biosensors for studies on adhesion-mediated cellular responses to their microenvironment. *Frontiers in Bioengineering and Biotechnology* 8: 597950.

Santhi, K., S. A. Dhanaraj, M. Koshy, S. Ponnusankar and B. Suresh. (2000). Study of biodistribution of methotrexate-loaded bovine serum albumin nanospheres in mice. *Drug Dev Ind Pharm* 26: 1293–1296.

Tallury, P., A. Malhotra, L. M. Byrne and S. Santra. (2010). Nanobioimaging and sensing of infectious diseases. *Adv Drug Deliv Rev* 62: 424–437.

Tasciotti, E., F. J. Cabrera, M. Evangelopoulos, J. O. Martinez, U. R. Thekkedath, M. Kloc, R. M. Ghobrial, X. C. Li, A. Grattoni and M. Ferrari. (2016). The emerging role of nanotechnology in cell and organ transplantation. *Transplantation* 100: 1629–1638.

Vo-Dinh, T. and P. Kasili. (2005). Fiber-optic nanosensors for single-cell monitoring. *Anal Bioanal Chem* 382: 918–925.

Wang, Y., L. Yu, X. Kong and L. Sun. (2017). Application of nanodiagnostics in point-of-care tests for infectious diseases. *Int J Nanomedicine* 12: 4789–4803.

Williams, R. M., C. Lee, T. V. Galassi, J. D. Harvey, R. Leicher, M. Sirenko, M. A. Dorso, J. Shah, N. Olvera, F. Dao, D. A. Levine and D. A. Heller. (2018). Noninvasive ovarian cancer biomarker detection via an optical nanosensor implant. *Sci Adv* 4: eaaq1090.

Wilson, M. R., S. N. Naccache, E. Samayoa, M. Biagtan, H. Bashir, G. Yu, S. M. Salamat, S. Somasekar, S. Federman, S. Miller, R. Sokolic, E. Garabedian, F. Candotti, R. H. Buckley, K. D. Reed, T. L. Meyer, C. M. Seroogy, R. Galloway, S. L. Henderson, J. E. Gern, J. L. DeRisi and C. Y. Chiu. (2014). Actionable diagnosis of neuroleptospirosis by next-generation sequencing. *N Engl J Med* 370: 2408–2417.

Yao, C. G. and P. N. Martins. (2020). Nanotechnology applications in transplantation medicine. *Transplantation* 104: 682–693.

Zenkl, G., T. Mayr and I. Klimant. (2008). Sugar-responsive fluorescent nanospheres. *Macromol Biosci* 8: 146–152.

Zhang, P., H. Lu, J. Chen, H. Han and W. Ma. (2014). Simple and sensitive detection of HBsAg by using a quantum dots nanobeads based dot-blot immunoassay. *Theranostics* 4: 307–315.

Zhu, Z., W. Song, K. Burugapalli, F. Moussy, Y. L. Li and X. H. Zhong. (2010). Nano-yarn carbon nanotube fiber based enzymatic glucose biosensor. *Nanotechnology* 21: 165501.

Zottel, A., A. Videtic Paska and I. Jovcevska. (2019). Nanotechnology meets oncology: nanomaterials in brain cancer research, diagnosis and therapy. *Materials* (Basel) 12.

CHAPTER 2

Nanosensor Technology in Biomarker Detection

Priyanka Mishra and *Shashi Kant Tiwari**

1. Introduction

Diseases are an integral part of human life, which includes simpler ones such as cold/influenza to more serious, deadly diseases which cause loss of lives every year. According to World Health Organization (WHO) data, the deadliest diseases in the world include ischemic heart diseases, cerebrovascular diseases, and respiratory infections, which together are responsible for almost a quarter of deaths annually (https://www.who.int/news-room/q-a-detail/what-is-the-deadliest-disease-in-the-world). Next, cancer plays a significant role in mortality worldwide and is the second leading cause of death globally. It is estimated that, globally, 9.6 million deaths in 2018 were due to cancer. By 2040, the number of new cancer cases per year is expected to rise to 29.5 million and the number of cancer-related deaths to 16.4 million (https://www.who.int/news-room/fact-sheets/detail/cancer and https://www.cancer.gov). Similarly, neurological diseases such as Alzheimer's, Parkinson's and other dementia cases keep rising worldwide (https://www.who.int/mental_health/neurology/neurological_disorders_report_web.pdf). Overall goal is to find cheaper and efficient treatment for these diseases. But the major challenges in disease control and prevention is early detection. An emerging way for earlier detection of disease progression is the use of disease specific biomarkers. Biomarkers are defined as an indicator of a particular disease state or some other physiological state of an organism. It can be a protein, or its fragment, nucleic acid (DNA/RNA), lipids, enzymes or a chemical substance secreted by abnormal cells. A biomarker is a 'molecular signature' of the physiological state of a disease in specific time frame, which is important for early detection and staging of disease. Thus, disease biomarkers provide an information on the underlying mechanism of the initiation of a disease and eventually offer powerful methods to

Department of Pediatrics, University of California San Diego USA.
* Corresponding author: sktiwari@health.ucsd.edu

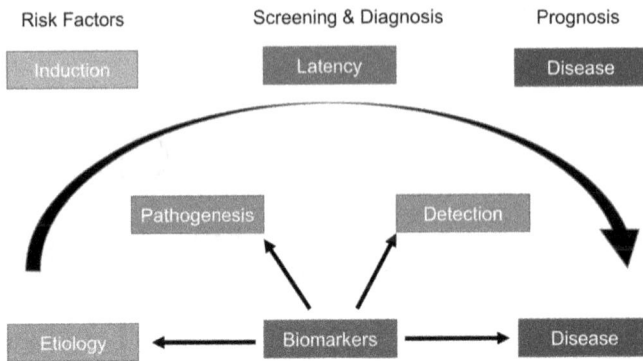

Fig. 1. Potential impact of biomarkers for understanding disease pathways. Source: Richard Mayeux, Biomarkers: Potential uses and limitations. *NeuroRx: The Journal of the American Society for Experimental NeuroTherapeutics* 1: 182–188, April 2004.

diagnose and treat the disease at an anticipated time. Broadly, biomarker is an umbrella coalescence term, which covers an immense number of disciplines that involve use and development of tools, technologies, examining novel drug development and further understanding the prediction, cause progression, regression, outcome, diagnosis and treatment of diseases (Figure 1). Biomarkers are widely applicable in the diagnosis and management of cancers, cardiovascular disease, infections, immunological, neurological and genetic disorders (Perera and Weinstein 2000, Mayeux 2004). The most common traditional investigative method for cancer includes computed tomography (CT), endoscopy, positron emission tomography (PET), mammography and magnetic resonance imaging (MRI). However, these technologies have major limitations such as they are not accessible to bigger population and also not practical for repeated screening at early stages of disease advancement. Nowadays, high throughput screening and detection of a specific biomarker is performed by utilizing enzyme-linked immunosorbent assay (ELISA)/enzyme immunoassay (EIA), in which specific antibodies against that biomarkers are used, and its signals detected by a chromogenic reporter and substrate. More recently, biomarkers have been detected by fluorescence and electrochemiluminescence with high sensitivity in picomole (pM, 10^{-12} M) concentrations. ELISA is a sensitive and well-established technique; however, it is time consuming, technically oppressive, and costly.

Nanotechnology is a fast moving field, having the ability to overcome the existing limitations and is already playing an increasingly important role in the improvement of biosensing. Nanosensors are devices having ability to sense any biological or chemical agents at the nanoscale. Usually, nanosensors are created by using nanoparticles, which are conjugated to a targeting ligand that binds to specific markers. Here nanoparticle acts as the generator or detector of signals and its specificity. Nanoparticles offer suitable and incomparable characteristics for detection such as amplified electrical conductivity, unique magnetic properties, high reactivity, strength, and increased surface area to volume ratio. High surface area to volume ratio of nanoparticle can detect a high concentration of markers at very low quantity of samples. In addition, nanosensors can be used for multi-parametric analysis in real time detection of signals. Moreover, nanoscale properties are tunable

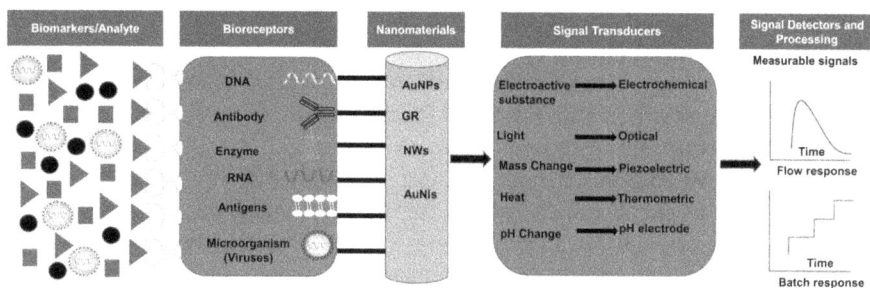

Fig. 2. Schematic representation showing biomarkers' detection using various nanomaterial-based affinity nano-biosensors. Abbreviations: AuNPs: gold nanoparticles, GR: graphene, NWs: nanowires, AuNIs: gold nanoislands.

by their shape such as nanotubes, thin films, and nanowires that give nanosensors versatility and great sensitivity detection. This sensitive approach can also be used to determine novel disease biomarkers. Till date, various studies have been conducted on developing high sensitivity nanosensors for biomarkers detection that opens up new era for early disease diagnosis and their treatment. In this chapter, we will familiarize various novel types of high sensitivity nanosensors, classified by different signal detection approaches like optical, electrical, mechanical, and magnetic relaxation (Figure 2). Thus, nanotechnology can provide unique detection capabilities for high sensitivity biomarker sensing that was earlier not possible. Thus, the aim of this chapter is to analyze the role of nanosensors in biomarker detection and further to understand the nanomaterials in sensor design and relate the difference in methods for the detection and determination of the biomarkers. Overall, this chapter is divided into four major heads: first, different type of nanomaterials and nanotechnology, which are exploited to design biomarkers detection; second, about biomarkers and their diagnostic role in various diseases; third, about methodology for detection of biomarkers, including mechanical, electrical, optical and magnetic sensors; and fourth, discussion about future vision of nanosensor technology in biomarker detection.

2. Role of Nanomaterials to Design Sensors for Biomarkers Detection

Sensor can be defined as "whatever that responds to an input of interest". Thus, sensor is a device that is used for detection and measurement of specific property. Nowadays, sensors are used in biomedical sciences for performing clinical research and development of unique drug discovery and diagnostic purposes. In addition, introduction of nanotechnology plays a crucial role in making very sensitive and specific biosensors, having potential to detect abnormal health condition, drug efficacy, infections and diagnosis of diseases. In brief, nanosensor can be defined as biological, chemical or physical sensors whose sensitivity belongs to nanometer range (10^{-9} nm) (Dahman 2017). In this chapter, we discuss about the nanomaterials used to design the sensor that are of size smaller than 100 nm. Nanomaterials can be classified based on their properties and composition such as carbon based (carbon nanotubes, fullerenes), metal based (Ag, Au nanoparticles and titanium oxide), polymers (dendrimers), semiconductor (quantum dots) and composite based material. Another way is to classify based on the origin of nanomaterials such as natural (found in environment), incidental (from wastes of industrial processes), and engineered nanomaterials (produced in lab conditions with specific shape and size) (Khanna 2012). Besides this, nanosensors can be classified according to their ability of sensing such as chemical nanosensors (sensing molecules, pH, atoms), physical nanosensors (detects optical properties, mass, charge and thermal properties) and nanobiosensors (detects enzymes, DNA, biomarkers—antigens, vitamins, antibodies, viral proteins, and cytotoxicity).

2.1 Carbon Nanotubes

Carbon nanomaterials have recently been preferred widely as an electrode material due to their multiple physical forms, such as fullerenes, graphene, carbon nanotubes, and carbon nanoribbons. While all these materials are composed of the same atoms, these materials have different physical and chemical properties, like electrical, optical and thermal qualities, and chemical reactivity (Huang 2015). Moreover, carbon nanomaterials are cheaper, have relatively inert electrochemistry, a rich surface chemistry, and strong electro-catalytic activity for several redox reactions (Pujado 2012). Carbon nanotubes (CNTs) are tubes made up of carbon with diameter ranges in nanometers, often called single-wall carbon nanotubes (SWCNTs). SWCNTs are one of the allotropes of carbon, in-between flat graphene and fullerene cages. Another carbon nanotube is called multi-wall carbon nanotubes (MWCNTs) because it is composed of nested single-wall carbon nanotubes bound together by van der Waals interactions in a tree ring-like fashion. CNTs exhibit unique features such as electrical conductivity, semiconductors, thermal conductivity and exceptional tensile strength (Hamada et al. 1992, Mintmire et al. 1992, Berber et al. 2000, Yu et al. 2000, Kim et al. 2001). These characteristic features are due to their nanostructure and strength of the bonds between carbon atoms. In addition to this, they can be chemically modified (Karousis et al. 2010). These unique properties enhance their application in wide areas of technology such as optics, nanotechnology, electronics for electrode design, composite materials and many other uses in material sciences. Due to greater surface area of CNTs have lower detection potential and also contribute to the emergence of higher currents that increases the analytical sensitivity and overall selectivity. In addition, CNTs are stable and resistant to passivation, which results in better reproducibility in CNT-based sensors. Further, other advantages are light weight, low toxicity, easy availability, high elasticity, and their biocompatibility, which make CNTs versatile materials for biosensor design (Mehra et al. 2016). In brief, CNTs based biosensor have improved sensitivity, selectivity, biocompatibility and reproducibility (Escarpa et al. 2015).

2.2 Graphene

Graphene is the thinnest and strongest material in the universe. Graphene is a single carbon layer made up of graphite (Chen et al. 2010). Graphene has been considered as the "mother of all graphitic forms" because it can take many forms such as spherical

C60 buckyball (when it is in zero dimensional) and CNTs (one-dimensional), graphene (two-dimensional sheets) or graphite (three-dimensional rolled). Moreover, graphene can be oxidized in fabrication process or in contact with air to another form called graphene oxide (GO) (Brownson and Banks 2014). Graphene can be synthesized by unique methods, such as arc discharge, chemical vapor deposition, and reduction of GO (Chen et al. 2010). Thus, graphene is a sheet of carbon atoms arranged in a hexagonal design. Graphene is considered as a "wonder material" due to their enormous group of properties. Graphene is considered as one of the toughest materials in universe, having excellent tensile strength, superior thermal and optical properties, incredible electrical conductivity, relative transparency, and impermeability to most gases and liquids. These properties make graphene a widely popular and very promising sensing material. Graphene-based sensors are commonly used for the detection of gases (CO_2 and NH_3), biomarkers (DNA, glucose, and dopamine), and chemical materials (cadmium and paracetamol) (Chen et al. 2010).

2.3 Magnetic Nanoparticles

Magnetic nanoparticles (MNs)-based biosensors have received significant attention because of unique advantages over other available strategies. MNs are cheap to produce, environmentally safe, chemically and physically stable and biocompatible. Currently, MNs are used in various biosensors such as magnetometers, SQUID, hall sensors and magnetic resonance devices such as MRI/NMR and diagnostic magnetic resonance (DMR). MNs utilized in DMR should be strong magnetic moment, have superparamagnetic behavior, and must be passivated with a hydrophilic and biocompatible coating to prevent aggregation in aqueous solution. The most common MNs utilized are cross-linked iron oxide (CLIO) nanoparticles, manganese-doped iron oxide magnetic nanoparticles, and elemental iron core/ferrite shell nanoparticles ('Cannonballs') (Josephson et al. 1999, Harisinghani et al. 2003, Peng et al. 2006, Lee et al. 2007, Miguel et al. 2007). CLIO nanoparticles have chiefly been used for DMR biosensing applications due to their excellent biological properties (Harisinghani et al. 2003). CLIO nanoparticles contain a superparamagnetic iron oxide core (monocrystalline iron oxide nanoparticles), which is composed of ferrimagnetic magnetite (Fe_3O_4)/maghemite (γFe_2O_3). Another amine conjugated CLIO nanoparticle (amino-CLIO) developed with an average hydrodynamic diameter of 25–40 nm and around 40–80 amines available per particle for binding of biomolecules via amine, anhydride, carboxyl, hydroxyl, thiol, or epoxide groups (Josephson et al. 1999, Sun et al. 2006). Another MNs used for DMR application is manganese-doped iron oxide magnetic nanoparticles. In this, the magnetic moment of iron oxide nanoparticles can be improved by doping the magnetite crystal with metal ions (manganese, cobalt, and nickel) (Lee et al. 2007a, 2009b). Next, elemental iron core/ferrite shell nanoparticles (Cannonballs, 16 nm) was extensively used for DMR applications have higher magnetization than metal oxides. This Fe-core magnetic nanoparticles with extremely high reflexivity and Fe-core are protected from oxidation using a ferrite shell (Peng et al. 2006, Miguel et al. 2007).

2.4 Metal Nanoparticles

Metallic nanoparticles have enthralled scientist for over a period and are now profoundly utilized in biomedical sciences and engineering. They are a center of interest because of their huge potential in nanotechnology. Nowadays, these materials can be synthesized and modified with various chemical functional groups, which allow them to be conjugated with drugs, ligands, and antibodies of interest, thus opening enormous potential applications in biomedical technology, magnetic separation, targeted drug delivery and preconcentration of target analytes, and vehicles for gene and drug delivery and, more importantly, diagnostic imaging (Mody et al. 2010). Metallic nanoparticles are meticulously used in electrochemistry for electrode and biosensor design. The specific properties of metallic nanoparticle include electro-conductive and thermal. In addition, they can catalyze the electrochemical reactions of target analytes on their surface, avoidance of undesirable interference from other electroactive compounds and also reduce the working potential (Pingarron et al. 2013). These are used in biosensor design and function as biomolecule immobilizers, electron transfer enhancers, reaction catalysts, and labelling the biomolecules. Due to these unique functions, metal nanoparticle-based biosensors have improved stability, selectivity and sensitivity (Pingarron et al. 2013). Other features like small particle size and high surface area allow metal nanoparticles to provide more active sites to catalyze a specific chemical reaction (Lawrence and Liang 2008). Most commonly used metallic nanoparticles are gold nanoparticles (AuNPs) and silver nanoparticles (AgNPs). AuNPs are very stable nanoparticles and having chemical stability and resistance to surface oxidation makes them one of the most important nanosensor material (Oliveira et al. 2015). Nowadays, AuNPs have been widely utilized in biosensor design due to their high biocompatibility and easy functionalization. Particularly, when DNA, chemicals or enzymes are immobilized on AuNPs, very efficient genosensor, chemical sensors, and immunosensors may be designed. Low toxicity potential of AuNPs make them an important tool in bio-nanotechnology (Ravalli et al. 2015) and are also used in electrochemical detection in very small amounts. Recently, AuNPs have been utilized for detection of dopamine (Park et al. 2017), protein kinase enzyme (Sun et al. 2017), and norepinephrine (Lee et al. 2017). In addition, AuNPs are used to determine pollutants (arsenic and nitrites), and so are widely used in the determination of arsenic in water and nitrites in body tissues (Lawrence and Liang 2008). Similarly, silver nanoparticles (AgNPs) have the highest conductivity both thermally and electrically and it is cheaper than gold due to its abundance. AgNPs are attached to electrode surface through electrochemical methods and can be used for both the detection of analytes and as a template in a biosensor (Lawrence and Liang 2008). The unique electrical, optical, thermal, and biological properties of AgNPs allow them to be applied in wide areas of applications, such as medical device coatings, optical sensors, diagnostics and anticancer agents. Moreover, biologically, AgNPs are used as antibacterial, anti-inflammatory, antiviral, antitumor, antifungal agents, and antioxidative effects, drug carriers and biosensors for biomarker detection, and for air and water disinfection as well (Zhang et al. 2016, Tran et al. 2013).

3. Biomarkers and Their Diagnostic Role in Diseases

Biological markers (biomarkers) have been defined by Hulka and colleagues as "cellular, biochemical or molecular alterations that are measurable in biological media such as human tissues, cells, or fluids" (Hulka et al. 1993). More recently, the definition has been broadened to include biological characteristics that can be objectively measured and evaluated as an indicator of normal biological processes, pathogenic processes, or pharmacological responses to a therapeutic intervention. Schulte has outlined the role of biomarkers in biomedical research (Table 1).

Biomarkers can also provide insight into disease progression, prognosis, and response to therapy. Simply, a biomarker can be defined as an objective indication of a medical state, which is measured accurately and reproducibly from outside the patient (Strimbu and Tavel 2010). Thus, biomarkers are important for demonstrating biological processes of health and disease states, as well as biological responses in order determine the most effective therapeutic interventions (Neumann 2015). An ideal biomarker should have high specificity, sensitivity and a great prognostic value (Fallaha et al. 2007). Based on the functional characteristics, biomarkers can be classified into three categories (Table 2). Further, biomarkers can be also classified based on many biological substances' uses such as gene, protein, enzyme lipid and metabolite biomarkers as summarized in Table 3. New types of studies, called OMICs, such as genomics, metabolomics, proteomics, and lipidomics, have been employed to develop and understand the dynamics of these biomarkers. Few studies validated the range of some biomarkers in human serum which is shown in Table 4. Altogether, biomarkers have been quite influential in detecting and diagnosing a particular disease or in predicting the risks attached to that disease at an early stage. Moreover, they can also be used for understanding the prognosis and outcome of the disease as well. Various advantages and disadvantages as discussed in Table 5.

Table 1. Significant role of biomarkers in clinical research.

I.	Delineation of events between exposure and disease
II.	Establishment of dose–response
III.	Identification of early events in the natural history
IV.	Identification of mechanisms by which exposure and disease are related
V.	Reduction in misclassification of exposures or risk factors and disease
VI.	Establishment of variability and effect modification
VII.	Enhanced individual and group risk assessments

4. Type of Nano-Biosensors for Detection of Biomarkers

The major challenges in biomedicine are the rapid and precise measurement of various biomarkers such as antigens, proteins, cell types, and pathogens in biological samples. Many new analytical platforms have recently been developed to measure biomarkers with high sensitivity that might allow immediate disease detection or provide valuable understandings into biology at the systems level. As earlier sections described about numerous nanomaterials (NMs) such as magnetic

Table 2. Types of biomarkers.

S.No.	Biomarker's Categories	Applications	Examples	References
1	Type 0 (i) Diagnostic/prognostic biomarkers (ii) Predictive biomarkers	Allow to understand the progression of disease and their mechanisms. For early detection of specific diseases, their progression over time and its outcome. To measure the risk for a disease and predicting diseases at presymptomatic stage.	Carcinoma antigen 125 for ovarian cancer PGC1α for development of diabetes. Cytokines, antibodies again amyloid-beta protein can be perceived as predictive biomarkers for AD.	Cho 2010 Sharma et al. 2016 Jain 2010 Galaske 2007
2	Type 1 biomarker (i) Efficacy biomarkers (ii) Mechanistic biomarkers (iii) Toxicity biomarkers	Used for drug discovery and therapeutic development and determination of biological effects of drugs like response, non-response and toxicity. Positive effects of drugs. Downstream mechanisms of drugs. Negative effects of particular drugs *in vitro* and *in vivo*.	HbA1C for indication of glycemic control in diabetic patients. Elevated concentrations of creatinine or urea nitrogen demonstrate kidney damage.	Dakubo 2016 Bakhtiar 2017 Jain 2010 Hall and Everds 2008
3	Type 2 biomarker (Surrogate endpoints)	For prediction based on epidemiological, therapeutic, pathophysiological, or survival of patient.	PSA is studied as a surrogate biomarker for measuring the failure to respond to endocrine therapy in advanced prostate carcinoma.	Pivac et al. 2013 Gion and Gasparini 2002

Table 3. Types of biomarkers based on biological molecules.

Biomarkers	Example	References
Protein Biomarkers	Proteomic methods to detect specific protein present in fluids or tissues to understand the pathology. GFAP increased in CSF of AD patients.	Diez et al. 2014 Diez et al. 2015, Teunissen 2018
Genomic Biomarkers (DNA/RNA)	Genome sequencing methods used to know the impaired genes. UGT1A1 gene used for assessing irinotecan therapy in colon cancer patient. Similarly, miRNAs are used for understanding the progression of PD because they regulate neuronal development, mitochondrial dysfunction, and oxidative stress.	Novelli et al. 2008 Arshad et al. 2017
Metabolite Biomarkers	Metabolomics help to understand the importance of metabolites in biochemical pathways. Creatinine and bilirubin are used for indicating kidney and liver dysfunctions, respectively.	Gant et al. 2015
Enzymes	PSA, which is a glycoprotein enzyme produced by the epithelial cells of the prostate gland, has been used to detect prostate cancer and it is an excellent example of an enzyme biomarker.	Liang and Chan 2007
Lipid Biomarkers	For early diagnosis of various cancers (breast, prostate, endometrium, ovary, and colon).	Fernandis and Wenk 2009

Table 4. Concentration of usual biomarkers in humans.

Biomarker	Concentration in Humans	References
Prostate-specific antigen (PSA)	0.5–2 ng/mL	Polascik et al. 1999, Kulasingam and Diamandis 2008
Interleukin 6 (IL-6)	< 6 pg/mL	Riedel et al. 2005
Carcinoembryonic antigen (CEA)	3–5 ng/mL	Mujagic et al. 2004
α-Fetoprotein (AFP)	0–20 ng/mL	Blohm et al. 1998

Source: Data from Ruitao Liu et al. Recent progress of biomarker detection sensors. *AAAS Research* Vol. 2020, Article ID 7989037.

Table 5. Advantages and disadvantages of biomarkers.

Advantages	Disadvantages
Objective assessment	Timing is critical
Precision of measurement	Expensive (costs for analyses)
Reliable; validity can be established	Storage (longevity of samples)
Less biased than questionnaires	Laboratory errors
Disease mechanisms often studied	Normal range difficult to establish
Homogeneity of risk or disease	Ethical responsibility

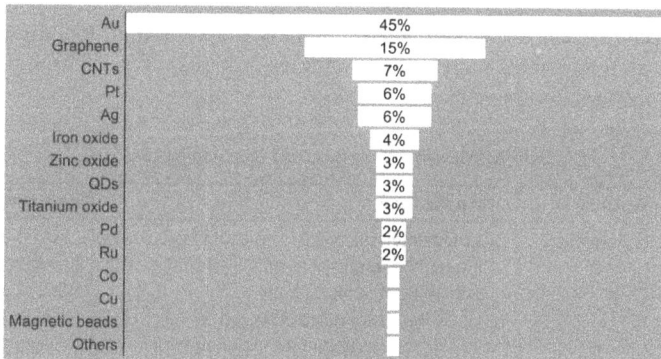

Fig. 3. The most commonly used nanomaterial in biosensors for detection of cancer. (Reprint with permission from Sharif, M. and M. R. Avadi et al. Cancer diagnosis using nanomaterials based electrochemical nanobiosensors. *Biosens Bioelectron* 126(2018): 773–784. Available from: https://doi.org/10/1016/j.bios.2018.11.026).

NPs (iron oxide NPs), metal NPs (gold and silver NPs), carbon-based nanotubes and carbon allotropes, quantum dots and nanowire with distinctive biological recognition elements (enzymes, DNA/RNA, antibodies, antigens, peptide, viruses) provide countless opportunities for enhancing the performance of nano-biosensor (Golichenari et al. 2018, Waifalkar et al. 2018). As for example here providing in figure the commonly used various nanomaterials in biosensors for cancer detection (Figure 3). Biosensor devices are broadly categorized based on the different types of nanomaterials currently being utilized for analytical purposes of biomarkers, which include magnetic, optical, carbon nanotube and electrochemical nano-biosensor devices.

4.1 Magnetic Nano-Biosensors

Till date various nanosensors have been developed to sense biomolecules using magnetic nanoparticles such as magnetometers, Superconducting quantum interference devices (SQUID), and hall sensors (Baselt et al. 1998, Chemla et al. 2000, Parkin et al. 2004, Aytur et al. 2006, Osterfeld et al. 2008, Tamanaha et al. 2008). Another most successful method based on magnetic resonance is MRI/NMR, in which magnetic nanoparticles as proximity sensor to accelerate the relaxation rate of water molecules are used. The technique based on MRI is utilized to see inside the human body known as diagnostic magnetic resonance (DMR). Currently, DMR has been widely used to detect various molecular targets (RNA, proteins, DNA metabolites, enzymes, drugs, pathogens and cancerous cells) with high specificity and sensitivity using clinical MRI scanners or benchtop NMR systems (Hogemann et al. 2002, Perez et al. 2002, Lee et al. 2008, 2009a, 2009b). Another technology like magnetoelectronics is applicable in various magnetic biosensors for identification, localization, and detection of wide range of chemical, physical and biological agents (Table 6). In addition, DMR utilized the ferric based NMs having magnetic properties for biomedical applications (Issa et al. 2013). Similarly, SQUID method is used for quick inspection of precise antigens from mixture by utilizing antibodies, which employ the super paramagnetic nature of these MNs (Chemla et al. 2000). Thus, magnetic nanoparticle based guided drug supply, hyperthermia treatment, and the use of MNs as MRI contrast agents are examples of highly efficacious application in biomedical technology (Nabaei et al. 2018).

Table 6. Magnetic biosensors and their applications.

Magnetic Biosensors	Application	References
Giant magnetoresistance (GMR)	Detect interleukin-6 (IL-6), a potential cancer biomarker, in unprocessed human serum with only 4 µL of serum sample. GMR biosensor detected as low as 1.5×10^2 $TCID_{50}$/mL virus and the signal intensity increased with increasing concentration of virus up to 1.0×10^5 $TCID_{50}$/mL.	Li et al. 2010 Ennen et al. 2016 Rizzi et al. 2017 Xu et al. 2016 Krishna et al. 2016
Anisotropic magnetoresistance (AMR) effect	Biomarker detection	Ren et al. 2020
Superconducting quantum interference device (SQUID)	Detect antibodies labelled magnetic nanoparticles	Lenz 1990, Kotitz et al. 1997

4.2 Carbon Nanotubes (CNTs) Based Biosensors

Carbon nanotubes (CNTs) have some extraordinary properties like flexible physic-geometric features, electrical conductivity, optical and dynamic physicochemical properties along with high mechanical strength (Nabaei et al. 2018). Due to these unique properties, CNTs are sensitive to be affected by contact with biomolecules, which leads to being widely used as biosensing materials. This category of biosensor was most popularly used for advancing the design of glucose biosensor, which involved use of nanotube for immobilization of glucose oxidase enzyme. This development led to the estimation of glucose from body fluids like tears and saliva, which were limited to major body tissues in previous undeveloped model. In this case, use of single walled nanotube has significantly increased the activity of enzyme (Azamian et al. 2002). These CNTs are also used for electrical detection of sensing apparatus. These nanotubes have potential of rapid friendliness with biological molecules like oligonucleotide probes, DNA and proteins, cytokines, and viruses (Table 7). Thus, the major advantages of these CNTs based biosensors are: (a) highly sensitive and fast detection at low concentration of biomolecules due to high surface-to-volume ratio, (b) can be applied to immobilize the enzyme to keep elevated biological activity, (c) quick response time due to incredible fast electron-transfer kinetics, (d) lower surface fouling and less potential of redox reaction, (e) high stability and durability.

Table 7. Carbon nanotube-based biosensors and their applications.

Analyte	Functionalized CNT for Biosensors	Detection Range	Sensitivity	References
Glutamate	Glutamate dehydrogenase (GLDH) immobilized on thionine–SWNTs	0.5–400 μM	137.3 ± 15.7 μA mM^{-1} cm^{-2}	Meng et al. 2009
BSA	Self-assembly of oxidative SWCNTs on Au	0.1–1.2 mM	3.89 uA uM^{-1}	Chen and Huang 2009
IL-6	SWCNT forests modified electrodes	40–150 pg mL^{-1}	3 nAcm^{-2} (pg mL^{-1})	Munge et al. 2009
Virus NS1	Anti-NS1 antibodies were covalently linked to CNT–SPE	40 ng mL^{-1}–2 μg mL^{-1}	85.59 uA mM^{-1} cm^{-2}	Dias et al. 2013
Dengue virus	Anti-NS1 antibodies are immobilized on the electrode surface	0.1–2.5 μg mL^{-1}	0.035 μg mL^{-1}	Mízia et al. 2014
Glucose	GOx immobilized on a hybrid film of AgNPs/CNTs/chitosan	0.5–50 M	135.9 A mM^{-1}	Jiehua et al. 2009
	GOx bound to ZnO NPs/ MWCNTs	0.1–16 mM	12.5 mA M^{-1} cm^{-2}	Wang et al. 2009
DNA/gene	Probe DNA immobilized on aligned MWCNT thin films	50–200 nM	0.02 AnM^{-1}	Shumyantseva et al. 2020

4.3 Optical Biosensors

Optical biosensor is the most diverse class of biosensor due to its application in different spectroscopy like adsorption, fluorescence, Raman, SERS, etc. Optical detection is achieved by using the optical field with biorecognition elements and these biosensors work on total internal reflection of light and surface characterization by difference in angle of incidence of reflected light. The surface modification is based on optical fibers and surface plasmon resonance (SPR). These sensors detect binding between analytes without any involvement of electrical connection, which make them fast, sensitive and reliable. Moreover, optical biosensing can be divided into two groups: first, like label free in which signal detected is generated by interaction of transducer with analyte. Second group is label based that uses optical signals generated by colorimetric, luminescent and fluorescent methods. As for example, glucose can be detected by enzymatic oxidation using label-based sensing. Currently, colorimetric biosensor is the more popular optical biosensor because presence of pathogenic microorganisms in sample can be instantly observed with the naked eye through a color change (Yoo and Lee 2016). In addition, various type of optical biosensors have been developed overtime having wide range of application in biomedical sciences. These optical biosensors are surface plasmon resonance biosensors, localized surface plasmon resonance biosensors, interferometric, ellipsometric, evanescent wave fluorescence and bioluminescent optical fiber biosensors, reflectometric interference spectroscopy and surface enhanced Raman scattering biosensors. These biosensors are applicable to rapid and accurate pathogen diagnosis, which is important for saving lives from infection. Various optical sensors have been developed for the detection of pathogens and viral infections (Table 8). Microfluidic-integrated optical sensors are useful for point-of-care diagnostics. Smartphone-based optical sensors provide a simple user interface for rapid sensing at reduced cost with ubiquitous capabilities. Overall optical biosensors allow the sensitive and selective detection

Table 8. Optical biosensors and their applications.

Optical Biosensors	Label Free	Biological Applications	References
SPR	Yes	Kinetic analysis of biointeractions Antigens in clinical samples Proteins in biological samples Xenobiotics and toxins in food	Riedel et al. 2014 Pimkova et al. 2012 Fernandez et al. 2010 Safina 2012
SPRi	Yes	Screening of biomarkers and therapeutic targets Screening of drug–target protein interactions	Safina 2012
LSPR	Yes	Detection of DNA hybridization Screening of antigen–antibody interactions Cancer biomarker and toxin detection	Piliarik et al. 2012 Endo et al. 2006 Yuan et al. 2012
Evanescent wave fluorescence	No	Clinical diagnostics, biodefence, food testing Clinical biomarkers Toxin screening	Taitt et al. 2016 Lochhead et al. 2011 Yildirim et al. 2012
Bioluminescent optical fiber	No	Response of cells to genotoxic agents Multidetection of genotoxins by live cell array	Jia et al. 2012 Biran et al. 2003
Waveguide interferometric	Yes	Study of cellular responses and processes Virus detection	Xu et al. 2007
Ellipsometric	Yes	Characterizing viral receptor profiles Detection of serum tumor biomarker	Fei et al. 2015
RIfS	Yes	Xenobiotics in food Detection of circulating tumor cells	Rau et al. 2014 Kumeria et al. 2012
SERS	Yes	Detection of cancer proteins Protein biomarker in environment	U et al. 2012 Srivastava et al. 2014

of a wide spectrum of analytes such as viruses, drugs, toxins, antibodies, and tumor biomarkers. Thus, major advantages of the optical biosensors over conventional analytical methods are: (a) they will enable direct and label free detection of biomolecules and chemicals in analytes (Damborsky et al. 2016), (b) they are highly specific, sensitive and cheaper, (c) they provide new analytical tools (micro/nanotechnologies, microelectromechanical systems (MEMSs) and microelectronics) with smallest size for large scale high throughput sensitive screening of diverse array of samples with variable parameters.

4.4 Electrochemical Biosensors

Electrochemical biosensors are invented for analysis of biochemical reactions through developed electrical device. In this electro-biochemistry, reactions that are measured will either generate a measurable potential or charge accumulation (potentiometric), a quantifiable current (amperometric), measurable conductive properties (conductometric), measurable impedance like resistance and reactance (impedimetric) between electrode (Mirsky et al. 1997, Chaubey and Malhotra 2002, Bonanni et al. 2006, Thevenot et al. 2001) as shown in Table 9.

Table 9. Types of electrochemical biosensors.

Measurement Type	Transducer	Analyte
Potentiometric	Ion selective electrode (Parkin et al. 2004) Glass Electrode Gas Electrode Metal Electrode	K^+, Cl^-, Ca^{2+}, F^- H^+, Na^+ CO_2, NH_3 Redox species
Amperometric	Metal or carbon electrode Chemically modified electrode (CME)	O_2, sugars, alcohols sugars, alcohols, phenols, oligonucleotides
Conductometric, impedimetric	Interdigitated electrode, metal electrode	Rea, charged species, oligonucleotides
Ion charge or field effect	Ion selective field effect transistor (ISFET), Enzyme FET (ENFET)	H^+, K^+

These biosensors mostly utilize the metallic nanoparticles that facilitate immobilization of reactants in the reaction between biomolecules. Colloidal gold-based nanoparticles were used to enhance immobilization of DNA on gold electrodes, which increased the efficiency of biosensor. Metallic nanoparticles have also been used for their coupling with biological probes to carry out detection reaction from a mixture (Gonzalez-Garcia et al. 2000). Here are the brief details of these different types of biosensing devices utilized in biomedical sciences for detection of analytes:

Amperometric Biosensors: These are a type of electrochemical sensor that measures current generated from the oxidation or reduction reactions of an electroactive species in any biochemical reaction (Luppa et al. 2001). If the current is measured at a constant potential, it is called amperometry, while if it is measured in controlled variations of the potential, it is known as voltammetry.

Table 10. Electrochemical nanobiosensor for virus detection.

Viral Target	Biosensor Type	Nanomaterial	LOD	Application	References
Influenza virus M1 protein	Electrochemical impedance	Nanocrystalline boron doped diamond	5×10^{-4} g/mL	Saliva	Siuzdak et al. 2019
Avian influenza virus H5N1	Voltammetric	Porous AuNPs	1 pM	Chicken serum	Lee et al. 2019
Chikungunya virus DNA	Electrochemical paper analytical device	Gold shells coated magnetic nanocubes	0.1 nM	Serum	Singhal et al. 2018
Zika virus-specific antibodies	Electrochemical impedance, voltammetric	Carboxylated CNTs	–	Blood and saliva	Palomar et al. 2018
Hepatitis C virus	Amperometric	Nanoliposomes	1.9 pM	Human serum	Tu et al. 2018
HIV envelope glycoprotein	Amperometric	GO	8.3 fM	Human serum	Nehra et al. 2017
Japanese encephalitis virus	Electrochemical impedance	AuNPs	167PFU/mL	–	Geng et al. 2016
Dengue virus DNA	Electrochemical impedance	Nanoporous alumina	2.7310212 M	–	Deng and Toh 2013
HIV	Voltammetric	AuNPs	600 fg/mL	-	Lee et al. 2013
Dengue virus NS1 protein	Voltammetric	Carboxylated CNTs	0.035 µg/mL	Serum	Mízia et al. 2014
Hepatitis C virus core antigen	Voltammetric	MWCNTs	0.01 pg/mL	Serum	Ma et al. 2013
Dengue type 2 virus	Voltammetric	Nanoporous alumina	1 PFU/mL	Infected Aedes Aegypti mosquito	Cheng et al. 2012
HIV-1	Voltammetric	Fe_3O_4 nanoparticles	50 pM	–	Tran et al. 2011
HBV DNA	Electrochemical impedance	Zeolite nanocrystals and MWCNT nanocomposite	50 copies/mL	Patient samples	Narang et al. 2016
HBV surface antigen	Voltammetric	Fe_3O_4 magnetic nanoparticles	0.19 pg/mL	Human serum	Alizadeh et al. 2017
Zika virus	Voltammetric	GO	2×10^{-2} PFU/mL	Blood serum	Tancharoen et al. 2019
HIV p24	Voltammetric	MWCNTs	0.083 pg/cm^3	Human serum	Ma et al. 2017

Source: Data from Kaya et al. (2020). Chapter 18 - Electrochemical virus detections with nanobiosensors. pp. 303–326. *In*: Baoguo Han, Vijay K. Tomer, Tuan Anh Nguyen, Ali Farmani and Pradeep Kumar Singh (eds.). Micro and Nano Technologies, Nanosensors for Smart Cities. Elsevier. https://doi.org/10.1016/B978-0-12-819870-4.00017-7.

Potentiometric Biosensors: They measure an accumulation of a charge potential at the working electrode compared to the reference electrode in an electrochemical cell when no significant current flows between them (Chaubey and Malhotra 2002, Murugaiyan et al. 2014). The lowest detection limits for these biosensors are currently achieved with ion-selective electrodes (Parkin et al. 2004).

Conductometric Biosensors: This is utilized to measure the ability of an analyte or a medium to conduct an electrical current between electrodes or reference nodes. Mostly, these biosensors are strongly coupled with enzymes, where the ionic strength, and thus the conductivity, of a solution between two electrodes changes due to an enzymatic reaction. Thus, this can be used to analyze enzymatic reactions that induce changes in the concentration of charged species in a solution (Murugaiyan et al. 2014).

Thus, based on these diverse electro-biochemical reactions, the electrochemical nanobiosensors were used in versatile areas of cancer diagnostics and detection of infectious microorganisms and viruses (Table 10).

5. Advantages and Disadvantages of Nano-Biosensors

Current nanotechnology offers the opportunity to fabricate the nanostructures with wide range of properties like magnetic, electrical, and optical to design the biosensors for detection of biomarkers. These nanoscale properties are extensively applicable in diverse fields, which include optics, mechanical, electronics and in biomedicine. In addition to this, it is crucial to know that this technology responds to the detection of biomolecules, which are physically attached to their surface due to some physical properties such as refractive index and mass. These nano-biosensors have wide applicability, and also various advantages and disadvantages which are summarized as: Advantages: (i) Quick response time, (ii) Working at atomic scale with highest efficacy, (iii) Fast and continuous measurement, (iv) Improved surface to volume ratio, (v) Calibration required a very small number of reagents, (vi) Applicable to measure non-polar molecules. Disadvantages: (i) Currently progress of nano-biosensors field are still under infancy stage, (ii) Due to high sensitivity, they may be error prone.

6. Conclusions and Future Vision

Globally, the most extensively researched area is the human health and associated diseases. Detection of biomarkers is the most effective method for early detection of diseases' progress and their prognosis. Recently, numerous nano sensors have been developed such as optical, electrochemistry, magnetic-based biosensors for analyzing biomarkers. Although currently enormous variety of biosensors have been developed, performance and efficiency of commercial biosensors in biomedical fields has not upgraded significantly. However, integration of nanomaterials into biosensing applications allows quick and sensitive detection of infectious viruses (SARS-CoV-2, HIV, MERS) and various biomarkers in fluid, blood and saliva. Various nanomaterials such as CNTs, graphene, magnetic, In_2O_3, gold nanoparticles and electrochemical biosensors were used in detection of glucose, viruses, and

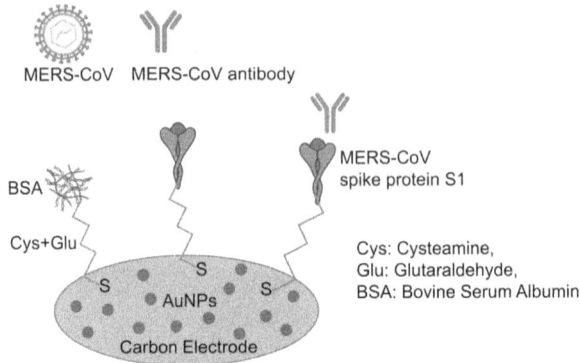

Fig. 4. Schematic representation showing application of AuNPs biosensor for MERS-CoV virus detection. Source: Modified from Riccarda Antiochia. (2020). Nanobiosensors as new diagnostic tools for SARS, MERS and COVID-19: From past to perspectives. *Microchimica Acta* 187: 639. https://doi.org/10.1007/s00604-020-04615-x and virus icons adapted from BioRender (https://biorender.com).

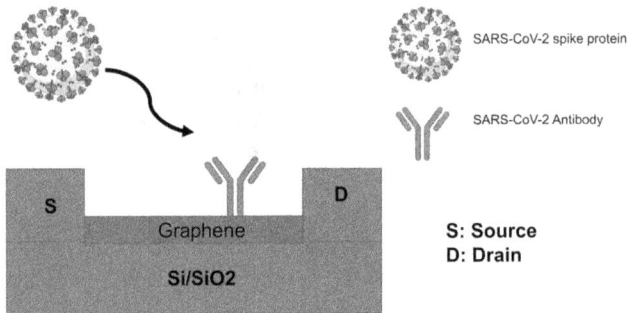

Fig. 5. Schematic representation showing application of graphene FET based biosensor for SARS-CoV-2 virus detection. Source: Modified from Riccarda Antiochia. (2020). Nanobiosensors as new diagnostic tools for SARS, MERS and COVID-19: From past to perspectives. *Microchimica Acta* 187: 639. https://doi.org/10.1007/s00604-020-04615-x and SARS-CoV-2 virus icons adapted from BioRender (https://biorender.com).

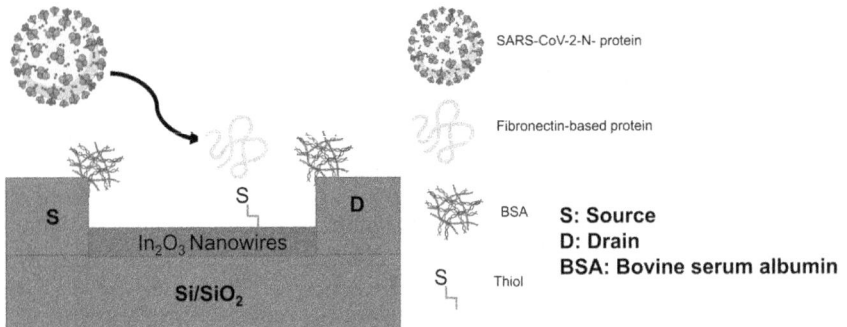

Fig. 6. Schematic representation showing application of In₂O₃ FET-based biosensor for SARS-CoV-2 virus detection. Source: Modified from Riccarda Antiochia. (2020). Nanobiosensors as new diagnostic tools for SARS, MERS and COVID-19: From past to perspectives. *Microchimica Acta* 187: 639. https://doi.org/10.1007/s00604-020-04615-x and SARS-CoV-2 virus icons adapted from BioRender (https://biorender.com).

other cancer biomarkers with detection limit in fempto molar range. In future, nanobiosensors with electrochemical and FET transduction can be miniaturized into inexpensive and integrated platforms. Furthermore, nanobiosensor design involves combination of extraction system into projected biosensor to develop wearable and user friendly biosensors. In addition, also we can develop microneedle-based biosensor for monitoring of viruses, antigen, antibodies and DNA or RNA in real time condition in the symptomatic and asymptomatic individuals.

References

Alizadeh, N., R. Hallaj and A. Salimi. (2017). A highly sensitive electrochemical immunosensor for hepatitis B virus surface antigen detection based on Hemin/G-quadruplex horseradish peroxidase-mimicking DNAzyme-signal amplification. *Biosens Bioelectron* 94: 184–192.

Aytur, T., J. Foley, M. Anwar, B. Boser, E. Harris and P. R. Beatty. (2006). A novel magnetic bead bioassay platform using a microchip-based sensor for infectious disease diagnosis. *J Immunol Methods* 314(1-2): 21–9.

Azamian, B. R., J. J. Davis, K. S. Coleman, C. B. Bagshaw and M. L. Green. (2002). Bioelectrochemical single-walled carbon nanotubes. *J Am Chem Soc* 124(43): 12664–5.

Bakhtiar, R. (2017). Translational application of biomarkers. pp. 17–34. *In*: Weng, N. and W. Jian (eds.). Targeted Biomarker Quantitation by LC-MS. Hoboken, NJ, USA: John Wiley and Sons, Inc.

Baselt, D. R., G. U. Lee, M. Natesan, S. W. Metzger, P. E. Sheehan and R. J. Colton. (1998). A biosensor based on magnetoresistance technology. *Biosens Bioelectron* 13(7-8): 731–9.

Berber, S., Y. K. Kwon and D. Tomanek. (2000). Unusually high thermal conductivity of carbon nanotubes. *Phys Rev Lett* 84(20): 4613–6.

Biran, I., D. M. Rissin, E. Z. Ron and D. R. Walt. (2003). Optical imaging fiber-based live bacterial cell array biosensor. *Anal Biochem* 315(1): 106–13.

Blohm, M. E., D. Vesterling-Horner, G. Calaminus and U. Gobel. (1998). Alpha 1-fetoprotein (AFP) reference values in infants up to 2 years of age. *Pediatr Hematol Oncol* 15(2): 135–42.

Bonanni, A., M. J. Esplandiu, M. I. Pividori, S. Alegret and M. del Valle. (2006). Impedimetric genosensors for the detection of DNA hybridization. *Anal Bioanal Chem* 385(7): 1195–201.

Brownson, D. A. C. and C. E. Banks. (2014). The Handbook of Graphene Electrochemistry. Heidelberg: Springer-Verlag.

Chaitali Singhal, Amidha Dubey, Ashish Mathur, C. S. Pundir and Jagriti Narang. (2018). Paper based DNA biosensor for detection of chikungunya virus using gold shells coated magnetic nanocubes. *Process Biochemistry* 74: 35–42, ISSN 1359–5113.

Chaubey, A. and B. D. Malhotra. (2002). Mediated biosensors. *Biosens Bioelectron* 17(6-7): 441–56.

Chemla, Y. R., H. L. Grossman, Y. Poon, R. McDermott, R. Stevens, M. D. Alper and J. Clarke (2000). Ultrasensitive magnetic biosensor for homogeneous immunoassay. *Proc Natl Acad Sci USA* 97(26): 14268–72.

Chen, D., L. Tang and J. Li. (2010). Graphene-based materials in electrochemistry. *Chemical Society Reviews* 39(8): 3157–3180.

Cheng, M. S., J. S. Ho, C. H. Tan, J. P. Wong, L. C. Ng and C. S. Toh. (2012). Development of an electrochemical membrane-based nanobiosensor for ultrasensitive detection of dengue virus. *Anal Chim Acta* 725: 74–80.

Cho, W. C. S. (2010). Cancer biomarkers (an overview). pp. 21–40. *In*: Hayat, M. A. (ed.). Methods of Cancer Diagnosis, Therapy, and Prognosis Volume 7: General Overviews, Head and Neck Cancer and Thyroid Cancer. New York: Springer.

Dahman, Y. (2017). Nanotechnology and functional materials for engineers. *In*: Amsterdam. Netherlands: Elsevier.

Dakubo, G. D. (2016). Cancer Biomarkers in Body Fluids: Principles. Switzerland: Springer.

Damborsky, P., J. Svitel and J. Katrlik. (2016). Optical biosensors. *Essays Biochem* 60(1): 91–100.

Deng, J. and C. S. Toh. (2013). Impedimetric DNA biosensor based on a nanoporous alumina membrane for the detection of the specific oligonucleotide sequence of dengue virus. *Sensors* (Basel) 13(6): 7774–85.

Dias, A. C., S. L. Gomes-Filho, M. M. Silva and R. F. Dutra. (2013). A sensor tip based on carbon nanotube-ink printed electrode for the dengue virus NS1 protein. *Biosens Bioelectron* 44: 216–21.

Diez, P., M. Gonzalez-Gonzalez, N. Dasilva, R. Jara-Acevedo, A. Orfao and M. Fuentes. (2014). Serum profiling by targeted proteomics for biomarker discovery. pp. 19–33. *In*: Fuentes, M. and J. LaBaer (eds.). Proteomics: Targeted Technology, Innovations and Applications. Norfolk, UK: Caister Academic Press.

Diez, P., R. M. Degano, N. Ibarrola, J. Casado and M. Fuentes. (2015). Genomics and proteomics for biomarker validation. pp. 231–242. *In*: Seitz, H. and S. Schumacher (eds.). Biomarker Validation: Technological, Clinical and Commercial Aspect. Weinheim, Germany: Wiley-VCH Verlag GmbH and Co. KGaA.

Endo, T., K. Kerman, N. Nagatani, H. M. Hiepa, D. K. Kim, Y. Yonezawa, Koichi Nakano and Eiichi Tamiya. (2006). Multiple label-free detection of antigen-antibody reaction using localized surface plasmon resonance-based core-shell structured nanoparticle layer nanochip. *Anal Chem* 78(18): 6465–75.

Escarpa, A., M. C. González and M. A. López. (2015). Electrochemical sensing on microfluidic chips. pp. 331–356. *In*: Escarpa, A., M. C. González and M. A. López (eds.). Agricultural and Food Electroanalysis. London: John Wiley and Sons, Ltd.

Fallaha, D., G. Hillis and B. H. Cuthbertson. (2007). Novel biomarkers and the outcome from critical illness and major surgery. pp. 32–43. *In*: Vincent, J.-L. (ed.). Intensive Care Medicine. New York, NY: Springer.

Fei, Y., Y. S. Sun, Y. Li, H. Yu, K. Lau, J. P. Landry, Zeng Luo, Nicole Baumgarth, Xi Chen and Xiangdong Zhu. (2015). Characterization of receptor binding profiles of influenza A viruses using an ellipsometry-based label-free glycan microarray assay platform. *Biomolecules* 5(3): 1480–98.

Fernandez, F., K. Hegnerova, M. Piliarik, F. Sanchez-Baeza, J. Homola and M. P. Marco. (2010). A label-free and portable multichannel surface plasmon resonance immunosensor for on site analysis of antibiotics in milk samples. *Biosens Bioelectron* 26(4): 1231–8.

Fernandis, A. Z. and M. R. Wenk. (2009). Lipid-based biomarkers for cancer. *Journal of Chromatography B* 877(26): 2830–2835.

Gant, T. W., E. L. Marczylo and M. O. Leonard. (2015). Discovery and application of novel biomarkers. pp. 129–150. *In*: Pfannkuch, F. and L. Suter-Dick (eds.). Predictive Toxicology: From Vision to Reality. Weinheim, Germany: Wiley-VCH Verlag GmbH and Co. KGaA.

Geng, X., F. Zhang, Q. Gao and Y. Lei. (2016). Sensitive impedimetric immunoassay of japanese encephalitis virus based on enzyme biocatalyzed precipitation on a gold nanoparticle-modified screen-printed carbon electrode. *Anal Sci* 32(10): 1105–1109.

Gion, M. and G. Gasparini. (2002). Biomarkers as therapeutic targets: Toward personalized treatments in oncology. pp. 151–162. *In*: Diamandis, E. P., H. A. Fritsche, H. Lilja, D. W. Chan and M. K. Schwartz (eds.). Tumor Markers: Physiology, Pathobiology, Technology, and Clinical Applications. Washington DC: AACC Press.

Golichenari, B., R. Nosrati, A. Farokhi-Fard, K. Abnous, F. Vaziri and J. Behravan. (2018). Nano-biosensing approaches on tuberculosis: Defy of aptamers. *Biosens Bioelectron* 117: 319–331.

Gonzalez-Garcia, M. B., C. Fernandez-Sanchez and A. Costa-Garcia. (2000). Colloidal gold as an electrochemical label of streptavidin-biotin interaction. *Biosens Bioelectron* 15(5-6): 315–21.

Hall, R. L. and N. E. Everds. (2008). Principles of clinical pathology for toxicology studies. pp. 1317–1358. *In*: Hayes, A. W. (ed.). Principles and Methods of Toxicology (5th ed.). New York, London: Informa Healthcare.

Hamada, N., S. Sawada and A. Oshiyama. (1992). New one-dimensional conductors: Graphitic microtubules. *Phys Rev Lett* 68(10): 1579–1581.

Harisinghani, M. G., J. Barentsz, P. F. Hahn, W. M. Deserno, S. Tabatabaei, C. H. van de Kaa, J. Rosette and R. Weissleder. (2003). Noninvasive detection of clinically occult lymph-node metastasis in prostate cancer. *N Engl J Med* 348(25): 2491–9.

Hogemann, D., V. Ntziachristos, L. Josephson and R. Weissleder. (2002). High throughput magnetic resonance imaging for evaluating targeted nanoparticle probes. *Bioconjug Chem* 13(1): 116–21.

Issa, B., I. M. Obaidat, B. A. Albiss and Y. Haik. (2013). Magnetic nanoparticles: surface effects and properties related to biomedicine applications. *Int J Mol Sci* 14(11): 21266–305.

Jain, K. K. (2010). The Handbook of Biomarkers. https://doi.org/10.1007/978-1-60761-685-6.

Jia, K., E. Eltzov, T. Toury, R. S. Marks and R. E. Ionescu. (2012). A lower limit of detection for atrazine was obtained using bioluminescent reporter bacteria via a lower incubation temperature. *Ecotoxicol Environ Saf* 84: 221–6.

Jiehua Lin, Chunyan He, Yue Zhao and Shusheng Zhang. (2009). One-step synthesis of silver nanoparticles/ carbon nanotubes/chitosan film and its application in glucose biosensor. *Sensors and Actuators B: Chemical* 137(2): 768–773, ISSN 0925-4005.

Josephson, L., C. H. Tung, A. Moore and R. Weissleder. (1999). High-efficiency intracellular magnetic labeling with novel superparamagnetic-Tat peptide conjugates. *Bioconjug Chem* 10(2): 186–91.

Karousis, N., N. Tagmatarchis and D. Tasis. (2010). Current progress on the chemical modification of carbon nanotubes. *Chem Rev* 110(9): 5366–97.

Khanna, V. K. (2012). *In*: Jones, B. and W. B. Spillman (eds.). Physical, Chemical, and Biological. Boca Raton, FL: CRC Press.

Kim, P., L. Shi, A. Majumdar and P. L. McEuen. (2001). Thermal transport measurements of individual multiwalled nanotubes. *Phys Rev Lett* 87(21): 215502.

Kulasingam, V. and E. P. Diamandis. (2008). Strategies for discovering novel cancer biomarkers through utilization of emerging technologies. *Nat Clin Pract Oncol* 5(10): 588–99.

Kumeria, T., M. D. Kurkuri, K. R. Diener, L. Parkinson and D. Losic. (2012). Label-free reflectometric interference microchip biosensor based on nanoporous alumina for detection of circulating tumour cells. *Biosens Bioelectron* 35(1): 167–173.

Lam Dai Tran, Binh Hai Nguyen, Nguyen Van Hieu, Hoang Vinh Tran, Huy Le Nguyen and Phuc Xuan Nguyen. (2011). Electrochemical detection of short HIV sequences on chitosan/Fe$_3$O$_4$ nanoparticle-based screen-printed electrodes. *Materials Science and Engineering: C* 31(2): 477–485, ISSN 0928-4931.

Lawrence, N. S. and H.-P. Liang. (2008). Metal nanoparticles: Applications in electroanalysis. pp. 435–457. *In*: Eftekhari, A. (ed.). Nanostructured Materials in Electrochemistry. Weinheim, Germany: Wiley-VCH Verlag GmbH and Co. KGaA.

Lee, E. J., J.-H. Choi, S. H. Um and B.-K. Oh. (2017). Electrochemical sensor for selective detection of norepinephrine using graphene sheets-gold nanoparticle complex modified electrode. *Korean Journal of Chemical Engineering* 34(4): 1129–1132.

Lee, H., E. Sun, D. Ham and R. Weissleder. (2008). Chip-NMR biosensor for detection and molecular analysis of cells. *Nat Med* 14(8): 869–74.

Lee, H., T. J. Yoon and R. Weissleder. (2009). Ultrasensitive detection of bacteria using core-shell nanoparticles and an NMR-filter system. *Angew Chem Int Ed Engl* 48(31): 5657–60.

Lee, H., T. J. Yoon, J. L. Figueiredo, F. K. Swirski and R. Weissleder. (2009). Rapid detection and profiling of cancer cells in fine-needle aspirates. *Proc Natl Acad Sci USA* 106(30): 12459–64.

Lee, J. H., Y. M. Huh, Y. W. Jun, J. W. Seo, J. T. Jang, H. T. Song, S. Kim, E. Chao, H. Yoon, J. S. Suh and J. Cheon. (2007). Artificially engineered magnetic nanoparticles for ultra-sensitive molecular imaging. *Nat Med* 13(1): 95–9.

Lee, J. H., B. K. Oh and J. W. Choi. (2013). Electrochemical sensor based on direct electron transfer of HIV-1 virus at Au nanoparticle modified ITO electrode. *Biosens Bioelectron* 49: 531–5.

Lee, T., S. Y. Park, H. Jang, G. H. Kim, Y. Lee, C. Park, M. Mohammadniaei, M. Lee and J. Min. (2019). Fabrication of electrochemical biosensor consisted of multi-functional DNA structure/porous au nanoparticle for avian influenza virus (H5N1) in chicken serum. *Mater Sci Eng C Mater Biol Appl* 99: 511–519.

Liang, S.-L. and D. W. Chan. (2007). Enzymes and related proteins as cancer biomarkers: A proteomic approach. *Clinica Chimica Acta* 381(1): 93–97.

Lochhead, M. J., K. Todorof, M. Delaney, J. T. Ives, C. Greef, K. Moll, K. Rowley, K. Vogel, C. Myatt, X. Zhang, C. Logan, C. Benson, S. Reed and R. T. Schooley. (2011). Rapid multiplexed immunoassay for simultaneous serodiagnosis of HIV-1 and coinfections. *J Clin Microbiol* 49(10): 3584–90.

Luppa, P. B., L. J. Sokoll and D. W. Chan. (2001). Immunosensors—principles and applications to clinical chemistry. *Clin Chim Acta* 314(1-2): 1–26.

Ma, C., M. Liang, L. Wang, H. Xiang, Y. Jiang, Y. Li and G. Xie. (2013). MultisHRP-DNA-coated CMWNTs as signal labels for an ultrasensitive hepatitis C virus core antigen electrochemical immunosensor. *Biosens Bioelectron* 47: 467–74.

Ma, Y., X. L. Shen, Q. Zeng, H. S. Wang and L. S. Wang. (2017). A multi-walled carbon nanotubes based molecularly imprinted polymers electrochemical sensor for the sensitive determination of HIV-p24. *Talanta* 164: 121–127.

Mayeux, R. (2004). Biomarkers: potential uses and limitations. *NeuroRx* 1(2): 182–8.

Mehra, N. K., K. Jain and N. K. Jain. (2016). Multifunctional carbon nanotubes in cancer therapy and imaging. pp. 421–453. *In*: Grumezescu, A. M. (ed.). Nanobiomaterials in Medical Imaging: Applications of Nanobiomaterials. Oxford, UK: Elsevier.

Meng, L., P. Wu, G. Chen, C. Cai, Y. Sun and Z. Yuan. (2009). Low potential detection of glutamate based on the electrocatalytic oxidation of NADH at thionine/single-walled carbon nanotubes composite modified electrode. *Biosens Bioelectron* 15; 24(6): 1751–6.

Miguel, O. B., Y. Gossuin, M. P. Morales, P. Gillis, R. N. Muller and S. Veintemillas-Verdaguer. (2007). Comparative analysis of the 1H NMR relaxation enhancement produced by iron oxide and core-shell iron-iron oxide nanoparticles. *Magn Reson Imaging* 25(10): 1437–41.

Mintmire, J. W., B. I. Dunlap and C. T. White. (1992). Are fullerene tubules metallic? *Phys Rev Lett* 68(5): 631–634.

Mirsky, V. M., M. Riepl and O. S. Wolfbeis. (1997). Capacitive monitoring of protein immobilization and antigen-antibody reactions on monomolecular alkylthiol films on gold electrodes. *Biosens Bioelectron* 12(9-10): 977–89.

Mízia, M. S., C. M. S. Silva Ana, V. M. Dias Bárbara, L. R. Silva Sérgio, Gomes-Filho Lauro, M. O. F. Goulart and R. F. Dutra. (2014). Electrochemical detection of dengue virus NS1 protein with a poly(allylamine)/carbon nanotube layered immunoelectrode.

Mody, V. V., R. Siwale, A. Singh and H. R. Mody. (2010). Introduction to metallic nanoparticles. *J Pharm Bioallied Sci* 2(4): 282–9.

Mujagic, Z., H. Mujagic and B. Prnjavorac. (2004). The relationship between circulating carcinoembryonic antigen (CEA) levels and parameters of primary tumor and metastases in breast cancer patients. *Med Arh* 58(1): 23–6.

Munge, B. S., C. E. Krause, R. Malhotra, V. Patel, J. S. Gutkind and J. F. Rusling. (2009). Electrochemical immunosensors for interleukin-6. Comparison of carbon nanotube forest and gold nanoparticle platforms. *Electrochem commun* 11(5): 1009–1012.

Murugaiyan, S. B., R. Ramasamy, N. Gopal and V. Kuzhandaivelu. (2014). Biosensors in clinical chemistry: An overview. *Adv Biomed Res* 3: 67.

Nabaei, V., R. Chandrawati and H. Heidari. (2018). Magnetic biosensors: Modelling and simulation. *Biosens Bioelectron* 103: 69–86.

Narang, J., C. Singhal, N. Malhotra, S. Narang, A. K. Pn, R. Gupta, R. Kansal and C. S. Pundir. (2016). Impedimetric genosensor for ultratrace detection of hepatitis B virus DNA in patient samples assisted by zeolites and MWCNT nano-composites. *Biosens Bioelectron* 86: 566–574.

Nehra, A., W. Chen, D. S. Dimitrov, A. Puri and K. P. Singh. (2017). Graphene oxide-polycarbonate track-etched nanosieve platform for sensitive detection of human immunodeficiency virus envelope glycoprotein. *ACS Appl Mater Interfaces* 9(38): 32621–32634.

Neumann, S. (2015). Biomarkers—Past and future. pp. 1–22. *In*: Seitz, H. and S. Schumacher (eds.). Biomarker Validation: Technological, Clinical and Commercial Aspects. Weinheim, Germany: Wiley-VCH Verlag GmbH and Co. KGaA.

Novelli, G., C. Ciccacci, P. Borgiani, M. P. Amati and E. Abadie. (2008). Genetic tests and genomic biomarkers: Regulation, qualification and validation. *Clinical Cases in Mineral and Bone Metabolism* 5(2): 149–154.

Osterfeld, S. J., H. Yu, R. S. Gaster, S. Caramuta, L. Xu, S. J. Han, D. A. Hall, R. J. Wilson, S. Sun, R. L. White, R. W. Davis, N. Pourmand and S. X. Wang. (2008). Multiplex protein assays based on real-time magnetic nanotag sensing. *Proc Natl Acad Sci USA* 105(52): 20637–40.

Park, D.-J., J.-H. Choi, W.-J. Lee, S. H. Um and B.-K. Oh. (2017). Selective electrochemical detection of dopamine using reduced graphene oxide sheets-gold nanoparticles modified electrode. *Journal of Nanoscience and Nanotechnology* 17(11): 8012–8018.

Parkin, S. S. S., C. Kaiser, A. Panchula, P. M. Rice, B. Hughes, M. Samant and S-H. Yang. (2004). Giant tunnelling magnetoresistance at room temperature with MgO (100) tunnel barriers. *Nat Mater* 3(12): 862–7.

Peng, S., C. Wang, J. Xie and S. Sun. (2006). Synthesis and stabilization of monodisperse Fe nanoparticles. *J Am Chem Soc* 128(33): 10676–7.

Perera, F. P. and I. B. Weinstein. (2000). Molecular epidemiology: recent advances and future directions. *Carcinogenesis* 21(3): 517–24.

Perez, J. M., L. Josephson, T. O'Loughlin, D. Hogemann and R. Weissleder. (2002). Magnetic relaxation switches capable of sensing molecular interactions. *Nat Biotechnol* 20(8): 816–20.

Piliarik, M., H. Sipova, P. Kvasnicka, N. Galler, J. R. Krenn and J. Homola. (2012). High-resolution biosensor based on localized surface plasmons. *Opt Express* 20(1): 672–80.

Pimkova, K., M. Bockova, K. Hegnerova, J. Suttnar, J. Cermak, J. Homola and Jan E. Dyr. (2012). Surface plasmon resonance biosensor for the detection of VEGFR-1—a protein marker of myelodysplastic syndromes. *Anal Bioanal Chem* 402(1): 381–7.

Pingarron, J. M., R. Villalonga and P. Yanez-Sedeno. (2013). Nanoparticle-modified electrodes for sensing. pp. 47–87. *In*: Pumera, M. (ed.). Nanomaterials for Electrochemical Sensing and Biosensing. Boca Raton, FL: CRC Press.

Pivac, N., G. Nedic, D. Kozaric-Kovacic, M. Nikolac, M. Grusibic-Ilic, M. Mustapic and D. Muck-Seler. (2013). Biomarkers as new tools to improve the diagnosis and treatment of PTSD. pp. 21–72. *In*: Wiederhold, B. K. (ed.). New Tools to Enhance Posttraumatic Stress Disorder Diagnosis and Treatment: Invisible Wounds of War. Amsterdam, Netherlands: IOS Press.

Polascik, T. J., J. E. Oesterling and A. W. Partin. (1999). Prostate specific antigen: a decade of discovery—what we have learned and where we are going. *J Urol* 162(2): 293–306.

Pujado, M. P. (2012). Carbon Nanotubes as Platforms for Biosensors with Electrochemical and Electronic Transduction. Heidelberg: Springer.

Quentin Palomar, Chantal Gondran, Robert Marks, Serge Cosnier and Michael Holzinger. (2018). Impedimetric quantification of anti-dengue antibodies using functional carbon nanotube deposits validated with blood plasma assays. *Electrochimica Acta* 274: 84–90, ISSN 0013-4686.

Rau, S., U. Hilbig and G. Gauglitz. (2014). Label-free optical biosensor for detection and quantification of the non-steroidal anti-inflammatory drug diclofenac in milk without any sample pretreatment. *Anal Bioanal Chem* 406(14): 3377–86.

Ravalli, A., C. G. da Rocha, H. Yamanaka and G. Marrazza. (2015). A label-free electrochemical affisensor for cancer marker detection: The case of HER2. *Bioelectrochemistry* 106: 268–275.

Ren, C., Q. Bayin, S. Feng, Y. Fu, X. Ma and J. Guo. (2020). Biomarkers detection with magnetoresistance-based sensors. *Biosens Bioelectron* 165: 112340.

Riedel, F., I. Zaiss, D. Herzog, K. Gotte, R. Naim and K. Hormann. (2005). Serum levels of interleukin-6 in patients with primary head and neck squamous cell carcinoma. *Anticancer Res* 25(4): 2761–5.

Riedel, T., C. Rodriguez-Emmenegger, A. de los Santos Pereira, A. Bedajankova, P. Jinoch, P. M. Boltovets and E. Brynda. (2014). Diagnosis of Epstein-Barr virus infection in clinical serum samples by an SPR biosensor assay. *Biosens Bioelectron* 55: 278–84.

Safina, G. (2012). Application of surface plasmon resonance for the detection of carbohydrates, glycoconjugates, and measurement of the carbohydrate-specific interactions: a comparison with conventional analytical techniques. A critical review. *Anal Chim Acta* 712: 9–29.

Sharma, A., E. Abdelfatah, M. Al Eissa and N. Ahuja. (2016). Prognostic epigenetics. pp. 177–195. *In*: Tollefsbol, T. O. (ed.). Medical Epigenetics. Amsterdam, Netherlands: Elsevier.

Shumyantseva, V. V., T. V. Bulko, E. G. Tikhonova, M. A. Sanzhakov, A. V. Kuzikov, R. A. Masamrekh, D. V. Pergushov, F. H. Schacher and L. V. Sigolaeva. (2020). Electrochemical studies of the interaction of rifampicin and nanosome/rifampicin with dsDNA. *Bioelectrochemistry* 140: 107736.

Siuzdak, K., P. Niedziałkowski, M. Sobaszek, T. Łęga, M. Sawczak, E. Czaczyk, K. Dziąbowska, T. Ossowski, D. Nidzworski and R. Bogdanowicz. (2019). Biomolecular influenza virus detection based on the electrochemical impedance spectroscopy using the nanocrystalline boron-doped diamond electrodes with covalently bound antibodies. *Sensors and Actuators B: Chemical* 280: 263–271, ISSN 0925-4005.

Srivastava, S. K., A. Shalabney, I. Khalaila, C. Gruner, B. Rauschenbach and I. Abdulhalim. (2014). SERS biosensor using metallic nano-sculptured thin films for the detection of endocrine disrupting compound biomarker vitellogenin. *Small* 10(17): 3579–87.

Strimbu, K. and J. A. Tavel. (2010). What are biomarkers? *Curr Opin HIV AIDS* 5(6): 463–6.

Sun, E. Y., L. Josephson, K. A. Kelly and R. Weissleder. (2006). Development of nanoparticle libraries for biosensing. *Bioconjug Chem* 17(1): 109–13.

Sun, K., Y. Chang, B. Zhou, X. Wang and L. Liu. (2017). Gold nanoparticles-based electrochemical method for the detection of protein kinase with a peptide-like inhibitor as the bioreceptor. *International Journal of Nanomedicine* 12: 1905–1915.

Taitt, C. R., G. P. Anderson and F. S. Ligler. (2016). Evanescent wave fluorescence biosensors: Advances of the last decade. *Biosens Bioelectron* 76: 103–12.

Tamanaha, C. R., S. P. Mulvaney, J. C. Rife and L. J. Whitman. (2008). Magnetic labeling, detection, and system integration. *Biosens Bioelectron* 24(1): 1–13.

Tancharoen, C., W. Sukjee, C. Thepparit, T. Jaimipuk, P. Auewarakul, A. Thitithanyanont and C. Sangma. (2019). Electrochemical biosensor based on surface imprinting for Zika virus detection in serum. *ACS Sens* 4(1): 69–75.

Teunissen, C. E., C. Verheul and E. A. J. Willemse. (2018). The use of cerebrospinal fluid in biomarker studies. pp. 3–20. *In*: Aminoff, M. J., F. Boller and D. F. Swaab (eds.). Cerebrospinal Fluid in Neurological Disorders. Amsterdam, Netherlands: Elsevier.

Thevenot, D. R., K. Toth, R. A. Durst and G. S. Wilson. (2001). Electrochemical biosensors: recommended definitions and classification. *Biosens Bioelectron* 16(1-2): 121–31.

Tran, Q. H., V. Q. Nguyen and A.-T. Le. (2013). Silver nanoparticles: Synthesis, properties, toxicology, applications and perspectives. *Advances in Natural Sciences: Nanoscience and Nanotechnology* 4(3): 033001.

Tu, H., K. Lin, Y. Lun and L. Yu. (2018). Magnetic bead/capture DNA/glucose-loaded nanoliposomes for amplifying the glucometer signal in the rapid screening of hepatitis C virus RNA. *Anal Bioanal Chem* 410: 3661–3669.

U, S. D., C. Y. Fu, K. S. Soh, B. Ramaswamy, A. Kumar and M. Olivo. (2012). Highly sensitive SERS detection of cancer proteins in low sample volume using hollow core photonic crystal fiber. *Biosens Bioelectron* 33(1): 293–8.

Waifalkar, P. P., A. D. Chougale, P. Kollu, P. S. Patil and P. B. Patil. (2018). Magnetic nanoparticle decorated graphene based electrochemical nanobiosensor for H_2O_2 sensing using HRP. *Colloids Surf B Biointerfaces* 167: 425–431.

Xu, J., D. Suarez and D. S. Gottfried. (2007). Detection of avian influenza virus using an interferometric biosensor. *Anal Bioanal Chem* 389(4): 1193–9.

Yan-Shi Chen and Jin-Hua Huang. (2009). Electrochemical sensing of bovine serum albumin at self-assembled SWCNTs on gold. *Diamond and Related Materials* 18(2–3): 516–519, ISSN 0925-9635.

Yi-Ting Wang, Lei Yu, Zi-Qiang Zhu, Jian Zhang, Jian-Zhong Zhu and Chun-hai Fan. (2009). Improved enzyme immobilization for enhanced bioelectrocatalytic activity of glucose sensor. *Sensors and Actuators B: Chemical* 136(2): 332–337, ISSN 0925-4005.

Yildirim, N., F. Long, C. Gao, M. He, H. C. Shi and A. Z. Gu. (2012). Aptamer-based optical biosensor for rapid and sensitive detection of 17beta-estradiol in water samples. *Environ Sci Technol* 46(6): 3288–94.

Yoo, S. M. and S. Y. Lee. (2016). Optical biosensors for the detection of pathogenic microorganisms. *Trends Biotechnol* 34(1): 7–25.

Yu, M. F., O. Lourie, M. J. Dyer, K. Moloni, T. F. Kelly and R. S. Ruoff. (2000). Strength and breaking mechanism of multiwalled carbon nanotubes under tensile load. *Science* 287(5453): 637–40.

Yuan, J., R. Duan, H. Yang, X. Luo and M. Xi. (2012). Detection of serum human epididymis secretory protein 4 in patients with ovarian cancer using a label-free biosensor based on localized surface plasmon resonance. *Int J Nanomedicine* 7: 2921–8.

Zhang, S., N. Huang, Q. Lu, M. Liu, H. Li and Y. Zhang. (2016). A double signal electrochemical human immunoglobulin G immunosensor based on gold nanoparticles-polydopamine functionalized reduced graphene oxide as a sensor platform and AgNPs/carbon nanocomposite as signal probe and catalytic substrate. *Biosensors and Bioelectronics* 77: 1078–1085.

CHAPTER 3

Innovative Nanobiosensors for Infectious Disease Diagnosis

Amitesh Anand[1],* and *Deependra Kumar Ban*[2],*

1. Introduction

We are living in an overcrowded world with accelerated changes. New scientific discoveries and inventions revolutionized our living standards, health care services, and communication. Globalization has also increased the risk of spreading microbial infection faster than any other time in human history. Unfortunately, the clinical diagnostic methods for infectious diseases have failed to match the pace of development. It still takes days to culture, perform a test, identify specific microbial agents (e.g., virus, bacteria, and fungus) and report to health management authorities to start the right treatment or precautionary measures. In case of critical situations such as epidemic/pandemic, bioterrorism, and emergency cases, rapid and accurate results become critical to start the right treatment and to initiate essential measures to prevent or slow down the spread.

These diagnoses are based on identification of specific biomolecules and every life form is composed of a common set of biomolecules (Figure 1); however, they often differ in the exact chemical compositions or arrangements. The standard pathogen diagnostics rely on the characteristic features of the organism and corresponding detection methods (Figure 2) (Table 1).

Each of these methods has its own merits and weaknesses and often a combination of these methods produces reliable results. Morphological analysis by microscopic examination and laboratory cultures are the most widely used methods (Table 1). There has been a substantial improvement in these methods catalyzed by technological advancements like electron microscopy, and high-throughput culturing. However,

[1] Department of Bioengineering, University of California San Diego, 9500 Gilman Drive, La Jolla, CA 92093, USA.
[2] Department of Mechanical and Aerospace Engineering University of California San Diego, 9500 Gilman Drive, La Jolla, CA 92093, USA.
* Corresponding author: amiteshanand@eng.ucsd.edu; dban@ucsd.edu

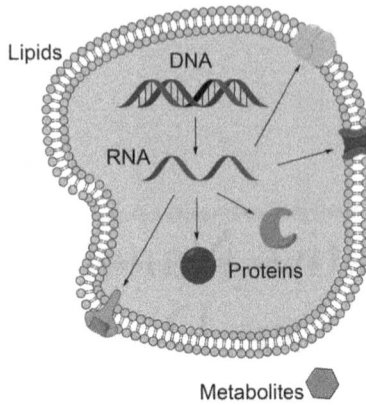

Fig. 1. Basic cellular constituents which are targeted for the microbial diagnosis.

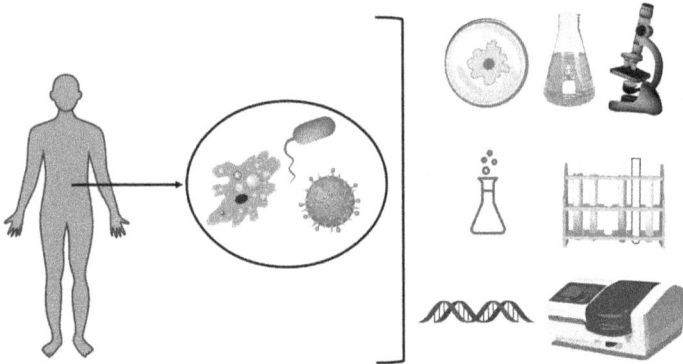

Fig. 2. Schematic representing standard clinical diagnosis methods for infectious diseases.

Table 1. Standard clinical method of pathogen detection.

S. No.	Method	Principle	Limitation
1	Microscopic examination	Gross morphology often aided by specific stains	Morphological redundancies, the limit of detection
2	Culture-based identification	Specific growth/culture condition	Cultivability, redundancies in nutrient requirements
3	Immunologic tests	Surrogate markers like antibodies or antigens	Less conclusive about the active presence of the pathogen
4	Nucleic acid-based identification methods	Genetic material identification	Expensive, technically challenging

the fraction of microbes is cultivable in the laboratory, which limits the diagnosis potential of standard techniques. Besides, these laboratories require multiple types of equipment, and the reliability of test results is highly dependent on the technical expertise of the laboratory personnel.

Several of these limitations are overcome by adopting molecular biology-based detection methods. In the COVID-19 era, PCR (Polymerase chain reaction) based infection diagnostics has become a household name (Wang et al. 2020). This method relies on the fact that every organism has a unique genetic material that can be extracted and analyzed to determine the presence or absence of a microorganism in a sample. An alternate molecular biology technique is the serological test which examines specific antibodies produced by the host immune system in response to a specific pathogen. Since serological tests are dependent on surrogate markers, there are significant chances of false-negative and false-positive results. There are several limitations to all these diagnostic methods, like technical know-how, extensive laboratory set-up, low throughput, etc. (Table 1). Therefore, despite decades of practical application and success in mitigating infectious disease burden, there is a need for next-generation diagnostics that can be easily upscaled without compromising precision and accuracy. Another highly desired feature is the easy deployment of the method to lesser equipped point-of-care facilities.

These features can be achieved by the integration of nanomaterials with the biosensor. Integration of nanomaterials makes the biosensor a portable and sensitive handheld device for rapid detection of pathogen-specific biomolecules as well as whole cell. In this chapter, we briefly describe the working principle of different biosensors, their components, and different nanomaterials utilized to prepare nanobiosensors. The later part of the chapter is focused on recent advancements in nanomaterials-based biosensors for virus and bacteria detection, which can be implemented for infectious disease diagnosis, and we conclude the chapter with future prospect of nanobiosensors.

2. Principle, Components, and Nanomaterials

A general working principle of nanobiosensor involves target molecules or cells and receptor interaction mediated modulation of the physiochemical signal into an electrical, optical, and mechanical signal. A nanobiosensor is developed to detect disease-specific antibodies, pathogens, nucleic acids (DNA and RNA), and metabolites (Prasad 2014). A nano-biosensor contains three major components: (i) bioreceptor, (ii) transducer and (iii) detector. Based on the interaction of the biomolecule with bio-receptor, signal generation, and detection, nano-biosensors can be further classified as optical (Figure 3A), electrochemical (Figure 3B), electrical (Figure 3C), and nanomechanical sensors (Figure 3D) (Table 2).

Optical biosensors: Working principle of optical sensors is based on interaction mediated stoke-shift, surface plasmon resonance shift, and color change, which can be detected by optical devices or mobile-based application (Figure 3A).

Electrochemical and electrical biosensors: Electrochemical sensors are based on the redox reaction mediated current (amperometry sensor), voltage (voltammetry sensor), and impedance (impedimetric sensor) modulation. Major components of electrochemical sensors include a working electrode, counter electrode, and a reference electrode (Figure 3B).

Electric sensors: Moreover, an electrical sensor detects modulation of electric properties such as resistance, conductivity, and current due to interaction between

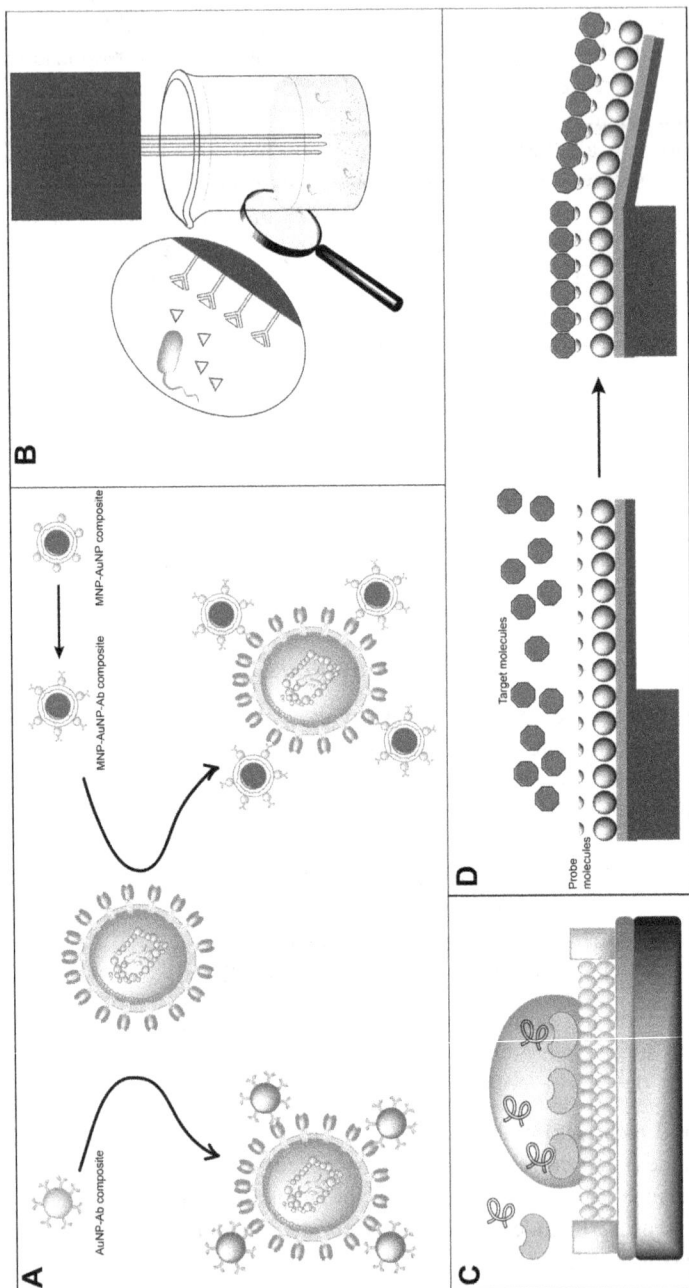

Fig. 3. Basic design and mechanism of different types of nanobiosensors. (A) Nanoparticle-based sensors; (B) Electrochemical sensors; (C) Electrical sensors; (D) Nanomechanical sensors.

Table 2. Different type of sensors and their advantages and limitations.

Method	Advantages	Limitations	References
Optical	Variety of dye and quantum dots available, b inding specific visible color detection	Stability of dye, limited conjugation chemistry, optical instrument	(Wilson 2012, Zhao et al. 2007, Zhang et al. 2009, Sanvicens et al. 2009)
Electrical and electrochemical	Label-free, sensitive Based on oxidation and reduction reaction, the interaction of the molecule	Cross-reactivity, limited shelf life and working temperature, and molecule stability	(Pumera et al. 2007, Islam et al. 2019, Monzó et al. 2015)
Nanomechanical	Based on the change in resonance frequency, amplitude	Sensitivity restricted by molecular weight, concentration	(Datar et al. 2009, Boisen and Thundat 2009, Ziegler 2004)

receptor and biomolecules. The major components of electric sensors contain bioreceptor, electric signal transducer, and detector (Figure 3C).

Nanomechanical sensors: Nanomechanical sensors' working principle is based on the variation in its resonance frequencies, and amplitude due to the interaction of a receptor molecule with a target on the sensor surface. It contains a piezoelectric signal generator, a cantilever with known resonance frequency/amplitude, and a detector (e.g., optical or electric detector). Interaction of target with receptor changes the resonance frequency of cantilever, which is detected by a detector (Figure 3D). Some sensors also contain additional parts such as a signal amplifier, filters, and actuators.

Nanomaterials: Nanomaterials play important role in the development of nano-biosensors (NBS) based handheld, highly sensitive devices, which are utilized for the development of nanobiosensors for rapid detection of infectious disease markers and whole microbial cells. Integration of different inorganic (e.g., metallic (Au, Ag), and semiconductor (ZnO)), and organic nanomaterials improve the NBS sensitivity and specificity severalfold (Abdel-Karim et al. 2020). It also reduced the volume of the sample used for detection to few microliters. Utilization of nanomaterials' NBS enhances the detection limits to a few hundred molecules per microliter of samples. Carbon-based nanomaterials such as a single atomic layer of graphene, graphene oxide, carbon nanotube, and fullerene were utilized to develop highly sensitive nano-biosensors (Jariwala et al. 2013). Quantum dots such as ZnSe and ZnS were used for optical biosensors (Koch 1993, Wegner and Hildebrandt 2015). Silicon and silicon nitride-based materials were bio-functionalized to build nanomechanical sensors (Datar et al. 2009).

3. Nanobiosensors for Virus Detection

Every diagnostic method is being put to test during the ongoing pandemic caused by the severe acute respiratory syndrome coronavirus 2 (SARS-CoV-2). Among many, one challenge the entire scientific community is facing is in scaling up the testing without compromising the detection sensitivity. The rapid antigen test has serious limitations of relatively high false negatives and therefore requires further confirmation by the PCR based molecular test. This has taken its toll on the mechanisms of curbing the spread of the virus. This bottleneck has motivated the scientific community to develop simpler and reliable detection methods.

Towards this, a Swiss group has developed a localized surface plasmon resonance (LSPR) biosensing system by combining the photothermal effect and plasmonic sensing transduction to detect the virus with high precision (Qiu et al. 2020). The sensor consists of a gold-nanoisland on a glass substrate and uses an artificially produced DNA receptor, which is complementary to the unique RNA sequence of the virus. Markita P. Landry and colleagues have developed a single-walled carbon nanotube (SWCNT)-based optical sensor, which has been noncovalently functionalized with a host protein, ACE2, that has high binding affinity for the SARS-CoV-2 spike protein (Pinals et al. 2020). This technique is claimed to be easily deployable into a point-of-care device to rapidly separate the patient samples into definite negatives and potential positives for further testing. Electrochemical sensors in conjugation with nanomaterials and the development of smart portable point of care devices will help to prevent the spread of infection (Bukkitgar et al. 2020).

Respiratory virus infections such as influenza (H1Nx), and severe acute respiratory syndrome (SARS) infection were detected by gold, silver, and magnetic nanoparticle, and quantum dot-based methods (Griffin et al. 2009). The gold nanoparticle-based colorimetric sensor was developed by functionalization of antisense oligonucleotides (ASOs), which is specific for N-gene (nucleocapsid phosphoprotein) of SARS-CoV-2. Gold nanoparticle-based ASO sensor can detect and discriminate 0.18 ng/uL of SARS-Cov2 in presence of MERS-COV viral RNA by the naked eye (Perez et al. 2003). SARS-CoV-2 spike, envelop and membrane protein-specific antibodies were derivatized on the gold nanoparticle surface. Antibody functionalized AuNP showed redshift in its optical density at 560 nm within 3 min (Ahmed et al. 2017). Magnetic nanoparticle (MNP) sensor was developed for rapid detection of SARS-CoV2. The change in magnetic response in the A.C. magnetic field due to the interaction of functionalized MNP was detected up to 0.084 nM (Zhong et al. 2020).

The rapid development of several portable diagnostic methods for the current SARS COV2 pandemic relies on the past continuous progress in nanotechnology and parallel advancement in the development of nanosensors. Gold nanoparticle size and distance-dependent surface energy transfer (SET) property were utilized to recognize highly contagious hepatitis C (HCV) RNA at 300 fM concentration (Griffin et al. 2009). A different strain of influenza infects millions of people every year. Norovirus is highly contagious and causes vomiting and diarrhea. Gold/magnetic nanoparticle decorated carbon nanotube-based sensor was developed to detect ~ 8 pM of H1N1 and Norovirus (Lee et al. 2018). Gold nanoparticle film on the different substrate was used to detect Caledonia/H1N1/1999 influenza virus in the limit of detection (LOD) of 50.5 pg/ml, while LOD for H3N2 detection was 24.3 PFU/ml, which is 100 times lower than traditional ELISA and sensitivity is greater than 500 times of immunochromatography (Ahmed et al. 2017).

Gold nanoparticle spatial assembly on a virus-like particle (VLP) and redshift in surface plasmon resonance due to plasmon coupling was utilized to optically detect

VLP (Niikura et al. 2009). Core-shell Ag-SiO$_2$ was derivatized using a specific guanine-richen anti-rHA protein aptamer for the H5N1 influenza virus. Thiazole orange fluorescence signal was used as a reporter molecule for target aptamer binding mediated G-quadruplex. Viral load of 2 ng/mL was detected using metal enhanced fluorescence (MEF) within 30 min (Pang et al. 2015).

Electrochemical nanosensors based on a nanoparticle-enabled smartphone (NES) nanoprobe were developed for rapid and sensitive detection of hepatitis B virus (HBV), HCV, and Zika virus (ZIKV). Viruses were captured on a microchip with specially labeled platinum nanoprobe, which induced gas bubble formation in presence of hydrogen peroxide. The distinctly visible pattern was formed by a clinically relevant sample containing 250 copies/ml of viruses captured in smartphones (Draz et al. 2020). Virus-specific antibody functionalized magnetic nanoparticles create a supramolecular assembly in presence of virus particles. The analysis of magnetic relaxation of these assemblies detects the 5 viral particles/10 μl without the need for extensive sample preparation (Perez et al. 2003). Graphene, a wonder material, was used to develop a field-effect transistor (GFET) based electrical sensor for detection of human immune deficiency virus (HIV) and associated disorders such as cardiovascular disorders and rheumatoid arthritis at fg/ml of LOD. This type of electrical nanosensors can be commissioned as a point of care diagnosis device (Islam et al. 2019).

Every year, dengue infection mediated hemorrhagic fever or dengue shock syndrome causes a large number of morbidity and mortality in developing countries. Developing a rapid and highly sensitive nano-biosensor for different dengue strain is an effective strategy to prevent and start early treatment. In this regard, graphene oxide and silicon dioxide nanocomposite-based electrochemical paper nanosensor was developed to detect genetic materials of four different dengue strains at pM to μM concentration (Singhal et al. 2020). Hepatitis B virus (HBV) causes hepatic infection and remains a constant threat to the world. Quantum dot (CdSe/ZnS QDs)-DNA nanosensor based on fluorescence resonance energy transfer (FRET) detected HBV's DNA and its single mismatch in one pot at the sensitivity of nM (Wang et al. 2010). An electrochemical sensor based on the electric pulse was developed to detect the hepatitis E virus (HEV) genotype. The sensitivity of detection is similar to RT-qPCR but the advantage is its portability and easy handling (Chowdhury et al. 2019). Polymer and graphene oxide composite based electrochemical sensors have been developed to detect ZIKV genetic materials. The detection limit of the developed nanosensor was equivalent to RT-PCR based testing (Tancharoen et al. 2019). Nanomechanical biosensors based on Boron Nitrite Nanotube detect a change in resonance frequency and bending deformation due to the interaction of viruses (Chaudhary et al. 2014). Silicon-based nanosensors were developed based on the shift of resonance frequency due to interaction of the virus with a specific antibody (Karamrezaei and Mokhtarian 2020). All the reported nanomaterial-based sensors described here showed very high sensitivity and work as a point of care of devices, which can be commissioned for a variety of viral strain detection without the need for a costly clinical setup.

4. Nanobiosensors for Bacteria Detection

A pathogenic bacterial infection causes several diseases and misdiagnosis or delay in diagnosis causes mortality and morbidity worldwide. Portable, sensitive and low-cost nanobiosensors are particularly useful in critical illness and analyze antibiotic resistance (Ahmed et al. 2014). Metal nanoparticles such as gold and silver were used to detect different bacterial pathogen (e.g., *Escherichia coli* (Joung et al. 2008), *Staphylococcus aureus* (Zhu et al. 2009)) by analyzing the shift in surface plasmon resonance. Nanoparticle-based bio-barcode nano-sensors not only amplify the signal but are also used in multiplexing. These nanosensors contain two major components: one, biomolecule (e.g., DNA, proteins, etc.) functionalized nanoparticle (e.g., AuNP, AgNP) bio-barcode, and second is receptor (e.g., antibody, antigen, peptide, and oligonucleotide) functionalized nanoparticle (e.g., magnetic nanoparticle). Interaction of two nanoparticle assemblies was separated using a magnetic field and further used to detect enhanced signals (e.g., color change, fluorescence, and biochip) (Figure 3) (Zhou et al. 2014, Xu et al. 2012).

Antibiotic-resistant *S. aureus* and *K. pneumoniae* were detected using bio-barcode-based fluorescent quantum dots and magnetic nanoparticles at 102 CFU/mL concentration using bacteria-specific fnbA, mecA, and wcaG genes (Cihalova et al. 2017). Similarly, fluorescent molecule tagged DNA bio-barcode sensors were used to detect 86 CFU/mL of *S. aureus* (Amini et al. 2019). *B. subtilis* and *S. enteritidis* were detected at fM concentration using bio-barcode nanosensor (Hill et al. 2007, Zhang et al. 2009). Another method of nanoparticle-based sensing is based on the conjugation of fluorescence and colorimetric molecules and analyze the evolution of fluorescence, luminescence, and color due to interaction with target molecules. The signal readout can be detected by a photometric instrument, mobile app-based scanning, and visible color difference. Bioligand functionalized magneto-fluorescent nanobeads were used for the detection of *Legionella pneumophila* bacteria as well as its separation and enrichment (Martynenko et al. 2019). Dye doped and antibody functionalized silica nanoparticles were used to detect disease-causing *E. coli* O157:H7 and its genetic marker detection within 20 min (Zhao et al. 2007). A rapid (< 30 min), sensitive, and multiplexed silica nanoparticle-based multicolored FRET (fluorescence resonance energy transfer) method was developed to detect disease-causing *E. coli*, *S. typhimurium*, and *S. aureus* (Wang et al. 2007). Luminescent gold nanocluster (Au NC) mediated rapid detection of gram-positive and negative bacteria as well as kanamycin resistance bacteria were performed as low as 600–700 CFU/mL (Goswami et al. 2018). Food born pathogen such as *Bacillus cereus, Hepatitis A, Shigella* causes disease; rapid identification of these bacteria in food is highly essential to prevent morbidity, and mortality. Nanoparticle nano-sensing improves the sensitivity and speed of detection of food borne pathogens.

Multiplexed magnetic nanoparticle-based rapid detection of specific DNA marker of *E. coli* O157:H7, *S. enterica, V. cholera,* and *C. jejuni* was performed by hybridization of PCR product and microscope or CCD or by visible eye based detection up to 316 CFU/mL (Li et al. 2013). Engineered phage M13 conjugated with gold nanoparticles was used for colorimetric detection of approximately 100 cells of pathogens (e.g., two strains of *E. coli, P. aeruginosa,* and *V. cholerae*) within an hour and without cross-reactivity (Peng and Chen 2019).

Microimpedance biosensor was developed by conjugation of *E. coli* specific antibody. Based on the change in impedance, different concentration of *E. coli* was discriminated within 5 min (Radke and Alocilja 2004). Urease functionalized magnetic nanoparticle (MNP) based electrochemical nanoarray sensor was developed for detection of *S. aureus, A. junii, V. harveyi, M. luteus, E. tarda, V. Parahemolyticus,* and *E. coli.* Nanosensor is based on the hydrolysis of urea and change in pH. MNP-urease based electrochemical sensor measured the bacteria with 90.7% accuracy within 30 min. Interfacing of nanomaterials' biomolecules enhances its detection potential. Derivatization of graphene surface with bacteria-specific peptides can be utilized to develop a highly sensitive detection platform. Mansoor et al. were able to develop a highly sensitive graphene-based nanosensor with the potential of detecting a single cell. The developed graphene sensors were implanted for remote monitoring of bacteria in saliva (Mannoor et al. 2012). Low cost, rapid, and sensitive bacteria diagnostic sensors are boon for mass scale deployment, so that it will be accessible to the large population. Zinc oxide nanorods based electrochemical sensor based on voltammetry analysis showed great potential in this regard. It showed response within 15 min at a limit of 103 CFU/ml of bacteria (Al-Fandi Mohamed Ghazi et al. 2018). Food mediated infection is another problem. Electrochemical genosensors are highly sensitive nanosensors for the detection of *Escherichia coli* O157:H7 in food samples (Abdalhai et al. 2015).

5. Future of Nano-Biosensor

These sensors hold immense potential in enabling better disease monitoring and preventive policy decisions. Various market analysts are already projecting the global nanosensor market to jump from the expected revenue generation worth $536.6 million in 2019 to reach $1,321.30 million by 2026 (Allied Analytics LLP 2019). Advancement in nanotechnology and fabrication facilities facilitated the manufacturing and testing of the performance of nanobiosensor for different applications. Miniaturization of the biochip, development of powerless or lower power devices, and integration of nanobiosensors with microfluidic-based devices provided the platform to develop automated lab-on-chip devices. Portability of nanobiosensors to the remote location, ability to work non-clinical set, and the need for no clinical expert for analysis make NBS a device with great future promise. In wake of the current pandemic and the prospect of rapid and portable diagnosis platform, nanobiosensor can be commissioned in passenger immigration points, clinics, and emergency wards of hospitals, trauma centers, and biohazard monitoring situations. It can be further developed as a wearable sensor for continuous monitoring and data transmission by Bluetooth and wifi-based devices to a remote location to an expert for a quick decision in case of an epidemic, bioterrorism, and pandemic.

References

Abdalhai, Mandour H., António Maximiano Fernandes, Xiaofeng Xia, Abubakr Musa, Jian Ji and Xiulan Sun. (2015). Electrochemical genosensor to detect pathogenic bacteria (*Escherichia coli* O157:H7) as applied in real food samples (Fresh Beef) to improve food safety and quality control. *Journal of Agricultural and Food Chemistry* 63(20): 5017–25. https://doi.org/10.1021/acs.jafc.5b00675.

Abdel-Karim, R., Y. Reda and A. Abdel-Fattah. (2020). Review—nanostructured materials-based nanosensors. *Journal of The Electrochemical Society* 167(3): 037554. https://doi.org/10.1149/1945-7111/ab67aa.

Ahmed, Asif, Jo V. Rushworth, Natalie A. Hirst and Paul A. Millner. (2014). Biosensors for whole-cell bacterial detection. *Clinical Microbiology Reviews* 27(3): 631. https://doi.org/10.1128/CMR.00120-13.

Ahmed, Syed Rahin, Jeonghyo Kim, Van Tan Tran, Tetsuro Suzuki, Suresh Neethirajan, Jaebeom Lee and Enoch Y. Park. (2017). *In situ* self-assembly of gold nanoparticles on hydrophilic and hydrophobic substrates for influenza virus-sensing platform. *Scientific Reports* 7(1): 44495. https://doi.org/10.1038/srep44495.

Al-Fandi Mohamed Ghazi, Alshraiedeh Nid'a Hamdan, Oweis Rami Joseph, Hayajneh Rawan Hassan, Alhamdan Iman Riyad, Alabed Rama Adel and Al-Rawi Omar Farhan. (2018). Direct electrochemical bacterial sensor using ZnO nanorods disposable electrode. *Sensor Review* 38(3): 326–34. https://doi.org/10.1108/SR-06-2017-0117.

Allied Analytics LLP. (2019). Nanosensors Market by Type and Application: Global Opportunity Analysis and Industry Forecast, 2018–2026, October 2019. 4989461. https://www.researchandmarkets.com/reports/4989461/nanosensors-market-by-type-and-application?utm_source=dynamic&utm_medium=BW&utm_code=9nmrm3&utm_campaign=1355042+-+Global+Nanosensors+Market+Expected+to+Grow+in+Value+to+%241%2c+321.30+Million+with+a+CAGR+of+11%25&utm_exec=anwr281bwd.

Amini, Ali, Mehdi Kamali, Bahram Amini, Azam Najafi, Asghar Narmani, Leila Hasani, Jamal Rashidiani, Hamid Kooshki and Narges Elahi. (2019). Bio-barcode technology for detection of *Staphylococcus aureus* protein a based on gold and iron nanoparticles. *International Journal of Biological Macromolecules* 124(March): 1256–63. https://doi.org/10.1016/j.ijbiomac.2018.11.123.

Boisen, Anja and Thomas Thundat. (2009). Design & fabrication of cantilever array biosensors. *Materials Today* 12(9): 32–38.

Bukkitgar, Shikandar D., Nagaraj P. Shetti and Tejraj M. Aminabhavi. (2020). Electrochemical investigations for COVID-19 detection-A comparison with other viral detection methods. *Chemical Engineering Journal*, November, 127575. https://doi.org/10.1016/j.cej.2020.127575.

Chaudhary, G., K. K. Singh, A. Mittal and N. Sood. (2014). Design and simulation of nano-mechanical resonator for virus detection. pp. 170–74. *In*: 2014 IEEE Sensors Applications Symposium (SAS). https://doi.org/10.1109/SAS.2014.6798940.

Chowdhury, Ankan Dutta, Kenshin Takemura, Tian-Cheng Li, Tetsuro Suzuki and Enoch Y. Park. (2019). Electrical pulse-induced electrochemical biosensor for hepatitis E virus detection. *Nature Communications* 10(1): 3737. https://doi.org/10.1038/s41467-019-11644-5.

Cihalova, Kristyna, Dagmar Hegerova, Ana Maria Jimenez, Vedran Milosavljevic, Jiri Kudr, Sylvie Skalickova, David Hynek, Pavel Kopel, Marketa Vaculovicova and Vojtech Adam. (2017). Antibody-free detection of infectious bacteria using quantum dots-based barcode assay. *Journal of Pharmaceutical and Biomedical Analysis* 134(February): 325–32. https://doi.org/10.1016/j.jpba.2016.10.025.

Datar, Ram, Seonghwan Kim, Sangmin Jeon, Peter Hesketh, Scott Manalis, Anja Boisen and Thomas Thundat. (2009). Cantilever Sensors: Nanomechanical Tools for Diagnostics.

Draz, Mohamed, S., Anish Vasan, Aradana Muthupandian, Manoj Kumar Kanakasabapathy, Prudhvi Thirumalaraju, Aparna Sreeram, Sanchana Krishnakumar, Vinish Yogesh, Wenyu Lin, Xu G. Yu, Raymond T. Chung and Hadi Shafiee. (2020). Virus detection using nanoparticles and deep neural network–enabled smartphone system. *Science Advances* 6(51): eabd5354. https://doi.org/10.1126/sciadv.abd5354.

Goswami, Upashi, Amaresh Kumar Sahoo, Arun Chattopadhyay and Siddhartha Sankar Ghosh. (2018). *In situ* synthesis of luminescent Au nanoclusters on a bacterial template for rapid detection, quantification, and distinction of kanamycin-resistant bacteria. *ACS Omega* 3(6): 6113–19. https://doi.org/10.1021/acsomega.8b00504.

Griffin, Jelani, Anant Kumar Singh, Dulal Senapati, Patsy Rhodes, Kanieshia Mitchell, Brianica Robinson, Eugene Yu and Paresh Chandra Ray. (2009). Size- and distance-dependent nanoparticle surface-energy transfer (NSET) method for selective sensing of hepatitis C virus RNA. *Chemistry – A European Journal* 15(2): 342–51. https://doi.org/10.1002/chem.200801812.

Hill, Haley D., Rafael A. Vega and Chad A. Mirkin. (2007). Nonenzymatic detection of bacterial genomic DNA using the bio bar code assay. *Analytical Chemistry* 79(23): 9218–23. https://doi.org/10.1021/ac701626y.

Hosseini-Ara, Reza, Amir Hossein Karamrezaei and Ali Mokhtarian. (2020). Exact analysis of antibody-coated silicon biological nano-sensors (SBNSs) to identify viruses and bacteria. *Microsystem Technologies* 26(2): 509–16. https://doi.org/10.1007/s00542-019-04533-w.

Islam, Saurav, Shruti Shukla, Vivek K. Bajpai, Young-Kyu Han, Yun Suk Huh, Ashok Kumar, Arindam Ghosh and Sonu Gandhi. (2019). A smart nanosensor for the detection of human immunodeficiency virus and associated cardiovascular and arthritis diseases using functionalized graphene-based transistors. *Biosensors and Bioelectronics* 126(February): 792–99. https://doi.org/10.1016/j.bios.2018.11.041.

Jariwala, Deep, Vinod K. Sangwan, Lincoln J. Lauhon, Tobin J. Marks and Mark C. Hersam. (2013). Carbon nanomaterials for electronics, optoelectronics, photovoltaics, and sensing. *Chemical Society Reviews* 42(7): 2824–60.

Joung, Hyou-Arm, Nae-Rym Lee, Seok Ki Lee, Junhyoung Ahn, Yong Beom Shin, Ho-Suk Choi, Chang-Soo Lee, Sanghyo Kim and Min-Gon Kim. (2008). High sensitivity detection of 16s RRNA using peptide nucleic acid probes and a surface plasmon resonance biosensor. *Analytica Chimica Acta* 630(2): 168–73. https://doi.org/10.1016/j.aca.2008.10.001.

Koch, Stephan W. (1993). Semiconductor Quantum Dots. Vol. 2. World Scientific.

Lee, Jaewook, Masahiro Morita, Kenshin Takemura and Enoch Y. Park. (2018). A multi-functional gold/iron-oxide nanoparticle-CNT hybrid nanomaterial as virus DNA sensing platform. *Biosensors and Bioelectronics* 102(April): 425–31. https://doi.org/10.1016/j.bios.2017.11.052.

Li, Song, Hongna Liu, Yan Deng, Lin Lin and Nongyue He. (2013). Development of a magnetic nanoparticles microarray for simultaneous and simple detection of foodborne pathogens. *Journal of Biomedical Nanotechnology* 9(7): 1254–60.

Mannoor, Manu S., Hu Tao, Jefferson D. Clayton, Amartya Sengupta, David L. Kaplan, Rajesh R. Naik, Naveen Verma, Fiorenzo G. Omenetto and Michael C. McAlpine. (2012). Graphene-based wireless bacteria detection on tooth enamel. *Nature Communications* 3(1): 763. https://doi.org/10.1038/ncomms1767.

Martynenko, Irina, V., Dragana Kusić, Florian Weigert, Shelley Stafford, Fearghal C. Donnelly, Roman Evstigneev, Yulia Gromova, Alexander V. Baranov, Bastian Rühle, Hans-Jörg Kunte, Yurii K. Gun'ko and Ute Resch-Genger. (2019). Magneto-fluorescent microbeads for bacteria detection constructed from superparamagnetic Fe_3O_4 nanoparticles and AIS/ZnS quantum dots. *Analytical Chemistry* 91(20): 12661–69. https://doi.org/10.1021/acs.analchem.9b01812.

Monzó, Javier, Ignacio Insua, Francisco Fernandez-Trillo and Paramaconi Rodriguez. (2015). Fundamentals, achievements and challenges in the electrochemical sensing of pathogens. *Analyst* 140(21): 7116–28. https://doi.org/10.1039/C5AN01330E.

Niikura, Kenichi, Keita Nagakawa, Noriko Ohtake, Tadaki Suzuki, Yasutaka Matsuo, Hirofumi Sawa and Kuniharu Ijiro. (2009). Gold nanoparticle arrangement on viral particles through carbohydrate recognition: a non-cross-linking approach to optical virus detection. *Bioconjugate Chemistry* 20(10): 1848–52. https://doi.org/10.1021/bc900255x.

Pang, Yuanfeng, Zhen Rong, Junfeng Wang, Rui Xiao and Shengqi Wang. (2015). A fluorescent aptasensor for H5N1 influenza virus detection based-on the core–shell nanoparticles metal-enhanced fluorescence (MEF). *Biosensors and Bioelectronics* 66(April): 527–32. https://doi.org/10.1016/j.bios.2014.10.052.

Peng, Huan and Irene A. Chen. (2019). Rapid colorimetric detection of bacterial species through the capture of gold nanoparticles by chimeric phages. *ACS Nano* 13(2): 1244–52. https://doi.org/10.1021/acsnano.8b06395.

Perez, J. Manuel, F. Joseph Simeone, Yoshinaga Saeki, Lee Josephson and Ralph Weissleder. (2003). Viral-induced self-assembly of magnetic nanoparticles allows the detection of viral particles in biological media. *Journal of the American Chemical Society* 125(34): 10192–93. https://doi. org/10.1021/ja036409g.

Pinals, Rebecca L., Francis Ledesma, Darwin Yang, Nicole Navarro, Sanghwa Jeong, John E. Pak, Lili Kuo, Yung-Chun Chuang, Yu-Wei Cheng and Hung-Yu Sun. (2020). Rapid SARS-CoV-2 detection by carbon nanotube-based near-infrared nanosensors. *MedRxiv*.

Prasad, Shalini. (2014). Nanobiosensors: The future for diagnosis of disease? *Configurations* 8(9).

Pumera, Martin, Samuel Sánchez, Izumi Ichinose and Jie Tang. (2007). Electrochemical nanobiosensors. *Sensors and Actuators B: Chemical* 123(2): 1195–1205. https://doi.org/10.1016/j.snb.2006.11.016.

Qiu, Guangyu, Zhibo Gai, Yile Tao, Jean Schmitt, Gerd A. Kullak-Ublick and Jing Wang. (2020). Dual-functional plasmonic photothermal biosensors for highly accurate severe acute respiratory syndrome coronavirus 2 detection. *ACS Nano* 14(5): 5268–77. https://doi.org/10.1021/acsnano.0c02439.

Radke, S. M. and E. C. Alocilja. (2004). Design and fabrication of a microimpedance biosensor for bacterial detection. *IEEE Sensors Journal* 4(4): 434–40. https://doi.org/10.1109/JSEN.2004.830300.

Sanvicens, Nuria, Carme Pastells, Nuria Pascual and M.-Pilar Marco. (2009). Nanoparticle-based biosensors for detection of pathogenic bacteria. *TrAC Trends in Analytical Chemistry* 28(11): 1243–52. https://doi.org/10.1016/j.trac.2009.08.002.

Singhal, Chaitali, Sudheesh K. Shukla, Akshay Jain, Chandrashekhar Pundir, Manika Khanuja, Jagriti Narang and Nagaraj P. Shetti. (2020). Electrochemical multiplexed paper nanosensor for specific dengue serotype detection predicting pervasiveness of DHF/DSS. *ACS Biomaterials Science & Engineering* 6(10): 5886–94. https://doi.org/10.1021/acsbiomaterials.0c00976.

Tancharoen, Chompoonuch, Wannisa Sukjee, Chutima Thepparit, Thitigun Jaimipuk, Prasert Auewarakul, Arunee Thitithanyanont and Chak Sangma. (2019). Electrochemical biosensor based on surface imprinting for zika virus detection in serum. *ACS Sensors* 4(1): 69–75. https://doi.org/10.1021/acssensors.8b00885.

Wang, Lin, Wenjun Zhao, Meghan B. O'Donoghue and Weihong Tan. (2007). Fluorescent nanoparticles for multiplexed bacteria monitoring. *Bioconjugate Chemistry* 18(2): 297–301. https://doi.org/10.1021/bc060255n.

Wang, Wenling, Yanli Xu, Ruqin Gao, Roujian Lu, Kai Han, Guizhen Wu and Wenjie Tan. (2020). Detection of SARS-CoV-2 in different types of clinical specimens. *Jama* 323(18): 1843–44.

Wang, Xiang, Xinhui Lou, Yi Wang, Qingchuan Guo, Zheng Fang, Xinhua Zhong, Hongju Mao, Qinghui Jin, Lei Wu, Hui Zhao and Jianlong Zhao. (2010). QDs-DNA nanosensor for the detection of hepatitis B virus DNA and the single-base mutants. *Biosensors and Bioelectronics* 25(8): 1934–40. https://doi.org/10.1016/j.bios.2010.01.007.

Wegner, K. David and Niko Hildebrandt. (2015). Quantum dots: bright and versatile *in vitro* and *in vivo* fluorescence imaging biosensors. *Chemical Society Reviews* 44(14): 4792–4834. https://doi.org/10.1039/C4CS00532E.

Wilson, Michael L. (2012). Malaria rapid diagnostic tests. *Clinical Infectious Diseases* 54(11): 1637–41. https://doi.org/10.1093/cid/cis228.

Xu, Jin, Bingying Jiang, Jiao Su, Yun Xiang, Ruo Yuan and Yaqin Chai. (2012). Background current reduction and biobarcode amplification for label-free, highly sensitive electrochemical detection of pathogenic DNA. *Chemical Communications* 48(27): 3309–11.

Zhang, Deng, David J. Carr and Evangelyn C. Alocilja. (2009). Fluorescent bio-barcode DNA assay for the detection of Salmonella Enterica Serovar Enteritidis. *Selected Papers from the Tenth World Congress on Biosensors Shangai, China, May 14–16, 2008* 24(5): 1377–81. https://doi.org/10.1016/j.bios.2008.07.081.

Zhao, Wenjun, Lin Wang and Weihong Tan. (2007). Fluorescent nanoparticle for bacteria and DNA detection. pp. 129–35. *In*: Warren C. W. Chan (ed.). Bio-Applications of Nanoparticles. New York, NY: Springer New York. https://doi.org/10.1007/978-0-387-76713-0_10.

Zhong, Jing, Enja Laureen Roesch, Thilo Viereck, Meinhard Schilling and Frank Ludwig. (2020). Rapid and sensitive detection of SARS-CoV-2 with functionalized magnetic nanoparticles. *ArXiv Preprint ArXiv:2010.03886*.

Zhou, Zhenpeng, Tian Li, Hongduan Huang, Yang Chen, Feng Liu, Chengzhi Huang and Na Li. (2014). A dual amplification strategy for DNA detection combining bio-barcode assay and metal-enhanced fluorescence modality. *Chemical Communications* 50(87): 13373–76.

Zhu, Shaoli, ChunLei Du and Yongqi Fu. (2009). Localized surface plasmon resonance-based hybrid Au–Ag nanoparticles for detection of *Staphylococcus aureus* Enterotoxin B. *Optical Materials* 31(11): 1608–13. https://doi.org/10.1016/j.optmat.2009.03.009.

Ziegler, Christiane. (2004). Cantilever-based biosensors. *Analytical and Bioanalytical Chemistry* 379(7–8): 946–59.

CHAPTER 4

Wearable Devices for Real-time Disease Monitoring in Healthcare

Pramila Jakhar,[1,]* *Pandey Rajagopalan,*[2] *Mayoorika Shukla*[3] and *Vipul Singh*[3]

1. Introduction

Currently, the development of Point-of-care technology (POCT) in medical industry is dramatically evolving with the rise in the number of elderly people. The focus is on the development of biosensors which can provide real-time monitoring and personalized healthcare. In this regard, point-of-care technology (POCT) facilitate easy and fast diagnostics for the patients who have limited access to health services, whereas conventional disease diagnostic tests in hospitals and laboratories are time-taking and costly, and require highly trained personnel. With the focus of technology development towards personalized medicine, wearable sensors will find progress rate of approximately 38% from 2017 to 2025 annually, among which the smart watch is anticipated to have a high growth rate (Guk et al. 2019).

Wearable biosensors are attached to the human body (Figure 1) and based on the physiological signals like heart rate, respiratory rate, blood pressure, temperature, etc., the physical signal is used to get the clinical information of the patient. These devices are noninvasive biosensors with real-time detection and facilitate continuous monitoring of patients which helps in preliminary medical diagnosis.

The wearable biosensors have evolved progressively in the form of pacemaker, watches, rings, glasses, clothing, bandages, and contact lenses that can be easily attached to individual's body, and provide the information (McCaul et al. 2017,

[1] Department of Electrical & Electronics Engineering, BITS Pilani K K Birla Goa Campus, Goa, 403726, India.
[2] College of Information Science & Electronic Engineering, Zhejiang University, Hangzhou 310027, China.
[3] Molecular and Nanoelectronics Research Group (MNRG), Discipline of Electrical Engineering, Indian Institute of Technology Indore, Simrol, Indore, 453-552, Madhya Pradesh, India.
* Corresponding author: pramila@goa.bits-pilani.ac.in

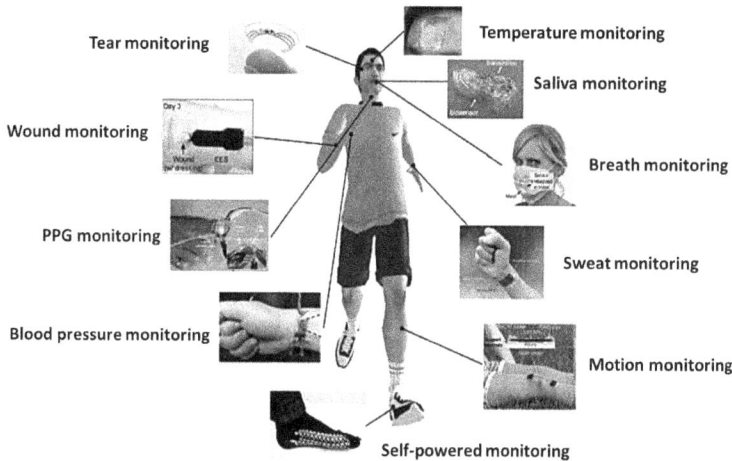

Fig. 1. Representative examples of wearable healthcare devices. Reprinted with permission from Zheng Lou (Materials Science & Engineering R 140 100523, 2020).

Yao et al. 2011). Due to easy usage and portability, these attributes differentiate the wearable biosensors from existing devices. Recently, many reports on the advanced achievements in wearable sensors for health monitoring have been reported (Lu et al. 2019). Yao et al. fabricated a contact lens integrated with amperometric glucose sensor for the monitoring of tear glucose level (Yao et al. 2011).

2. Common Diseases Where Real Time Monitoring Can be a Boon and How

Human health can be diagnosed based on the biological signal produced by human body. The wearable healthcare device can sense these biological signals for early diagnosis of diseases. Variety of wearable sensors have been developed to monitor the biological signals based on the characteristics of physical or chemical signals. Herein, the advancements in wearable healthcare devices that can be utilized for real-time monitoring of patients have been presented.

2.1 Glucose Monitoring

One of the popular wearable devices is smart watch. First commercially available Glucowatch is by biographer (Cygnus Inc., Redwood City, CA, USA). This Glucowatch is the first commercially accepted non-invasive glucowatch for glucose detection which was approved by the Food and Drug Administration (FDA). It actually provides the information based on the concentration of glucose obtained from skin interstitial fluid by electrochemical method. Pirovano et al. demonstrated a smart watch shown in Figure 2(a), which can monitor sodium and potassium constituents in the body from sweat in real time (Pirovano et al. 2020). Additionally, smart contact lenses can detect the physiological information in the tears. Variety of contact lenses have been developed using optical and electrical techniques to monitor the glucose concentration in tear. Yao et al. developed a contact lens for

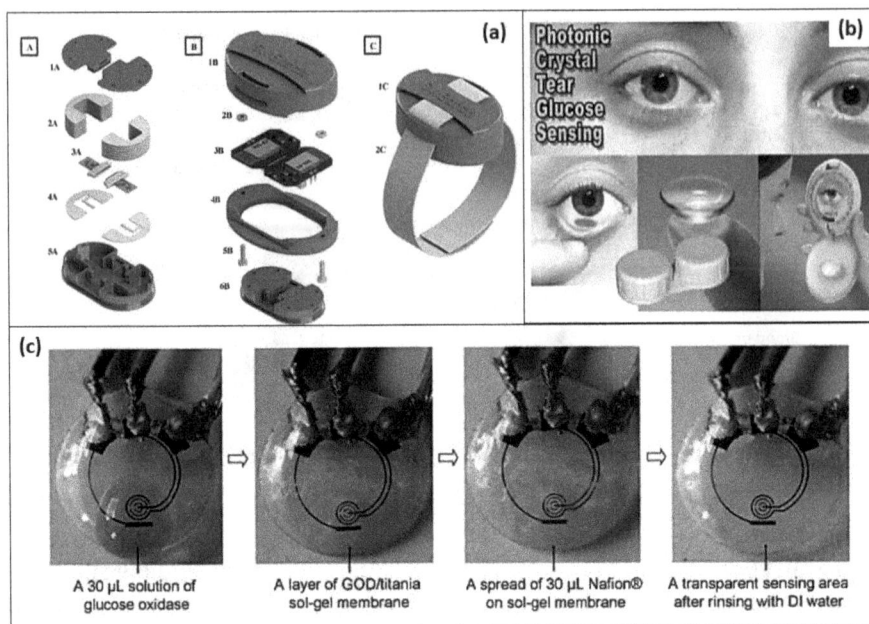

Fig. 2. Wearable glucose biosensor. (a) Smart watch A. Microfluidic unit; 1A. 3D printed lid using rigid polymer VEROBLACK with flexible polymer TANGOBLACK contact pads, 2A. Sorbent material, 3A. K+ and Na+ ISEs, 4A. Sorbent material contact layer, 5A. Sweat harvester. B. Platform body; 1B. 3D printed platform body, 2B and 5B. Nylon M3 × M12 mm nut and bolt arrangement 3B. Shimmer single channel PCBs, 2 × 3.7 V 155 mAh battery, 4B. Sealing and press fit connector, 6B. Microfluidic unit, C. 1C fully enclosed 3D printed SwEatch platform, 2C. Image adapted from Paolo Pirovano (Talanta 219, 121145, 2020). (b) Pictorial representation of use of glucose-sensing material to determine glucose concentration in tear fluid. Image reprinted with the permission from Alexeev (Clinical Chemistry 50, Vol 12, 2004). (c) Sequential images of sensor pre-treatment with GOD/titania/Nafion. Reprinted with the permission from Huanfen Yao (Biosensors and Bioelectronics 26: 3290–3296, 2011).

detection of glucose from tear presented in Figure 2(b) (Yao et al. 2011). This sensor was fabricated by developing the microstructures on a polymer substrate. Alexeev et al. fabricated photonic crystals based sensor for non-invasive detection of glucose in tears (Alexeev et al. 2004). The developed sensor is shown in Figure 2(c). Moreover, a fluorescent contact lens with hand-held photofluorometer was developed for the detection of glucose concentration non-invasively. The contact lenses were fabricated by using hydrogel nanospheres consisting of tetramethylrhodamine isothiocyanate concanavalin A and fluorescein isothiocyanate dextran (March et al. 2006).

The principle of detection of glucose was based on the fact that with the addition of glucose it causes displacement of FITC-dextran from the combined position on TRITC-Con A and thereby increased the fluorescent intensity with the increased glucose concentration. Additionally, optical sensor as wearable lens was introduced to continuously monitor glucose in physiological conditions. Moreover, with the increasing demand of glucose sensors, the development towards minimally invasive glucose monitoring devices has been focused, as illustrated in Table 1.

Table 1. Selected examples of commercial noninvasive or minimally invasive biosensors.

Product, Company	Analyte	Wearable Platform	Monitoring Mechanism	Website
Smart contact lens, Google and Novartis	Glucose in tears	Contact lens	Electrochemistry	https://verily.com/projects/sensors/smart-lens-program/
GlucoWatch, Cygnus Inc.	Glucose in ISF	Watch type	Electrochemistry	No longer available
GlucoWise, MediWise	Blood glucose	Finger clip	Radio frequency	http://www.gluco-wise.com/
Freestyle Libre, Abbott	Glucose in ISF	Patch	Electrochemistry	https://www.freestylelibre.us/

Source: Data from Kim et al. Wearable biosensors for healthcare monitoring. *Nature Biotechnology* 37(2019): 389–406.

2.2 *Parkinson's Disease*

Detection of motion in human body is important for monitoring of medical care, and sports performance, and is important to determine physical disability of individual. Frequently analyzing the body activities can detect abnormal walk patterns and sudden trembles in the hands, which are the signs of incurable diseases like diabetes, Parkinson's disease, and Alzheimer's disease. The detection of the signs facilitates the early diagnosis and treatment of these diseases. The movement recognition in human body (for example, bending of the hands, legs, arms, and spine are large scale movements and the delicate movements include movement of neck, chest and face during speaking and emotional expression) is based on the strain measurement. Several research articles reported the development of wearable sensors for recording human body motion by measuring real-time signals. Hands and limbs movements are important for most of the actions of human daily life. Yamada et al. fabricated a stretchable wearable device using SWCNT films and applying them with bandages and clothing. This group demonstrated that this sensor can monitor variety of human motion viz. movement, typing, breathing and speech. During walking, the skin on human body continuously gets stretched and contracts. The developed device could easily detect, and also distinguish, various human motions involving the movement of the knee while bending, running, squatting and jumping, and permutations of these motions (Yamada et al. 2011). Zhou et al. suggested a bio-inspired stretchable nanogenerator for underwater energy harvesting motivated by the structure of an electric eel (Zou et al. 2019).

Similarly, the detection of finger movement is critical as their motion is involved in many of the activities of hand. Several flexible strain measuring devices have been reported to observe fingers' movement. These sensors are usually integrated with gloves or wearing on target fingers. As the finger has movement, it changes the stretching or bending of the sensor, which relates the change in strain caused by the bending or curvature of finger with the electrical parameters (Kanaparthi and Badhulika 2016,

Li et al. 2017). Li et al. fabricated a superhydrophobic MWCNT/TPE smart material showing high-performance sensing capability for stretching, bending, and torsion. This bending and stretching of finger was monitored by the relative increase and decrease in the resistance, respectively. Additionally, the superhydrophobic property of MWCNT/TPE exhibited the reliable performance of sensor in wet environmental conditions. Chen et al. fabricated a highly flexible conductive fiber strain sensor, which can monitor different gestures of fingers based on the resistance change of the optical fiber sensor on each finger joint (Li et al. 2017).

2.3 *Monitoring of Respiration Rate*

Monitoring of the respiration rate is reliable information for routine health check-ups as the abnormal respiratory rate is indicative of patients' health at risk. It is the primary symptom of various diseases viz. asthma, anemia, chronic obstructive pulmonary disease, and sleep apnea (Jubran 2004, 2015, Massaroni et al. 2018). Diagnosis of symptoms of any disease is indispensable for curing the disease. The wearable technology in this respect has been explored widely and several pioneering research work based on pressure or strain sensors have been reported. These sensors are small and flexible that are appropriate for patients to put on (Han et al. 2019, Roy et al. 2019, Xiong et al. 2020). Tao et al. fabricated graphene-paper based pressure sensor for monitoring of pulse, respiratory rate, as well as voice recognition (Tao et al. 2017). Zhao et al. synthesized textile tri-boelectric nanogenerators (t-TENG) by using copper-coated PET and polyimide (PI)-coated Cu-PET (PI-Cu-PET) (Zhao et al. 2016). Kim and group demonstrated a process to develop a flexible, solderable electronic platform and its integration with soft adhesive films for breath sensing (Kim et al. 2017). Though, most of the reports have demonstrated only the detection of respiration rate but not volume. In this regard, Chu and group reported wearable sensor, which can altogether measure both respiration rate and volume with high reliability. The detection principle was based on the local strain measurement on chest and abdomen (Chu et al. 2019) as shown in Figure 3(A). Moreover, the flexible piezoelectric sensor was attached with conventional mask fabricated to monitor the continuous flow of respiration (Yaghouby et al. 2016). Additionally, the breathing can be detected by measuring the output voltage as presented in Figure 3(B) (Huang et al. 2014, Lei et al. 2015).

Also, by continuous measurement of human oral humidity, breathing can be monitored. Kano et al. fabricated a fast and flexible nanocrystal humidity sensor using PI substrate. The developed sensor measured the signal generated during the respiratory movements based on the changes in sensor signal attached on a human chest wall, causing the stretch in piezoelectric PVDF film to produce the corresponding electrical signals (Kano et al. 2017). Guder et al. fabricated a paper based humidity sensor and defined the measured electrical signal corresponding to inhalation and exhalation of human breathing, which was integrated with electronic device for the data display (Güder et al. 2016).

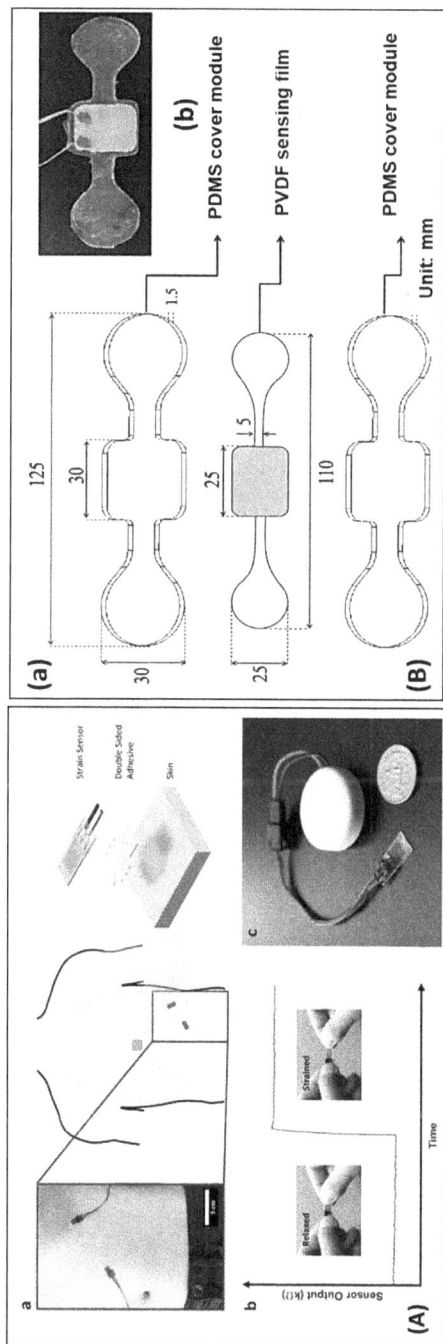

Fig. 3. (A) (a) The left image shows the strain sensors on the ribcage and abdomen. The middle schematics shows the placement of the accelerometer (purple square) in addition to the strain sensors (gray rectangles). The exploded schematic on the right shows the strain sensor and double-sided tape in order of attachment on the skin. (b) Change in resistance of the sensor, under strain, measured using the wireless Bluetooth unit. (c) Image of the wireless Bluetooth unit with a single strain sensor attached. (B) (a) Schematic illustrations of the structure of sensor patch, and (b) Image of sensor patch (upper right). Reprinted with the permission from {Yamada, Nature Nanotechnology 2011, 26}.

2.4 Cardiovascular Health Monitoring

Heart is one of the most important organs of human body, which plays important role in circulating blood, pumping nutrients to body cells and also oxygen containing blood and removes deoxygenated blood from body. Any dysfunction in heart can result in serious impact on human health. Hence, regular monitoring of heart working state is important for patient health. In this respect, the wearable devices for detecting cardiac disorders are playing potential role for the early detection of severe health conditions by regular monitoring of persons (Sana et al. 2020). For the non-invasive detection of cardiovascular signals which can be blood pressure, heart rate, many of the wearable sensors have been developed in the recent years (Murray et al. 2016, Shu et al. 2015, Trung and Lee 2016, Zang et al. 2015).

The wearable sensor for cardiac monitoring is first attached with the human body in the form of watch, chest band, clothing, etc. This attached sensor collects the information of cardiac condition based on the blood pressure, heart rate or electrocardiogram and thus collected data is processed for the display purpose. Moreover, this information is sent to individual via proper app or cloud based server (Luo et al. 2018). However, with the technological advancements, some of the nontechnical barriers concerning regulatory and privacy issues require to be considered to fully leverage the potential of these monitoring devices in providing high quality, reliable, and affordable health care (Sana et al. 2020).

To monitor the heart rate and blood pressure with accuracy, strain, optical and electronic sensors have been developed (Gong et al. 2014, Lin et al. 2020, Luo et al. 2018). Lin et al. developed a flexible pressure sensor based on conducting polymer PPy-cotton composite material, which has piezoresistive property. The fabricated piezoresistive sensor was attached to the wrist of individual to monitor wrist artery pulse signal (Lin et al. 2020). Dagdeviren et al. reported a PZT piezoelectric thin-film sensor which can be skin-mounted (wrist, throat) for blood pressure detection (Dagdeviren et al. 2014). Photoplethysmogram (PPG) is another important technique for measuring heart rate and blood pressure. In this regard, extensive research has been devoted in the field of flexible, stretchable and high performance PDs and LEDs for medical diagnostics applications. Someya's group fabricated a highly flexible and reliable PPG sensor using repetitive parylene layers and inorganic layer (SiON). The detection in these sensors is based on measuring the reflected signal when one part of the tissue is illuminated. The smart electronic skin (e-skin) device consisting of health-monitoring sensors is shown in Figure 4 (Yokota et al. 2016). Hence, individual's heart rate detected by PPG sensor can play vital role in healthcare monitoring devices.

2.5 Sweat Monitoring

The bio-fluids secreted from human body can be used for the monitoring of individual health. These epidermal fluids can be sweat or interstitial fluids. The selection and analysis of biomarkers in these epidermal bio fluids is done at the skin surface. Moreover, sweat embodies easily available bio-fluid, which is composed of electrolytes and metabolites, which can provide information regarding health status

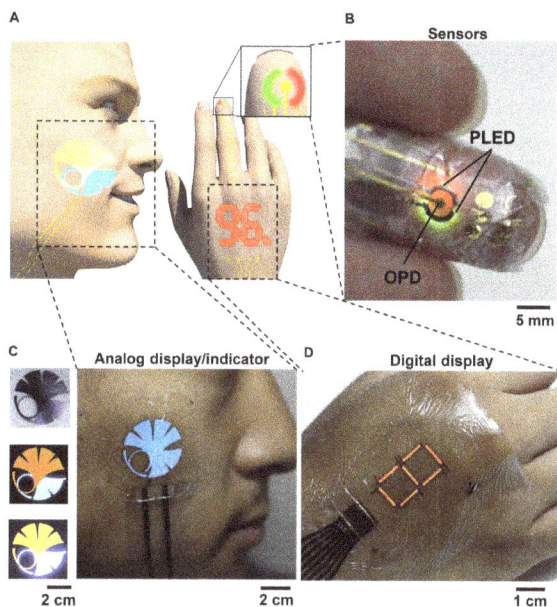

Fig. 4. Smart e-skin system comprising health-monitoring sensors, displays, and ultraflexible PLEDs. Image adapted from (Yokota et al. 2016).

and physical condition of the individual (Pirovano et al. 2020). For example, the electrolytes in sweat viz. sodium, and potassium are associated with the hydration status and can provide information regarding conditions like electrolyte imbalance and cystic fibrosis. Thus, by continuous monitoring of electrolyte concentration in sweat can be used for the medical diagnostics. The detection principle of these skin-worn sensors depends on different transduction methods viz. optical, electrochemical, mechanical. Finally, the sensor is integrated with data processing and transmission components.

Nyein et al. developed microfluidic sensing patches via roll-to-roll (R2R) processes for point-of-care medical applications. The patch permits sweat capture within a spiral microfluidic for real-time measurement of electrolytes concentration in sweat such as Na^+, K^+, glucose, and sweat rate in exercise and chemically induced sweat. The patch is demonstrated for investigating regional sweat composition, predicting whole-body fluid/electrolyte loss during exercise, uncovering relationships between sweat metrics, and tracking glucose dynamics to explore sweat-to-blood correlations in healthy and diabetic individuals (Nyein et al. 2019). Parrilla et al. fabricated wearable potentiometric ion sensors utilizing MWCNTs which can detect $[Na^+]$, $[K^+]$, $[Cl^-]$ (Parrilla et al. 2019). Glennon et al. demonstrated a watch named 'SwEatch' for sweat electrolyte monitoring (Glennon et al. 2016). Alizadeh et al. fabricated a patch system for continuous monitoring of fluid based biomarkers such as electrolytes, metabolites and proteins (Alizadeh et al. 2018). Table 2 shows the developed novel wearable and flexible sweat monitoring devices based on different detection methods, which have been developed for continuous health monitoring.

Table 2. Wearable and flexible chemical sensors for continuous sweat analysis.

Analyte	Analytical Technique	Recognition Element	Material and Platform
Lactate	Electrochemical-amperometry	Lactate oxidase (LOx)	Temporary transfer tattoo
pH	Electrochemical-potentiometry	Polaniline (PANI)	Temporary transfer tattoo
Ammonium	Electrochemical-potentiometry	Nonactin ionophore based ISE	Temporary transfer tattoo
Alcohol	Electrochemical-amperometry	Alcohol-oxidase	Temporary transfer tattoo
Sodium, potassium, lactate, glucose	Electrochemical-potentiometry and amperometry	Sodium and potassium ISEs; glucose oxidase (GOx); LOx	PET wristband/handband
Humidity, glucose, pH	Resistive for humidity; amperometry; potentiometry	PEDOT for humidity; GOx; PANI	Disposable silicone strip
Chloride	Electrochemical-potentiometry	Ag/AgCl electrodes with salt bridge	PDMS on wristband or adhesive bandage
Potassium, lactate	Electrochemical-amperometry and potentiometry	LOx; potassium ISE	PET sticker on eyeglasses

Source: Data from Kim et al. Wearable and flexible electronics for continuous molecular monitoring. *Chemical Society Reviews* 48(2019): 1465–1491.

3. Features for Wearable Healthcare Devices

The wearable biosensors in healthcare should be designed with some common characteristics such as flexibility, biocompatibility, biodegradability, ultrathin, and self-healing. All the required properties of theses biosensors are discussed in the present section.

3.1 Flexibility or Stretchability

In today's healthcare devices, a biosensor having high stretchability, i.e., elastic mechanical behavior, and high performance is required. With the high flexible properties of biosensor, it is possible to make a consistent and close contact between the sensor and the dynamic complex structured skin. This conformity of the sensor provides high spatial and temporal resolution, which gives more chances to collect the signal from the skin interface. The development in new flexible nanomaterials, new design concepts of devices, manufacturing technology and its wide range of applications in healthcare have accelerated research in the field of stretchable electronic technology.

Additionally, the evolutions in the synthesis of stretchable materials have motivated the realization of many stretchable devices. Oh et al. introduced the chemical groups for dynamic non-covalent crosslinked conjugated polymers and reported the new concept for scalable semiconductor polymers' development (Oh et al. 2016). Recently, the same group explored the manufacturing platform which produces intrinsically scalable transistor array (Wang et al. 2018) where they have shown 347 transistors per square centimeters. Likewise, more number of transistors (6300 transistors with an area of 4.4 × 4.4 cm²) can be developed on a large scale by this manufacturing process and also demonstrated the transistor array as the basic component of skin-like electronics. This had led the research in the area of intrinsically stretchable polymer materials and gave way to explore the next generation of stretchable wearable biosensors. Additionally, alternative ways have been demonstrated to increase the conductivity of the stretchable materials by fabricating the composite material with metal nanoparticles (NPs), and nanosheets. Nanowires (NWs), and carbon nanotubes (CNTs) can be combined with stretchable elastomers (Amjadi et al. 2014). Cao et al. demonstrated a transparent transistor, which was fabricated using PDMS as dielectric material and CNTs were used as both the electrodes and active material. This flexible transistor has shown maximum of 3% strain (Cao et al. 2006). Hu et al. synthesized a composite ink material by using adhesive rubber and soluble silver salts and fabricated a stretchable sensor by placing the ink directly on ballpoint pen which can write on stretchable substrate. Moreover, there are significant challenges in the stretchable technology that need to be addressed. Additionally, intrinsically fragile materials also offer exceptional properties and incorporating those with stretchable material can provide further development in the technology. These materials can facilitate excellent stretchable properties by using proper geometry and device design. Additionally, it is inspiring to develop personalized intelligent prosthesis for temperature or pressure sensing, which can be of significant interest due to its viability in wearable health monitoring technology.

3.2 Conformity and Ultra-thin

To develop wearable health care technology, large-scale integration and close contact of these devices with human body are important considerations. The conventional real-time monitoring devices have rigid structure, high power consumption, and restricted functions which are not suitable for human body and thus not applicable for healthcare application. In this respect, flexible biosensor is an alternative method that provides conformal and high-sensitive properties for interaction between individual's body and electronic devices. Various techniques have been amended to realize large-scale contact area and to get conformity of skin. In these techniques, the synthesis of active sensing matrices and the fabrication of light weight, ultra-thin, high flexible and stretchable membranes for conformal contact on human skin surfaces have been reported (Pang et al. 2015, Park et al. 2018a, 2017). One of the important factors for this is to develop ultra-thin devices so that it is possible to make the devices conformable and flexible, which is principally favorable for skin and implantable electronics. Regardless of these benefits, the polymer properties and ultra-thin geometry of the substrates are some of the limitations and challenges in fabrication of such devices. Lee et al. developed an integrated sensor matrix of 2 μm thickness for medical purpose and soft robotics (Lee et al. 2016b). Wagner et al. introduced a rolling backplane with amorphous silicon FETs on a 3 μm thick steel substrate (Ma and Wagner 1999). Lee et al. reported an ultra-thin, vibration sensitive e-skin, capable of detecting skin acceleration based on linear relation with the sound pressure of Ti/Cu film as a sacrificial layer (Lee et al. 2019).

3.3 Biocompatible and Biodegradable

For real-time health monitoring, the wearable device should not create any discomfort, health threats and should not restrict the daily activities of human life. Thus, biocompatibility is one of the important parameters to consider before developing wearable health devices. Usually, active materials are of higher risk as compared to matrix material. Hence, for application of nanostructured materials in future healthcare, it is required to have profound understanding of immune response and the exposure phenomenon under different conditions; for instance, skin contact, intake, inhalation and injection, etc. In this respect, GaN is an alternative for high-speed electronic and optoelectronic devices development as well as it has good biocompatibility. Moreover, several biosensors based on GaN have been reported so far. In this area, Rogers et al. developed a multifunctional brain sensor using silicon, which was completely biodegradable (Kang et al. 2016). Several immunohistochemical studies of brain tissue after implantation (2, 4 and 8 weeks) showed that the sensor and its by-products were biocompatible. Their report suggested that this conventional semiconductor is biocompatible and can be utilized for biomedical implants and health monitoring. Additionally, several number of organic active substances, viz. polyethylenedioxythiophene (PEDOT) and polypyrrole (PPy), have been reported to be biocompatible (Bhatt et al. 2016). Carbonized tissues and cotton also have huge prospective in developing biocompatible wearable sensors (Rahimi et al. 2017).

3.4 Self-healing

Currently, the wearable healthcare device performance is usually restricted by toughness as its components are simply scratched and impaired on the human body. An ideal wearable device can not only preserve its electronic function, but also heal itself by restoring the electrical and mechanical characteristics when some mechanical distortion happens. In this regard, recent developments have been attained for introducing self-healing property and a few self-healing polymer systems have been reported in electronic applications (Huynh and Haick 2016, Lu et al. 2019). Commonly used materials which exhibit self-healing property are composite materials which consist of conductive particles. Yu et al. demonstrated a scalable three-layer network structure using aerogel-based ternary network hydrogel. These composite hydrogels exhibit strong and fast self-healing capability (Lu et al. 2019). Bao et al. reported supramolecular organic and inorganic composite to form self-healing e-skin for the application of self-healing care equipment (Chen et al. 2012). Recently, the same group demonstrated the properties of self-reconfiguration of conductive nanostructures with dynamically cross-linked polymer networks. Moreover, the composite of conductive 1-D network with the tensile self-healing polymer, nano-materials or polymer composites can restructure and return to its mechanical and conductive properties. Utilizing the self-bonding properties involved in self-healing polymers, many of the electronic components like light-emitting capacitor cells, electrocardiogram or strain sensors and high-performance systems can be easily included and thus develop the highly suitable sensors in the field of robotics and prosthetics. Nakahata et al. fabricated a transparent, flexible, and tough hydrogel demonstrating the self-healing property in both wet and dry environment (Nakahata et al. 2016). Zhang et al. reported the self-healing property of ethylene propylene diene monomer rubber prepared by graft-polymerization of zinc dimethacrylate onto rubber chains to form a reversible ionic cross-linked network (Zhang et al. 2019). Li et al. demonstrated a novel variety of supramolecular elastomer by photopolymerizing two rationally designed polymerizable deep eutectic solvent (PDES) monomers. The synthesized network exhibited transparent, intrinsically conductive, stretchable, and self-healable properties. Specifically the healing property has been demonstrated in several organic solvents facilitating the improved adaptability of the network (Li et al. 2020). Cao et al. fabricated a bio-inspired skin-like material exhibiting the properties viz. transparent, electrically conductive and self-healing in both dry and wet surroundings. The synthesized material composed of a fluorocarbon elastomer and a fluorine-rich ionic liquid demonstrates ionic conductivity tuned to as high as 10^{-3} S cm^{-1} and can tolerate strains as high as 2,000% (Cao et al. 2019). The synthesized material demonstrated a touch, pressure and strain sensor platform to demonstrate the sensory capabilities. Additionally, high optical transparency of self-healing materials can be used in evolving opto-electronic human machine and machine–machine communication interfaces for underwater investigation. In accordance with the advantages of self-healing materials and developed devices, there are several challenges which need the attention before the integration of these electronic devices and applicability of these devices into human daily lives.

4. Recent and Advanced Trends

4.1 Smart Glasses

Glasses have been an intricate part of human civilizations of centuries. The earliest account of glasses goes back to 13th century. Glasses themselves improved the vision of billions of people and let them see the world even in their advanced ages. The modern opticians have surpassed all the previous development by introducing smart sensors and electronics in a simple glass. Such inventions have always been a part of sci-fi novels in the 19th and 20th century; however, recent development by big tech companies like Google and Bosch have made it in reality. The device from Google was released with lot of possibilities and holds a huge potential for the future industries. The company claimed that its glasses can help the industry assembly workers in complex tasks to finish in less time. So, in the recent times, several companies are coming up with their products to help the health care professionals. Companies like "*Augmedix*" and "*Vuzix*" have come up with certain glasses made for health care professionals. In this section, we will discuss the recent developments and future possibilities for the smart glasses in the health care systems.

Donald et al. published the preliminary studies on the head mounted display for surgeons and anesthesiologists during crucial surgeries at Ohio State University Hospital. Cecchini et al. found that transparent displays allow anesthesiologists to focus more on patient (Block et al. 1995). Identifying physical vital sign of patient health allows more chances to health professionals to save life in catastrophic events (Liu et al. 2009, Ormerod et al. 2002). In many emergency instances when anesthesiologists are not available, a surgeon has to complete several tasks and has to monitor both screen and physical discomforts of patients. It was observed that surgeons using such technologies have reduced the shifting of attention by whooping 90% from procedural field. This allowed a faster multitasking and better attention to the patients (Liebert et al. 2016). Moreover, it allowed surgeons to finish their task faster compared to those who were not using it. Moreover, these glasses have evolved to allow the surgeons to access the critical information of the patients like CT scan, X-rays, etc. during the surgeries (Muensterer et al. 2014, Scheck 2014). Readers can read the detailed review in this direction about Google glass in pediatric surgeries (Muensterer et al. 2014). Moreover, these devices can also be used as a training tool for junior surgeons. The device has inbuilt camera and Wi-Fi (Collman 2013). This allows junior doctors to train themselves where to look while performing the surgeries. Moreover, it can work in a reverse flow where an expert advice can be taken directly by the doctors performing the operation from senior specialists. This curtails the hectic documentation procedure, further increasing the efficiency of entire system (Armstrong et al. 2014, Kim et al. 2018).

With the surge of COVID-19 in the early months of 2020, several companies rolled out the products with no touch temperature monitoring (infrared signals). The initial reports from WHO suggested that almost 88% of all the symptomatic patients showed a high temperature. Hence, companies rolled out several products where people with high temperature can be easily located within no time. Ajna Lens from Mumbai rolled out their product where it can screen more than 100 people in a minute (Kaushal 2020). Moreover, the company added the artificial intelligence feature in the product to process the image and reduce the human error. Chinese startup company rokid visions also developed a similar product for the security professionals in the public places, which can potentially detect hundreds of people from 3 m distance (Smith 2020). Many international companies like Xiaomi, Samsung and Vuzix, etc. are also entering this field of smart wearables. A recent market report has predicted that smart wearable devices' industry will reach as high 24 billion $ industry by 2023 (Company 2020).

4.2 Textile and Fiber Based Wearable Biosensors

Wearable sensors come in variety of materials; however, many times it isn't comfortable to wear them. To solve this, many researchers tried to make wearables from textile fibers, which can be easily introduced to the users. This not only increases the comfort for the user but also increases larger area for sensing. Additionally, being natural they can sustain harsh weather conditions and various forms of strains. This area has seen an unprecedented growth in the past decade and to cover each and every development in this area demands a separate chapter. Readers can read the detailed reviews in this direction elsewhere (Ghahremani Honarvar and Latifi 2017, Hatamie et al. 2020). However, at this stage, we wish to briefly introduce the major recent development in this area in sensing technology.

A variety of materials have been utilized in plethora of applications. Materials from simple thread to high grade stretchable fabrics all have been utilized for sensing (Hatamie et al. 2020). In the broad categories, the sensor can be utilized to sense chemical species (like humidity, bio-molecules, sweat, etc.) and as physical sensor (like respiratory function, temperature, pressure sensor, etc.).

4.2.1 Chemical Sensors

Wang et al. have fabricated a wearable bandage type sensor. The sensor can measure the pH around a wound, which can detect the healing of the wound in real time. The device used the conducting polymer polyaniline (PANi), which was judiciously screen printed and it can work in pH range of 5–8. Textile sensors find a suitable application in athletic field for sport related activity. Lopezab et al. fabricated a microfluidic based device which was transferred to the wrist band (Curto et al. 2012). The device measured the sweat pH concentration in real time for the athletes to diagnose the symptoms of dehydration. The device incorporated the ionogels, which were equally efficient as the commercial pH sensors. Dendukuri et al. have fabricated a yarn based lactase sensor, which utilized the chrono-amperometric detection of lactase in sweat (Modali et al. 2016). The technique was non-invasive employing all biocompatible materials. Andrade et al. showed a generalized approach to fabricate the yarn based biosensor (Guinovart et al. 2013). Here, the clean yarn was coated with CNT ink. The conducting yarn was dip coated with ion selective membrane for the detection of various ions like potassium and ammonium ions. The device was demonstrated to be used as wound bandage. Brosseau et al. developed a surface enhanced Raman based flexible and wearable biosensor (Bindesri et al. 2018). In this research, they detected the concentrations of levofloxacin (a frequently prescribed antibiotic), in both 0.1 M sodium fluoride and artificial urine solution. Para-Aminothiophenol (pATP) was used as a preliminary molecule to assess the performance of device.

4.2.2 Physical Sensors

A wide variety of wearable textile based sensors are available where the researchers have made clothes to measure diverse functions like temperature, heart rate, pulse, blood pressure, etc. (Hatamie et al. 2020, Lee et al. 2016a). Interested readers can find a detailed review elsewhere (Pan et al. 2011, Van Langenhove 2007). Briefly, Takamatsu et al. fabricated a pressure sensor based on nylon fiber coated with PEDOT:PSS. The fibers were weaved and the joints act as a capacitor which changes its capacitance value upon applying the pressure. The device was carefully tested with a force sensor. Huang et al. fabricated breath rate sensor using Ag nanowires (Huang et al. 2008). Pradhan et al. fabricated a sensor for monitoring of neurological and cardiovascular disorders (Rai et al. 2013). Kalaoglu et al. fabricated an obstacle detection sensor for the visually impaired people (Bahadir et al. 2012). The majority of textile sensors are passive sensors where they need an external power to operate; however, only a few have been shown as active sensors. Nirmal et al. have fabricated the self-powered yarn type piezoelectric sensor for the breath sensing (Raj et al. 2018). The novel device was very cost effective, which was used on a simple cotton textile thread. Here, the authors used bismuth ferrite and PVDF composite and coated this composite onto the Ag electrode with a simple brush coating technique. The simple device was tested with a linear motor setup and could produce a high voltage output of 40 V and could power 5 green LEDs. Finally, the device was demonstrated to monitor the human breathing. Recently, several triboelectric based devices have been the hot topic of research and several researchers have contributed largely to textile based wearable triboelectric. Yang et al. have fabricated fiber based device which is self-powered. The team used a tribo-ferroelectric synergistic effect with a very high power density of 5 Wm^{-1} at low frequency. The device not only harvested energy but was also used for powering other small devices. A detailed review in this class of material can be found here (Hu and Zheng 2019).

4.3 E-skin/E-tattoos

Another interesting aspect of recent development in wearable devices are the body conformal tattoos or skins, popularly known as e-skin or e-tattoos. These devices are next generation devices, which work like the health monitoring system in the real time. The device is conveniently placed on human skin with improved level of comfort, which could monitor the physiological functions like temperature, pulse, electrooculography, ECG, etc. (Guo et al. 2019, Kim et al. 2018, Sadri et al. 2018, Zhang et al. 2015, Zhao et al. 2018). Zhang et al. fabricated an E-skin composed of polyurethane coated with PEDOT:PSS network (Zhang et al. 2015). The device could sense both temperature as well as pressure in the same device. The use of thermoelectric material like PEDOT:PSS changes the thermal conductivity with the change in temperature. Hence, this device can function as a human skin to measure functions simultaneously. The pressure signal was identified using the piezo-resistive signal from the PEDOT:PSS. The device was a complex set of pixelated network of devices, which was extremely sensitive to both temperature and pressure. Guo et al. fabricated the eco-flex based electronic tattoo using the silver nano-flakes (Guo et al. 2019) for the ECG monitoring. The composite device was fabricated using a simple screen printing technique which can be easily implemented for the large area applications. The device was optimized to have an improved conformal contact with the body by controlling the concentration of Triton X and cure temperature. The cell viability studies were also conducted to test the biocompatibility of the device. Chunhui et al. fabricated the porous Polypyrrole (ppy) based wearable sensor sensors to measure the movement of elbows (Wang et al. 2016). The fabrication process was quite interesting for this sensor. Here, the ppy was coated on the porous nickel foam using electrodeposition. Later, the nickle was eroded from the matrix using FeCl$_3$ solution and a 3D porous and hollow polymer structure was ready to use. Using the piezo resistive property of ppy, the strain was calculated. A similar metal organic framework based porous carbon device was fabricated by Zhao et al. (Zhao et al. 2018). Sadri et al. fabricate a simple paper based stretchable sensor using spray coatings and laser cutting technique (Sadri et al. 2018). The device can be used for ECGs, EMGs, EOGs, and thermo-therapeutic treatments to joints, etc. For a detailed review on such sensors, readers can find a good review article elsewhere (Zhang et al. 2020).

Most of the devices shown here require external source of complicated wires. Hence, there is a need of conformal battery also for such sensors. Recently, Amey et al. realized that these epidermal devices require power to consume; hence, the group fabricated a sweat activated biocompatible batteries (Bandodkar et al. 2020).

The device design along with the power management circuit was carefully optimized to power the wearable sensors. The battery can be easily detached from the module which contains wireless sensor node and power management systems, which enable it for continuous on-skin recording of physiological signals. The device can be used as both chemical and physical sensor.

4.4 Self-powered Systems

The device, which is actively producing electrical energy along with sensing the environmental moieties, can be classified under the self-powered systems. Such devices are taking up the space for the next generation wearable biosensors. Several techniques are implemented to make the devices self-powered; however, only a few of the technologies have addressed the issue fairly whereas many are still dependent upon other technologies. Electrochemical based approaches are able to produce sustainable power, sufficient for sensing purpose. Katz et al. gave the concept of self-powered enzymatic sensor for the first time two decade ago (Katz et al. 2001). Here, they used glucose oxidase as an enzymatic layer which reacts with glucose present in the solution, thus sensing as well as working as energy harvesting. Stramibini et al. fabricated a microneedle based self-powered biosensor for the measurement of glycaemia in the interstitial fluid. The microneedles could potentially absorb the analytes using the capillary action and can measure the glucose concentration in real time. The devices were used for the pain-free glucose detection. Jeerapan et al. fabricated a self-powered biofuel socks using simple printing technology (Jeerapan et al. 2016). The device showed high order of stability even after severe deformations. The device could power from glucose and lactate using a single enzyme. The fabrication process was intricate; however, the serpentine structure of the electrodes allowed high deformation without change in the contact resistance. Similar approaches have been utilized for other wearables like Band-Aid and diapers for elderly patients (Cho et al. 2017, Shitanda et al. 2019). The other transduction technique which has been used are piezoelectric materials for self-powered detection. Though it is not completely self-powered, some researchers have put efforts to realize such sensors. Here, in a piezoelectric material a mechanical energy is converted to electrical energy in certain materials which show non centro-symmetry. Here we discuss some of the major developments in this area; however, most of them are not wearable device at present stage. However, with little efforts, they can be improvised for wearable application. Majority of devices utilized the charge screening principle for the detection. In simple words, under periodic strain the charge is produced on the surface of material. When these charges are inhibited or screened by the biological entities present in the surrounding, the concentration is detected. Kim et al. fabricated a barium titanate (BTO) based first glucose biosensors using this principle (Selvarajan et al. 2016, 2017b). The barium titanate nanoparticles were mixed in aqueous PVA solution and drop casted on the glass slide, which was later heat treated to remove PVA. When the nanoparticle based device is exposed to the different concentrations of glucose (50 μM–1 mM), the load resistance value changes. Hence, the voltage produced across the device changes and sensing becomes so simple and easy to investigate without the use of complex electrochemistry. The authors cross verified the glucose sensing with electrochemical sensing platforms also. The device was shown to perform both energy harvesting and sensing purpose. A similar approach was used to detect the cysteine (Selvarajan et al. 2017a). Cysteine is a vital amino acid which contains thiol group, thus playing an important role in homeostasis. Here, in this research the BTO was functionalized with (3-aminopropyl) triethoxysilane which primarily contains the NH_2 groups, which was done using IV characterization for different concentrations (10 μM–1000 μM) with a limit of detection of 10 μM. Self-powered characteristics were also measured using the previously explained principle. A similar approach was recently published by the same group of functionalizing BTO with casein and its interaction with cysteamine was identified (Selvarajan et al. 2020). This anticipated research can work out to be an introductory apparatus for investigating the possible interactions and application in theranostics before performing invasive real-time experiment. Recently, Abisegpriyan et al. fabricated the hollow cylindrical bismuth ferrite based sensor and energy harvester (Abisegapriyan et al. 2020). The device was designed to detect the transitional flow along with the sensing. The device calculated the concentration of catechol in the water. The device could sense catechol concentration in a wide range from 25 μM to 250 mM concentration with a large linear range. The sensor has good selectivity and detection limits down to 10.2 μM. However, the device was tested only for 5 concentrations. Fu et al. fabricated another hollow PVDF/PANI (polyaniline) based biosensor suitable for gas sensing (Fu et al. 2018). A variety of gases like acetone, methane, carbon monoxide, ethanol, etc. were sensed at different concentrations (0–600 ppm). Various device configurations were tried to fabricate an alcohol based breadth sensor which could finally work. Pandey et al. fabricated some simple biomechanical energy based physical sensors using zinc oxide (Pandey et al. 2019, Rajagopalan et al. 2019). The device was used as smart gas purging station or smart urinals for societal benefits.

Recently, the research in the area of triboelectric energy based self-powered devices is also accelerating. Lin et al. fabricated a gelatin PDMS based triboelectric nanogenerator (TENG), which could detect the lactase through the sweat (Chen et al. 2017). The device was implanted in the socks, which could convert the mechanical energy of walking into electrical energy. The Au-Pd nanoparticle based sensing layer was used to detect the concentration of lactate with high selectivity. Xue et al. fabricated poly acrylonitrile PDMS based e-skin type self-powered sensor (He et al. 2018). The device was fabricated using the lithography technique. The device could harvest energy as well as work as sensor to detect various bio-molecules like lactase, glucose, urea, Na and K ions using various enzymes. The device was fabricated on a PDMS substrate worked in contact separation mode upon bending. A similar approach was used for the calcium detection by the same group with the different pattern and enzyme (Zhao et al. 2019). Gaurav et al. also fabricated a PVDF-TiO$_2$ based biosensor for the ammonia detection using a similar principle (Khandelwal et al. 2019). The TENG field is still growing and every year the average power density is increasing due to new researches being done in this field. Hence, we believe a lot of exciting opportunities are there in this field of research.

5. Other Factors Crucial in the Device Development and Healthcare System

5.1 Power Source Development in Battery Technology

The bottle neck for all the wearable devices is the portable power supply. One cannot expect the wearable with electrical wires hanging around. Hence, the requirement is that the device must be self-powered (using thermoelectric, triboelectric, etc.) or have a portable power source like battery or capacitor. We have already discussed the concept of self-powered wearable device in the previous section and more details will be discussed in the next section. Here, we will discuss various other options developed for such devices.

Batteries are the first choice for device designers as it can easily meet the fluctuating power demand of the device without much efforts. The devices are built in various shapes and sizes and thus the battery designers have to keep these things in their mind; hence, next generation batteries for wearables are coming in variety of sizes. Companies like Samsung, Panasonic, Grepow, etc. are working towards the miniaturizing of battery technology. Moreover, they have a separate R and D division for the customized shapes and sized for the wearable technologies. No doubt the research community is working in this area of research but the latest trends are coming for the industrial R and D sector. The other important factor in this area is the flexibility as many devices require a flexible battery. Several research articles have been published in the areas of flexible batteries using various process like electrospinning, hydrothermal, carbonization, etc. (Chen et al. 2016, Li et al. 2014, Zamarayeva et al. 2017, Zhao et al. 2014). In this research domain, carbon nanofibers are interesting candidates which are quite flexible in nature. Moreover, the porous electrode increases the surface area for the efficient charge transfer. The porous material also here behaves as a freestanding electrode; moreover, its biocompatibility can be used in the wearable biosensor devices. A detailed review in this direction of flexible batteries can be found elsewhere (Tao et al. 2018). Another important class of these batteries are the textile/fiber based batteries which can be easily coupled with textile based sensors (Kwon et al. 2012, Park et al. 2015, Weng et al. 2014). These cable or fiber type batteries can deform to a very large extent compared to flexible battery counterparts. Moreover, the flexible batteries many times can deform in only one plane but such restrictions are not present with fiber based batteries. The other important class of batteries are stretchable batteries, origami based batteries and Bridge Island based batteries (Gaikwad et al. 2012, Song et al. 2014, Xu et al. 2013). Even with continuous advancements, several researchers found these battery related components bulky; hence, several researchers tried to fabricate a thin film based solar cells which are quite flexible and lightweight (Park et al. 2018b). Moreover, this reduces the regular recharging feature along with the reduced contact resistance. Under the normal illumination, flexible solar cells are less efficient compared to rigid counterparts. However, here they have used a nano grating technique to improve the absorption of visible radiation from all angles of light. This in-built power supply enabled a better response from the organic electrochemical transistors attached to the skin by three times. Here, this type of device is more likely to work as self-powered class of sensor due to absence of storage unit and several of such devices will be discussed in the next section in detail. Supercapacitors are also interesting alternatives for the electrical storage. Recently, Park et al. fabricated the dynamically stretchable textile based all in one device (Park et al. 2019). The device was fabricated using molybdenum oxide and multi walled carbon nanotubes composite. The device was used to fabricate a strain sensor but shows the potential to be used in physical and chemical based wearable sensors. Recently, Christina et al. also fabricated a similar device with a potential to be used in glucose biosensing (Mary et al. 2020).

A lot of work has been done in this area and a lot has to be achieved, which still remains the bottle neck for the next generation devices.

5.2 Regulations

Apart from research institutions, a lot of devices from industries are coming to the market. To what extent a simple wearable device falls under the jurisprudence of the medical device and comes under the scanner of health regulatory authorities of the country? This topic requires an in-depth analysis to answer some of the interesting questions. Here, in this study we take FDA (Food and Drug Administration) as a regulatory which has set itself up as a leading body in the work regarding such concerns. So the regulations for the medical devices are different from the medical drug in several ways. The trail of any new drug or vaccines is done in phases and it takes several years to finish the trails (exceptions are in COVID-19 and H1N1). However, the medical devices do not undergo such stringent criteria for evaluation. A medical device may also reach the market without any clinical or pivotal trial. The most commonly cited example is the hip metal replacements (Ardaugh et al. 2013). Unlike the medical drug where prediction does not play a role, medical device regulations are more or less based on predictions about the safety concerns. This reduces the overall cost of the product and keeps the curiosity driven research alive. However, it does not mean that these devices are not regulated at all. The authorities divide such devices basically into three categories (class I – low risk; class II – medium risk; class III – high risk) based on the risk factors. Class I and II products usually get the nod from the regulatory authority through a 510 k protocol prevalent in states (Challoner and Senate 2011). Here, in this protocol the owner has to prove some simple points like how safe is the product, how better the product is compared to its predecessor, simple trials and benchmark reports, etc. However, it's easier said than done and the rules are applied differently to different class of products. Recently, some of the hybrid device products which are coming in the market are a combination of device and drug (products 2020). For instance, the drug delivering stent belongs to class III product, but at the same time requires clinical trials for the drugs. Or sometimes companies working in telemedicine combine the bio-absorbable transmitter with the drug to check the level of absorption in the body (Tong 2018). Researchers have argued that such products should be treated like drugs and not like wearable products. This makes the task of regulators even more difficult. So in a nutshell, the wearable products come under the vague category of health, whereas regulatory bodies like FDA are more concerned about the medics. The medical device and the health based devices are not clearly divided and hence it becomes difficult to regulate these devices in the market. Other important considerations are the softwares, which are used in the device design and maintenance.

6. Conclusion

In summary, wearable sensors in healthcare equipment have paved a feasible way to resolve the issues in efficient medical amenities in real-time. The evolution in the field of material science, manufacturing technology, integrated circuit construction, and structural engineering has led to the development in wearable healthcare technology. These advancements in wearable sensors have evolved as a new era for the analysis, cure and prevention of numerous diseases. The implementation of wearable health monitoring devices is user friendly and people can easily try without the user's acquaintance. Hence, it is possible to detect critical diseases at their early stage. Furthermore, future electronic devices have opened up the scope for the development of electronic device, which can predict the health before the detection of symptoms. The prediction and treatment utilizing wearable health monitoring devices give way to next category of electronic products and open new possibilities for personal point of care applications. Subsequently, wearable health monitoring system can reduce the burden of doctors and limit the increasing health expenses.

Acknowledgements

Pramila Jakhar is thankful to Department of Electrical & Electronics Engineering, BITS Pilani K K Birla Goa Campus for providing resources to write this book chapter. Authors are thankful to Indian Institute of Technology, Indore to carry out research during the stay at Indian Institute of Technology, Indore.

References

Abisegapriyan, K., P. M. J. R. Nirmal, N. R. Alluri, C. Arunkumar and K. Sang-Jae. (2020). All in one transitional flow-based integrated self-powered catechol sensor using $BiFeO_3$ nanoparticles. *Sensors and Actuators B: Chemical* 320: 128417.

Alexeev, V. L., S. Das, D. N. Finegold and S. A. Asher. (2004). Photonic crystal glucose-sensing material for noninvasive monitoring of glucose in tear fluid. *Clinical Chemistry* 50(12): 2353–2360.

Alizadeh, A., A. Burns, R. Lenigk, R. Gettings, J. Ashe, A. Porter, M. McCaul, R. Barrett, D. Diamond and P. White. (2018). A wearable patch for continuous monitoring of sweat electrolytes during exertion. *Lab on a Chip* 18(17): 2632–2641.

Amjadi, M., A. Pichitpajongkit, S. Lee, S. Ryu and I. Park. (2014). Highly stretchable and sensitive strain sensor based on silver nanowire–elastomer nanocomposite. *ACS Nano* 8(5): 5154–5163.

Ardaugh, B. M., S. E. Graves and R. F. Redberg. (2013). The 510 (k) ancestry of a metal-on-metal hip implant. *New England Journal of Medicine* 368(2): 97–100.

Armstrong, D. G., T. M. Rankin, N. A. Giovinco, J. L. Mills and Y. Matsuoka. (2014). A heads-up display for diabetic limb salvage surgery: a view through the google looking glass. *Journal of Diabetes Science and Technology* 8(5): 951–956.

Bahadir, S. K., V. Koncar and F. Kalaoglu. (2012). Wearable obstacle detection system fully integrated to textile structures for visually impaired people. *Sensors and Actuators A: Physical* 179: 297–311.

Bandodkar, A., S. Lee, I. Huang, W. Li, S. Wang, C.-J. Su, W. Jeang, T. Hang, S. Mehta and N. Nyberg. (2020). Sweat-activated biocompatible batteries for epidermal electronic and microfluidic systems. *Nature Electronics* 3: 1–9.

Bhatt, V. D., S. Teymouri, K. Melzer, A. Abdellah, Z. Guttenberg and P. Lugli. (2016). Biocompatibility tests on spray coated carbon nanotube and PEDOT:PSS thin films. *IEEE Transactions on Nanotechnology* 15(3): 373–379.

Bindesri, S. D., D. S. Alhatab and C. L. Brosseau. (2018). Development of an electrochemical surface-enhanced Raman spectroscopy (EC-SERS) fabric-based plasmonic sensor for point-of-care diagnostics. *Analyst* 143(17): 4128–4135.

Cao, Q., S. H. Hur, Z. T. Zhu, Y. G. Sun, C. J. Wang, M. A. Meitl, M. Shim and J. A. Rogers. (2006). Highly bendable, transparent thin-film transistors that use carbon-nanotube-based conductors and semiconductors with elastomeric dielectrics. *Advanced Materials* 18(3): 304–309.

Cao, Y., Y. J. Tan, S. Li, W. W. Lee, H. Guo, Y. Cai, C. Wang and B. C.-K. Tee. (2019). Self-healing electronic skins for aquatic environments. *Nature Electronics* 2(2): 75–82.

Challoner, D. R. and U. Senate. (2011). Medical devices and the public's health: the FDA 510 (k) clearance process at 35 years. Written Statement before the Committee on Health, Education, Labor, and Pensions US Senate, Institute of Medicine of the National Academics, Washington.

Chen, C.-H., P.-W. Lee, Y.-H. Tsao and Z.-H. Lin. (2017). Utilization of self-powered electrochemical systems: metallic nanoparticle synthesis and lactate detection. *Nano Energy* 42: 241–248.

Chen, L., M. Nakamura, T. D. Schindler, D. Parker and Z. Bryant. (2012). Engineering controllable bidirectional molecular motors based on myosin. *Nature Nanotechnology* 7(4): 252–256.

Chen, R., Y. Hu, Z. Shen, Y. Chen, X. He, X. Zhang and Y. Zhang. (2016). Controlled synthesis of carbon nanofibers anchored with $Zn_x Co_{3-x} O_4$ nanocubes as binder-free anode materials for lithium-ion batteries. *ACS Applied Materials & Interfaces* 8(4): 2591–2599.

Cho, E., M. Mohammadifar and S. Choi. (2017). A self-powered sensor patch for glucose monitoring in sweat. 2017 IEEE 30th International Conference on Micro Electro Mechanical Systems (MEMS).

Chu, M., T. Nguyen, V. Pandey, Y. Zhou, H. N. Pham, R. Bar-Yoseph, S. Radom-Aizik, R. Jain, D. M. Cooper and M. Khine. (2019). Respiration rate and volume measurements using wearable strain sensors. *NPJ Digital Medicine* 2(1): 1–9.

Collman, A. (2013). First ever surgery conducted by doctor wearing Google Glass. *Daily Mail*.

Company, T. B. R. (2020). Smart Wearables Global Market Report 2020–30: Covid 19 Growth and Change. https://www.reportlinker.com/p05935473/Smart-Wearables-Global-Market-Report-30-Covid-19-Growth-and-Change.html?utm_source=GNW.

Curto, V. F., C. Fay, S. Coyle, R. Byrne, C. O'Toole, C. Barry, S. Hughes, N. Moyna, D. Diamond and F. Benito-Lopez. (2012). Real-time sweat pH monitoring based on a wearable chemical barcode microfluidic platform incorporating ionic liquids. *Sensors and Actuators B: Chemical* 171: 1327–1334.

Dagdeviren, C., Y. Su, P. Joe, R. Yona, Y. Liu, Y.-S. Kim, Y. Huang, A. R. Damadoran, J. Xia and L. W. Martin. (2014). Conformable amplified lead zirconate titanate sensors with enhanced piezoelectric response for cutaneous pressure monitoring. *Nature Communications* 5(1): 1–10.

Block, F. E., D. O. Yablok and J. S. McDonald. (1995). Clinical evaluation of the 'head-up' display of anesthesia data. *International Journal of Clinical Monitoring and Computing* 12: 21–24.

Fu, Y., H. He, T. Zhao, Y. Dai, W. Han, J. Ma, L. Xing, Y. Zhang and X. Xue. (2018). A self-powered breath analyzer based on PANI/PVDF Piezo-gas-sensing arrays for potential diagnostics application. *Nanomicro Lett* 10(4): 76.

Gaikwad, A. M., A. M. Zamarayeva, J. Rousseau, H. Chu, I. Derin and D. A. Steingart. (2012). Highly stretchable alkaline batteries based on an embedded conductive fabric. *Advanced Materials* 24(37): 5071–5076.

Ghahremani Honarvar, M. and M. Latifi. (2017). Overview of wearable electronics and smart textiles. *The Journal of The Textile Institute* 108(4): 631–652.

Glennon, T., C. O'Quigley, M. McCaul, G. Matzeu, S. Beirne, G. G. Wallace, F. Stroiescu, N. O'Mahoney, P. White and D. Diamond. (2016). 'SWEATCH': A wearable platform for harvesting and analysing sweat sodium content. *Electroanalysis* 28(6): 1283–1289.

Gong, S., W. Schwalb, Y. Wang, Y. Chen, Y. Tang, J. Si, B. Shirinzadeh and W. Cheng. (2014). A wearable and highly sensitive pressure sensor with ultrathin gold nanowires. *Nature Communications* 5(1): 1–8.

Güder, F., A. Ainla, J. Redston, B. Mosadegh, A. Glavan, T. Martin and G. M. Whitesides. (2016). Paper-based electrical respiration sensor. *Angewandte Chemie International Edition* 55(19): 5727–5732.

Guinovart, T., M. Parrilla, G. A. Crespo, F. X. Rius and F. J. Andrade. (2013). Potentiometric sensors using cotton yarns, carbon nanotubes and polymeric membranes. *Analyst* 138(18): 5208–5215.

Guk, K., G. Han, J. Lim, K. Jeong, T. Kang, E.-K. Lim and J. Jung. (2019). Evolution of wearable devices with real-time disease monitoring for personalized healthcare. *Nanomaterials* 9(6): 813.

Guo, W., P. Zheng, X. Huang, H. Zhuo, Y. Wu, Z. Yin, Z. Li and H. Wu. (2019). Matrix-independent highly conductive composites for electrodes and interconnects in stretchable electronics. *ACS Applied Materials & Interfaces* 11(8): 8567–8575.

Han, Z., Z. Cheng, Y. Chen, B. Li, Z. Liang, H. Li, Y. Ma and X. Feng. (2019). Fabrication of highly pressure-sensitive, hydrophobic, and flexible 3D carbon nanofiber networks by electrospinning for human physiological signal monitoring. *Nanoscale* 11(13): 5942–5950.

Hatamie, A., S. Angizi, S. Kumar, C. M. Pandey, A. Simchi, M. Willander and B. D. Malhotra. (2020). Textile based chemical and physical sensors for healthcare monitoring. *Journal of The Electrochemical Society* 167(3): 037546.

He, H., H. Zeng, Y. Fu, W. Han, Y. Dai, L. Xing, Y. Zhang and X. Xue. (2018). A self-powered electronic-skin for real-time perspiration analysis and application in motion state monitoring. *Journal of Materials Chemistry C* 6(36): 9624–9630.

Hu, Y. and Z. Zheng. (2019). Progress in textile-based triboelectric nanogenerators for smart fabrics. *Nano Energy* 56: 16–24.

Huang, C.-T., C.-L. Shen, C.-F. Tang and S.-H. Chang. (2008). A wearable yarn-based piezo-resistive sensor. *Sensors and Actuators A: Physical* 141(2): 396–403.

Huang, C.-Y., M.-C. Chan, C.-Y. Chen and B.-S. Lin. (2014). Novel wearable and wireless ring-type pulse oximeter with multi-detectors. *Sensors* 14(9): 17586–17599.

Huynh, T. P. and H. Haick. (2016). Self-healing, fully functional, and multiparametric flexible sensing platform. *Advanced Materials* 28(1): 138–143.

Jeerapan, I., J. R. Sempionatto, A. Pavinatto, J.-M. You and J. Wang. (2016). Stretchable biofuel cells as wearable textile-based self-powered sensors. *Journal of Materials Chemistry A* 4(47): 18342–18353.

Jubran, A. (2004). Pulse oximetry. *Intensive Care Medicine* 30(11): 2017–2020.

Jubran, A. (2015). Pulse oximetry. *Critical Care* 19(1): 272.

Kanaparthi, S. and S. Badhulika. (2016). Solvent-free fabrication of a biodegradable all-carbon paper based field effect transistor for human motion detection through strain sensing. *Green Chemistry* 18(12): 3640–3646.

Kang, S.-K., R. K. Murphy, S.-W. Hwang, S. M. Lee, D. V. Harburg, N. A. Krueger, J. Shin, P. Gamble, H. Cheng and S. Yu. (2016). Bioresorbable silicon electronic sensors for the brain. *Nature* 530(7588): 71–76.

Kano, S., K. Kim and M. Fujii. (2017). Fast-response and flexible nanocrystal-based humidity sensor for monitoring human respiration and water evaporation on skin. *ACS Sensors* 2(6): 828–833.

Katz, E., A. F. Bückmann and I. Willner. (2001). Self-powered enzyme-based biosensors. *Journal of the American Chemical Society* 123(43): 10752–10753.

Kaushal, B. (2020). This Mumbai company's AR and AI-powered smart glasses can scan 300 people at a time for COVID-19 symptoms. https://yourstory.com/smbstory/pivot-persist-mumbai-company-ar-ai-smart-glasses-ajna-lens?utm_pageloadtype=scroll.

Khandelwal, G., A. Chandrasekhar, R. Pandey, N. P. M. J. Raj and S.-J. Kim. (2019). Phase inversion enabled energy scavenger: A multifunctional triboelectric nanogenerator as benzene monitoring system. *Sensors and Actuators B: Chemical* 282: 590–598.

Kim, J.-H., S.-R. Kim, H.-J. Kil, Y.-C. Kim and J.-W. Park. (2018). Highly conformable, transparent electrodes for epidermal electronics. *Nano Letters* 18(7): 4531–4540.

Kim, Y. S., J. Lu, B. Shih, A. Gharibans, Z. Zou, K. Matsuno, R. Aguilera, Y. Han, A. Meek and J. Xiao. (2017). Scalable manufacturing of solderable and stretchable physiologic sensing systems. *Advanced Materials* 29(39): 1701312.

Kwon, Y. H., S. W. Woo, H. R. Jung, H. K. Yu, K. Kim, B. H. Oh, S. Ahn, S. Y. Lee, S. W. Song and J. Cho. (2012). Cable-type flexible lithium ion battery based on hollow multi-helix electrodes. *Advanced Materials* 24(38): 5192–5197.

Lee, H., T. K. Choi, Y. B. Lee, H. R. Cho, R. Ghaffari, L. Wang, H. J. Choi, T. D. Chung, N. Lu and T. Hyeon. (2016a). A graphene-based electrochemical device with thermoresponsive microneedles for diabetes monitoring and therapy. *Nature Nanotechnology* 11(6): 566–572.

Lee, S., A. Reuveny, J. Reeder, S. Lee, H. Jin, Q. Liu, T. Yokota, T. Sekitani, T. Isoyama and Y. Abe. (2016b). A transparent bending-insensitive pressure sensor. *Nature Nanotechnology* 11(5): 472–478.

Lee, S., J. Kim, I. Yun, G. Y. Bae, D. Kim, S. Park, I.-M. Yi, W. Moon, Y. Chung and K. Cho. (2019). An ultrathin conformable vibration-responsive electronic skin for quantitative vocal recognition. *Nature Communications* 10(1): 1–11.

Lei, K.-F., Y.-Z. Hsieh, Y.-Y. Chiu and M.-H. Wu. (2015). The structure design of piezoelectric poly (vinylidene fluoride) (PVDF) polymer-based sensor patch for the respiration monitoring under dynamic walking conditions. *Sensors* 15(8): 18801–18812.

Li, L., Y. Bai, L. Li, S. Wang and T. Zhang. (2017). A superhydrophobic smart coating for flexible and wearable sensing electronics. *Advanced Materials* 29(43): 1702517.

Li, R. a., G. Chen, T. Fan, K. Zhang and M. He. (2020). Transparent conductive elastomers with excellent autonomous self-healing capability in harsh organic solvent environments. *Journal of Materials Chemistry A* 8(10): 5056–5061.

Li, W., Z. Yang, Y. Jiang, Z. Yu, L. Gu and Y. Yu. (2014). Crystalline red phosphorus incorporated with porous carbon nanofibers as flexible electrode for high performance lithium-ion batteries. *Carbon* 78: 455–462.

Liebert, C. A., M. A. Zayed, O. Aalami, J. Tran and J. N. Lau. (2016). Novel use of Google Glass for procedural wireless vital sign monitoring. *Surgical Innovation* 23(4): 366–373.

Lin, X., T. Zhang, J. Cao, H. Wen, T. Fei, S. Liu, R. Wang, H. Ren and H. Zhao. (2020). Flexible piezoresistive sensors based on conducting polymer-coated fabric applied to human physiological signals monitoring. *Journal of Bionic Engineering* 17(1): 55–63.

Liu, D., S. A. Jenkins and P. M. Sanderson. (2009). Clinical implementation of a head-mounted display of patient vital signs. 2009 International Symposium on Wearable Computers.

Lu, Y., Z. Liu, H. Yan, Q. Peng, R. Wang, M. E. Barkey, J.-W. Jeon and E. K. Wujcik. (2019). Ultrastretchable conductive polymer complex as a strain sensor with a repeatable autonomous self-healing ability. *ACS Applied Materials & Interfaces* 11(22): 20453–20464.

Luo, N., J. Zhang, X. Ding, Z. Zhou, Q. Zhang, Y. T. Zhang, S. C. Chen, J. L. Hu and N. Zhao. (2018). Textile-enabled highly reproducible flexible pressure sensors for cardiovascular monitoring. *Advanced Materials Technologies* 3(1): 1700222.

Ma, E. Y. and S. Wagner. (1999). Amorphous silicon transistors on ultrathin steel foil substrates. *Applied Physics Letters* 74(18): 2661–2662.

March, W., D. Lazzaro and S. Rastogi. (2006). Fluorescent measurement in the non-invasive contact lens glucose sensor. *Diabetes Technology & Therapeutics* 8(3): 312–317.

Mary, A. J. C., S. S. Shalini, R. Balamurugan, M. Harikrishnan and A. C. Bose. (2020). Supercapacitor and non-enzymatic biosensor application of the $Mn_2O_3/NiCo_2O_4$ composite material. *New Journal of Chemistry* 44: 11316.

Massaroni, C., C. Venanzi, A. P. Silvatti, D. Lo Presti, P. Saccomandi, D. Formica, F. Giurazza, M. A. Caponero and E. Schena. (2018). Smart textile for respiratory monitoring and thoraco-abdominal motion pattern evaluation. *Journal of Biophotonics* 11(5): e201700263.

McCaul, M., T. Glennon and D. Diamond. (2017). Challenges and opportunities in wearable technology for biochemical analysis in sweat. *Current Opinion in Electrochemistry* 3(1): 46–50.

Modali, A., S. R. K. Vanjari and D. Dendukuri. (2016). Wearable woven electrochemical biosensor patch for non-invasive diagnostics. *Electroanalysis* 28(6): 1276–1282.

Muensterer, O. J., M. Lacher, C. Zoeller, M. Bronstein and J. Kübler. (2014). Google Glass in pediatric surgery: an exploratory study. *International Journal of Surgery* 12(4): 281–289.

Murray, T. and S. Krishnan. (2018). Medical Wearables for Monitoring Cardiovascular Diseases.

Nakahata, M., Y. Takashima and A. Harada. (2016). Highly flexible, tough, and self-healing supramolecular polymeric materials using host–guest interaction. *Macromolecular Rapid Communications* 37(1): 86–92.

Nyein, H. Y. Y., M. Bariya, L. Kivimäki, S. Uusitalo, T. S. Liaw, E. Jansson, C. H. Ahn, J. A. Hangasky, J. Zhao and Y. Lin. (2019). Regional and correlative sweat analysis using high-throughput microfluidic sensing patches toward decoding sweat. *Science Advances* 5(8): eaaw9906.

Oh, J. Y., S. Rondeau-Gagné, Y.-C. Chiu, A. Chortos, F. Lissel, G.-J. N. Wang, B. C. Schroeder, T. Kurosawa, J. Lopez and T. Katsumata. (2016). Intrinsically stretchable and healable semiconducting polymer for organic transistors. *Nature* 539(7629): 411–415.

Ormerod, D., B. Ross and A. Naluai-Cecchini. (2002). Use of a see-through head-worn display of patient monitoring data to enhance anesthesiologists' response to abnormal clinical events. Proceedings. Sixth International Symposium on Wearable Computers.

Pan, C., Z. Li, W. Guo, J. Zhu and Z. L. Wang. (2011). Fiber-based hybrid nanogenerators for/as self-powered systems in biological liquid. *Angewandte Chemie International Edition* 50(47): 11192–11196.

Pandey, R., N. P. Maria Joseph Raj, V. Singh, P. Iyamperumal Anand and S.-J. Kim. (2019). Novel interfacial bulk heterojunction technique for enhanced response in ZnO nanogenerator. *ACS Applied Materials & Interfaces* 11(6): 6078–6088.

Pang, C., J. H. Koo, A. Nguyen, J. M. Caves, M. G. Kim, A. Chortos, K. Kim, P. J. Wang, J. B. H. Tok and Z. Bao. (2015). Highly skin-conformal microhairy sensor for pulse signal amplification. *Advanced Materials* 27(4): 634–640.

Park, H., J. W. Kim, S. Y. Hong, G. Lee, H. Lee, C. Song, K. Keum, Y. R. Jeong, S. W. Jin and D. S. Kim. (2019). Dynamically stretchable supercapacitor for powering an integrated biosensor in an all-in-one textile system. *ACS Nano* 13(9): 10469–10480.

Park, J., M. Park, G. Nam, J. s. Lee and J. Cho. (2015). All-solid-state cable-type flexible zinc–air battery. *Advanced Materials* 27(8): 1396–1401.

Park, S., K. Fukuda, M. Wang, C. Lee, T. Yokota, H. Jin, H. Jinno, H. Kimura, P. Zalar and N. Matsuhisa. (2018a). Ultraflexible near-infrared organic photodetectors for conformal photoplethysmogram sensors. *Advanced Materials* 30(34): 1802359.

Park, S., S. W. Heo, W. Lee, D. Inoue, Z. Jiang, K. Yu, H. Jinno, D. Hashizume, M. Sekino and T. Yokota. (2018b). Self-powered ultra-flexible electronics via nano-grating-patterned organic photovoltaics. *Nature* 561(7724): 516–521.

Park, Y., J. Shim, S. Jeong, G. R. Yi, H. Chae, J. W. Bae, S. O. Kim and C. Pang. (2017). Microtopography-guided conductive patterns of liquid-driven graphene nanoplatelet networks for stretchable and skin-conformal sensor array. *Advanced Materials* 29(21): 1606453.

Parrilla, M., I. Ortiz-Gomez, R. Canovas, A. Salinas-Castillo, M. Cuartero and G. A. Crespo. (2019). Wearable potentiometric ion patch for on-body electrolyte monitoring in sweat: toward a validation strategy to ensure physiological relevance. *Analytical Chemistry* 91(13): 8644–8651.

Pirovano, P., M. Dorrian, A. Shinde, A. Donohoe, A. J. Brady, N. M. Moyna, G. Wallace, D. Diamond and M. McCaul. (2020). A wearable sensor for the detection of sodium and potassium in human sweat during exercise. *Talanta* 219: 121145.

Products, U. F. a. D. A. O. o. c. (2020). Frequently Asked Questions About Combination Products. https://www.fda.gov/combination-products/about-combination-products/frequently-asked-questions-about-combination-products.

Rahimi, R., M. Ochoa, A. Tamayol, S. Khalili, A. Khademhosseini and B. Ziaie. (2017). Highly stretchable potentiometric pH sensor fabricated via laser carbonization and machining of Carbon–Polyaniline composite. *ACS Applied Materials & Interfaces* 9(10): 9015–9023.

Rai, P., S. Oh, P. Shyamkumar, M. Ramasamy, R. E. Harbaugh and V. K. Varadan. (2013). Nano-bio-textile sensors with mobile wireless platform for wearable health monitoring of neurological and cardiovascular disorders. *Journal of The Electrochemical Society* 161(2): B3116.

Raj, N. P. M. J., N. R. Alluri, N. Vivekananthan, A. Chandrasekhar, G. Khandelwal and S.-J. Kim. (2018). Sustainable yarn type-piezoelectric energy harvester as an eco-friendly, cost-effective battery-free breath sensor. *Applied Energy* 228: 1767–1776.

Rajagopalan, P., V. Singh, I. Palani and S.-J. Kim. (2019). Superior response in ZnO nanogenerator via interfaced heterojunction for novel smart gas purging system. *Extreme Mechanics Letters* 26: 18–25.

Roy, K., S. K. Ghosh, A. Sultana, S. Garain, M. Xie, C. R. Bowen, K. Henkel, D. Schmeißer and D. Mandal. (2019). A self-powered wearable pressure sensor and pyroelectric breathing sensor based on GO interfaced PVDF nanofibers. *ACS Applied Nano Materials* 2(4): 2013–2025.

Sadri, B., D. Goswami, M. Sala de Medeiros, A. Pal, B. Castro, S. Kuang and R. V. Martinez. (2018). Wearable and implantable epidermal paper-based electronics. *ACS Applied Materials & Interfaces* 10(37): 31061–31068.

Sana, F., E. M. Isselbacher, J. P. Singh, E. K. Heist, B. Pathik and A. A. Armoundas. (2020). Wearable devices for ambulatory cardiac monitoring: JACC state-of-the-art review. *Journal of the American College of Cardiology* 75(13): 1582–1592.

Scheck, A. (2014). Special report: Seeing the (google) glass as half full. *Emergency Medicine News* 36(2): 20–21.

Selvarajan, S., N. R. Alluri, A. Chandrasekhar and S.-J. Kim. (2016). BaTiO$_3$ nanoparticles as biomaterial film for self-powered glucose sensor application. *Sensors and Actuators B: Chemical* 234: 395–403.

Selvarajan, S., N. R. Alluri, A. Chandrasekhar and S.-J. Kim. (2017a). Direct detection of cysteine using functionalized BaTiO$_3$ nanoparticles film based self-powered biosensor. *Biosensors and Bioelectronics* 91: 203–210.

Selvarajan, S., N. R. Alluri, A. Chandrasekhar and S.-J. Kim. (2017b). Unconventional active biosensor made of piezoelectric BaTiO$_3$ nanoparticles for biomolecule detection. *Sensors and Actuators B: Chemical* 253: 1180–1187.

Selvarajan, S., N. R. Alluri, A. Chandrasekhar and S.-J. Kim. (2020). Biocompatible electronic platform for monitoring protein-drug interactions with potential in future theranostics. *Sensors and Actuators B: Chemical* 305: 127497.

Shitanda, I., Y. Fujimura, S. Nohara, Y. Hoshi, M. Itagaki and S. Tsujimura. (2019). Based disk-type self-powered glucose biosensor based on screen-printed biofuel cell array. *Journal of The Electrochemical Society* 166(12): B1063.

Shu, Y., C. Li, Z. Wang, W. Mi, Y. Li and T.-L. Ren. (2015). A pressure sensing system for heart rate monitoring with polymer-based pressure sensors and an anti-interference post processing circuit. *Sensors* 15(2): 3224–3235.

Smith, C. (2020). Crazy new glasses might be able to spot people with coronavirus. https://bgr.com/2020/04/16/coronavirus-symptoms-screening-rokid-glasses-perform-fever-readings/.

Song, Z., T. Ma, R. Tang, Q. Cheng, X. Wang, D. Krishnaraju, R. Panat, C. K. Chan, H. Yu and H. Jiang. (2014). Origami lithium-ion batteries. *Nature Communications* 5(1): 1–6.

Tao, L.-Q., K.-N. Zhang, H. Tian, Y. Liu, D.-Y. Wang, Y.-Q. Chen, Y. Yang and T.-L. Ren. (2017). Graphene-paper pressure sensor for detecting human motions. *ACS Nano* 11(9): 8790–8795.

Tao, T., S. Lu and Y. Chen. (2018). A review of advanced flexible lithium-ion batteries. *Advanced Materials Technologies* 3(9): 1700375.

Tong, R. (2018). Wearable Technology in Medicine and Health Care. Academic Press.

Trung, T. Q. and N. E. Lee. (2016). Flexible and stretchable physical sensor integrated platforms for wearable human-activity monitoring and personal healthcare. *Advanced Materials* 28(22): 4338–4372.

Van Langenhove, L. (2007). Smart Textiles for Medicine and Healthcare: Materials, Systems and Applications. Elsevier.

Wang, C., Y. Ding, Y. Yuan, A. Cao, X. He, Q. Peng and Y. Li. (2016). Multifunctional, highly flexible, free-standing 3D polypyrrole foam. *Small* 12(30): 4070–4076.

Wang, S., J. Xu, W. Wang, G.-J. N. Wang, R. Rastak, F. Molina-Lopez, J. W. Chung, S. Niu, V. R. Feig and J. Lopez. (2018). Skin electronics from scalable fabrication of an intrinsically stretchable transistor array. *Nature* 555(7694): 83–88.

Weng, W., Q. Sun, Y. Zhang, H. Lin, J. Ren, X. Lu, M. Wang and H. Peng. (2014). Winding aligned carbon nanotube composite yarns into coaxial fiber full batteries with high performances. *Nano Letters* 14(6): 3432–3438.

Xiong, J., J. Chen and P. S. Lee. (2020). Functional fibers and fabrics for soft robotics, wearables, and human–robot interface. *Advanced Materials* 2002640.

Xu, S., Y. Zhang, J. Cho, J. Lee, X. Huang, L. Jia, J. A. Fan, Y. Su, J. Su and H. Zhang. (2013). Stretchable batteries with self-similar serpentine interconnects and integrated wireless recharging systems. *Nature Communications* 4(1): 1–8.

Yaghouby, F., K. D. Donohue, B. F. O'Hara and S. Sunderam. (2016). Noninvasive dissection of mouse sleep using a piezoelectric motion sensor. *Journal of Neuroscience Methods* 259: 90–100.

Yamada, T., Y. Hayamizu, Y. Yamamoto, Y. Yomogida, A. Izadi-Najafabadi, D. N. Futaba and K. Hata. (2011). A stretchable carbon nanotube strain sensor for human-motion detection. *Nature Nanotechnology* 6(5): 296.

Yao, H., A. J. Shum, M. Cowan, I. Lähdesmäki and B. A. Parviz. (2011). A contact lens with embedded sensor for monitoring tear glucose level. *Biosensors and Bioelectronics* 26(7): 3290–3296.

Yokota, T., P. Zalar, M. Kaltenbrunner, H. Jinno, N. Matsuhisa, H. Kitanosako, Y. Tachibana, W. Yukita, M. Koizumi and T. Someya. (2016). Ultraflexible organic photonic skin. *Science Advances* 2(4): e1501856.

Zamarayeva, A. M., A. E. Ostfeld, M. Wang, J. K. Duey, I. Deckman, B. P. Lechêne, G. Davies, D. A. Steingart and A. C. Arias. (2017). Flexible and stretchable power sources for wearable electronics. *Science Advances* 3(6): e1602051.

Zang, Y., F. Zhang, C.-a. Di and D. Zhu. (2015). Advances of flexible pressure sensors toward artificial intelligence and health care applications. *Materials Horizons* 2(2): 140–156.

Zhang, F., Y. Zang, D. Huang, C.-a. Di and D. Zhu. (2015). Flexible and self-powered temperature–pressure dual-parameter sensors using microstructure-frame-supported organic thermoelectric materials. *Nature Communications* 6(1): 1–10.

Zhang, S., S. Li, Z. Xia and K. Cai. (2020). A review of electronic skin: soft electronics and sensors for human health. *Journal of Materials Chemistry B* 8(5): 852–862.

Zhang, Z.-F., K. Yang, S.-G. Zhao and L.-N. Guo. (2019). Self-healing behavior of ethylene propylene diene rubbers based on ionic association. *Chinese Journal of Polymer Science* 37(7): 700–707.

Zhao, C., J. Kong, X. Yao, X. Tang, Y. Dong, S. L. Phua and X. Lu. (2014). Thin MoS_2 nanoflakes encapsulated in carbon nanofibers as high-performance anodes for lithium-ion batteries. *ACS Applied Materials & Interfaces* 6(9): 6392–6398.

Zhao, T., C. Zheng, H. He, H. Guan, T. Zhong, L. Xing and X. Xue. (2019). A self-powered biosensing electronic-skin for real-time sweat Ca^{2+} detection and wireless data transmission. *Smart Materials and Structures* 28(8): 085015.

Zhao, X.-H., S.-N. Ma, H. Long, H. Yuan, C. Y. Tang, P. K. Cheng and Y. H. Tsang. (2018). Multifunctional sensor based on porous carbon derived from metal–organic frameworks for real time health monitoring. *ACS Applied Materials & Interfaces* 10(4): 3986–3993.

Zhao, Z., C. Yan, Z. Liu, X. Fu, L. M. Peng, Y. Hu and Z. Zheng. (2016). Machine-washable textile triboelectric nanogenerators for effective human respiratory monitoring through loom weaving of metallic yarns. *Advanced Materials* 28(46): 10267–10274.

Zou, Y., P. Tan, B. Shi, H. Ouyang, D. Jiang, Z. Liu, H. Li, M. Yu, C. Wang and X. Qu. (2019). A bionic stretchable nanogenerator for underwater sensing and energy harvesting. *Nature Communications* 10(1): 1–10.

CHAPTER 5

Nanocarbon-Based Sensor for Wearable Health Monitoring

Md. Milon Hossain,[1,3,]* *Abbas Ahmed*[2] *and Maliha Marzana*[3]

1. Introduction

Flexible and wearable electronics are developing rapidly and changing the typical medical diagnosis system by rendering it with integrated features of wearability, remote operation, comfortability, and real-time feedback. Therefore, wearable health and activity monitoring and personal health management are emerging and can be utilized for continuous, point-of-care, noninvasive and comfortable monitoring of critical health signs. The health information obtained by wearable electronics could be used for disease diagnosis, preventive health care and rehabilitation (Wang et al. 2020). The biophysical signs such as the electrocardiogram (ECG), electroencephalogram (EEG), electromyogram (EMG), skin temperature, and biochemical signs such as glucose, pH, lactate, etc. can be measured using wearable electronics (Choi et al. 2016). Therefore, different types of flexible and wearable sensors such as strain sensors, pressure sensors, electrophysiological and electrochemical sensors have been developed. Besides, flexible and stretchable interconnects and power devices have been developed to be integrated with wearable sensors (Wang et al. 2020). However, significant challenges exist due to the interface incompatibility between bulky and rigid electronics and soft skin tissue. Therefore, the development of body conformable soft, stretchable and comfortable electronic devices is surging to capture high-quality signals from the human body. These requirements can be met by either nanoengineering the existing materials or synthesis of new functional nanomaterials (Choi et al. 2016, Khan et al. 2016). In general, biophysical signs including pressure and temperature monitoring sensors work based on relative variations in their

---

[1] Department of Textile Engineering, Chemistry and Science, North Carolina State University, North Carolina-27606, USA.
[2] National Institute of Textile Engineering and Research, University of Dhaka, Dhaka 1000, Bangladesh.
[3] Department of Textile Engineering, Khulna University of Engineering and Technology, Bangladesh.
* Corresponding author: mhossai5@ncsu.edu

electrical parameters such as resistance, capacitance, and piezoelectricity. According to the active sensing material used, the sensing devices are classified as solid-state and liquid-state sensing devices. For solid-state sensors, different solid materials such as conductive metals, polymers, nanocarbons, semiconductors, and nanowires, etc. are widely used. Among many materials, nanocarbons are highlighted, considering their unique electrical, mechanical, chemical and thermal properties. Besides, nanocarbon materials have high mechanical flexibility and can be functionalized with different materials for target-specific sensing. While nanocarbon refers to different carbon allotropes such graphite, diamond, fullerene, this chapter is limited to carbon nanotubes (CNTs) and graphene including graphene oxide (GO) and reduced graphene oxide (rGO) (Saba et al. 2019, Ahmed et al. 2020a). CNTs are cylindrical carbonaceous materials produced from rolled-up planar graphene sheets and they might be single-walled (SWCNT) or multi-walled (MWCNT) depending on the number of concentric layers of graphene (Wang et al. 2017). This chapter discusses different structures of CNTs and graphene and their application in wearable health monitoring. However, the wearable health monitoring applications focus mainly on biopotential monitoring, skin temperature and body motion monitoring. The requirements of the wearable sensor and mechanism of sensing are also elucidated.

2. Nanostructured Carbon Materials

Among different nanocarbon materials, CNTs and graphene are dominant and different from conventional carbon materials. The nanocarbon materials can be categorized into three generations as shown in Figure 1.

First-generation: The three basic nanocarbon materials are included in the first generation such as fullerene, graphene and CNTs. The various novel derivatives of nanocarbon like carbon quantum dots, nanofibers, nanohorns, nanoribbons, nanocapsulates, nanocages, etc. are also included. Generally, these nanocarbons are characterized by morphology and low dimensionality.

Second generation: The electronic structure of first-generation nanocarbons is tailored to obtain second-generation nanocarbons. Different heteroatoms/dopants are introduced to tune their properties and they undergo a controlled synthesis process.

Third generation: The synthesis process and properties are extremely controlled by the development of hybrid and/or hierarchical systems at the nano-level (Perathoner and Centi 2018).

The properties of CNTs depend on their structure. The structure of SWCNTs are zigzag, armchair, and chiral and they might be either metallic or semiconducting according to their chirality and diameter. The basic unit of two-dimensional (2D) carbon materials is graphene and can be wrapped into zero-dimensional (0D) fullerene, rolled up into one-dimensional (1D) CNTs and stacked into three-dimensional (3D) graphite (Saba et al. 2019). The primary properties of different dimensional nanocarbon are presented in Table 1. The atomic-scale dimensions of the materials greatly affect the flexural rigidity of devices. Materials with 0D demonstrate unique characteristics due to quantum confinement and large surface

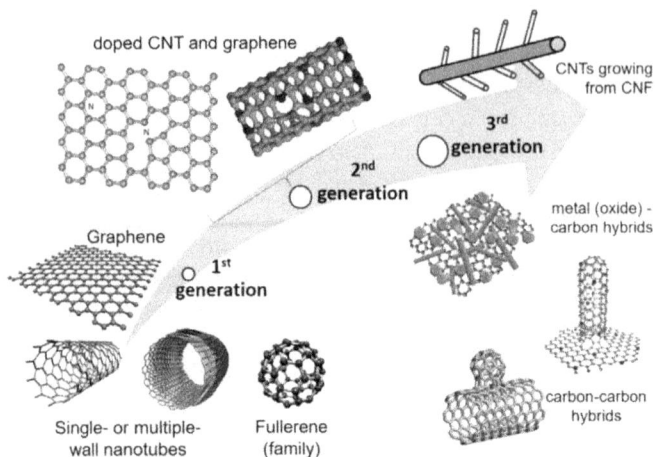

Fig. 1. Three generations of nanocarbon (Reprinted with permission from Perathoner and Centi 2018).

Table 1. The primary properties of different dimensional nanocarbon materials.

	0D	1D	2D	3D
Typical materials	Fullerene, carbon quantum dots, etc.	Carbon nanotube (CNT), carbon nanowires, etc.	Graphene, graphdiyne, etc.	Carbon sphere, carbon shell, graphite, porous carbon, etc.
Basic properties	Nonlinear optical property, low density, etc.	Good conductivity, hydrogen storage capacity, diathermancy, etc.	High specific surface area, electronic transmission capacity, etc.	High specific surface area, designability, multihole, etc.

Source data obtained from He et al. 2021

area. Compared to 0D materials, carrier transport is more efficient in 1D materials; for example, nanowires and carbon nanotubes. For planar device structures, 2D materials exhibit enhanced charge transport behavior. However, for materials with higher dimensions such as 3D, the performance directly depends on the uniformity of the device. With the increase of dimensional scales, defects, and rigidity of the devices increase (Choi et al. 2016).

3. Fabrication of Electrodes for Wearable Health Monitoring

Wearable electronics have made the health care system more user-friendly and provided easy access to real-time monitoring of an individual's health status. Using sophisticated wearables, it is possible to measure different physiological (e.g., body motion and body temperature), (bio) chemical signals (blood pressure, glucose, sweat, etc.) and bioelectrical signals (ECG, EEG, etc.). Such technology makes it possible for systematic homecare medicine practices by connecting to the Internet of Things (Wang et al. 2020). Importantly, the key component of these devices is the electrode. The output of a sensor device largely depends on electrode characteristics, any noise associated with the electrode may affect the result significantly. Notably, in wearable electronics, electrodes or conductors should have the ability to maintain high conductance over a larger strain which is required for transmitting signals.

Generally, there are two different electrode systems involved in wearable health monitoring, which are (i) passive and (ii) active. Both electrodes perform based on the similar operation principle of conventional electrodes. In passive electrodes, the sensor can acquire an electrical signal. Such a type of electrode is often applied in cardiac and muscle assessments by detecting electrical signals obtained from the heart and muscles. Besides, passive electrodes are applied in smart textile-based wearable and wireless biomonitoring systems. On the other hand, active electrodes are the type of electrodes used for transcutaneous electrical stimulation. In this case, by transducing the applied electrical current to a tissue probe on a skin surface, the targeted nerve cells, skin receptors and other sensory units can be activated and pain relief through nerve electrical stimulation can also be achieved (Keller and Kuhn 2008, Li et al. 2010, Mečnika and Dipl 2014). Using the above-mentioned mechanism, several electrodes can be fabricated. However, passive textiles electrodes are widely employed for the acquisition of different electrophysiological signals (Acar et al. 2019).

4. Requirements for Sensing

Understanding the major requirements for wearable sensing would impart the appropriate designation of healthcare devices and assess their usability in particular applications. Based on the application types, several performance metrics are essential to consider when developing a sensing system. Overall, the selection of sensing material for electrode development is important because it mostly affects the functionality of the electrodes. Wearable electronic devices typically require several components in their structure including modules for the sensor, processing, communication and powering that are needed to be mounted on skin, curved surfaces and movable joints of the body. Hence, flexibility and stretchability are key aspects of skin-conformability and wearability (Ahmed et al. 2020c, Wang et al. 2020).

As mentioned above, sensing requirements vary according to the applications. For instance, in biopotential monitoring, several factors should be considered during/ after electrode fabrication that influence biopotential acquisition significantly, which are as follows-

i) Skin-electrode contact impedance: When recording biopotential signals, skin-electrode contact impedance is important as it affects the signal obtained from the body. It was found that decreasing the skin-electrode impedance improves the quality of signals.

ii) Motion-artifacts: These are undesirable signals in biopotential monitoring that arise from motion caused by the movement of one part to another. In such cases, proper electrode placement or placing a preamplifier next to the active electrode may reduce motion artifacts (Acar et al. 2019).

Apart from flexibility and stretchability, several other key requirements are essential in the fabrication of sensing electrodes for body motion monitoring-

iii) Sensitivity or gauge factor (GF): This is the foremost parameter of monitoring body motion. For effective functionality, the strain/pressure electrode should possess high sensitivity or GF. Besides, according to the type of motions (small and broad range of strain/pressure), sensitivity may vary. Therefore, the proper selection of active materials for sensing is a prerequisite.

iv) Linearity and response time: The stretchable sensor should possess linearity in performance because a very large strain should be accommodated by the strain sensor. The nonlinearity of the sensor causes issues in calibration and operation. Furthermore, the sensor should be fast-responsive to reach a steady-state condition.

v) Durability: For long-time operation, strain sensors should have dynamic durability as they experience several mechanical deformations. This parameter is important for all wearable and skin-mountable devices (Amjadi et al. 2016, Hossain and Bradford 2021).

5. Mechanism of Sensing

The mechanism of sensing such as piezoresistive, capacitative, piezoelectric and triboelectric is discussed in the following sections. The optical transduction mechanism of sensing is also highlighted.

5.1 Piezoresistive

Piezoresistivity is the most used mechanism to fabricate wearable health monitoring sensors due to its simplicity, cost-effectiveness, large ranges of pressure, suitability, and high pixel density (Zang et al. 2015). Moreover, the working principle of these sensors is straightforward (Homayounfar and Andrew 2020). The piezoresistive sensor converts the electric resistance of a system into an electrical signal when a mechanical stimulus is applied (Figure 2a) (Wan et al. 2017). It can measure the strains and stresses to monitor the health signal, to build a relationship between the human and computer, and may function as electronic skin (Yang et al. 2017).

The resistance R of the operative component can be demonstrated as (Homayounfar and Andrew 2020)

$$R = (\rho L)/A,$$

where, ρ = Electrical resistivity, r = Resistivity, L = Length, A = Cross-sectional area.

Resistance can be varied by the deformity of the factors ρ, r, L, and A (any three of them or all of them), when any stress (compression or tensile) is applied on the piezoresistive sensors. Depending on the type of stress, these devices can be of two types—pressure sensor and strain sensor (Homayounfar and Andrew 2020).

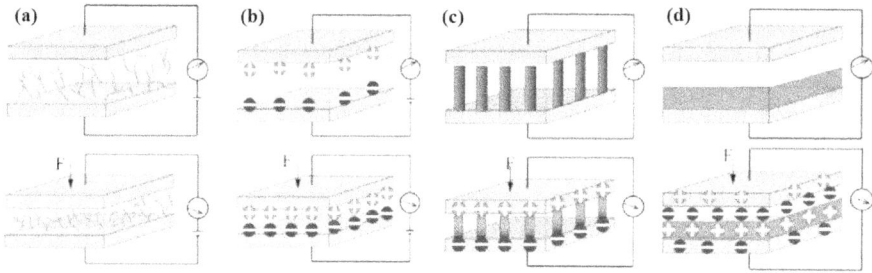

Fig. 2. Illustration of four sensing mechanisms (a) piezoresistive, (b) piezocapacitive, (c) piezoelectric, (d) triboelectric (Reprinted with permission from Wan et al. 2017).

For both piezoresistive and piezocapacitive sensors, the resistance variation is based on geometry and resistivity (Yang et al. 2017) and can be stated as

$$\Delta R/R = (1 + 2v)\epsilon + \Delta\rho/\rho,$$

where, v = Poisson's ratio, ϵ = Poisson's strain, ΔR = the change in resistance.

Physiological actions can deform the structure, which will create a minor change in R and a low GF value (Li et al. 2020b). Therefore, the materials should be chosen carefully to fabricate these sensors. Therefore, the design strategy significantly affects sensor performance and with system design, a highly sensitive sensor can be fabricated. Recently, some novel design methods are used; for instance, crack extension in the slender films, tunneling effect, separation of the sensing substances, etc. Some elements exhibit high sensitivity due to the nook of the channel or diversion of the morphology inside the sensing components or between the sensing component and the electrodes (Yang et al. 2017).

5.2 Capacitive

The ability to store electrical charge is known as capacitance. The capacitive sensor exhibits a parallel-plate arrangement where the dielectric surface is sandwiched between the coupled electrodes (Li et al. 2020b). The conventional representation of capacitance for the parallel plate is given by the equation

$$C = \varepsilon_r\varepsilon_0 A/d,$$

where ε_r relative permittivity of the materials between the plates, ε_0 is the electric constant, A is the area of the plates, d is the distance between the plates, and C is the capacitance.

The external force applied to the pressure sensors affects variables A and d, and causes the plate to deflect, leading to variations in capacitance (Zang et al. 2015) (Figure 2b). Generally, the changes in parameters A and d are caused by the shear force and perpendicular pressure applied to the electrodes, respectively.

Capacitive sensors have high sensitivity, low hysteresis, high response time, thermal stability, less power consumption, higher linearity, etc. Different materials are used to improve their performance such as carbon nanotubes, graphene and its derivatives, elastic polydimethylsiloxane, stretchable conductive materials, ecoflex silicone, pyramidal elastomer, etc. Besides, the ionic conductors are used as an alternative to the electrodes to impart biocompatibility, optical transparency, etc. These ionic conductors include hydrogels, ionogels, and many more superior elements (Yang et al. 2017).

5.3 Piezoelectric

Facile material preparation, inexpensive, ease of data acquisition attracts the researcher for the piezoelectric sensor. When an external force is applied, charge separation takes place inside the materials (Xu et al. 2018). One surface of the materials is negatively charged, and the other surface is positively charged and thus creates a potential difference as shown in Figure 2c. The efficiency of energy conversion of the piezoelectric material is given by the physical quantity known as a piezoelectric coefficient (d33). Traditional inorganic materials used for piezoelectric sensors demonstrated high sensitivity due to their high d33 and have been widely explored. However, inorganic materials are less preferred for flexible pressure sensing applications due to their high crystallinity and rigidity. Therefore, carbon nanotube, graphene, and their derivatives have received significant attention to synthesize the wearable health monitoring piezoelectric sensors, owing to their acoustic vibration detection properties, high response time, sensitivity, and many more features (Li et al. 2020b). However, recently, ultrathin structures of inorganic materials such as films and nanowires demonstrated good mechanical flexibility and are being extensively studied for flexible and stretchable pressure sensors.

5.4 Triboelectric

The triboelectric phenomenon caused by friction is common in our daily life. This sensing mechanism is a relatively new area compared to other sensing mechanisms and was discovered in 2012 (Tao et al. 2019). The triboelectric mechanism is also known as contact electrification and sensors based on this mechanism transform mechanical energy to an electrical signal by a coupling of triboelectrification and electrostatic induction (Yu et al. 2019) (Figure 2d). Triboelectrification facilitates the generation of high output charge/voltage and electrostatic induction can amplify the reaction signal to the mechanical motion, thus enhancing the sensitivity of the sensor device (Ren et al. 2018). The charge generation is directly related to the triboelectric polarities difference between two materials. In normal conditions, the small gap separates the two layers but when force is applied, this gap diminishes, and the triboelectric effect induces opposite charges on each of the two materials. At relaxed conditions, surfaces containing opposite charges are separated and compensating charges are generated on the top of each side (Wan et al. 2017). There are two mechanisms involved: electron transfer and ion transfer. For polymers, mobile ions are contained or hydronium and hydroxide ions from the atmosphere are introduced. Electron transfer involves the electrification between metal and metal, semiconductors and insulators (Tao et al. 2019). Compared to other transduction mechanisms, triboelectric sensors can detect tiny pressure or motion in a self-power mode. However, most self-powered sensors that have high sensitivity in monitoring pulse waveforms fail to identify precise actions such as gesture recognition, physical rehabilitation and human-machine interaction due to the narrow measurement range and poor linearity (Yu et al. 2019).

5.5 Optical Transduction

The optical transduction method is another mechanism of sensing in which the mechanical incentive is converted into diversified signals. Optical fibers play a significant role in this method (Pinet et al. 2007). Light is generated by a LED that falls out on a force-sensitive transmission medium such as optical fibers, which is an elastomeric waveguide. Then it travels through the system and gets picked by a photodetector (Yang et al. 2017). This process has several advantages, for example, it can exhibit high resolution, repeatability, lower attenuation coupled with the vision sensing methodology (Chi et al. 2018). Besides, it links mechanical reports to the light, which is susceptible to detach the health monitoring sensor from electronics technology (Yang et al. 2017). Moreover, they are independent of electrical trespass. However, the optical transduction mechanism requires a large amount of power to produce the light and needs transparency, interferences, labels, etc., which makes it challenging for different applications (Zhang and Hoshino 2014). The comparison of the different sensing mechanisms in terms of their advantages and disadvantages is presented in Table 2.

Table 2. Advantages and disadvantages of the different mechanisms of sensing.

Mechanism of Sensing	Advantages	Disadvantages
Piezoresistive	• High sensitivity • Higher pressure range • High pixel density	• Poor stability • Lower thermal stability • Hysteresis effect
Capacitive	• High sensitivity • Low hysteresis • High response time • Thermal stability • Low power consumption • Higher linearity	• Parasitical dominance • Low signal-to-noise ratio • Electromagnetic influence • Complex structure • Crosstalk between the sensing elements
Piezoelectric	• High sensitivity • Fleeting characteristics • Reversibility • Low power consumption • Good dynamic response	• Inappropriate for static sensing • Less mechanical stability • Complex circuit structure • Susceptibility to temperature • Drift of sensor output over time
Triboelectric	• High sensitivity • Lightweight • Fast response • Low detection limit	• Decoupling of interfered signals is difficult • Low reliability for long term • Lack of positive triboelectric materials
Optical transduction	• High resolution • Lower attenuation • Electrical trespass	• Higher power consumption • Higher cost • Complexity

Source data obtained from Li et al. 2017b, Yang et al. 2017, Chi et al. 2018, Kim et al. 2020

6. Applications of Graphene for Wearable Health Monitoring

6.1 Graphene for Biopotential Monitoring

Recently, monitoring health conditions and human performance during physically demanding activities have received increased interest. Wearable health monitoring devices are desired for improved personal health care, long term monitoring of known health issues and rehabilitation. Currently, electrodes are being developed to monitor human vital signs continuously, noninvasively and comfortably (Khan et al. 2016). The traditional bioelectric electrode includes silver/silver-chloride (Ag/Ag-Cl), but epidermal inflammation and allergic reaction inhibit their prolonged uses for biopotential monitoring. Moreover, poor bioelectric signal quality may arise because the dry electrode shows less adherence to the curved surface with human skin (Das et al. 2020). Graphene-based sensors are well suited for bioelectric signal acquisition because of their high conductivity, biocompatibility and skin conformity when integrated with a flexible substrate such as textiles. Besides, graphene-based materials for biopotential monitoring electrodes are regarded as efficient and cost-effective. For example, Das et al. (2020) fabricated a thermally reduced graphene oxide and nylon-based wearable electrode for the recording of three different bioelectric signals: ECG, EEG, and EMG. The sensor electrode showed excellent epidermal conformity, high throughput signals, and long-time monitoring. A similar study was reported by Yapici et al. (2015), where authors developed a graphene-clad nylon textile electrode for cardiac biosignal recording. The graphene-clad electrode was compared with the conventional Ag/Ag-Cl electrode and showed that graphene-clad electrode-based ECG recordings were not influenced by any unwanted distortions as evident by the discrete wavelet transform procedure. It was shown that both filtered waveforms from conventional and graphene-clad textile electrodes perfectly overlapped and demonstrated distinguishable P-QRS-T ECG morphology.

Xu et al. designed a screen-printed graphene textile electrode for ECG measurement (Xu et al. 2020). The electrode showed promise in performance over the conventional wet electrode. Besides, the electrode exhibited superior signals even after washing and bending 9 and 2000 times, respectively. Using graphene as an active material, similar approaches have been employed to fabricate electrodes for biopotential monitoring, revealing outstanding ECG signal acquisition capacity (Sinha et al. 2019). Wearable biopotential monitoring electrodes should have high signal quality, low skin-electrode contact impedance and high signal-to-noise ratio (SNR). A graphene-based electrode for cardiac ECG measurement by the coating of graphene on top of a metallic layer of a Ag/AgCl electrode was developed (Celik et al. 2016). The electrode showed reliability in biopotential monitoring and improved performance in terms of signal quality, skin-contact impedance, and SNR.

6.2 Graphene for Skin Temperature Monitoring

In the medical health assessment of a human body, temperature is the first key physiological health indicator determined by the physician. By measuring the skin temperature of a body, doctors may assess the initial health conditions of a patient and thereby assist in taking preventive or precautionary decisions as well as predict potential implications. Therefore, real-time and continuous monitoring of skin temperature is of significant importance. Traditional temperature measurement system involves mercury thermometer and infrared thermometer and often fails to provide long-term and continuous temperature monitoring, whereas wearable skin-mountable sensors are feasible and advantageous in terms of high accuracy and sensitivity, prolonged performance and real-time monitoring (Liu et al. 2019).

Due to the excellent electrical and thermal properties, nanocarbon materials (graphene and CNTs) were extensively explored to develop skin attachable temperature sensors. However, the sensitivity of pure graphene-based temperature sensors is not satisfactory (typically less than 1.00% $°C^{-1}$) (Kabiri Ameri et al. 2017, Nguyen and Kim 2018) and therefore, they are often coupled with other materials such as PEDOT:PSS (Soni et al. 2020), textiles (Rajan et al. 2020) and polydimethylsiloxane (PDMS) (Wang et al. 2019), etc. Recently, Liu et al. (2019) fabricated a skin-attachable temperature sensor by a facile spray dipping approach on a polyimide (PI) substrate. In the composite structure, rGO was utilized as an active temperature sensing element and polyethyleneimine (PEI) was employed as an adhesive. The flexible sensor demonstrated high sensitivity at a temperature range of 25–45°C and was able to detect the tiny temperature change (0.1°C) and durability up to 120 days. The sensor has the potential to be used in healthcare and disease diagnosis. The performance summary of the different graphene-based temperature sensors is provided in Table 3.

A resistive temperature sensor by placing micro-patterned single-layer graphene on silicon dioxide/Si, a silicon nitride membrane (SiN), and a Si wafer with etched rectangular pits was produced (Davaji et al. 2017). The resistive sensor demonstrated a quadratic dependence of resistance on the temperature between 283 K and 303 K. As shown in Figure 3(a-b), the device resistance depends on the electron conductance. It was found that the graphene sensor placed on SiN membrane exhibited high sensitivity which could be useful for various sensing applications.

Temperature sensors with excellent and reliable sensing properties under external deformations are essential for artificial electronic e-skins. Wang et al. prepared a cellular solid structure graphene/PDMS composites by 3D printing technique (Wang et al. 2019). Unlike traditional sensors, where skin conformability is an issue, the as-obtained cellular composites exhibited improved conformability with the curved surfaces of the human body. The real-time monitoring of normal skin temperature

Table 3. Performance summary of graphene-based temperature sensors.

Material	Range of Measurement (°C)	Sensitivity (°C⁻¹)	Linearity	References
rGO coated polyamide	25–45	1.30	yes	(Liu et al. 2019)
rGO	30–100	0.6345%	yes	(Liu et al. 2018b)
rGO filled cellulose films	25–80	-	yes	(Sadasivuni et al. 2015)
rGO/PU	30–80	1.34	yes	(Trung et al. 2016)
Graphene/PDMS	25–75	0.8	yes	(Wang et al. 2019)
Graphene woven fabric	25–50	1.34	yes	(Zhao et al. 2017)

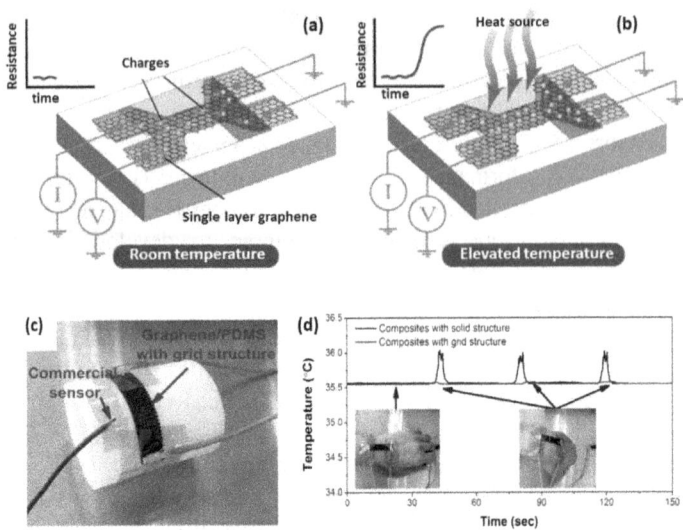

Fig. 3. Schematic representation of a graphene sensor. The increase in temperature from room temperature (a) to a high temperature (b) causes the increase in electron while a decrease in electron mobility (reprinted with permission from Davaji et al. 2017). (c) Experimental set up of the temperature sensor and (d) monitoring of wrist skin temperature and joint bending (reprinted with permission from Wang et al. 2019).

was found to be ~ 35.6°C, while the bending of the wrist to a larger angle caused an increase in temperature by 0.5°C. This could be due to the strain-sensitivity effect of the sensor. However, the graphene/PDMS sensor showed exceptional temperature sensing properties even at large deformation states, Figure 3(c-d). In another study, an ink-jet printed temperature sensor was developed which was compatible with an epidermal electronic system. The sensors were fabricated from graphene and PEDOT:PSS ink and screen-printed silver ink. Since the sensor was printed on a skin-conformable flexible bandage type substrate, it showed excellent adhesion to the skin. The sensor can measure the temperature under normal conditions 34–35°C with higher temperature coefficient resistance of 0.06% °C^{-1} (Vuorinen et al. 2016).

6.3 Graphene for Body Motion Monitoring

Typically, the change of physical position of the human body can be termed as human motion. To measure and monitor a subject's real-time three-dimensional (3D) kinematics, an analysis of human motion is becoming an indispensable diagnostic technique. It is a biomedical model that deals with classical forward kinematics by which a particular position or orientation of a body segment can be measured. Furthermore, by analyzing human body motion, the recovery condition of a patient from the rehabilitation treatments can be elucidated (Huang et al. 2019, Xie et al. 2020). To date, several human motion monitoring systems have been designed and developed exploiting numerous types of materials (Ahmed et al. 2020b). However, the ideal selection of sensing materials, creating an unobtrusive sensing platform and reliable sensing performance, is still challenging.

The key requirements for a strain sensor to function appropriately ranges from high sensitivity to flexibility and adaptability to configure with various body postures. Importantly, recent research focuses on enabling these aspects in strain sensors, thus adopting wide varieties of nanomaterials which includes conducting polymers (Hatamie et al. 2020) metallic nanoparticles (Lee et al. 2014a, Chen et al. 2016) and carbonaceous materials (CNT and graphene) (Molina 2016, Li et al. 2018, Huang et al. 2019). One of the ways of achieving a high-performance strain sensor is by coupling these nanomaterials with textile structures, which are termed as smart textiles. Since smart textiles are flexible and have excellent adaptability with the human body, they are often referred to as intelligent materials that can sense and respond in a specific environment (Faruk et al. 2021). Recently, graphene-enabled smart textiles gained tremendous attention owing to their high sensitivity and excellent conformability to human skin, offering real-time tracking of different dynamic body movements. Based on different functional materials and diverse sensing mechanisms, different strain sensors can be developed such as resistive, capacitive and piezoelectric, etc. However, resistive and capacitive sensors are predominant because of their facile fabrication strategy and direct data acquisition system (Ray et al. 2019). In such cases, emphasis has been put on the utilization of graphene as an active sensing material.

Graphene-enabled wearable strain sensors are utilized mainly for two different types of motion monitoring: (i) large motion detection, e.g., hands and arms bending, muscle movements, different gestures and (ii) subtle motion detection, e.g., eye blinking, speech recognition, etc. For example, Zheng et al. (2020) prepared graphene-based smart textiles for low limit detection of strain by the deposition of graphene onto cotton fabrics, followed by encapsulation of polydimethylsiloxane. The sensor exhibited successful detection of various human motions including bending of fingers, wrist, elbow and squatting at a very low strain (~ 0.4%). Such sensors may find applications in areas where very low range motion detection is required. A similar work demonstrated that the graphene textiles strain sensor can detect the respiration rate with various breath patterns along with measuring various large and subtle human motions. Besides, the sensor is endowed with close-fitting to the human body, showing excellent wearability (Yang et al. 2018).

Graphene and its derivatives have also been employed with other materials forming composites to serve as flexible strain sensors. For example, a wearable strain sensor was developed by Liu et al. (2018a), combining cracked graphene sheets with glycerol/potassium chloride ionic conductor. The sensor demonstrated high stretchability (300%), gauge factor (25.2) and fast response time (80 ms), thus enabling monitoring of both large range and subtle human body motions. Human skin is comparatively rough with dermatoglyphic and getting ultra-conformal contact with skin is a challenging task. This issue was addressed by developing an ultra-conformal strain sensor based on a double transfer process utilizing graphene as an active material (Wan et al. 2018). The sensor performed through sliding of nanosheets in fish scale graphene layers when strain is applied, thus showing ultra-high sensitivity (GF = 502), skin level stretchability (35%) and fast response. This sensor can be employed for monitoring several physiological signals as well as vocal cord movement, radial artery waves and jugular venous pulses.

The effective detection and monitoring of different dynamic deformations of a body require the strain sensor to be highly sensitive, flexible, stretchable, fast responsive, durable and linear in performance. Although graphene-based strain sensors are widely appreciated for their high conductivity, gauge factors and low cost, they showed compromise in stretchability. Furthermore, encapsulation is required for some graphene-based sensors to prevent delamination because of the week interface between fiber and graphene (Chatterjee et al. 2019). However, these limitations are expected to be overcome in the forthcoming years with further research.

7. Application of CNT for Wearable Health Monitoring

CNTs are widely used in wearable electronics due to their excellent mechanical flexibility and superior conductive properties. CNTs usually exhibit an entangled network structure rendering CNT-based sensors highly stretchable (Bansal and Gandhi 2019). This section discusses the applications of CNTs for wearable health monitoring such as biopotential monitoring, skin temperature monitoring, and body motion monitoring.

7.1 CNT for Biopotential Monitoring

CNTs have been utilizing in the medical field as cardiovascular diseases are the most common cause of higher death rates, and CNT-based ECG sensors monitor cardiac activities regularly to reduce sudden heart attacks and many more diseases (Bansal and Gandhi 2019). Moreover, CNT-based ECG sensors are comfortable to wear. A facile dipping and drying method (Zhao et al. 2018) was used to produce SWCNT coated cotton fabric sensor for real-time monitoring of ECG. The flexible fabric sensor exhibited efficient transmission of weak bio-signals showing significant distortion or attenuation and can be employed as a promising wearable device for ECG monitoring. In a recent study, Li et al. (2020a) adopted a "cut" and "paste" technique to design a CNT-based ECG monitoring sensor. In this study, a laser cutter was used to cut CNT-PU thin film to get the desired design of electrode and then heat pressed to paste the electrode pattern onto the textile substrate. The sensor demonstrated lower skin-electrode impedance, which can transduce high-quality ECG signal and SNR similar to that of conventional Ag/AgCl wet electrodes.

CNT-based electrodes were also used to design an Ear-EEG to monitor the bioelectrical brain activity of human beings through several personal devices such as a smartphone (Fu et al. 2020), as shown in Figure 4. Although the use of CNT based devices for wearable applications is debatable due to their toxicity concern, recent studies show that CNT exhibit low cytotoxicity and good biocompatibility (Jung et al. 2012, Lee et al. 2014b). This illustrates that CNT has no adverse effect on the wearer's skin and is promising for long-term wearable applications.

7.2 CNT for Skin Temperature Monitoring

The excellent thermal conductivity of CNTs is used for long-term and continuous temperature monitoring. The 1D and 2D fibrous structures produced using CNTs are suitable for wearable skin-mountable sensors and are considered more feasible

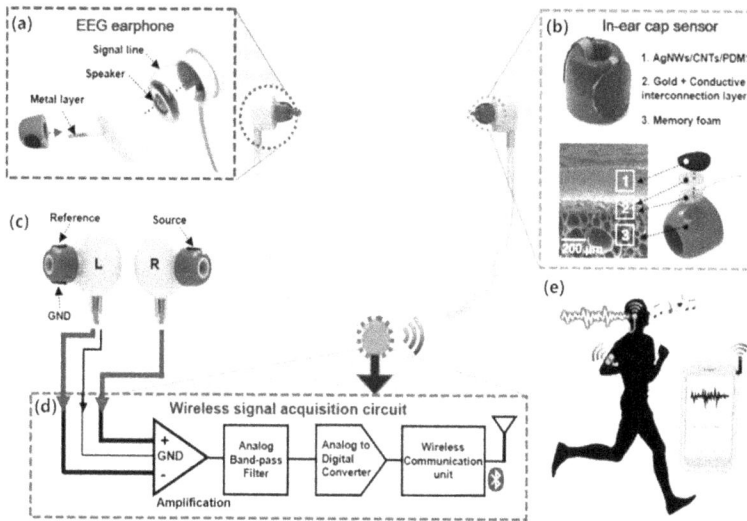

Fig. 4. Segments and diagram of an intrinsic earphone for bioelectrical brain activity recording. (a) Structural formation of portable earphones electrode, (b) figures of circumstantial arrangements and components of the synthesized earbud, (c) illustration of the placement of each EEG electrodes, (d) schematic form of a signal processing model and (e) EEG data from the daily activity of a person is shown in real-time on a personal smartphone (Reprinted with permission from Fu et al. 2020).

and advantageous in terms of high accuracy and sensitivity, prolonged performance and real-time monitoring (Liu et al. 2019). An MWCNT-based skin temperature monitoring device was developed to treat chronic wounds such as a diabetic foot ulcer caused by the increment of the temperature at the risky areas of the foot (Matzeu et al. 2011). This sensor can be located on the calf and held by a light bandage. A radio frequency identification (RFID) tag was used to connect the sensors with the leg motion. In another work, Karimov et al. (2012) demonstrated an MWCNT-based flexible Al-CNT-Al temperature sensor model, which was synthesized by the deposition of MWCNTs on the adhesive elastic polymer tape to measure the skin temperature effectively. The sensor exhibited an average temperature sensitivity of -1.26% °C^{-1}.

7.3 CNT for Body Motion Monitoring

Skin-mountable and stretchable wearable body motion monitoring devices are increasing due to the increasing demand in personalized health monitoring, human motion detection, human-machine interfaces, soft robotics, etc. (Zhang et al. 2017). Besides, wearable health monitoring devices can significantly improve the well-being of humans. For example, the health safety of people working in a harsh and dangerous environment can be ensured by monitoring their physiological performance in real-time. Importantly, wearable health monitoring can reduce the health care cost and hospital visits for the elderly by providing self-care at home. Besides, the performance of the athletes can be improved by quantifying and analyzing their motions, speed, and daily exercise to protect them from injury (Wang and Loh 2017).

The mechanical flexibility and excellent conductivity of CNTs are advantageous for body motion monitoring. The interconnected network structure of CNTs allows them to be highly stretched and CNT-based elastomeric nanocomposites can be stretched up to 500% (Dahiya et al. 2020).

A CNT thin film was fabricated by spray coating technique and integrated with fabric to be used as wearable body motion monitoring applications (Wang and Loh 2017). The sensor was used to detect the finger movements as a proof-of-concept device and it was found that the different bending motions of the finger can be detected with relatively good stability. In another study, PDMS and CNTs or silver/CNTs composites were prepared by the template method (Zhang et al. 2017). The result demonstrated successful detection of arm motion, wrist bending, and elbow bending. However, the CNT-based sensor also exhibits very high sensitivity to detect voice. CNT-based resistive force wearable sensors are capable of altering the resistance variation of any sensitive elements resulting from an external evocation into an electrical signal output to exhibit effective sensitivity in voice recognition and pulse detection (Gu et al. 2019). Wang et al. designed an E-skin device with MWCNT/Poly(di-methylsiloxane) composite film, which can track pressure and strain dynamically when joined with a human finger or a human throat (Wang et al. 2017). Furthermore, CNT/PDMS nanocomposites were used to develop a wearable strain sensor that exhibits superior competency to monitor multiple human body motion, for instance, bending of a finger and elbow, speaking, drinking, breathing, etc. (Li et al. 2017a).

However, the large deformation of the body may induce crack, break, or damage the sensors leading to significant changes in the output signal. As a result, maintaining linearity, which is a vital parameter of strain sensor over an extensive strain range, is difficult. Therefore, Li et al. developed a self-repairing CNT/ethylene vinyl acetate fiber strain sensor with a high stretchability of 190%, large linear working range up to 88% strain, excellent dynamic durability (5000 cycles) and fast response speed (312 ms) (Li et al. 2020c). The sensor was able to detect the movement of finger, wrist and elbow bending as shown in Figure 5.

8. Conclusion

The structure and properties of carbon nanotube and graphene have been discussed in this chapter. Different efforts have been made to utilize the unique characteristics of nanocarbon for wearable health monitoring applications. Both graphene and carbon nanotube have been extensively used for biopotential and body motion monitoring compared to skin temperature monitoring. To develop highly stretchable body movement tracking devices, elastomeric composites of nanocarbon are preferred. While wearable health monitoring is expanding rapidly, still challenges remain to use carbon nanomaterials for particular health monitoring applications. The toxicity and health safety of nanocarbon require further understanding. Besides, most of the development of wearable health monitoring focuses on functionality and the accuracy and security of the health information are yet to be properly addressed.

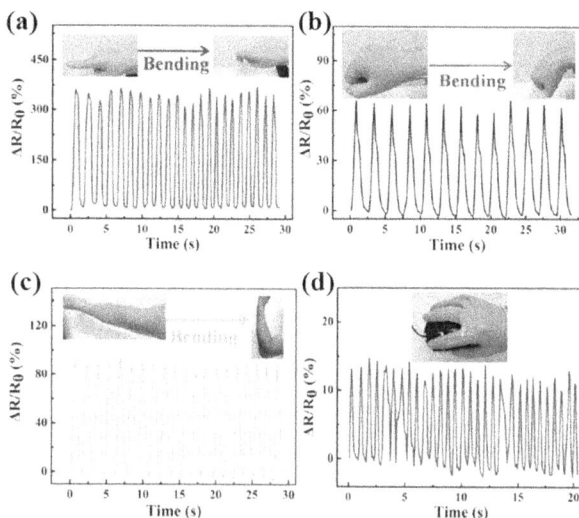

Fig. 5. Application of CNTs/EVA fiber strain sensors. Movement tracking of (a) the finger, (b) the wrist, and (c) the elbow under bending. (d) Response curve of CNTs/EVA fiber strain sensors to a mouse click (Reprinted with permission from Li et al. 2020c).

References

Acar, G., O. Ozturk, A. J. Golparvar, T. A. Elboshra, K. Böhringer and M. Kaya Yapici. (2019). Wearable and flexible textile electrodes for biopotential signal monitoring: A review. *Electronics (Switzerland)* 8: 1–25.

Ahmed, A., B. Adak, T. Bansala and S. Mukhopadhyay. (2020a). Green solvent processed cellulose/ graphene oxide nanocomposite films with superior mechanical, thermal, and ultraviolet shielding properties. *ACS Applied Materials and Interfaces* 12: 1687–1697.

Ahmed, A., M. M. Hossain, B. Adak and S. Mukhopadhyay. (2020b). Recent advances in 2D MXene integrated smart-textile interfaces for multifunctional applications. *Chemistry of Materials* 10296–10320.

Ahmed, A., M. A. Jalil, M. M. Hossain, M. Moniruzzaman, B. Adak, M. T. Islam, M. S. Parvez and S. Mukhopadhyay. (2020c). A PEDOT:PSS and graphene-clad smart textile-based wearable electronic Joule heater with high thermal stability. *Journal of Materials Chemistry C* 8: 16204–16215.

Amjadi, M., K. U. Kyung, I. Park and M. Sitti. (2016). Stretchable, skin-mountable, and wearable strain sensors and their potential applications: a review. *Advanced Functional Materials* 26: 1678–1698.

Bansal, M. and B. Gandhi. (2019). CNT based wearable ECG sensors. pp. 208–213. *In*: 2019 3rd International Conference on Recent Developments in Control, Automation and Power Engineering, RDCAPE 2019 (Institute of Electrical and Electronics Engineers Inc.).

Celik, N., N. Manivannan, A. Strudwick and W. Balachandran. (2016). Graphene-enabled electrodes for electrocardiogram monitoring. *Nanomaterials* 6: 1–16.

Chatterjee, K., J. Tabor and T. K. Ghosh. (2019). Electrically conductive coatings for fiber-based E-textiles. *Fibers* 7: 1–45.

Chen, S., Y. Wei, X. Yuan, Y. Lin and L. Liu. (2016). A highly stretchable strain sensor based on a graphene/silver nanoparticle synergic conductive network and a sandwich structure. *Journal of Materials Chemistry C* 4: 4304–4311.

Chi, C., X. Sun, N. Xue, T. Li and C. Liu. (2018). Recent progress in technologies for tactile sensors. *Sensors (Switzerland)* 18.

Choi, S., H. Lee, R. Ghaffari, T. Hyeon and D. H. Kim. (2016). Recent advances in flexible and stretchable bio-electronic devices integrated with nanomaterials. *Advanced Materials* 28: 4203–4218.

Dahiya, A. S., T. Gil, J. Thireau, N. Azemard, A. Lacampagne, B. Charlot and A. Todri-Sanial. (2020). 1D nanomaterial-based highly stretchable strain sensors for human movement monitoring and human–robotic interactive systems. *Advanced Electronic Materials* 2000547: 1–13.

Das, P. S., S. H. Park, K. Y. Baik, J. W. Lee and J. Y. Park. (2020). Thermally reduced graphene oxide-nylon membrane based epidermal sensor using vacuum filtration for wearable electrophysiological signals and human motion monitoring. *Carbon* 158: 386–393.

Davaji, B., H. D. Cho, M. Malakoutian, J. K. Lee, G. Panin, T. W. Kang and C. H. Lee. (2017). A patterned single layer graphene resistance temperature sensor. *Scientific Reports* 7: 1–10.

Faruk, M. O., A. Ahmed, M. A. Jalil, M. T. Islam, B. Adak, M. M. Hossain and S. Mukhopadhyay. (2021). Functional textiles and composite based wearable thermal devices for Joule heating: progress and perspectives. *Applied Materials Today* 23: 101025.

Fu, Y., J. Zhao, Y. Dong and X. Wang. (2020). Dry electrodes for human bioelectrical signal monitoring. *Sensors (Switzerland)* 20: 1–30.

Gu, Y., T. Zhang, H. Chen, F. Wang, Y. Pu, C. Gao and S. Li. (2019). Mini review on flexible and wearable electronics for monitoring human health information. *Nanoscale Research Letters* 14: 263.

Hatamie, A., S. Angizi, S. Kumar, C. M. Pandey, A. Simchi, M. Willander and B. D. Malhotra. (2020). Review—textile based chemical and physical sensors for healthcare monitoring. *Journal of The Electrochemical Society* 167: 037546.

He, B., M. Feng, X. Chen and J. Sun. (2021). Multidimensional (0D-3D) functional nanocarbon: Promising material to strengthen the photocatalytic activity of graphitic carbon nitride. *Green Energy and Environment*.

Homayounfar, S. Z. and T. L. Andrew. (2020). Wearable sensors for monitoring human motion: a review on mechanisms, materials, and challenges. *SLAS Technology* 25: 9–24.

Hossain, M. M. and P. D. Bradford. (2021). Durability of smart electronic textiles. pp. 27–53. *In*: Nanosensors and Nanodevices for Smart Multifunctional Textiles (Elsevier).

Huang, H., S. Su, N. Wu, H. Wan, S. Wan, H. Bi and L. Sun. (2019). Graphene-based sensors for human health monitoring. *Frontiers in Chemistry* 7: 1–26.

Jung, H. C., J. H. Moon, D. H. Baek, J. H. Lee, Y. Y. Choi, J. S. Hong S. H. Lee. (2012). CNT/PDMS composite flexible dry electrodes for long-term ECG monitoring. *IEEE Transactions on Biomedical Engineering* 59: 1472–1479.

Kabiri Ameri, S., R. Ho, H. Jang, L. Tao, Y. Wang, L. Wang and N. Lu. (2017). Graphene electronic tattoo sensors. *ACS Nano* 11: 7634–7641.

Karimov, K. H. S., F. A. Khalid, M. Tariq Saeed Chani, A. Mateen, M. Asif Hussain and A. Maqbool. (2012). Carbon nanotubes based flexible temperature sensors. *Optoelectronics and Advanced Materials, Rapid Communications* 6: 194–196.

Keller, T. and A. Kuhn. (2008). Electrodes for transcutaneous (surface) electrical stimulation. *Journal of Automatic Control* 18: 35–45.

Khan, Y., A. E. Ostfeld, C. M. Lochner, A. Pierre and A. C. Arias. (2016). Monitoring of vital signs with flexible and wearable medical devices. *Advanced Materials* 28: 4373–4395.

Kim, D. W., J. H. Lee, J. K. Kim and U. Jeong. (2020). Material aspects of triboelectric energy generation and sensors. *NPG Asia Materials* 12.

Lee, J., S. Kim, J. Lee, D. Yang, B. C. Park, S. Ryu and I. Park. (2014a). A stretchable strain sensor based on a metal nanoparticle thin film for human motion detection. *Nanoscale* 6: 11932–11939.

Lee, S. M., H. J. Byeon, J. H. Lee, D. H. Baek, K. H. Lee, J. S. Hong and S. H. Lee. (2014b). Self-adhesive epidermal carbon nanotube electronics for tether-free long-term continuous recording of biosignals. *Scientific Reports* 4: 6074.

Li, B. M., O. Yildiz, A. C. Mills, T. J. Flewwellin, P. D. Bradford and J. S. Jur. (2020a). Iron-on carbon nanotube (CNT) thin films for biosensing E-textile applications. *Carbon* 168: 673–683.

Li, L., Wai Man au, Y. Li, K. M. Wan, Sai ho Wan and K. S. Wong. (2010). Design of intelligent garment with transcutaneous electrical nerve stimulation function based on the intarsia knitting technique. *Textile Research Journal* 80: 279–286.

Li, Q., J. Li, D. Tran, C. Luo, Y. Gao, C. Yu and F. Xuan. (2017a). Engineering of carbon nanotube/polydimethylsiloxane nanocomposites with enhanced sensitivity for wearable motion sensors. *Journal of Materials Chemistry C* 5: 11092–11099.

Li, T., J. Zou, F. Xing, M. Zhang, X. Cao, N. Wang and Z. L. Wang. (2017b). From dual-mode triboelectric nanogenerator to smart tactile sensor: a multiplexing design. *ACS Nano* 11: 3950–3956.

Li, Y., W. Chen and L. Lu. (2020b). Wearable and biodegradable sensors for human health monitoring. *ACS Applied Bio Materials*.

Li, Y., B. Zhou, G. Zheng, X. Liu, T. Li, C. Yan and Z. Guo. (2018). Continuously prepared highly conductive and stretchable SWNT/MWNT synergistically composited electrospun thermoplastic polyurethane yarns for wearable sensing. *Journal of Materials Chemistry C* 6: 2258–2269.

Li, Z., X. Qi, L. Xu, H. Lu, W. Wang, X. Jin and Y. Dong. (2020c). Self-repairing, large linear working range shape memory carbon nanotubes/ethylene vinyl acetate fiber strain sensor for human movement monitoring. *ACS Applied Materials & Interfaces* 12: 42179–42192.

Liu, C., S. Han, H. Xu, J. Wu and C. Liu. (2018a). Multifunctional highly sensitive multiscale stretchable strain sensor based on a graphene/glycerol-KCl synergistic conductive network. *ACS Applied Materials and Interfaces* 10: 31716–31724.

Liu, G., Q. Tan, H. Kou, L. Zhang, J. Wang, W. Lv, H. Dong and J. Xiong. (2018b). A flexible temperature sensor based on reduced graphene oxide for robot skin used in internet of things. *Sensors (Switzerland)* 18.

Liu, Q., H. Tai, Z. Yuan, Y. Zhou, Y. Su and Y. Jiang. (2019). A high-performances flexible temperature sensor composed of polyethyleneimine/reduced graphene oxide bilayer for real-time monitoring. *Advanced Materials Technologies* 4: 1–9.

Matzeu, G., M. Losacco, E. Parducci, A. Pucci, V. Dini, M. Romanelli and F. Di Francesco. (2011). Skin temperature monitoring by a wireless sensor. pp. 3533–3535. *In*: IECON Proceedings (Industrial Electronics Conference).

Mečņika, V. and M. H. Dipl. (2014). Smart textiles for healthcare: applications and technologies. *Rural Environment, Education, Personality*, 7–8. Available at: https://pdfs.semanticscholar.org/b536/fd03cfb4b9a39e1d78e746a5f39cd3f358d5.pdf [Accessed October 12, 2020].

Molina, J. (2016). Graphene-based fabrics and their applications: A review. *RSC Advances* 6: 68261–68291.

Nguyen, D. K. and T. Y. Kim. (2018). Graphene quantum dots produced by exfoliation of intercalated graphite nanoparticles and their application for temperature sensors. *Applied Surface Science* 427: 1152–1157.

Perathoner, S. and G. Centi. (2018). Advanced nanocarbon materials for future energy applications. pp. 305–325. *In*: Emerging Materials for Energy Conversion and Storage (Elsevier Inc.).

Pinet, É., C. Hamel, B. Glišić, D. Inaudi and N. Miron. (2007). Health monitoring with optical fiber sensors: from human body to civil structures. *In*: Health Monitoring of Structural and Biological Systems 2007 (SPIE), 653219.

Rajan, G., J. J. Morgan, C. Murphy, E. Torres Alonso, J. Wade, A. K. Ott, S. Russo, H. Alves, M. F. Craciun and A. I. Neves. (2020). Low operating voltage carbon-graphene hybrid E-textile for temperature sensing. *ACS Applied Materials and Interfaces* 12: 29861–29867.

Ray, T. R., J. Choi, A. J. Bandodkar, S. Krishnan, P. Gutruf, L. Tian, R. Ghaffari and J. A. Rogers. (2019). Bio-integrated wearable systems: A comprehensive review. *Chemical Reviews* 119: 5461–5533.

Ren, Z., J. Nie, J. Shao, Q. Lai, L. Wang, J. Chen, X. Chen and Z. L. Wang. (2018). Fully elastic and metal-free tactile sensors for detecting both normal and tangential forces based on triboelectric nanogenerators. *Advanced Functional Materials* 28: 1–9.

Saba, N., M. Jawaid, H. Fouad and O. Y. Alothman. (2019). Nanocarbon: Preparation, properties, and applications. pp. 327–354. *In*: Nanocarbon and its Composites (Elsevier).

Sadasivuni, K. K., A. Kafy, H. C. Kim, H. U. Ko, S. Mun and J. Kim. (2015). Reduced graphene oxide filled cellulose films for flexible temperature sensor application. *Synthetic Metals* 206: 154–161.

Sinha, S. K., F. A. Alamer, S. J. Woltornist, Y. Noh, F. Chen, A. McDannald, C. Allen, R. Daniels, A. Deshmukh, M. jain and K. Chon. (2019). Graphene and Poly(3,4-ethylene dioxythiophene):Poly(4-styrenesulfonate) on nonwoven fabric as a room temperature metal and its application as dry electrodes for electrocardiography. *ACS Applied Materials and Interfaces* 11: 32339–32345.

Soni, M., M. Bhattacharjee, M. Ntagios and R. Dahiya. (2020). Printed temperature sensor based on PEDOT: PSS-graphene oxide composite. *IEEE Sensors Journal* 20: 7525–7531.

Tao, J., R. Bao, X. Wang, Y. Peng, J. Li, S. Fu, C. Pan and Z. L. Wang. (2019). Self-powered tactile sensor array systems based on the triboelectric effect. *Advanced Functional Materials* 29: 1–23.

Trung, T. Q., S. Ramasundaram, B. U. Hwang and N. E. Lee. (2016). An all-elastomeric transparent and stretchable temperature sensor for body-attachable wearable electronics. *Advanced Materials* 28: 502–509.

Vuorinen, T., J. Niittynen, T. Kankkunen, T. M. Kraft and M. Mäntysalo. (2016). Inkjet-printed graphene/PEDOT:PSS temperature sensors on a skin-conformable polyurethane substrate. *Scientific Reports* 6: 1–8.

Wan, S., Z. Zhu, K. Yin, S. Su, H. Bi, T. Xu, H. Zhang, Z. Shi, L. He and L. Sun. (2018). A highly skin-conformal and biodegradable graphene-based strain sensor. *Small Methods* 2: 1700374.

Wan, Y., Y. Wang and C. F. Guo. (2017). Recent progresses on flexible tactile sensors. *Materials Today Physics* 1: 61–73.

Wang, L., J. A. Jackman, E. L. Tan, J. H. Park, M. G. Potroz, E. T. Hwang and N. J. Cho. (2017). High-performance, flexible electronic skin sensor incorporating natural microcapsule actuators. *Nano Energy* 36: 38–45.

Wang, L. and K. J. Loh. (2017). Wearable carbon nanotube-based fabric sensors for monitoring human physiological performance. *Smart Materials and Structures* 26.

Wang, S., Y. Bai and T. Zhang. (2020). Materials, systems, and devices for wearable bioelectronics. pp. 1–48. *In*: Wearable Bioelectronics.

Wang, Z., W. Gao, Q. Zhang, K. Zheng, J. Xu, W. Xu, E. Shang, J. Jiang, J. Zhang and Y. Liu. (2019). 3D-printed graphene/polydimethylsiloxane composites for stretchable and strain-insensitive temperature sensors. *ACS Applied Materials and Interfaces* 11: 1344–1352.

Xie, J., Q. Chen, H. Shen and G. Li. (2020). Review—wearable graphene devices for sensing. *Journal of The Electrochemical Society* 167: 037541.

Xu, F., X. Li, Y. Shi, L. Li, W. Wang, L. He and R. Liu. (2018). Recent developments for flexible pressure sensors: a review. *Micromachines* 9: 580.

Xu, X., M. Luo, P. He and J. Yang. (2020). Washable and flexible screen printed graphene electrode on textiles for wearable healthcare monitoring. *Journal of Physics D: Applied Physics* 53.

Yang, T., D. Xie, Z. Li and H. Zhu. (2017). Recent advances in wearable tactile sensors: Materials, sensing mechanisms, and device performance. *Materials Science and Engineering R: Reports* 115: 1–37.

Yang, Z., Y. Pang, X. L. Han, Y. Yang, J. Ling, M. Jian, Y. Zhang, Y. Yang and T. L. Ren. (2018). Graphene textile strain sensor with negative resistance variation for human motion detection. *ACS Nano* 12: 9134–9141.

Yapici, M. K., T. Alkhidir, Y. A. Samad and K. Liao. (2015). Graphene-clad textile electrodes for electrocardiogram monitoring. *Sensors and Actuators, B: Chemical* 221.

Yu, J., X. Hou, M. Cui, S. Zhang, J. He, W. Geng, J. Mu and X. Chou. (2019). Highly skin-conformal wearable tactile sensor based on piezoelectric-enhanced triboelectric nanogenerator. *Nano Energy* 64: 103923.

Zang, Y., F. Zhang, C. A. Di and D. Zhu. (2015). Advances of flexible pressure sensors toward artificial intelligence and health care applications. *Materials Horizons* 2: 140–156.

Zhang, J. X. J. and K. Hoshino. (2014). Optical transducers. pp. 233–320. *In*: Molecular Sensors and Nanodevices (Elsevier).

Zhang, Q., L. Liu, D. Zhao, Q. Duan, J. Ji, A. Jian, W. Zhang and S. Sang. (2017). Highly sensitive and stretchable strain sensor based on Ag@CNTs. *Nanomaterials* 7.

Zhao, X., Y. Long, T. Yang, J. Li and H. Zhu. (2017). Simultaneous high sensitivity sensing of temperature and humidity with graphene woven fabrics. *ACS Applied Materials and Interfaces* 9: 30171–30176.

Zhao, Y., Y. Cao, J. Liu, Z. Zhan, X. Li and W. J. Li. (2018). Single-Wall carbon nanotube-coated cotton yarn for electrocardiography transmission. *Micromachines* 9: 1–10.

Zheng, Y., Y. Li, Y. Zhou, K. Dai, G. Zheng, B. Zhang, C. Liu and C. Shen. (2020). High-performance wearable strain sensor based on graphene/cotton fabric with high durability and low detection limit. *ACS Applied Materials and Interfaces* 12: 1474–1485.

CHAPTER 6

Electrochemical Wearable Sensor for Biomedical and Healthcare Applications

Yugender Goud Kotagiri,[1,] Shekher Kummari,[2] Roger Narayan,[3]
Vinay Sharma[4] and Rupesh Kumar Mishra[4,5,6,]**

1. Introduction

Wearable sensors have attracted immense attention over the last decade as a result of their enormous promise to track the health, fitness and surroundings of the wearer (Campbell et al. 2018, Miller et al. 2016). However, scant attention has been paid to the production of wearable chemical/biosensors that provide more detailed information on the well-being of the wearer. The creation of wearable chemical sensors faces several obstacles on different fronts. The scale, rigidity and operating specifications of the existing chemical sensors are incompatible with the wearable technologies. Similarly, due to their poor energy capacity and slow recharging, modern wearable power supplies are unable to fulfill the criteria for wearable electronics. Thus, the problems facing the field of wearable chemical sensors and wearable sensors in general can only be tackled by a multidisciplinary approach in which researchers from a range of fields work in unison.

[1] Department of Nanoengineeringa, University of California San Diego, La Jolla, CA 92093, USA.
[2] Department of Chemistry, National Institute of Technology, Warangal, Telangana, 506004, India.
[3] Joint Department of Biomedical Engineering, University of North Carolina and Carolina State University, Raleigh, North Carolina 27695-7115, United States.
[4] Amity Institute of Biotechnology, Amity University Rajasthan, Jaipur, 303002, India.
[5] School of Materials Engineering, Purdue University, West Lafayette, IN 47907-2045, USA.
[6] Bindley Bio-science Center, Lab 222, 1203 W. State St., Purdue University, West Lafayette, IN 47907.
Emails: yugenderkotagiri@gmail.com, ykotagiri@eng.ucsd.edu; shekar.kummari1@gmail.com; roger_narayan@unc.edu; vsharma4@jpr.amity.edu; rkmishra@purdue.edu
* Corresponding author: ykotagiri@eng.ucsd.edu; rupeshmishra02@gmail.com, rkmishra@purdue.edu

Different wearable platforms have been used to analyze various analytes such as flexible and stretchable tattoo-like biosensors (Kim et al. 2015a, Bandodkar et al. 2014, Guinovart et al. 2013, Mishra et al. 2018, Wang et al. 2019, Sempionatto et al. 2020, De Guzman et al. 2019, Insights et al. 2014, Bandodkar et al. 2015), microneedle based platform (Teymourian et al. 2020, Ciui et al. 2018, Goud et al. 2019b, Anastasova et al. 2017, Mishra et al. 2017a, 2020, Yin et al. 2019, Parrilla et al. 2019, Li et al. 2019, Nakanishi et al. 2005, Ren et al. 2017, Wang and Mintchev 2013, O'Mahony et al. 2016, Bollella et al. 2019a, b), mouth guard sensors (Arakawa and Mitsubayashi 2017, Arakawa et al. 2016, Mattson et al. 2012, Kim et al. 2015b), bandage based wearable sensors (Guinovart et al. 2014, Ciui et al. 2018, Agarwala et al. 2019, Yang 2019, Takei 2015), eye-glass based sensors (Chen et al. 2014, Zhang et al. 2016, Sempionatto et al. 2017, 2019), glove-based sensors (Barfidokht et al. 2019, Da Silva et al. 2011, Nishiyama and Watanabe 2009, Lee et al. 2010, Lee and Chung 2017, Shen et al. 2016a, b, Roy et al. 2015, Carbonaro et al. 2014, Kang et al. 2016), and textile sensors (Lee et al. 2015, Liu et al. 2017, Gualandi et al. 2016, Parrilla et al. 2016, Meyer et al. 2006, Yang et al. 2018, Salvado et al. 2012, Zhou et al. 2017, Matsuhisa et al. 2015, Afroj et al. 2019, Lai et al. 2017, Enokibori et al. 2013) fabricated via conventional screen printing methods, which conform to the contours of the body and display resilience toward mechanical stresses expected from the wearer's activity. Non-invasive sensing of sweat lactate (Currano et al. 2018) and sodium (Bandodkar et al. 2014), alcohol (Kim et al. 2016) and glucose (Kim et al. 2018) have thus been demonstrated using wearable devices in recent times.

2. Tattoo-based Wearable Sensors

Bandodkar et al. have developed a potentiometric tattoo sensor combined with a miniaturized wireless transceiver for real-time monitoring of sodium in human perspiration (Bandodkar et al. 2014). Here, the authors initially fabricated the solid-state non-invasive tattoo sensor through screen printing process, laser printing, fluidics and wireless technologies. They have prepared the selective sodium ionophore membrane cocktail solution by using the components of Na ionophore X, polyvinyl chloride (PVC), bis(2-ethylhexyl) sebacate (DOS). They employed potentiometric technique to analyze the sodium ion concentration and the sensor displayed near-Nernstian response and carryover effects. Moreover, the authors carried-out the mechanical resilience studies of the tattoo sensor, and the sensor exhibits excellent resistance against the deformation studies.

In another report, a potentiometric sensor has been developed for the specific detection of ammonia ion in sweat (Guinovart et al. 2013). Here the authors initially fabricated the solid-state tattoo sensor with the screen printing technology. They have modified the working and reference electrodes with the respective cocktail solution. They have tested the printed tattoo sensor for robust mechanical resilience studies such as bending and stretching experiments; interestingly, the senor exhibits good response even after the deformation studies. They have tested the tattoo sensor performance in the concentration ranges between 10^{-4} M to 0.1 M, well within the physiological levels of ammonium in sweat. The have claimed that the sensor exhibited the near-Nernst response and carryover studies.

Mishra et al. have developed the wearable potentiometric tattoo biosensor for the specific detection of G-type nerve agents (Mishra et al. 2018). The authors initially prepared the pH sensitive poly aniline modified the working electrodes; this electrode monitored the proton concentration change, which was specifically released during the OPH hydrolysis of fluorine-containing organophosphates nerve agent simulant diisopropyl fluorophosphate (DFP). The tattoo sensor exhibits the good response in the concentration range of DFP from 10 to 120 mM. Moreover, they have tested the mechanical resilience studies of the tattoo sensor.

In another report, a wearable skin-worn non-invasive electrochemical biosensor has been developed for the specific detection of vitamin C in sweat (Sempionatto et al. 2020). The authors initially printed flexible tattoo patch electrodes with the screen-printing process. They changed the operating electrode surface with the ascorbate oxidase enzyme (AAOx) to track changes in the reduction current of the oxygen co-substrate. They reported that the tattoo sensor enzyme biosensor provides a highly selective response relative to typical direct (non-enzymatic) voltammetric measurements, with no effect on electroactive interference species such as uric acid or acetaminophen. Moreover, they have conducted the mechanical resilience studies of the tattoo senor path, which has exhibited excellent deformation stability. Finally, they have tested the real time measurement of vitamin C in the sweat before and after intake of vitamin C tablets with the different human subjects.

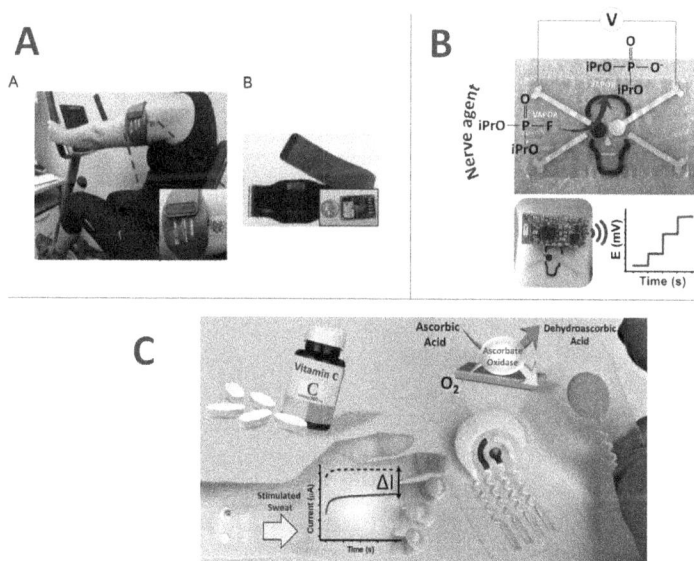

Fig. 1. Demonstration of epidermal sensors for biomedical and security applications: (A) Epidermal tattoo potentiometric sodium sensors with wireless signal transduction for continuous non-invasive sweat monitoring (Reprinted with permission from Bandodkar et al. 2014, Elsevier 2014). (B) Wearable potentiometric tattoo biosensor for on-body detection of G-type nerve agents simulants (Reprinted with permission from Mishra et al. 2018, Elsevier 2018). (C) Epidermal enzymatic biosensors for sweat Vitamin C: Toward personalized nutrition (Reprinted with permission from Sempionatto et al. 2020, American Chemical Society).

3. Microneedle-based Wearable Sensors

Goud et al. constructed a microneedle sensing platform for the continuous monitoring of levodopa (L-Dopa) for Parkinson's disease management (Goud et al. 2019a). Here, the authors have developed the parallel simultaneous sensing of the independent enzymatic and non-enzymatic voltammetric detection of L-Dopa using square wave voltammetry method. They claimed that this type of duel sensing method offers a built-in redundancy and enhances the information content of the microneedle sensor arrays. They have performed the L-Dopa sensing in the artificial ISF medium in the presence of possible albumin and globulins proteins. Moreover, they have demonstrated a skin-mimicking phantom gel as well as upon penetration through mice skin.

In another report, Mishra et al. constructed the microneedle sensor array for minimally invasive continuous electrochemical detection of opioid (OPi) and organophosphate (OP) nerve agents (Mishra et al. 2020). Multiplexed microneedle sensor array relies on simple carbon paste and OPH modified electrode for simultaneous detection of fentalyl and nerve agents, respectively. They have claimed that the microneedle sensor array is capable of continuously monitoring the fentanyl at nanomolar level and has exhibited a distinguishable performance in the presence of morphine and non-fentanyl. Finally, they said that the developed sensor array can be transfered for the real time monitoring of OPi-OP with an attractive analytical performance, including high sensitivity, selectivity, and stability.

In another report, a skin-worn microneedle sensing device has been constructed for the minimal invasive continuous monitoring of subcutaneous alcohol (Mohan et al. 2017). The authors initially constructed the pyramidal shape microneedle structures integrated with Pt and Ag wires, each with a microcavity opening, then the working electrode surfaces modified with the electropolymerizing of o-phenylene diamine, following the immobilization of alcohol oxidase (AOx) in the intermediate layer of chitosan, together with the outer layer of Nafion. They have tested the microneedle sensing performance in the artificial ISF medium, then they have demonstrated *ex vivo* mice skin model analysis as well for the evaluation of penetration of needle tips.

Teymourian et al. constructed a minimally invasive microneedle electrochemical sensor for continuous monitoring of glucose (CGM) and ketone (CKM) (Teymourian et al. 2020). The authors initially fabricated the microneedle sensor with hydroxybutyrate dehydrogenase (HBD) and glucose oxidase enzymes for the specific detection of ketone and glucose, respectively. It was reported that the CKM microneedle system exhibits attractive analytical efficiency, high sensitivity (with a low detection limit of 50 µM), high selectivity in the presence of potential interference, and good stability during prolonged activity in the artificial ISF. In addition, the possible applicability of this microneedle sensor to minimally invasive ketone body tracking has been demonstrated in a phantom gel skin-mimicking model.

4. Glove-based Wearable Sensors

Glove based sensor platforms are becoming very popular in recent days due to their simplicity and easy to use property (Barfidokht et al. 2019). Various applications

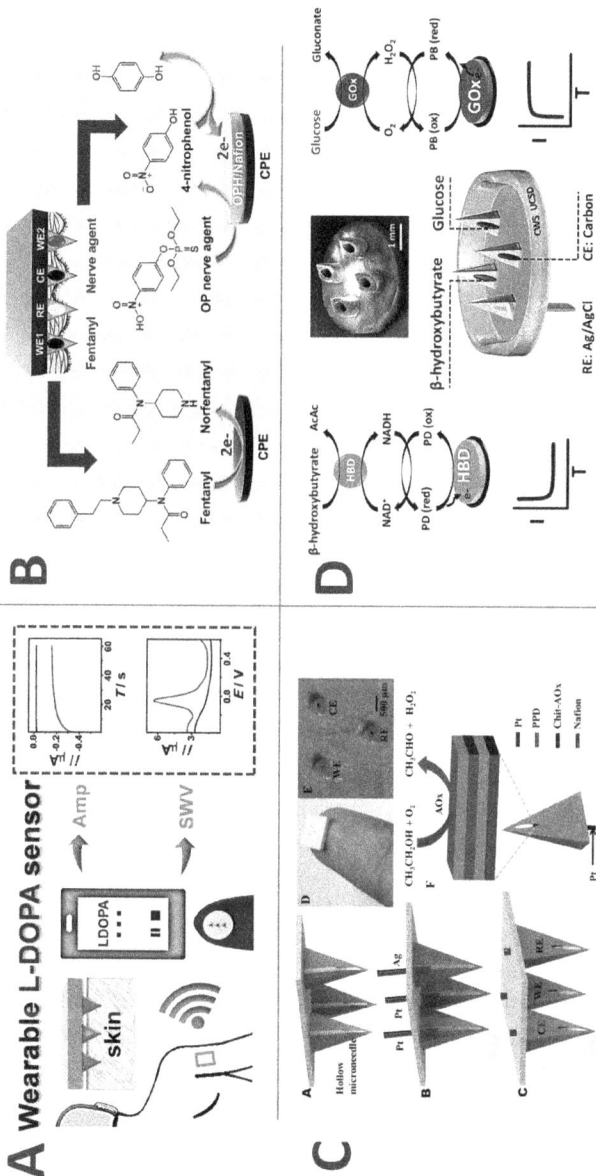

Fig. 2. Wearable microneedle sensing platform for biomedical and security applications (A) wearable electrochemical microneedle sensor platform for continuous screening of levodopa (Reprinted with permission from Goud et al. 2019a, American Chemical Society 2019). (B) Simultaneous monitoring of opioid and nerve agents on a wearable microneedle sensor array (Reprinted with permission from Mishra et al. 2020, American Chemical Society 2020). (C) Minimally-invasive alcohol monitoring using microneedle sensor arrays (Reprinted with permission from Mohan et al. 2017, Elsevier 2017) and (D) microneedle-based detection of ketone bodies along with glucose and lactate toward real-time interstitial fluid monitoring of diabetic ketosis and ketoacidosis (Reprinted with permission from Teymourian et al. 2020, American Chemical Society 2020).

of glove-based sensors have been demonstrated. Barfidokht et al. have presented a wearable "Lab-on-a-Glove" electrochemical sensor as a significant tool to detect several opioids simultaneously in powder and liquid samples. The glove-based sensor contains screen printed electrodes modified by mixture of an ionic liquid (4-(3-butyl-1-imidazolio)-1-butanesulfonate) and MWCNT. This combination showed oxidation of fentanyl using square wave voltammetric technique. The glove sensor achieved a detection level of 10 μM. The index finger sensor enables a rapid screening of fentanyl with considerable promise for point of use screening and timely analysis for first responders.

In another example of glove-based sensing, a flexible and stretchable glove-based electrochemical biosensor system has been shown as a wearable point-of-need screening tool to detect nerve agents and pesticides in food and vegetables. An organophosphate hydrolase enzyme was immobilized to a carbon-based working electrode sensor which specifically detects organophosphorus compounds. This robust and disposable "lab-on-a-glove" could be used a biosensing system for swiping on different fingers and integrated with a portable electronic device for electrochemical analysis and real-time wireless data broadcast to a smart host device. New ink formulations were used to print the sensors on index finger and various bending and stretching studies were performed for real and in-field glove application in the defense sector and in agriculture. The glove sensor was applied to test apples, tomatoes and various hard surfaces including glasses, wood and plastics for nerve agent's detection (Mishra et al. 2017).

Hubble and Wang (2019) have published a comprehensive review on glove-based sensors and discussed the applicability of this platform. It was discussed that how a fingertip can be deployed as a sensor platform and become a technology. Recent developments have been made for translation of glove based wearable electrochemical sensors to upsurge the sensual view of robotics, signifying the development of robotic skin with biochemical examination ability suitable for translation to remote chemical analysis in hazardous scenarios. The customization potential of wearable sensors is inordinate, yet challenges persist for evolving these arrangements from prototypes to more pervasive devices readily deployed in the field. We trust that these challenges can be overawed soon.

The expansion of robotic sensors that mimic the human sensing abilities is critical for the interface and intellectual abilities of modern robots. Robotic sensors have been developed for pressure and temperature sensing; however, they were never exploited for chemical analysis due to the challenges associated with realizing chemical sensing modalities on robotic platforms. Wang et al. (2019), Sempionatto et al. (2020), group at UCSD have demonstrated a first example of robotic sensing fingers for taste analysis in food and beverages. They have used gloves to print serpentine long structures on different fingers and used them for detection of sweetness, sourness and spiciness in food items. As per their published work, glucose was used as a marker to detect sweetness, ascorbic acid for sourness and capsaicin for spiciness. The printed sensors on gloves were connected to lab made PCB board and obtained data was wirelessly transmitted to mobile device. Printed middle, index, and ring robotic fingers allow accurate discrimination between sweetness, sourness, and spiciness by direct electrochemical glucose, ascorbic acid, and capsaicin detection.

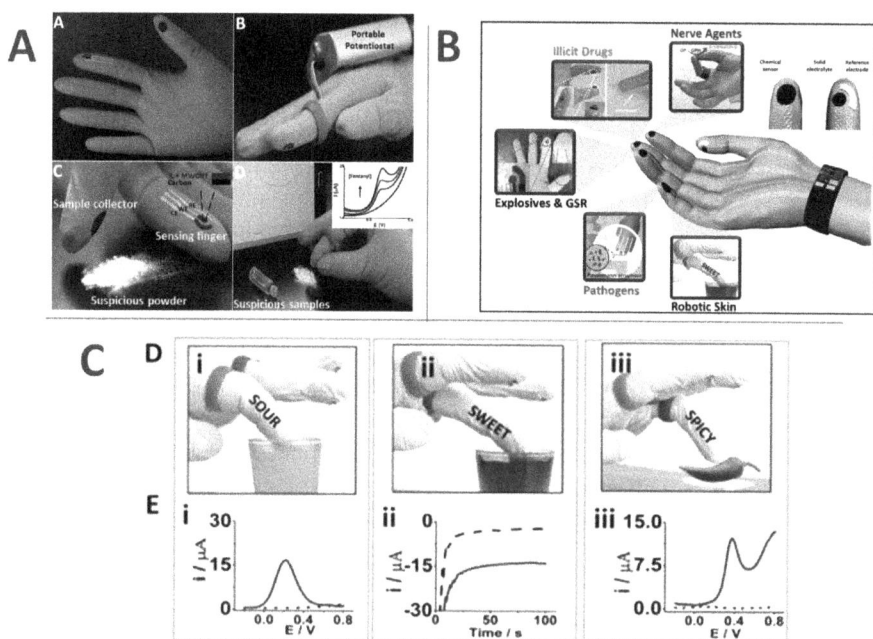

Fig. 3. Glove-based wearable platform for various sensing applications (A) Wearable electrochemical glove-based sensor for rapid and on-site detection of fentanyl (Reprinted with permission from Barfidokht et al. 2019, Elsevier 2019). (B) Examples of various fingertip based sensing using glove (Reprinted with permission from Hubble and Wang 2019, Wiley-VCH Verlag GmbH & Co. KGaA, Weinheim, 2019). (C) Illustration of chemical sensing at the robotic hand, wherein detection of sweetness, sourness and spiciness is confirmed toward automated taste discrimination in food samples (Reprinted with permission from Ciui et al. 2018, American Chemical Society 2018). (Reprinted with permission from De Jong et al. 2016, Royal Society of Chemistry 2016).

The analysis was further extended to detect caffeine in coffee samples alongside 3 tested molecules. The "sense of taste" robotic technology therefore allows correct discernment of various flavors, as shown in various experiments covering a wide variety of liquid and solid food samples. Such recognition of progressive wearable taste-sensing modules at the robot fingertips would pave the way to automatic chemical sensing machinery, enabling robotic decision making for applied food assistance with extensive implications to a varied range of robotic detecting applications (Ciui et al. 2018).

In another glove-based sensing application, an electrochemical fingerprint approach was used to recognize cocaine and its cutting agents in street samples. In this work, screen printed electrodes were coupled to glove and direct analysis with minimal sample treatment was required; this is all done due to the addition of screen-printed sensors. A semi-solid conductive gelatin gel was applied on the sensor surface, which acted as a conductive medium to diffuse the analyte from a thumb collector. Such novel methodologies allow on-site detection of chemical threats like cocaine at the airports, harbors and borders. It is noteworthy to mention that this glove-based sensor is the first example of sensing on fingertip for dangerous analytes (De Jong et al. 2016).

5. Translation to Commercial Market

The universal wearable sensors market was valued at crudely $150 million in 2016 and is expected to reach $2.86 billion by 2025 as per a recent market survey (https://www.grandviewresearch.com/industry-analysis/global-wearable-sensor-market).

A large amount of this future market segment is predicted to be made up of wearable sensors, primarily for noninvasive glucose analysis (Heikenfeld et al. 2018). The growth-rate of non-invasive bio-sensing platforms are slow at the moment due to depressed health market; however, with the fast progression of research outcomes and proof-of-concept demonstrations, many are still optimistic about the market prospects of wearable biosensors. The efficacious translation of wearable sensors and proof-of-concept demonstrations to the commercial market face various footraces connected to their essential operation. For entirely unswerving use, wearable sensors must have overawed stability issues instigated by extended operation under unrestrained conditions, biofouling from constituents present in the sampled biofluids, and intrinsic instabilities of the biological recognition element themselves. Furthermore, the devices must be capable of robust operation without the prerequisite for constant recalibration (usual in lab setting). The sensor research thus must ensure high bimolecular stability to preserve precision and reliability of the response (Elsherif et al. 2018, Sung et al. 2017, Yetisen et al. 2014).

Correspondingly, an appropriate fluid sampler structure, for instance microfluidics, is required to offer operative and quick transport of the biofluid on the sensor and warrant a reproducible and precise signal with insignificant sample contamination and carryover. Such progressive wearable microfluidic setup might also simplify multistep bioaffinity analyses, for antibody-based assays. The renewal of affinity sensor is an additional, foremost challenge to overcome, if extended usage of wearable sensor is to be achieved. Entirely combined wearable sensing platforms would need combination of electronics rich with wireless capabilities to simplify the data processing and protected signal communication. Furthermore, the usability of mobile devices and smartphone-assisted microscopes, combined with artificial intelligence and algorithm-based statistical applications, is anticipated to ease the data of optical biosensors. Seeing the foresaid challenges, we are at the initial phase of understanding how these wearable sensors can advance our well-being and performance.

6. Outlook and Perspectives

Wearable biosensors are projected to become more efficient and replaces from the wrist and into textiles such as socks, t-shirts and fashion accessories such as rings, eyeglasses that merge into a wearer's day-to-day life. Most of these devices will necessitate single use components to discourse sensor fouling matters. Future mode of wearable biosensors will certainly monitor a diverse range of biomarkers such as glucose, lactate, proteins, etc. in noninvasive manner, eventually permitting an inclusive medical diagnostics and performance assessment. The adaptation of such noninvasive sensors by the medical communal will necessitate widespread and efficacious endorsement in human testing and upgraded understanding of the scientific relevancy of the device evidence. Considering the viable research and incredible saleable opportunities in wearable biosensors, we anticipate stimulating novel expansions very soon. The wearable biosensors market is therefore projected to continue its speedy progress and its route to improve people's lives.

7. Conclusions

We have witnessed a tremendous progress in the field of electrochemical wearable sensor for biomedical and security applications using nanotechnological tools through various developments. These wearable sensors have great potential to provide an alternate solution for personalized healthcare monitoring, safety and security analysis and so on. They are small, well-lit, rapid in their performance and provides continuous monitoring of metabolites and toxic chemicals. They can be developed as a highly sensitive and selective device with low sample requirement. Development of various wearables such as gloves, tattoo, microneedles, eyeglasses, bandages, textiles and e-skin based sensor platforms has made people's life easier and they can have their healthcare related data on their mobile devices. Such technologies have changed the lifestyle and thinking patterns of human beings. We are confident that the pace which science is following will lead to even greater heights of technology development and beyond. To advance the efficacy and expand the superiority of such devices towards bioanalysis and chemical assessment, role of nanotechnology is immense and crucial.

References

Afroj, Shaila, Nazmul Karim, Zihao Wang, Sirui Tan, Pei He, Matthew Holwill, Davit Ghazaryan, Anura Fernando and Kostya S. Novoselov. (2019). Engineering graphene flakes for wearable textile sensors via highly scalable and ultrafast yarn dyeing technique. *ACS Nano*. https://doi.org/10.1021/acsnano.9b00319.

Agarwala, Shweta, Guo Liang Goh, Truong Son Dinh Le, Jianing An, Zhen Kai Peh, Wai Yee Yeong and Young Jin Kim. (2019). Wearable bandage-based strain sensor for home healthcare: combining 3D aerosol jet printing and laser sintering. *ACS Sensors*. https://doi.org/10.1021/acssensors.8b01293.

Anastasova, Salzitsa, Blair Crewther, Pawel Bembnowicz, Vincenzo Curto, Henry MD Ip, Bruno Rosa and Guang Zhong Yang. (2017). A wearable multisensing patch for continuous sweat monitoring. *Biosensors and Bioelectronics*. https://doi.org/10.1016/j.bios.2016.09.038.

Arakawa, Takahiro, Yusuke Kuroki, Hiroki Nitta, Koji Toma, Kohji Mitsubayashi, Shuhei Takeuchi, Toshiaki Sekita and Shunsuke Minakuchi. (2016). Mouth guard type biosensor cavitous sensor for monitoring of saliva glucose with telemetry system. *In*: Proceedings of the International Conference on Sensing Technology, ICST. https://doi.org/10.1109/ICSensT.2015.7438362.

Arakawa, Takahiro and Kohji Mitsubayashi. (2017). Cavitas sensors (soft contact lens type biosensor, mouth-guard type sensor, etc.) for daily medicine. *In*: Smart Sensors, Measurement and Instrumentation. https://doi.org/10.1007/978-3-319-47319-2_3.

Bandodkar, Amay J., Denise Molinnus, Omar Mirza, Tomás Guinovart, Joshua R. Windmiller, Gabriela Valdés-Ramírez, Francisco J. Andrade, Michael J. Schöning, and Joseph Wang. (2014). Epidermal tattoo potentiometric sodium sensors with wireless signal transduction for continuous non-invasive sweat monitoring. *Biosensors and Bioelectronics* 54(April): 603–9. https://doi.org/10.1016/j.bios.2013.11.039.

Bandodkar, Amay J., Wenzhao Jia and Joseph Wang. (2015). Tattoo-based wearable electrochemical devices: a review. *Electroanalysis*. https://doi.org/10.1002/elan.201400537.

Barfidokht, Abbas, Rupesh K. Mishra, Rajesh Seenivasan, Shuyang Liu, Lee J. Hubble, Joseph Wang and Drew A. Hall. (2019). Wearable electrochemical glove-based sensor for rapid and on-site detection of fentanyl. *Sensors and Actuators, B: Chemical*. https://doi.org/10.1016/j.snb.2019.04.053.

Bollella, Paolo, Sanjiv Sharma, Anthony E. G. Cass, Federico Tasca and Riccarda Antiochia. (2019a). Minimally invasive glucose monitoring using a highly porous gold microneedles-based biosensor: characterization and application in artificial interstitial fluid. *Catalysts*. https://doi.org/10.3390/catal9070580.

Bollella, Paolo, Sanjiv Sharma, Anthony E. G. Cass, Federico Tasca and Riccarda Antiochia. (2019b). Highly porous gold microneedles-based biosensor: characterization and application in artificial interstitial fluid. *Catalysts*.

Carbonaro, Nicola, Gabriele Dalle Mura, Federico Lorussi, Rita Paradiso, Danilo De Rossi and Alessandro Tognetti. (2014). Exploiting wearable goniometer technology for motion sensing gloves. *IEEE Journal of Biomedical and Health Informatics*. https://doi.org/10.1109/JBHI.2014.2324293.

Chen, Guo Zhen, Ion Seng Chan, Leo K. K. Leung and David C. C. Lam. (2014). Soft wearable contact lens sensor for continuous intraocular pressure monitoring. *Medical Engineering and Physics*. https://doi.org/10.1016/j.medengphy.2014.06.005.

Ciui, Bianca, Aida Martin, Rupesh K. Mishra, Barbara Brunetti, Tatsuo Nakagawa, Thomas J. Dawkins, Mengjia Lyu, Cecilia Cristea, Robert Sandulescu and Joseph Wang. (2018). Wearable wireless tyrosinase bandage and microneedle sensors: toward melanoma screening. *Advanced Healthcare Materials* 7(7): 1701264. https://doi.org/10.1002/adhm.201701264.

Ciui, Bianca, Aida Martin, Rupesh K. Mishra, Tatsuo Nakagawa, Thomas J. Dawkins, Mengjia Lyu, Cecilia Cristea, Robert Sandulescu and Joseph Wang. (2018). Chemical sensing at the robot fingertips: toward automated taste discrimination in food samples. *ACS Sensors*. https://doi.org/10.1021/acssensors.8b00778.

Currano, Luke J., F. Connor Sage, Matthew Hagedon, Leslie Hamilton, Julia Patrone and Konstantinos Gerasopoulos. (2018). Wearable sensor system for detection of lactate in sweat. *Scientific Reports* 8(1): 15890. https://doi.org/10.1038/s41598-018-33565-x.

de Jong, M., N. Sleegers, J. Kim, F. Van Durme, N. Samyn, J. Wang and K. De Wael. (2016). Electrochemical fingerprint of street samples for fast on-site screening of cocaine in seized drug powders. *Chem. Sci.* 7(3): 2364–2370. https://doi.org/10.1039/c5sc04309c.

Elsherif, Mohamed, Mohammed Umair Hassan, Ali K. Yetisen and Haider Butt. (2018). Wearable contact lens biosensors for continuous glucose monitoring using smartphones. *ACS Nano*. https://doi.org/10.1021/acsnano.8b00829.

Enokibori, Yu, Akihisa Suzuki, Hirotaka Mizuno, Yuuki Shimakami and Kenji Mase. (2013). E-textile pressure sensor based on conductive fiber and its structure. *In*: UbiComp 2013 Adjunct - Adjunct Publication of the 2013 ACM Conference on Ubiquitous Computing. https://doi.org/10.1145/2494091.2494158.

Goud, K. Yugender, Chochanon Moonla, Rupesh K. Mishra, Chunmei Yu, Roger Narayan, Irene Litvan and Joseph Wang. (2019a). Wearable electrochemical microneedle sensor for continuous monitoring of levodopa: toward parkinson management. *ACS Sensors* 20(3): acssensors.9b01127. https://doi.org/10.1021/acssensors.9b01127.

Goud, K. Yugender, Chochanon Moonla, Rupesh K. Mishra, Chunmei Yu, Roger Narayan, Irene Litvan and Joseph Wang. (2019b). Wearable electrochemical microneedle sensor for continuous monitoring of levodopa: toward parkinson management. *ACS Sensors*, August, acssensors.9b01127. https://doi.org/10.1021/acssensors.9b01127.

Gualandi, I., M. Marzocchi, A. Achilli, D. Cavedale, A. Bonfiglio and B. Fraboni. (2016). Textile organic electrochemical transistors as a platform for wearable biosensors. *Scientific Reports*. https://doi.org/10.1038/srep33637.

Guinovart, Tomàs, Amay J. Bandodkar, Joshua R. Windmiller, Francisco J. Andrade and Joseph Wang. (2013). A potentiometric tattoo sensor for monitoring ammonium in sweat. *Analyst*. https://doi.org/10.1039/c3an01672b.

Guinovart, Tomàs, Gabriela Valdés-Ramírez, Joshua R. Windmiller, Francisco J. Andrade and Joseph Wang. (2014). Bandage-based wearable potentiometric sensor for monitoring wound pH. *Electroanalysis* 26(6): 1345–53. https://doi.org/10.1002/elan.201300558.

Guzman, Keana De, Ghayadah Al-Kharusi, Tanya Levingstone and Aoife Morrin. (2019). Robust epidermal tattoo electrode platform for skin physiology monitoring. *Analytical Methods*. https://doi.org/10.1039/c8ay02678e.

Heikenfeld, J., A. Jajack, J. Rogers, P. Gutruf, L. Tian, T. Pan, R. Li, M. Khine, J. Kim, J. Wang and J. Kim. (2018). Wearable sensors: modalities, challenges, and prospects. *Lab on a Chip*. https://doi.org/10.1039/c7lc00914c.

Hubble, Lee J. and Joseph Wang. (2019). Sensing at your fingertips: glove-based wearable chemical sensors. *Electroanalysis*. https://doi.org/10.1002/elan.201800743.

Amay, J. Bandodkar, Denise Molinnus, Omar Mirza, Tomás Guinovart, Joshua R. Windmiller, Gabriela Valdés-Ramírez, Francisco J. Andrade, Michael J. Schöning and Joseph Wang. (2014). Epidermal tattoo potentiometric sodium sensors with wireless signal transduction for continuous non-invasive sweat monitoring. *Biosensors and Bioelectronics*.

Jong, Mats De, Nick Sleegers, Jayoung Kim, Filip Van Durme, Nele Samyn, Joseph Wang and Karolien De Wael. (2016). Electrochemical fingerprint of street samples for fast on-site screening of cocaine in seized drug powders. *Chemical Science*. https://doi.org/10.1039/c5sc04309c.

Kang, Brian Byunghyun, Haemin Lee, Hyunki In, Useok Jeong, Jinwon Chung and Kyu Jin Cho. (2016). Development of a polymer-based tendon-driven wearable robotic hand. *In*: Proceedings - IEEE International Conference on Robotics and Automation. https://doi.org/10.1109/ICRA.2016.7487562.

Kim, Jayoung, William R. De Araujo, Izabela A. Samek, Amay J. Bandodkar, Wenzhao Jia, Barbara Brunetti, Thiago R. L. C. Paixão and Joseph Wang. (2015a). Wearable temporary tattoo sensor for real-time trace metal monitoring in human sweat. *Electrochemistry Communications*. https://doi.org/10.1016/j.elecom.2014.11.024.

Kim, Jayoung, Somayeh Imani, William R. de Araujo, Julian Warchall, Gabriela Valdés-Ramírez, Thiago R. L. C. Paixão, Patrick P. Mercier and Joseph Wang. (2015b). Wearable salivary uric acid mouthguard biosensor with integrated wireless electronics. *Biosensors and Bioelectronics*. https://doi.org/10.1016/j.bios.2015.07.039.

Kim, Jayoung, Itthipon Jeerapan, Somayeh Imani, Thomas N. Cho, Amay Bandodkar, Stefano Cinti, Patrick P. Mercier and Joseph Wang. (2016). Noninvasive alcohol monitoring using a wearable tattoo-based iontophoretic-biosensing system. *ACS Sensors*. https://doi.org/10.1021/acssensors.6b00356.

Kim, Jayoung, Alan S. Campbell and Joseph Wang. (2018). Wearable non-invasive epidermal glucose sensors: a review. *Talanta*. https://doi.org/10.1016/j.talanta.2017.08.077.

Lai, Ying Chih, Jianan Deng, Steven L. Zhang, Simiao Niu, Hengyu Guo and Zhong Lin Wang. (2017). Single-thread-based wearable and highly stretchable triboelectric nanogenerators and their applications in cloth-based self-powered human-interactive and biomedical sensing. *Advanced Functional Materials*. https://doi.org/10.1002/adfm.201604462.

Lee, Boon Giin and Wan Young Chung. (2017). Wearable glove-type driver stress detection using a motion sensor. *IEEE Transactions on Intelligent Transportation Systems*. https://doi.org/10.1109/TITS.2016.2617881.

Jaehong Lee, Hyukho Kwon, Jungmok Seo, Sera Shin, Ja Hoon Koo, Changhyun Pang, Seungbae Son, Jae Hyung Kim, Yong Hoon Jang, Dae Eun Kim and Taeyoon Lee. (2015). Conductive fiber-based ultrasensitive textile pressure sensor for wearable electronics. *Advanced Materials*. https://doi.org/10.1002/adma.201500009.

Lee, Youngbum, Byungwoo Lee and Myoungho Lee. (2010). Wearable sensor glove based on conducting fabric using electrodermal activity and pulse-wave sensors for e-health application. *Telemedicine and E-Health*. https://doi.org/10.1089/tmj.2009.0039.

Li, Yan, Hang Zhang, Ruifeng Yang, Yohan Laffitte, Ulises Schmill, Wenhan Hu, Moufeed Kaddoura, Eric J. M. Blondeel and Bo Cui. (2019). Fabrication of sharp silicon hollow microneedles by deep-reactive ion etching towards minimally invasive diagnostics. *Microsystems and Nanoengineering*. https://doi.org/10.1038/s41378-019-0077-y.

Liu, Mengmeng, Xiong Pu, Chunyan Jiang, Ting Liu, Xin Huang, Libo Chen, Chunhua Du, Jiangman Sun, Weiguo Hu and Zhong Lin Wang. (2017). Large-area all-textile pressure sensors for monitoring human motion and physiological signals. *Advanced Materials*. https://doi.org/10.1002/adma.201703700.

Matsuhisa, Naoji, Martin Kaltenbrunner, Tomoyuki Yokota, Hiroaki Jinno, Kazunori Kuribara, Tsuyoshi Sekitani and Takao Someya. (2015). Printable elastic conductors with a high conductivity for electronic textile applications. *Nature Communications*. https://doi.org/10.1038/ncomms8461.

Mattson, J., R. Shultz, J. Goodman, S. Anderson and D. Garza. (2012). Validation of a novel mouth guard for measurement of linear and rotational accelerations during head impacts. *Clinical Journal of Sport Medicine*.

Meyer, Jan, Paul Lukowicz and Gerhard Tröster. (2006). Textile pressure sensor for muscle activity and motion detection. *In*: Proceedings - International Symposium on Wearable Computers, ISWC. https://doi.org/10.1109/ISWC.2006.286346.

Mishra, Rupesh K., Lee J. Hubble, Aida Martín, Rajan Kumar, Abbas Barfidokht, Jayoung Kim, Mustafa M. Musameh, Ilias L. Kyratzis and Joseph Wang. (2017). Wearable flexible and stretchable glove biosensor for on-site detection of organophosphorus chemical threats. *ACS Sensors* 2(4): 553–61. https://doi.org/10.1021/acssensors.7b00051.

Mishra, Rupesh K., A. M. Vinu Mohan, Fernando Soto, Robert Chrostowski and Joseph Wang. (2017). A microneedle biosensor for minimally-invasive transdermal detection of nerve agents. *Analyst*. https://doi.org/10.1039/c6an02625g.

Mishra, Rupesh K., Abbas Barfidokht, Aleksandar Karajic, Juliane R. Sempionatto, Joshua Wang and Joseph Wang. (2018). Wearable potentiometric tattoo biosensor for on-body detection of G-type nerve agents simulants. *Sensors and Actuators, B: Chemical*. https://doi.org/10.1016/j.snb.2018.07.001.

Mishra, Rupesh K., K. Yugender Goud, Zhanhong Li, Chochanon Moonla, Mona A. Mohamed, Farshad Tehrani, Hazhir Teymourian and Joseph Wang. (2020). Continuous opioid monitoring along with nerve agents on a wearable microneedle sensor array. *Journal of the American Chemical Society* 142(13): 5991–95. https://doi.org/10.1021/jacs.0c01883.

Mohan, A. M. Vinu, Joshua Ray Windmiller, Rupesh K. Mishra and Joseph Wang. (2017). Continuous minimally-invasive alcohol monitoring using microneedle sensor arrays. *Biosensors and Bioelectronics* 91(January): 574–79. https://doi.org/10.1016/j.bios.2017.01.016.

Nakanishi, Naoyuki, Hidetake Yamamoto, Kazuyoshi Tsuchiya, Yasutomo Uetsuji and Eiji Nakamachi. (2005). Development of wearable medical device for bio-MEMS. *In*: BioMEMS and Nanotechnology II. https://doi.org/10.1117/12.638162.

Nishiyama, Michiko and Kazuhiro Watanabe. (2009). Wearable sensing glove with embedded hetero-core fiber-optic nerves for unconstrained hand motion capture. *IEEE Transactions on Instrumentation and Measurement*. https://doi.org/10.1109/TIM.2009.2021640.

O'Mahony, Conor, Andrea Bocchino, Eleonora Sulas, Antonio Ciarlone, Guiseppe Giannoni, Suzanne O'Callaghan, Anan Kenthao, A. James, P. Clover, Danilo Demarchi, Paul Galvin and Konstantin Grygoryev. (2016). Embedded sensors for micro transdermal interface platforms (MicroTIPs). *In*: Symposium on Design, Test, Integration and Packaging of MEMS/MOEMS, DTIP 2016. https://doi.org/10.1109/DTIP.2016.7514859.

Parrilla, Marc, Rocío Cánovas, Itthipon Jeerapan, Francisco J. Andrade and Joseph Wang. (2016). A textile-based stretchable multi-ion potentiometric sensor. *Advanced Healthcare Materials* 5(9): 996–1001. https://doi.org/10.1002/adhm.201600092.

Parrilla, Marc, María Cuartero, Sara Padrell Sánchez, Mina Rajabi, Niclas Roxhed, Frank Niklaus and Gastón A. Crespo. (2019). Wearable all-solid-state potentiometric microneedle patch for intradermal potassium detection. *Analytical Chemistry* 91(2): 1578–86.

Ren, Lei, Qing Jiang, Zhipeng Chen, Keyun Chen, Shujia Xu, Jie Gao and Lelun Jiang. (2017). Flexible microneedle array electrode using magnetorheological drawing lithography for bio-signal monitoring. *Sensors and Actuators, A: Physical*. https://doi.org/10.1016/j.sna.2017.10.042.

Roy, Kathika, Durga Prasad Idiwal, Annapurna Agrawal and Bani Hazra. (2015). Flex sensor based wearable gloves for robotic gripper control. *In*: ACM International Conference Proceeding Series. https://doi.org/10.1145/2783449.2783520.

Salvado, Rita, Caroline Loss, Gon and Pedro Pinho. (2012). Textile materials for the design of wearable antennas: a survey. *Sensors (Switzerland)*. https://doi.org/10.3390/s121115841.

Sempionatto, Juliane R., Tatsuo Nakagawa, Adriana Pavinatto, Samantha T. Mensah, Somayeh Imani, Patrick Mercier and Joseph Wang. (2017). Eyeglasses based wireless electrolyte and metabolite sensor platform. *Lab on a Chip*. https://doi.org/10.1039/c7lc00192d.

Sempionatto, Juliane R., Laís Canniatti Brazaca, Laura García-Carmona, Gulcin Bolat, Alan S. Campbell, Aida Martin, Guangda Tang, Rushabh Shah, Rupesh K. Mishra, Jayoung Kim, Valtencir Zucolotto, Alberto Escarpa and Joseph Wang. (2019). Eyeglasses-based tear biosensing system: non-invasive detection of alcohol, vitamins and glucose. *Biosensors and Bioelectronics*. https://doi.org/10.1016/j. bios.2019.04.058.

Sempionatto, Juliane R., Ahmed A. Khorshed, Aftab Ahmed, Andre N. De Loyola e Silva, Abbas Barfidokht, Lu Yin, K. Yugender Goud, Mona A. Mohamed, Eileen Bailey, Jennifer May, Claude Aebischer, Claire Chatelle and Joseph Wang. (2020). Epidermal enzymatic biosensors for sweat vitamin C: toward personalized nutrition. *ACS Sensors*. https://doi.org/10.1021/acssensors.0c00604.

Shen, Zhong, Juan Yi, Xiaodong Li, Mark Hin Pei Lo, Michael Z. Q. Chen, Yong Hu and Zheng Wang. (2016). A soft stretchable bending sensor and data glove applications. *Robotics and Biomimetics*. https://doi.org/10.1186/s40638-016-0051-1.

Shen, Zhong, Juan Yi, Xiaodong Li, Lo Hin Pei Mark, Yong Hu and Zheng Wang. (2016). A soft stretchable bending sensor and data glove applications. *In*: 2016 IEEE International Conference on Real-Time Computing and Robotics, RCAR 2016. https://doi.org/10.1109/RCAR.2016.7784006.

Silva, Alexandre Ferreira Da, Anselmo Filipe Gonçalves, Paulo Mateus Mendes and José Higino Correia. (2011). FBG sensing glove for monitoring hand posture. *IEEE Sensors Journal*. https://doi. org/10.1109/JSEN.2011.2138132.

Sung, Yulung, Fernando Campa and Wei-Chuan Shih. (2017). Open-source do-it-yourself multi-color fluorescence smartphone microscopy. *Biomedical Optics Express*. https://doi.org/10.1364/ boe.8.005075.

Takei, Kuniharu. (2015). Wearable and fexible sensor sheets toward periodic health monitoring. *In*: *Smart Sensors, Measurement and Instrumentation*. https://doi.org/10.1007/978-3-319-18191-2_7.

Teymourian, Hazhir, Chochanon Moonla, Farshad Tehrani, Eva Vargas, Reza Aghavali, Abbas Barfidokht, Tanin Tangkuaram, Patrick P. Mercier, Eyal Dassau and Joseph Wang. (2020). Microneedle-based detection of ketone bodies along with glucose and lactate: toward real-time continuous interstitial fluid monitoring of diabetic ketosis and ketoacidosis. *Analytical Chemistry* 92(2): 2291–2300. https://doi.org/10.1021/acs.analchem.9b05109.

Wang, Gang and Martin P. Mintchev. (2013). Development of wearable semi-invasive blood sampling devices for continuous glucose monitoring: a survey. *Engineering*. https://doi.org/10.4236/ eng.2013.55b009.

Wang, Qi, Shengjie Ling, Xiaoping Liang, Huimin Wang, Haojie Lu and Yingying Zhang. (2019). Self-healable multifunctional electronic tattoos based on silk and graphene. *Advanced Functional Materials*. https://doi.org/10.1002/adfm.201808695.

Yang, Bin, Jilie Kong and Xueen Fang. (2019). Bandage-like wearable flexible microfluidic recombinase polymerase amplification sensor for the rapid visual detection of nucleic acids. *Talanta*. https://doi. org/10.1016/j.talanta.2019.06.031.

Yang, Zhen, Yu Pang, Xiao Lin Han, Yi Yang, Yifan Yang, Jiang Ling, Muqiang Jian, Yingying Zhang and Tian Ling Ren. (2018). Graphene textile strain sensor with negative resistance variation for human motion detection. *ACS Nano*. https://doi.org/10.1021/acsnano.8b03391.

Yetisen, Ali K., J. L. Martinez-Hurtado, Angel Garcia-Melendrez, Fernando Da Cruz Vasconcellos and Christopher R. Lowe. (2014). A smartphone algorithm with inter-phone repeatability for the analysis of colorimetric tests. *Sensors and Actuators, B: Chemical*. https://doi.org/10.1016/j.snb.2014.01.077.

Yin, Mengtian, Li Xiao, Qingchang Liu, Sung Yun Kwon, Yi Zhang, Poonam R. Sharma, Li Jin, Xudong Li and Baoxing Xu. (2019). 3D printed microheater sensor-integrated, drug-encapsulated microneedle patch system for pain management. *Advanced Healthcare Materials*. https://doi.org/10.1002/ adhm.201901170.

Zhang, Yu Shrike, Fabio Busignani, João Ribas, Julio Aleman, Talles Nascimento Rodrigues, Seyed Ali Mousavi Shaegh, Solange Massa, Camilla Baj Rossi, Irene Taurino, Su-Ryon, Giovanni Calzone, Givan Mark Amaratunga, Douglas Leon Chambers, Saman Jabari, Yuxi Niu, Vijayan Manoharan, Mehmet Remzi Dokmeci, Sandro Carrara9, Danilo Demarchi and Ali Khademhosseini. (2016). Google glass-directed monitoring and control of microfluidic biosensors and actuators. *Scientific Reports*. https://doi.org/10.1038/srep22237.

Zhou, Gengheng, Joon Hyung Byun, Youngseok Oh, Byung Mun Jung, Hwa Jin Cha, Dong Gi Seong, Moon Kwang Um, Sangil Hyun and Tsu Wei Chou. (2017). Highly sensitive wearable textile-based humidity sensor made of high-strength, single-walled carbon nanotube/poly(vinyl alcohol) filaments. *ACS Applied Materials and Interfaces*. https://doi.org/10.1021/acsami.6b12448.

CHAPTER 7

Smart Textile-Based Interactive, Stretchable and Wearable Sensors for Healthcare

Abbas Ahmed,[1] *Bapan Adak*[2,]* *and Samrat Mukhopadhyay*[3]

1. Introduction

The demand for non-invasive monitoring of human health has generated the idea of developing smart skin mountable devices. The flexible, stretchable, and smart electronic textiles (e-textiles) based wearable electronic sensors hold great promise in advanced healthcare systems, providing access to accurate and user-friendly interface to monitor patients' certain physiological aspects (Yetisen et al. 2016, Yang et al. 2018, Das et al. 2020, Karim et al. 2020, Esfahani 2021). With real-time monitoring, the unobtrusive features, good service life and cost effectiveness of smart-textile based sensors will further boast their importance in health monitoring. Smart textile based electrodes can be easily integrated into different parts of clothing, thereby endowing comfortability to wearer. Furthermore, textile based wires and electrodes may replace the traditional hard electrical wires or electrodes. Such textile electrode enables measurement of different biopotentials such as electrocardiogram (ECG) (Li et al. 2020a, Ruiz et al. 2020), electromyogram (EMG) (Acar et al. 2019) and respiratory activity (Yang et al. 2017).

One of the widely explored strategy to develop smart textile electrodes is coating them with conductive nanomaterials. This strategy enables textile fibers/yarns/fabrics to coat inherently with nanomaterials achieving a conductive surface (Shahariar

[1] National Institute of Textile Engineering and Research, University of Dhaka, Dhaka 1000, Bangladesh.
[2] Product Development Department, Kusumgar Corporates Pvt. Ltd., Vapi, Valsad, Gujarat, India, 396195.
[3] Department of Fibre and Textile Engineering, Indian Institute of Technology Delhi, Hauz Khas, New Delhi, India, 110016.
Email: samrat@textile.iitd.ac.in
* Corresponding author: bapan.iitd15@gmail.com

et al. 2019, Ahmed et al. 2020c, Islam et al. 2020, Hossain and Bradford 2021). Over the last decade, intensive research endeavors have been executed to fabricate smart textile materials and diverse range of applications have been realized. Specially, for health care systems, smart textile electrode devices have been extensively applied to monitor different biomedical health parameters of human such as body motion and temperature, sweat level, sleep and respiration. Besides, smart textile sensors are also utilized in infants' health supervisions (Zhu et al. 2015). In addition, smart textile fiber based artificial muscles may be applied in humanoid robots for wearables.

In this chapter, efforts have been made to summarize the current scenario in smart textile-based devices for healthcare monitoring. The chapter starts with highlighting different textile architectural designs and incorporation of different nanomaterials into textile structures. In the next sections, current development within smart textile based healthcare systems such as wearable motion and biomonitoring, thermotherapy, artificial muscles, etc. has been presented. The chapter concludes outlining some future research directions.

2. Intrinsic Textile Structures for Wearable Electronics

Textiles have a long history with human beings which not only provided for civilizations, but also enormously made human life easy and comfortable, thus facilitating thermal comfort and shielding us from unfavorable hot or cold environments. Interestingly, the recent advances in modern technologies have made human life more proactive than before. Notably, textiles are not only being used for apparel and aesthetics, but also for more sophisticated and advanced purposes. The widespread applications of nanotechnology in daily life have emerged and boosted up the concept of smart textiles, finding enormous application in areas such as wearable sensing, thermal management, real-time healthcare operations, energy harvesting as well as storing, and in many other intelligent systems. There are mainly two basic textile architectures (fibers/yarns and fabric) that are used intensively in wearable smart textile applications including healthcare sensing purposes, which are presented briefly in the following sections.

2.1 Fibers or Yarns for Smart Sensors

Fiber or yarns are the basic structural units for making textiles. Electro-conductive fiber or yarns are one of the key materials for the fabrication of textile based sensors. Fiber-based sensors can be integrated into garments to fabricate smart textiles, cloths and smart gloves for personal healthcare in order to monitor numerous physiological and biological signals of human body. Such fiber/yarns can also be used in wearable electronics either by sewing or simply attaching in garments.

Electronic properties can be imparted in non-conductive textile fibers by exploiting various means. In this regard, coating and spinning are the two most explored approaches through which a conductive nanomaterial can be deposited on fiber/yarn surfaces or may be doped with fiber spinning solution, respectively, thus achieving conductive properties. For instance, two conductive cotton threads were prepared by (Zhong et al. 2015) via coating of carbon nanotubes (CNTs) and polytetrafluoroethylene (PTFE) onto the surface of cotton threads, and wound helically around an elastic fiber. The produced flexible fiber was used in strain sensing performance. In another study, a fiber-based sensor was developed in matrix form by coating of poly(styrene-block-butadienstyrene) (SBS) polymer on the surface of poly(p-phenylene terephthalamide) (Kevlar) fiber (Lee et al. 2015). The fiber-based smart sensor was embedded in gloves and cloths for human-machine interfaces. Using a wet spinning method (Trung et al. 2019), designed a stretchable free-standing elastomeric fiber-based temperature monitoring sensor to detect changes in the human skin temperature.

2.2 Fabric Structures for Smart Sensors

Apart from fiber or yarn based smart sensors where textile integration is necessary using some supportive substrate, smart fabric-based sensors can be worn/used directly without any supporting substrates and accessories. The manufacturing of conductive fabric-based sensor involves coating of fabric with functional materials or integrating the sensing part into fabric structures. Basically, two textile fabric structures are used—knitted and woven. These knitted or woven smart devices can be prepared utilizing number of approaches, which include coating, printing, dying, etc. Besides these, by varying the fabric pattern in smart textile sensors, the desired sensing property, stretchability and durability can be achieved (Hossain and Bradford 2021).

Smart sensors developed from knitted textiles offer flexibility and easy fit-ability to human body. Importantly, due to the superior elasticity and tailorable sensing properties of knitted fabric, this structure is a widely preferred platform for customized wearable devices. For example (Seyedin et al. 2015), fabricated an all-polymeric strain sensor by co-knitting of as-developed polyurethane (PU)/poly(3,4-ethylenedioxythiophene) polystyrene sulfonate (PEDOT:PSS) fibers with commercial spandex yarns. The knitted strain sensor exhibited cyclically repeatable sensing response at applying strain up to 160% and the sensor was demonstrated to monitor joint motion sensing. Recently (Fan et al. 2020), developed a triboelectric all-textile sensor array by knitting with conductive nylon yarns in a full cardigan stitch. The developed sensor demonstrated simultaneous supervision of arterial pulse waves and respiratory signals. The smart textile sensor further broadens its application area by facilitating extended and noninvasive evaluation of sleep apnea syndrome and cardiovascular disease.

3. Materials for Smart Electronic Textile Manufacturing

3.1 Two-dimensional (2D) Nanomaterials

3.1.1 Graphene

The two-dimensional material graphene has been enormously used in nanofabrication of smart and wearable electronic materials. Graphene possesses extraordinary properties such as high surface area, electrical and thermal conductivity, excellent carrier mobility and mechanical flexibility (Ahmed et al. 2020a). These make graphene a promising nanomaterial in development of wearable electronics including smart electronic textiles for diverse applications. The excellent compatibility between graphene and textile substrates has promulgated to the development of smart textile based sensors, which benefits from the range of functionalities, especially for non-invasive (wearable) and invasive (implantable) devices for healthcare (Hatamie et al. 2020, Hossain and Bradford 2021). The intensive use of graphene in biomedical sensors may be attributed to its high sensitivity, which is because of the high specific surface area of graphene layer that facilitates carbon atom to get in touch directly with analytes. Moreover, because of the excellent mechanical robustness and low atomic thickness of graphene, several conformal and close contact with organ of interest such as brain (Masvidal-Codina et al. 2019), skin (Huang et al. 2019) and eyes can be obtained. Such features are essential for acquisition of superior biosignals preventing skin irritation and motion artifacts. To date, extensive research work has been carried out for the development of graphene enabled smart textile products for wearable health monitoring and biomedical engineering; for example, real time monitoring of heart/pulse-rate (Yang et al. 2017), motion monitoring (Wang et al. 2015, Afroj et al. 2020, Xu et al. 2020b, Esfahani 2021), bio-signals (ECG, EMG) and body temperature (Li et al. 2020c, Soni et al. 2020). Despite these advancements in graphene based smart materials for health-monitoring, safety assessment is a key issue to consider. Therefore, the impact of graphene based materials on human body should be systematically executed in terms of their biocompatibility, toxicity and potential to skin irritation.

3.1.2 Metal Carbides/Nitrides (MXenes)

The 2D transitional metal carbide or nitride (MXene) is relatively a new material in smart textile research. However, enormous progress has been observed in MXene enabled smart textiles for wearable smart textile applications, including smart health monitoring systems. MXene based smart textiles are enormously applied in real-time monitoring of human health such as thermotherapy, humidity sensing, pressure sensing, biomedical sensing, wound dressing and tissue engineering (Ahmed et al. 2020b, Huang et al. 2020b, Zhang et al. 2020, Zhao et al. 2020, Hasan et al. 2021). Several MXene based smart sensors for biomedical applications have been discussed later.

3.2 Intrinsically Conducting Polymers (ICPs)

Intrinsically conducting polymers are widely used to apply on textile fibers/yarns or fabric surfaces or in fiber spinning solution to obtain conducting fibers for making electroactive textile structures. ICPs are highly conductive and lightweight materials, having low-temperature processing routes, and can be applied on textile substrates without compromising its flexibility. Due to the ease of solution processibility, such conductive polymers are feasible for roll-to-roll processing in bulk, integrating widely approached textile manufacturing techniques including dyeing and printing (Sinha et al. 2019, Ahmed et al. 2020c). The most commonly employed ICPs into textiles are PEDOT:PSS, polypyrrole (PPy), and polyaniline (PANI) (Akter Shathi et al. 2020, Barakzehi et al. 2020). Electronic textiles developed from these polymers are fabricated via a wide range of approaches as follows:

1) Coating/dying (this technique is suitable for solution processed ICPs, where textile substrate is immersed and dried at convenient temperature. Importantly, there are variety of coating techniques available)

2) Printing (this process involves preparation of an ink solution which is applied on textiles through various printing techniques)

3) Spinning (ICPs are often spun into fibers or can be doped with other electroactive materials to achieve composite conductive fibers)

4) Polymerization [(3,4-Ethylenedioxythiophene (EDOT) can be polymerized onto the textile substrate by vapor phase, *in-situ* and electrochemical polymerization forming PEDOT)]

To date, a good number of works have been reported on ICP based conductive textiles, realizing their application in diverse fields including healthcare and biomedicine. Notable advancement has been made particularly for monitoring of human motion and heart rate (Tseghai et al. 2020), acquisition of bio-signals (ECG, EMG), etc. (Acar et al. 2019, Sinha et al. 2020).

3.3 *Metallic Nanostructures*

Development of smart textiles derived from metallic nanostructures has received a great focus because of their higher electrical conductivity. Several low dimensional metal nanowires/nanorods (Ag, ZnO) and nanoparticles (Ag, Au and Pt) have been found promising for the fabrication of fiber or fabric based smart textile electronics (Kapoor et al. 2020, Rauf et al. 2020). Such materials can be deposited on textiles using various techniques such as electroless-plating (Taghavi Pourian Azar et al. 2020), sputtering (Huang et al. 2020a), coating (Li et al. 2020b), etc. The developed smart textiles depicted excellent properties and have become an ideal material for biomedical and health monitoring. However, metal based smart textiles suffer from a serious issue of instability because the metallic nanoparticles often tend to be leached out from fabric surfaces. Besides, skin irritation is still a great concern for full implementation of such devices as wearables.

4. Smart Textile Electrode for Wearable Thermotherapy

Maintaining thermal comfort is very important because thermal states of the human body are important for psychological health, and could be life-threatening for humans if the core body temperature reaches to the conditions of hyperthermia (above 37.5–38.3°C) or hypothermia (below 35°C) (Peng and Cui 2020). Personalized heating devices with adaptable temperature features could be a great solution to overcome these issues. Moreover, localized thermotherapy has been found as an attractive way of physiotherapy to alleviate or relieve pain in injured joints, muscles and skins (Ahmed et al. 2020c). Fortunately, smart textile based electric heater is extensively applied for such purposes due to its long term heating stability and ability to comply with human skin. A good number of studies have been reported on flexible smart textile based electric heater, over the past years (Pakdel et al. 2019, Bhattacharjee et al. 2020, Faruk et al. 2021).

Traditional thermotherapy pad focuses mainly on orthopedics and muscles inside the skin. However, the therapeutic treatments to epidermal trauma and bacterial infectious diseases are generally ignored in traditional personal healthcare systems. Therefore, recently, researchers are trying to fill this gap by developing flexible smart textile based electric heater with high efficiency and ease of use. Zhao et al. (2020) reported a smart textile based thermotherapy pad which showed promise in thermotherapy when attached in neck guarding pad [Figure 1(a,b)]. Intriguingly, the generated heat can be transferred homogeneously to the bacterial affected skin, accelerating the healing of the infected wounds.

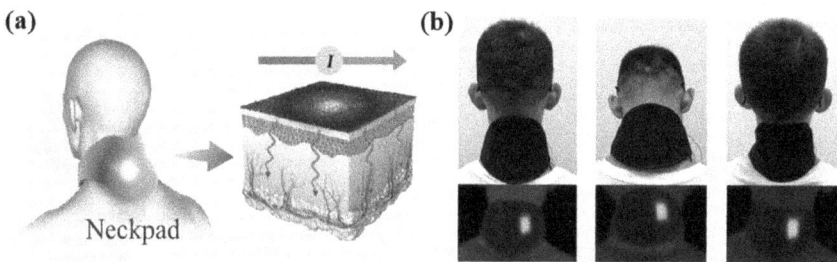

Fig. 1. (a) Schematics showing thermotherapy application of MXene coated smart-textile when attached in a neck-guarding pad. (b) Photographs showing various positions of head with smart fabric based neckpad and the respective IR thermal images. Reprinted with permission from (Zhao et al. 2020), American Chemical Society (2020): 8793–8805.

5. Artificial Muscles for Healthcare

Enormous advances have been realized in smart-textile based mechanically actuating devices or artificial muscles. Artificial muscles derived from smart fibers have found many applications, which include humanoid robots (soft robotics, prosthetic limbs), comfortable clothing and miniaturized actuator for microfluidics (Haines et al. 2014, Kanik et al. 2019). Smart textile structures are attractive for artificial muscles because of the following two reasons (Cochrane et al. 2016):

i) Since the muscles are prepared from muscles fiber and by biomimicry, they can be reproduced.

ii) Textile based structures provide external support to muscles due to their enhanced comfort and wearability to the wearer.

In a recent study (Foroughi et al. 2016), prepared a scalable actuating 3D textile fabric from spandex-multiwall carbon nanotube (MWCNT) yarns. The smart textile demonstrated excellent strain sensing performance when a knee-sleeve prototype was designed to aid personal rehabilitation after an injury, illustrating its possible application in soft robotics and medical devices. Using two traditional textile processing techniques (knitting and weaving) (Maziz et al. 2017), found an excellent way of producing soft wearable muscles. Firstly, the researchers prepared conducting polymer coated cellulose yarns and assembled them into fabrics by both knitting and weaving, and then a metal free deposition of conducting polymers was performed. The developed textile actuator showed good feasibility to integrate in simple soft-robotic devices.

Roach et al. (2019) fabricated a long liquid crystal elastomer (LCE) fiber for artificial muscles. The developed fiber showed the ability to knit, sew and weave to fabricate smart textiles. The excellent reversible actuation performance of fiber was utilized to mimic human fiber-based bicep muscle. Figure 2(a,b) shows the mimic of relaxation and contraction of single LCE bicep muscle fiber. Besides, the activation force increased as the number of fibers increased, revealing fibers' ability to lift up 250 times higher than its weight (Figure 2c).

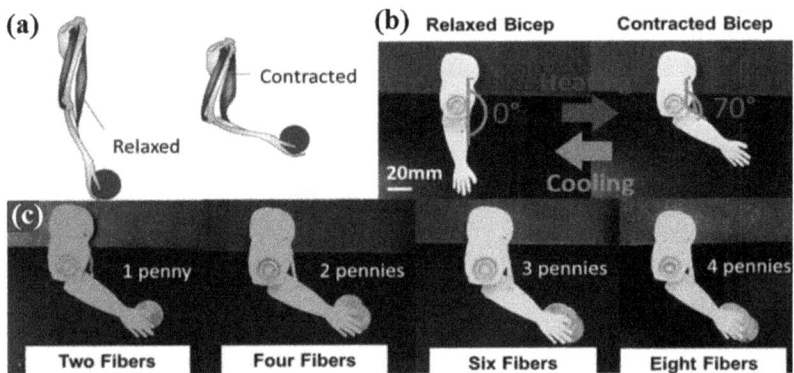

Fig. 2. (a) Anatomy showing a relaxed and contracted bicep muscle fiber. (b) Activation of a single relaxed and contracted bicep muscle fiber. (c) Demonstration of lifting number of pennies by integrating multiple fibers without compromising activation strain. Reprinted with permission from (Roach et al. 2019), American Chemical Society 2019: 19514–19521.

6. Wearable Smart Monitoring System for Infant's Health

The real time health status of infant is important to parents and clinicians so that they can identify infants' physiological condition and ensure the care or therapy at the onset of implications. Smart textile based wearable sensor system paves the way for early indication of infant's health issues through constant health supervision, thereby warning parents and clinicians. Once the signal is transmitted to clinicians, they can take effective measures to treat the infant patients. Smart textile based monitoring of infants has seen many advancements in recent years. One of the notable applications of smart textile based electrodes in infants' health monitoring is acquisition of biopotential such as ECG (Zhu et al. 2015). The traditional method of measuring ECG signal uses gel electrode, which has the potential to cause skin irritation and allergic reaction to infant.

Integrated measurement system offers multiple sensor or wearable electronic devices to be in a single platform, making the monitoring system more comfortable, small in size and convenient. For example (Bouwstra et al. 2009), has developed a smart jacket for integrated monitoring of a newborn's vital signals. The integrated textile electrode system is enabled for easy measurement of ECG and respiration parameters. The designed open chest and open fabric structure on the back of the jacket has made clinical observation effective. Moreover, the produced movement artifacts can be reduced by distributing six electrodes in different position of the jacket. Hopefully, in near future, the technological development of smart textile electronics could be at the forefront in health monitoring of patients, especially for infants.

7. Broad Range Human Motion Monitoring by Strain Sensor

One of the crucial goal in the application of robotics and prosthetics is monitoring of human body motions. Wearable sensors are extensively implemented in this field for continuous detection and monitoring of activities when a body is subject to mechanical strain or deformations. Strain sensors are the main form of sensors that can monitor various motions of human body such as hand motion, joint motion (scapular movement, knee movement, and wrist bending) and several subtle motions (blink, pulse, and respire), etc. The strain sensitivity of an object can be realized through different mechanisms such as resistive, piezoresistive, piezoelectric, capacitive, and triboelectric effects, while resistive and capacitive sensors are predominantly used in skin-interfaced practices ascribed to their simplicity in design and capability of measuring both static and dynamic data (Wang et al. 2020, Xu et al. 2020a).

Smart-textile based wearable sensors can be worn or integrated into fabrics or can be knitted in different textiles structures. Mainly two type of textile structures can be employed in strain sensitivity purposes, which are fibers/yarns and fabrics. Functional electronic fibers can be made out of strain sensitive materials (metals, electro-conductive materials, and conducting polymers). The same is applicable for yarns, which later can be given to various textile forms through knitting, weaving or by other special manufacturing facilities. These functional textiles are intensively applied in real-time detection of human motions in small (subtle) to broad ranges.

Stretchable and flexible strain sensors have been widely used in tracking several vital physiological signals of human body upon movement of body for different operations. Several works have been reported for accurate and simple detection of human hand motions; for example, a multi-functional sensor and user-interface device in textile platform has been developed by *in-situ* polymerization of PEDOT onto fibers (Eom et al. 2017). The integration of conductive fibers in fabric showed high sensitivity towards different hand movements. Notably, the strain sensor was able to express American Sign Language through pre-defined hand gestures by utilizing a user-interfaced wireless system. In an interesting study (Chen et al. 2020), prepared a conductive hydrogel hybrid fiber with core-shell segmental configuration for strain sensing. As demonstrated in Figure 3(a,b), the core-shell segmental structure was obtained by extrusion of conductive hydrogel precursor solution as core layers via dual-core coaxial wet spinning technique. The strain sensor illustrated accurate sensing of two adjacent joint motions (finger and wrist) at a time, Figure 3(c,d).

Recently (Cai et al. 2018), prepared a strain sensing stretchable composite yarn through dip-coating of cotton/spandex into CNT dispersion. The sensor demonstrated strain sensing in the range of 0–350% with sensitivity towards repetitive bending/releasing of fingers at certain angles. By introducing Ag nanoparticles into elastomeric fibers composed coalesced multi-microfilaments (Lee et al. 2018), synthesized a

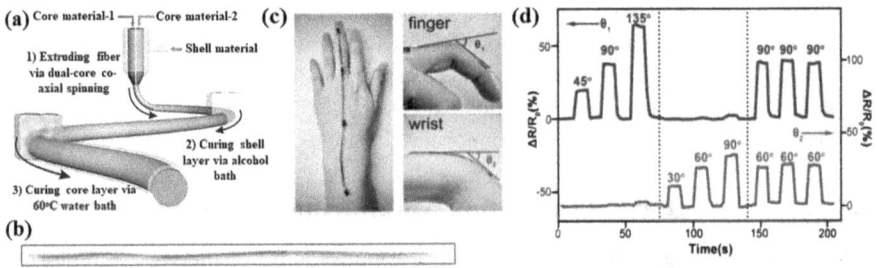

Fig. 3. (a) Fabrication process of core-shell segmental fiber. (b–c) Segmental conductive hydrogel fiber for monitoring hand movement. (d) Monitoring of signals during the movement of hand. Reprinted with permission from (Chen et al. 2020), American Chemical Society 2020: 7565–7574.

highly stretchable fiber based strain sensor for broad range sensing (gauge factor ~ 9.3 × 10⁵ and strain sensing range 450% in the 1st stretching strain). The strain sensor was mounted on a glove, which showed promise in detecting hand movement and had the potential of controlling the hand of a robot, paving the way to be used as flexible electronics in biomedical engineering. Similar type of strain-sensing glove has been developed by Baughman and co-workers with incorporation of stretchable hierarchical buckled fibers for monitoring a broad range of strain induced by human hands (Liu et al. 2015, Choi et al. 2016).

Zheng et al. (2020) had taken an approach to fabricate a smart strain sensor for low range strain detection, by depositing graphene nanosheets onto the cotton fabric surfaces through multiple dipping cycles followed by encapsulation with polydimethylsiloxane (PDMS). The strain sensor demonstrated a very fast and linear current-voltage response, excellent durability and very small range strain (~ 0.4%) detectability.

Smart textile based strain sensors are also utilized in certain complex human motion monitoring such as motion induced by joint movement of body. Ideally, a typical strain sensor is not able to monitor different motions at the same time because of the multidirectional movement of human body. Therefore, well monitoring of strain in various axes requires several strain sensors to be combined into multi-axial strain sensors (Wang et al. 2020). In a recent study, a composite core-spun polyurethane/cotton/CNT yarn was prepared to detect several human joint motions such as elbow flexion and knee bending (Wang et al. 2016b). An ultra-stretchable and sensitive strain sensor was prepared by (Zhang et al. 2017) to detect the strain at both ultra-low detection limit of 0.2% and large body movement. The authors also designed a similar type of strain sensor from carbonized silk fabrics to detect large and complex human body motions including the wrist bending and rotation, and knee movement (Wang et al. 2016a).

Smart textile sensors are also utilized to detect certain subtle human motion signals such as eye blinking, vocal vibration, and change of facial expressions with pulse or phonation of different words (Yang et al. 2018). For instance (Cheng et al. 2015), detected subtle physiological signals by mounting strain sensors onto throat, chest and wrist. The strain sensors were capable of differing pronunciation of diverse words such as "hi," "hello," and "sensor". Moreover, under regular exercise conditions, the sensor recorded the wrist pulse and respiration signals of humans. These sensors could potentially serve as early warning systems for sudden infant death syndrome or adult sleep apnea.

8. Smart Textile Sensors for Dynamic Bio-Monitoring

8.1 Sleep Monitoring

Sleep constitutes one-third of our lives and is the means of rest, adjustments and revitalize to start again. However, in this high-pace of modern society, sleeping disorder is considered to be one of the major issues and billions of people suffer from it (Yamamoto et al. 2016). Sleeping quality is an important sign that is related to certain health-related diseases such as diabetes, coronary heart disease, pressure ulcer and obesity sleep disorder (Lin et al. 2018). Thus, this calls for the development of effective real-time sleep behavior monitoring warning system for early medical diagnosis.

Recently, researchers devised a soft smart textile prepared from washable fibers to continuously monitor sleep and early intervention to sleep related diseases (Zhou et al. 2020). The smart fibers are fabricated by enwrapping an inner core of twisted conductive yarn with an outer sheath of ultra-thin silicone fiber and woven into a bedsheet, Figure 4(a,b). With the aid of a customized multi-channel data acquisition

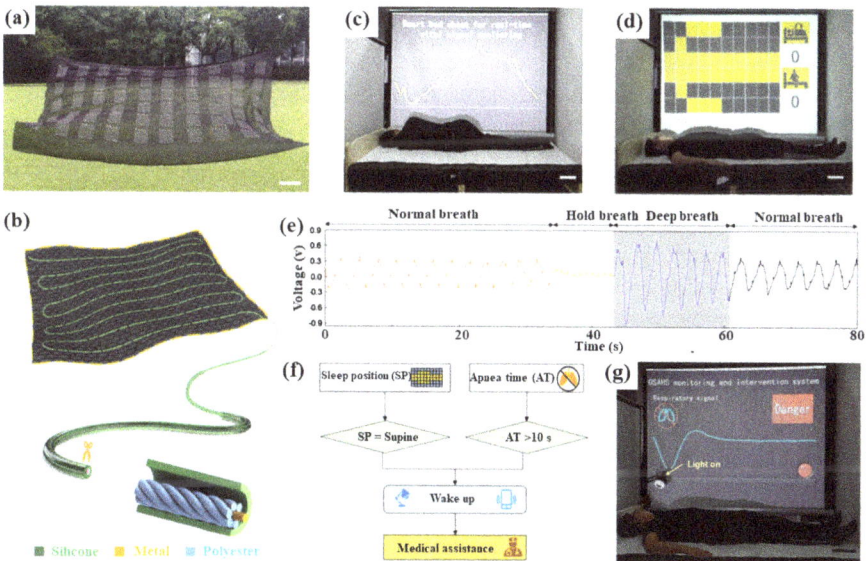

Fig. 4. (a) Photograph of a smart textile for sleep monitoring. (b) Demonstration of a sensing unit designed by weaving a washable functional fiber. (c) Real-time demonstration for monitoring of sleeping posture and body motion and (d) subtle physiological signal monitoring at the time of sleeping. (e) Recorded physiological signal at various breath conditions. (f) Flowchart showing smart textile based OSAHS supervision and intervention system. (g) Photograph showing an alarming system to wake up patients by an illuminated lamp in order to minimize the occurrence of sudden death during sleep. Reprinted with permission from (Zhou et al. 2020), Elsevier 2020: 112064.

circuit, the corresponding signals generated from each sensing unit are recorded, ensuring the final analog output from the smart textile which could accurately express the sleeping posture and physiological signals of a subject during sleep, as depicted in Figure 4(c–e). In addition, the smart textile was employed to reduce sudden death by monitoring obstructive sleep apnea-hypopnea syndrome (OSAHS). The smart textile embedded OSAHS supervision and intervening system could automatically interfere if the duration of apnea episodes exceeds 10 s along with a supine position, and based on the actual needs the time duration can be regulated. Moreover, a bulb or audible alarm will be activated to quickly wake up the patient and to breathe normally, Figure 4(f,g). Additionally, an alert may automatically be transferred to doctors for timely treatment, if the patient couldn't manage to wake up in time.

A similar smart textile based wireless bio-monitoring sensor with high sensitivity of 3.88 V/kPa has been developed for personalized healthcare through a commercially viable way (Meng et al. 2020). The smart textile device can measure human (elderly and weak people) pulse wave in accurate, continuous and noninvasive way. Moreover, the textile based wireless biomonitoring system can be utilized precisely to diagnose the OSAHS syndrome throughout the whole night, even with body movement. In essence, these smart textiles based wearable sleep monitoring system could intensively be utilized in personalized healthcare.

8.2 Respiratory Supervision

In critical care unit, respiration supervision is one of the crucial tasks by which patients' mortality rate and need for ventilation can be prognosticated. Breathing is controlled by the medulla oblongata together with other autonomic functions. Respiratory system sends information of the oxygen concentration, carbon dioxide exchange and the volume of inhaled air (Folke et al. 2003, Quandt et al. 2015). Several approaches have been adopted for respiration monitoring and the common sensors can respond to the flow of breath, while the expansion and contraction of the chest during breathing can be monitored by textile based sensors. Strain sensor can provide information of these motion activities and the obtained data can be exploited for respiration rate supervision.

The conductive nature of electronic textiles augmented the development of sensing elements that can be directly incorporated to garments or inherently weaved or knitted into textiles for respiration sensors. Recently, a fiber-based stretchable strain sensor was developed by (Jang et al. 2019) for several bio-signal detection including respiration supervision. The highly sensitive fibers were fabricated via coating of a carbon based active materials onto the single layer of fiber. The fiber sensor was horizontally sewed onto an electrical fabric-band characterized to supervision of waveforms of respirations during various breathing conditions. Moreover, the fiber sensor is comparable to commercially available biopac breath monitoring sensor (Figure 5). It was found that during the chest expansion, the pressure applied on the compressible textile reduces its thickness, thereby increasing the capacitance between the electrodes.

A specific example of textile based respiratory sensor is its utilization in supervision of respiration for patients suffering from sleep apnea syndrome. As

(a)

(b)

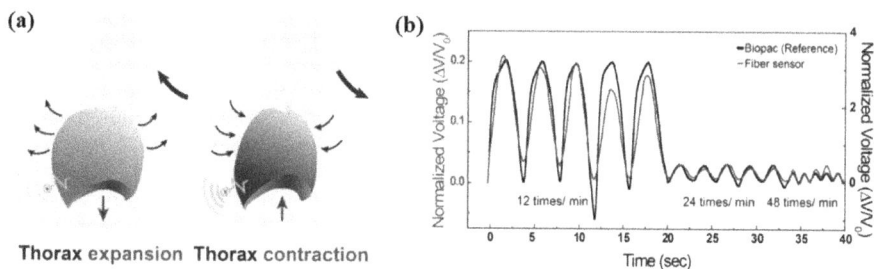

Thorax expansion Thorax contraction

Fig. 5. (a) Schematic showing mechanism of human respiration. (b) The relative changes in voltage for human respiration as function of time. Reprinted with permission from (Jang et al. 2019), American Chemical Society 2019: 15079–15087.

discussed in the Section 8.1, sleep is essential to regain the energy. Sleep apnea can be caused due to the reduced and irregular breathing pattern of patients which even leads to death, and therefore, the full sleep cycle is important to regulate patient's health. In this regard, bed sensors for long-term monitoring of breathing pattern can be a solution. Taking these into consideration, recently researchers fabricated a smart fabric that can be used as bed sheet for real-time and continuous supervision of patients' sleep conditions. The ultimate advantage of using such kind of sensors can be ascribed to the wireless monitoring of patients' sleep conditions and simply by transferring that generated data to a computer or smartphone simulated system even when the doctor is not around. Besides these, textile based respiratory rate sensors also find application in protective clothing, monitoring of vital signs for disabled patients and precautions for sudden infant death syndrome (SIDS) (Zhu et al. 2015, Zhou et al. 2020).

8.3 Sweat/Humidity Sensors

Wearable smart textile based chemical sensors have attracted huge attention to research community to aid in healthcare providing non-invasive and continuous tracking of human health. The health situation of a patient can be analyzed by the data obtained from body-fluids such as sweat, tears and urine. To detect these biomarkers, two distinct principles can be applied, which are resistive and capacitive sensor mechanism.

For resistive humidity sensor, the changes in electrical impedance in hygroscopic medium is measured. With the increase in humidity, the resistance of the material decreases. Employing different fabrication scenarios, resistive humidity sensor can be transferred into smart textiles; for example (Liu et al. 2019), fabricated MXene decorated silk textiles with excellent breathability. The MXene decorated textiles exhibited rapid changes in resistance by responding the blow containing subtle moistures, while no obvious resistance change was found upon dry wind blowing, enabling MXene textiles to monitor the changes in human humidity (i.e., sweating level).

On the other hand, capacitive humidity sensor is composed of two electrodes: a conductive plate and a dielectric placed between the plates. The principle of capacitive humidity sensors is based on the changes of capacitance of the dielectric constant, by which relative humidity is measured. Based on this mechanism, several smart textile-enabled capacitive humidity sensors have been designed for efficient healthcare to monitor human body sweat rate and moisture in wound. Recently (Rauf et al. 2020), have developed metal-organic framework (MOF) coated interdigitated textile electrodes for effective humidity monitoring. A wide range of humidity (3.7%–90%) was employed to observe the sensing capability of smart fabric constructed from cotton and linen.

Recently (He et al. 2019), developed a flexible sweat analysis patch sensor from silk-fabric and carbon based textile. The silk-based electrode worked as wearable electrochemical sensor arrays, enabling real-time and multiplex sweat detection of 6 biomarkers (glucose, lactate, ascorbic acid, Na^+, uric acid and K^+).

8.4 Temperature Sensors

One of the vital signal of health monitoring is body temperature. The typical human body temperature is stable around 36.5–37.5°C. However, an increase in temperature is the sign of bacterial infection or inflammatory conditions, while the decrease in body temperature can be due to the reduced blood flow or even organ failure. For medical assessment, the body temperature is very informative and an essential physiological health parameter. Generally, human body temperature is measured from the arm or chest (Hatamie et al. 2020, Xu et al. 2020a).

Smart textile based temperature sensor provides assessment of temperature changes on skin surfaces and in the near-body environment. These data can be utilized for several purposes such as physiological assessment maintaining patient's comfort, and supervision of wound healing (Mečņika and Dipl 2014). Several textile manufacturing approaches facilitated the development of temperature sensor embedding fiber/yarn and fabrics by integrating them on textiles through knitting, weaving, braiding and printing. Such textile-based sensor can be employed as a temperature probe based on thermistor (thermistor detects temperature variation according to the changes in electrical resistivity), resistive sensor (detects temperature values with respect to the changes in electrical resistance of metals), thermocouple (based on the thermoelectrical effect produced in the junction of two dissimilar metals), and optical sensor or fiber Bragg grating (FBG) sensor (that is sensitive to light, reflecting or transmitting particular wavelength of light) (Mečņika and Dipl 2014, Cochrane et al. 2016).

As an example, a smart textile yarn based temperature sensor was developed by (Hughes-Riley et al. 2017), integrating into a textile remote system. A commercial thermistor chip was implanted into the fibers or yarn, followed by encapsulation in a resin to protect the thermistor. The yarn based wearable temperature enabled the long-term monitoring of human skin, showing a very fast response (between 0.01–0.35 s) over the desired temperature range. In addition, such smart textile can be used to make socks which is promising for patients with, for example, diabetes.

Recently (Lugoda et al. 2020), explored three different industrially viable textile yarn preparation techniques such as knitting, braiding, and double covering to flexible temperature sensing yarn. Among these three techniques, double covering method was found to have the least impact on the sensors' performance because

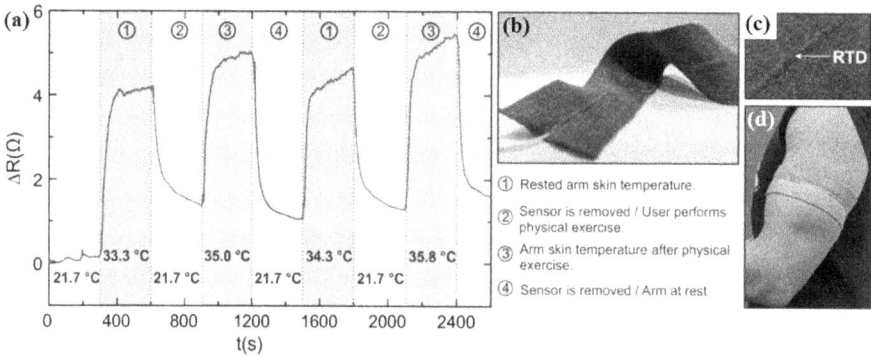

Fig. 6. (a) Temperature measurement results gathered from the temperature sensing yarn located within the armband. (b) Prototype armband showing temperature-sensing yarn located in the center. (c) Close up image of the sensing area of the armband. (d) The armband worn on the upper arm. Reprinted with permission from (Lugoda et al. 2020), MDPI 2020: 73.

of yarn's smaller dimensions. Moreover, the smart sensing yarns were attached in an armband and employed to evaluate changes in skin temperature during physical activity (Figure 6).

9. Smart Textile Electrodes for Acquision of Bio-Potentials

Smart textile enabled biosignal monitoring (cardiac, neural, muscular and ocular) is tremendously growing with the pace of the developments in wearable electronics and the Internet of Things (IoT). The advancement in wearable technology immensely augmented the prospect of real-time health monitoring, brain-computer interfaces and human-machine/computer interaction. Textile electronics is playing a vital role in these respect. This section will provide an overview of smart textile devised electrodes that are intensively utilized in different biopotential monitoring.

9.1 Electrocardiogram (ECG) Monitoring

Electrocardiogram is an important technique to monitor electrophysiological signal of the heart. Wearable ECGs have been intensively used in healthcare for early detection of heart disease and long term health monitoring. An electrocardiogram reveals the change in magnitude of the cardiac vector and its direction as the heart pumps the blood throughout the body. Usually, ECG electrodes are mounted on the chest and limbs to detect the potential differences in the body that exist as a result of different electric field on the skin epidermis (Sinha et al. 2017).

Currently, Ag/AgCl electrodes with a conductive gel layer are the mostly used electrodes in the ECG monitoring system. The pre-gelled Ag/AgCl electrode can achieve high quality ECG signals. However, skin inflammation is one of the key issues which may occur during prolonged ECG operation. Besides, the performance of the electrode may become unstable as the gel dries out over time (Pani et al. 2016). Therefore, current research is focusing on the development of ECG system excluding gel from electrode in order to overcome the skin allergy issues. Consequently, textiles have emerged as a promising substrate for dry ECG electrodes due to the inherent flexibility, stretchability and the possibility of scaling-up by direct weaving of medical clothing with biosensing capability (Yapici et al. 2015). Moreover, smart textile based electrode can measure ECG signals in a more non-invasive and customized way by maintaining conformal contact with human body. A schematic illustration of skin-electrode interface system is presented in Figure 7, showing how a textile electrode is different from wet electrode, essentially overcoming the issue of skin irritations caused by conventional gel electrode (Beckmann et al. 2010).

To fabricate textile based ECG electrode, a number of electroactive materials such as graphene (Yapici et al. 2015), conducting polymers (Pani et al. 2016, Sinha et al. 2019) and metallic nanoparticles (Boehm et al. 2016) have been widely integrated into textiles using several methodologies including screen printing of conductive pastes (Sinha et al. 2017), electroplating and deposition of metals and sewing or embroidery of conductive metallic thread into garments (Marozas et al. 2011), knitting (Jang et al. 2007) and weaving. Notably, PEDOT:PSS is the most widely explored polymer for biopotential ECG monitoring techniques, and has been successfully integrated into textiles. These could be due to the biocompatibility and rheological properties of PEDOTs, and above all the decrease in electrical impedance over classical electrodes (Bolin et al. 2009).

The designed textile based ECG electrodes embedded into smart garments have been applied to monitor elderly heart failure prevention and sports or extreme conditions (Baig et al. 2013), tele-health and infants monitoring (Coosemans et al. 2006). For example (Yapici et al. 2015), designed a graphene-clad, conductive textile ECG electrode for acquisition of biosignal specifically in cardiac assessment. A dip-coated strategy was devised for conformal coating of reduced graphene oxide (rGO) into the nylon fabrics. The graphene-clad electrode exhibited excellent conformity and cross correlation of 97% than those of Ag/AgCl electrodes.

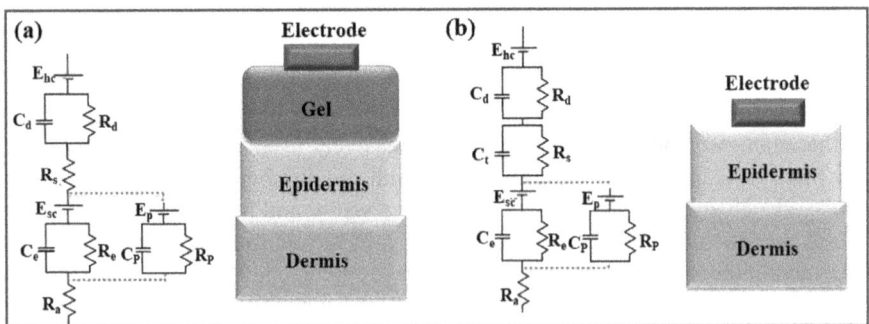

Fig. 7. Circuit of skin-interface electrode for (a) conventional wet electrode and (b) textile electrode.

9.2 *Electromyogram (EMG)*

Electromyography (EMG) provides muscle function assessment by examining the electrical signals driven by the central nervous system to the muscles. By this technique, the electrical signals that occur during the contraction and relaxation of muscle cycles can be measured. Moreover, in clinical system, EMG is also utilized as a diagnostic tool for nerve and muscle injury. EMG analyzes electrical signals within the muscle with two electrodes (Acar et al. 2019). Importantly, myoelectric signals are found to be entirely localized on the skin surface over the muscle of interest. Generally, EMG electrode is smaller in size but creates higher skin-electrode impedance. Smart textile enabled EMG electrode can acquire data in a more realistic way, thus providing the proof-of-concept, facilitating the way of embedding new textile electrode. The wearable EMG acquisition with textile electrode has found many applications including muscle status tracking (Pino et al. 2018), rehabilitation (Guo et al. 2018), prosthetics (Farina et al. 2010), and running leggings for muscle fatigue detection (Shafti et al. 2017). Notably, two companies, namely Athos (Mad Apparel Inc., CA, US) and Myontec (Myontec Ltd., Kuopio, Finland), have already brought textile integrated EMG electrodes in markets focusing on athlete training.

10. Summary

Textile plays an inevitable role in human civilization. Besides, the enormous progress in nanotechnology has rendered more functionalities in clothing systems, thereby serving many spheres in life including in healthcare. Smart e-textiles are flexible, stretchable and widely recognized as compatible with human body. The growing interest in the development of smart textiles for healthcare applications is being driven by the need for continuous monitoring of physiological parameters and timely treatment of patients. Nanosensor embedded smart-textiles provide wearable and non-invasive supervision of health. Therefore, the research interest combining textiles and sensors into a single structure to develop wearable sensors is growing rapidly and it is expected that in the coming days this wearable technology may become more accessible for personal medicare systems.

References

Acar, G., O. Ozturk, A. J. Golparvar, T. A. Elboshra, K. Böhringer and M. Kaya Yapici. (2019). Wearable and flexible textile electrodes for biopotential signal monitoring: A review. *Electronics (Switzerland)* 8: 1–25.

Afroj, S., S. Tan, A. M. Abdelkader, K. S. Novoselov and N. Karim. (2020). Highly conductive, scalable, and machine washable graphene-based E-textiles for multifunctional wearable electronic applications. *Advanced Functional Materials* 30: 2000293.

Ahmed, A., B. Adak, T. Bansala and S. Mukhopadhyay. (2020a). Green solvent processed cellulose/graphene oxide nanocomposite films with superior mechanical, thermal, and ultraviolet shielding properties. *ACS Applied Materials and Interfaces* 12: 1687–1697.

Ahmed, A., M. M. Hossain, B. Adak and S. Mukhopadhyay. (2020b). Recent advances in 2D MXene integrated smart-textile interfaces for multifunctional applications. *Chemistry of Materials* 32: 10296–10320.

Ahmed, A., M. A. Jalil, M. M. Hossain, M. Moniruzzaman, B. Adak, M. T. Islam, M. S. Parvez and S. Mukhopadhyay. (2020c). A PEDOT:PSS and graphene-clad smart textile-based wearable electronic Joule heater with high thermal stability. *Journal of Materials Chemistry C* 8: 16204–16215.

Akter Shathi, M., C. Minzhi, N. A. Khoso, H. Deb, A. Ahmed and W. Sai Sai. (2020). All organic graphene oxide and Poly (3, 4-ethylene dioxythiophene) - Poly (styrene sulfonate) coated knitted textile fabrics for wearable electrocardiography (ECG) monitoring. *Synthetic Metals* 263: 116329.

Baig, M. M., H. Gholamhosseini and M. J. Connolly. (2013). A comprehensive survey of wearable and wireless ECG monitoring systems for older adults. *Medical and Biological Engineering and Computing* 51: 485–495.

Barakzehi, M., M. Montazer, F. Sharif, T. Norby and A. Chatzitakis. (2020). MOF-modified polyester fabric coated with reduced graphene oxide/polypyrrole as electrode for flexible supercapacitors. *Electrochimica Acta* 336: 135743.

Beckmann, L., C. Neuhaus, G. Medrano, N. Jungbecker, M. Walter, T. Gries and S. Leonhardt. (2010). Characterization of textile electrodes and conductors using standardized measurement setups. *Physiological Measurement* 31: 233–247.

Bhattacharjee, S., C. R. Macintyre, P. Bahl, U. Kumar, X. Wen, A. A. Chughtai and R. Joshi. (2020). Reduced graphene oxide and nanoparticles incorporated durable electroconductive silk fabrics. *Advanced Materials Interfaces* 2000814: 1–13.

Boehm, A., X. Yu, W. Neu, S. Leonhardt and D. Teichmann. (2016). A novel 12-lead ECG T-shirt with active electrodes. *Electronics (Switzerland)* 5.

Bolin, M. H., K. Svennersten, X. Wang, I. S. Chronakis, A. Richter-Dahlfors, E. W. H. Jager and M. Berggren. (2009). Nano-fiber scaffold electrodes based on PEDOT for cell stimulation. *Sensors and Actuators, B: Chemical* 142: 451–456.

Bouwstra, S., W. Chen, L. Feijs and S. B. Oetomo. (2009). Smart jacket design for neonatal monitoring with wearable sensors. pp. 162–167. *In*: Proceedings - 2009 6th International Workshop on Wearable and Implantable Body Sensor Networks, BSN 2009.

Cai, G., M. Yang, J. Pan, D. Cheng, Z. Xia, X. Wang and B. Tang. (2018). Large-scale production of highly stretchable CNT/cotton/spandex composite yarn for wearable applications. *ACS Applied Materials and Interfaces* 10: 32726–32735.

Chen, J., H. Wen, G. Zhang, F. Lei, Q. Feng, Y. Liu, X. Cao and H. Dong. (2020). Multifunctional conductive hydrogel/thermochromic elastomer hybrid fibers with a core-shell segmental configuration for wearable strain and temperature sensors. *ACS Applied Materials and Interfaces* 12: 7565–7574.

Cheng, Y., R. Wang, J. Sun and L. Gao. (2015). A stretchable and highly sensitive graphene-based fiber for sensing tensile strain, bending, and torsion. *Advanced Materials* 27: 7365–7371.

Choi, C., J. M. Lee, S. H. Kim, S. J. Kim, J. Di and R. H. Baughman. (2016). Twistable and stretchable sandwich structured fiber for wearable sensors and supercapacitors. *Nano Letters* 16: 7677–7684.

Cochrane, C., C. Hertleer and A. Schwarz-Pfeiffer. (2016). Smart textiles in health: An overview. pp. 9–32. *In*: Smart Textiles and Their Applications.

Coosemans, J., B. Hermans and R. Puers. (2006). Integrating wireless ECG monitoring in textiles. *Sensors and Actuators, A: Physical* 130–131: 48–53.

Das, P. S., S. H. Park, K. Y. Baik, J. W. Lee and J. Y. Park. (2020). Thermally reduced graphene oxide-nylon membrane based epidermal sensor using vacuum filtration for wearable electrophysiological signals and human motion monitoring. *Carbon* 158: 386–393.

Eom, J., R. Jaisutti, H. Lee, W. Lee, J. S. Heo, J. Y. Lee, S. K. Park and Y. H. Kim. (2017). Highly sensitive textile strain sensors and wireless user-interface devices using all-polymeric conducting fibers. *ACS Applied Materials and Interfaces* 9: 10190–10197.

Esfahani, M. I. M. (2021). Smart textiles in healthcare: a summary of history, types, applications, challenges, and future trends. pp. 93–107. *In*: Nanosensors and Nanodevices for Smart Multifunctional Textiles (Elsevier).

Fan, W., Q. He, K. Meng, X. Tan, Z. Zhou, G. Zhang, J. Yang and Z. L. Wang. (2020). Machine-knitted washable sensor array textile for precise epidermal physiological signal monitoring. Available at: http://advances.sciencemag.org/ [Accessed October 25, 2020].

Farina, D., T. Lorrain, F. Negro and N. Jiang. (2010). High-density EMG E-textile systems for the control of active prostheses. pp. 3591–3593. *In*: 2010 Annual International Conference of the IEEE Engineering in Medicine and Biology Society, EMBC'10.

Faruk, M. O., A. Ahmed, M. A. Jalil, M. T. Islam, A. M. Shamim, B. Adak, M. M. Hossain and S. Mukhopadhyay. (2021). Functional textiles and composite based wearable thermal devices for Joule heating: progress and perspectives. *Applied Materials Today* 23: 101025.

Folke, M., L. Cernerud, M. Ekström and B. Hök. (2003). Critical review of non-invasive respiratory monitoring in medical care. *Medical and Biological Engineering and Computing* 41: 377–383.

Foroughi, J., G. M. Spinks, S. Aziz, A. Mirabedini, A. Jeiranikhameneh, G. G. Wallace, M. E. Kozlov and R. H. Baughman. (2016). Knitted carbon-nanotube-sheath/spandex-core elastomeric yarns for artificial muscles and strain sensing. *ACS Nano* 10: 9129–9135.

Guo, J., S. Yu, Y. Li, T. H. Huang, J. Wang, B. Lynn, J. Fidock, C. L. Shen, D. Edwards and H. Su. (2018). A soft robotic exo-sheath using fabric EMG sensing for hand rehabilitation and assistance. pp. 497–503. *In*: 2018 IEEE International Conference on Soft Robotics, RoboSoft 2018.

Haines, C. S., M. D. Lima, N. Li, G. M. Spinks, J. Foroughi, J. D. W. Madden, S. H. Kim, S. Fang, M. J. De Andrade, F. Göktepe and O. Göktepe. (2014). Artificial muscles from fishing line and sewing thread. *Science* 343: 868–872.

Hasan, M. M., M. M. Hossain and H. K. Chowdhury. (2021). Two-dimensional MXene-based flexible nanostructures for functional nanodevices: a review. *Journal of Materials Chemistry A* 9: 3231–3269.

Hatamie, A., S. Angizi, S. Kumar, C. M. Pandey, A. Simchi, M. Willander and B. D. Malhotra. (2020). Review—textile based chemical and physical sensors for healthcare monitoring. *Journal of The Electrochemical Society* 167: 037546.

He, W., C. Wang, H. Wang, M. Jian, W. Lu, X. Liang, X. Zhang, F. Yang and Y. Zhang. (2019). Integrated textile sensor patch for real-time and multiplex sweat analysis. *Science Advances* 5.

Hossain, M. M. and P. D. Bradford. (2021). Durability of smart electronic textiles. pp. 27–53. *In*: Nanosensors and Nanodevices for Smart Multifunctional Textiles (Elsevier).

Huang, H., S. Su, N. Wu, H. Wan, S. Wan, H. Bi and L. Sun. (2019). Graphene-based sensors for human health monitoring. *Frontiers in Chemistry* 7: 1–26.

Huang, M. L., Z. Cai, Y. Z. Wu, S. G. Lu, B. S. Luo and Y. H. Li. (2020a). Metallic coloration on polyester fabric with sputtered copper and copper oxides films. *Vacuum* 178: 109489.

Huang, R., X. Chen, Y. Dong, X. Zhang, Y. Wei, Z. Yang, W. Li, Y. Guo, J. Liu, Z. Yang and H. Wang. (2020b). MXene composite nanofibers for cell culture and tissue engineering. *ACS Applied Bio Materials* 3: 2125–2131.

Hughes-Riley, T., P. Lugoda, T. Dias, C. L. Trabi and R. H. Morris. (2017). A study of thermistor performance within a textile structure. *Sensors (Switzerland)* 17.

Islam, G. M. N., A. Ali and S. Collie. (2020). Textile sensors for wearable applications: a comprehensive review. *Cellulose* 27: 6103–6131.

Jang, S., J. Cho, K. Jeong and G. Cho. (2007). Exploring possibilities of ECG electrodes for bio-monitoring smartwear with Cu sputtered fabrics. pp. 1130–1137. *In*: Lecture Notes in Computer Science (including subseries Lecture Notes in Artificial Intelligence and Lecture Notes in Bioinformatics) (Springer Verlag).

Jang, S., J. Kim, D. W. Kim, J. W. Kim, S. Chun, H. J. Lee, G. R. Yi and C. Pang. (2019). Carbon-based, ultraelastic, hierarchically coated fiber strain sensors with crack-controllable beads. *ACS Applied Materials and Interfaces* 11: 15079–15087.

Kanik, M., S. Orguc, G. Varnavides, J. Kim, T. Benavides, D. Gonzalez, T. Akintilo, C. C. Tasan, A. P. Chandrakasan, Y. Fink and P. Anikeeva. (2019). Strain-programmable fiber-based artificial muscle. *Science* 365: 145–150.

Kapoor, A., P. Shankar and W. Ali. (2020). Green synthesis of metal nanoparticles for electronic textiles. pp. 81–97. *In*: Advanced Structured Materials (Springer).

Karim, N., S. Afroj, K. Lloyd, L. C. Oaten, D. V. Andreeva, C. Carr, A. D. Farmery, I. D. Kim and K. S. Novoselov. (2020). Sustainable personal protective clothing for healthcare applications: A review. *ACS Nano* 14: 12313–12340.

Lee, J., H. Kwon, J. Seo, S. Shin, J. H. Koo, C. Pang, S. Son, J. H. Kim, Y. H. Jang, D. E. Kim and T. Lee. (2015). Conductive fiber-based ultrasensitive textile pressure sensor for wearable electronics. *Advanced Materials* 27: 2433–2439.

Lee, J., S. Shin, S. Lee, J. Song, S. Kang, H. Han, S. Kim, S. Kim, J. Seo, D. Kim and T. Lee. (2018). Highly sensitive multifilament fiber strain sensors with ultrabroad sensing range for textile electronics. *ACS Nano* 12: 4259–4268.

Li, B. M., O. Yildiz, A. C. Mills, T. J. Flewwellin, P. D. Bradford and J. S. Jur. (2020a). Iron-on carbon nanotube (CNT) thin films for biosensing E-textile applications. *Carbon* 168: 673–683.

Li, G. P., F. Cao, K. Zhang, L. Hou, R. C. Gao, W. Y. Zhang and Y. Y. Wang. (2020b). Design of anti-UV radiation textiles with self-assembled metal–organic framework coating. *Advanced Materials Interfaces* 7: 1901525.

Li, Z., W. Guo, Y. Huang, K. Zhu, H. Yi and H. Wu. (2020c). On-skin graphene electrodes for large area electrophysiological monitoring and human-machine interfaces. *Carbon* 164: 164–170.

Lin, Z., J. Yang, X. Li, Y. Wu, W. Wei, J. Liu, J. Chen and J. Yang. (2018). Large-scale and washable smart textiles based on triboelectric nanogenerator arrays for self-powered sleeping monitoring. *Advanced Functional Materials* 28: 1704112.

Liu, L. X., W. Chen, Zhang, H. Bin, Q. W. Wang, F. Guan and Z. Z. Yu. (2019). Flexible and multifunctional silk textiles with biomimetic leaf-like MXene/silver nanowire nanostructures for electromagnetic interference shielding, humidity monitoring, and self-derived hydrophobicity. *Advanced Functional Materials* 29: 1–10.

Liu, Z. F., S. Fang, F. A. Moura, J. N. Ding, N. Jiang, J. Di, M. Zhang, X. Lepró, D. S. Galvao, C. S. Haines and N. Y. Yuan. (2015). Hierarchically buckled sheath-core fibers for superelastic electronics, sensors, and muscles. *Science* 349: 400–404.

Lugoda, P., J. C. Costa, C. Oliveira, L. A. Garcia-Garcia, S. D. Wickramasinghe, A. Pouryazdan, D. Roggen, T. Dias and N. Münzenrieder. (2020). Flexible temperature sensor integration into e-textiles using different industrial yarn fabrication processes. *Sensors (Switzerland)* 20.

Marozas, V., A. Petrenas, S. Daukantas and A. Lukosevicius. (2011). A comparison of conductive textile-based and silver/silver chloride gel electrodes in exercise electrocardiogram recordings. *Journal of Electrocardiology* 44: 189–194.

Masvidal-Codina, E., X. Illa, M. Dasilva, A. B. Calia, T. Dragojević, E. E. Vidal-Rosas, E. Prats-Alfonso, J. Martínez-Aguilar, M. Jose, R. Garcia-Cortadella and P. Godignon. (2019). High-resolution mapping of infraslow cortical brain activity enabled by graphene microtransistors. *Nature Materials* 18: 280–288.

Maziz, A., A. Concas, A. Khaldi, J. Stålhand, N. K. Persson and E. W. H. Jager. (2017). Knitting and weaving artificial muscles. *Science Advances* 3.

Mečņika, V. and M. H. Dipl. (2014). Smart textiles for healthcare : applications and technologies. *Rural Environment, Education, Personality*, 7–8. Available at: https://pdfs.semanticscholar.org/b536/fd03cfb4b9a39e1d78e746a5f39cd3f358d5.pdf [Accessed October 12, 2020].

Meng, K., S. Zhao, Y. Zhou, Y. Wu, S. Zhang, Q. He, X. Wang, Z. Zhou, W. Fan, X. Tan and J. Yang. (2020). A wireless textile-based sensor system for self-powered personalized health care. *Matter* 2: 896–907.

Pakdel, E., M. Naebe, L. Sun and X. Wang. (2019). Advanced functional fibrous materials for enhanced thermoregulating performance. *ACS Applied Materials and Interfaces* 11: 13039–13057.

Pani, D., A. Dessi, J. F. Saenz-Cogollo, G. Barabino, B. Fraboni and A. Bonfiglio. (2016). Fully textile, PEDOT:PSS based electrodes for wearable ECG monitoring systems. *IEEE Transactions on Biomedical Engineering* 63: 540–549.

Peng, Y. and Y. Cui. (2020). Advanced textiles for personal thermal management and energy. *Joule* 4: 724–742.

Pino, E. J., Y. Arias and P. Aqueveque. (2018). Wearable EMG shirt for upper limb training. pp. 4406–4409. *In*: Proceedings of the Annual International Conference of the IEEE Engineering in Medicine and Biology Society, EMBS.

Quandt, B. M., L. J. Scherer, L. F. Boesel, M. Wolf, G. L. Bona and R. M. Rossi. (2015). Body-monitoring and health supervision by means of optical fiber-based sensing systems in medical textiles. *Advanced Healthcare Materials* 4: 330–355.

Rauf, S., M. T. Vijjapu, M. A. Andrés, I. Gascón, O. Roubeau, M. Eddaoudi and K. N. Salama. (2020). Highly selective metal-organic framework textile humidity sensor. *ACS Applied Materials and Interfaces* 12: 29999–30006.

Roach, D. J., C. Yuan, X. Kuang, V. C. F. Li, P. Blake, M. L. Romero, I. Hammel, K. Yu and H. J. Qi. (2019). Long liquid crystal elastomer fibers with large reversible actuation strains for smart textiles and artificial muscles. *ACS Applied Materials and Interfaces* 11: 19514–19521.

Ruiz, L. L., M. Ridder, D. Fan, J. Gong, B. M. Li, A. Myers, E. Cobarrubias, J. Strohmaier, J. S. Jur and J. Lach. (2020). Self-powered cardiac monitoring: maintaining vigilance with multi-modal harvesting and E-textiles. *IEEE Sensors Journal* XX: 1–1.

Seyedin, S., J. M. Razal, P. C. Innis, A. Jeiranikhameneh, S. Beirne and G. G. Wallace. (2015). Knitted strain sensor textiles of highly conductive all-polymeric fibers. *ACS Applied Materials and Interfaces* 7: 21150–21158.

Shafti, A., R. B. Ribas Manero, A. M. Borg, K. Althoefer and M. J. Howard. (2017). Embroidered electromyography: a systematic design guide. *IEEE Transactions on Neural Systems and Rehabilitation Engineering* 25: 1472–1480.

Shahariar, H., I. Kim, H. Soewardiman and J. S. Jur. (2019). Inkjet printing of reactive silver ink on textiles. *ACS Applied Materials and Interfaces* 11: 6208–6216.

Sinha, S. K., Y. Noh, N. Reljin, G. M. Treich, S. Hajeb-Mohammadalipour, Y. Guo, K. H. Chon and G. A. Sotzing. (2017). Screen-printed PEDOT:PSS electrodes on commercial finished textiles for electrocardiography. *ACS Applied Materials and Interfaces* 9: 37524–37528.

Sinha, S. K., F. A. Alamer, S. J. Woltornist, Y. Noh, F. Chen, A. McDannald, C. Allen, R. Daniels, A. Deshmukh, M. Jain, K. Chon and G. A. Sotzing. (2019). Graphene and Poly(3,4-ethylene dioxythiophene):Poly(4-styrenesulfonate) on nonwoven fabric as a room temperature metal and its application as dry electrodes for electrocardiography. *ACS Applied Materials and Interfaces* 11: 32339–32345.

Sinha, S. K., H. F. Posada-Quintero, Y. Noh, C. Allen, R. Daniels, K. H. Chon, L. Sloan and G. A. Sotzing. (2020). Integrated dry poly(3,4-ethylenedioxythiophene):polystyrene sulfonate electrodes on finished textiles for continuous and simultaneous monitoring of electrocardiogram, electromyogram and electrodermal activity. *Flexible and Printed Electronics* 5: 035009.

Soni, M., M. Bhattacharjee, M. Ntagios and R. Dahiya. (2020). Printed temperature sensor based on PEDOT:PSS-graphene oxide composite. *IEEE Sensors Journal* 20: 7525–7531.

Taghavi Pourian Azar, G., D. Fox, Y. Fedutik, L. Krishnan and A. J. Cobley. (2020). Functionalised copper nanoparticle catalysts for electroless copper plating on textiles. *Surface and Coatings Technology* 396: 125971.

Trung, T. Q., T. M. L. Dang, S. Ramasundaram, P. T. Toi, S. Y. Park and N. E. Lee. (2019). A stretchable strain-insensitive temperature sensor based on free-standing elastomeric composite fibers for on-body monitoring of skin temperature. *ACS Applied Materials and Interfaces* 11: 2317–2327.

Tseghai, G. B., D. A. Mengistie, B. Malengier, K. A. Fante and L. Van Langenhove. (2020). PEDOT:PSS-based conductive textiles and their applications. *Sensors (Switzerland)* 20: 1881.

Wang, C., X. Li, E. L. Gao, M. Q. Jian, K. L. Xia, Q. Wang, Z. Xu, T. Ren and Y. Zhang. (2016a). Carbonized silk fabric for ultrastretchable, highly sensitive, and wearable strain sensors. *Advanced Materials* 28: 6640–6648.

Wang, J., C. Lu and K. Zhang. (2020). Textile-based strain sensor for human motion detection. *Energy & Environmental Materials* 3: 80–100.

Wang, X., Y. Qiu, W. Cao and P. Hu. (2015). Highly stretchable and conductive core-sheath chemical vapor deposition graphene fibers and their applications in safe strain sensors. *Chemistry of Materials* 27: 6969–6975.

Wang, Z., Y. Huang, J. Sun, Y. Huang, H. Hu, R. Jiang, W. Gai, G. Li and C. Zhi. (2016b). Polyurethane/cotton/carbon nanotubes core-spun yarn as high reliability stretchable strain sensor for human motion detection. *ACS Applied Materials and Interfaces* 8: 24837–24843.

Xu, C., Y. Yang and W. Gao. (2020a). Skin-interfaced sensors in digital medicine: from materials to applications. *Matter* 2: 1414–1445.

Xu, L., Z. Liu, H. Zhai, X. Chen, R. Sun, S. Lyu, Y. Fan, Y. Yi, Z. Chen, L. Jin and J. Zhang. (2020b). Moisture-resilient graphene-dyed wool fabric for strain sensing. *ACS Applied Materials and Interfaces* 12: 13265–13274.

Yamamoto, U., M. Nishizaka, C. Yoshimura, N. Kawagoe, A. Hayashi, T. Kadokami and S. I. Ando. (2016). Prevalence of sleep disordered breathing among patients with nocturia at a urology clinic. *Internal Medicine* 55: 901–905.

Yang, T., X. Jiang, Y. Zhong, X. Zhao, S. Lin, J. Li, X. Li, J. Xu, Z. Li and H. Zhu. (2017). A wearable and highly sensitive graphene strain sensor for precise home-based pulse wave monitoring. *ACS Sensors* 2: 967–974.

Yang, Z., Y. Pang, X. L. Han, Y. Yang, Y. Yang, J. Ling, M. Jian, Y. Zhang, Y. Yang and T. L. Ren. (2018). Graphene textile strain sensor with negative resistance variation for human motion detection. *ACS Nano* 12: 9134–9141.

Yapici, M. K., T. Alkhidir, Y. A. Samad and K. Liao. (2015). Graphene-clad textile electrodes for electrocardiogram monitoring. *Sensors and Actuators, B: Chemical* 221.

Yetisen, A. K., H. Qu, A. Manbachi, H. Butt, M. R. Dokmeci, J. P. Hinestroza, M. Skorobogatiy, A. Khademhosseini and S. H. Yun. (2016). Nanotechnology in textiles. *ACS Nano* 10: 3042–3068.

Zhang, M., C. Wang, H. Wang, M. Jian, X. Hao and Y. Zhang. (2017). Carbonized cotton fabric for high-performance wearable strain sensors. *Advanced Functional Materials* 27: 1604795.

Zhang, X., X. Wang, Z. Lei, L. Wang, M. Tian, S. Zhu, H. Xiao, X. Tang and L. Qu. (2020). Flexible MXene-decorated fabric with interwoven conductive networks for integrated joule heating, electromagnetic interference shielding, and strain sensing performances. *ACS Applied Materials and Interfaces* 12: 14459–14467.

Zhao, X., L. Y. Wang, C. Y. Tang, X. J. Zha, Y. Liu, B. H. Su, K. Ke, R. Y. Bao, M. B. Yang and W. Yang. (2020). Smart Ti3C2TxMXene fabric with fast humidity response and joule heating for healthcare and medical therapy applications. *ACS Nano* 14: 8793–8805.

Zheng, Y., Y. Li, Y. Zhou, K. Dai, G. Zheng, B. Zhang, C. Liu and C. Shen. (2020). High-performance wearable strain sensor based on graphene/cotton fabric with high durability and low detection limit. *ACS Applied Materials and Interfaces* 12: 1474–1485.

Zhong, J., Q. Zhong, Q. Hu, N. Wu, W. Li, B. Wang, B. Hu and J. Zhou. (2015). Stretchable self-powered fiber-based strain sensor. *Advanced Functional Materials* 25: 1798–1803.

Zhou, Z., S. Padgett, Z. Cai, G. Conta, Y. Wu, Q. He, S. Zhang, C. Sun, J. Liu, E. Fan and K. Meng. (2020). Single-layered ultra-soft washable smart textiles for all-around ballistocardiograph, respiration, and posture monitoring during sleep. *Biosensors and Bioelectronics* 155: 112064.

Zhu, Z., T. Liu, G. Li, T. Li and Y. Inoue. (2015). Wearable sensor systems for infants. *Sensors (Switzerland)* 15: 3721–3749.

CHAPTER 8

E-Skin for Futuristic Nanosensor Technology for the Healthcare System

Venkateswaran Vivekananthan,[1] *Gaurav Khandelwal,*[2]
Nagamalleswara Rao Alluri[1] and *Sang-Jae Kim*[1,2,3,]*

1. Introduction

Skin is the sensory organ of the human body which protects the entire body and acts as the direct interface between the body and external environment. It senses the external stimuli such as touch, heat, cold, and sends the respective signals to the brain. Inspired from the properties of skin, researchers have developed skin-based devices called electronic skins (e-skin), having a flexible nature which mimics the property of the human skin. Having a lot of advantages, e-skin has emerged as a futuristic portable and wearable device for health care monitoring and diagnostic applications. Conventional health care monitoring and bio-medical devices have certain drawbacks such as bigger instrument size, more instrumentation, trained professionals for operating the instrument, and radiation from instruments. On the other hand, wearable devices can collect a set of data like activity levels, sleep, heart rate and continuous monitoring. In addition, the device can communicate with the health care provider remotely via Internet of Things (IoT) based artificial intelligence (AI) technology (Hammock et al. 2013, Chortos et al. 2016).

Wearable devices have the potential to provide enormous benefits to healthcare providers by monitoring the patient data over a long period of time, paving the way for medical professionals getting a better view about the problems of patient. The

---
[1] Advanced Technology Institute, Department of Electrical and Electronics Engineering, University of Surrey, United Kingdom.
[2] Nanomaterials & System Laboratory, Major of Mechatronics Engineering, Faculty of Applied Energy System, Jeju National University, Jeju 63243, Republic of Korea.
[3] R&D center for Energy New Industry, Jeju National University, Jeju 63243, Republic of Korea.
* Corresponding author: kimsangj@jejunu.ac.kr

collected data helps the medical professionals to accurately diagnose the health issues of the patients. Wearables can also be used on patients when they return home after surgery or an operation to monitor their recovery and ensure no complications occur. This helps ease the burden on healthcare systems by letting the patients leave the hospital and return home, but still keeping an eye on their conditions using wearable devices. Emergencies can also be recognized as soon as they occur. This system can be set up to notify others, such as family members or healthcare professionals. This more proactive approach to healthcare can be very beneficial, as it can catch problems early before they develop into larger issues that could have dangerous health consequences (Trung and Lee 2016, Miyamoto et al. 2017).

The recent past has seen a wide range of wearable sensors to quantify and measure the human activities and the corresponding signals generated from those activities. For example, skin pressure (Luo et al. 2017), electrocardiograms (Luo et al. 2016), metabolite levels (Li et al. 2018), body temperature (Yeo et al. 2013), biomarkers (Yang et al. 2018), and electrolyte balance (Nakata et al. 2017) which is shown in Figure 1.

The advancement has now made drug delivery possible by integrating wearable devices, which can be worn on the human skin. Several aspects need to be considered in developing an e-skin based device with high flexibility, stretchability, self-healing ability, conductivity, stability, and reproducibility. Also, key features such as designing integrated circuits, embedding wireless communication, low power consuming circuits are also very important to facilitate the development of these devices. In addition, developing a technique which could able to fabricate the wearable devices in a cost-effective approach in bulk quantitates is highly desirable.

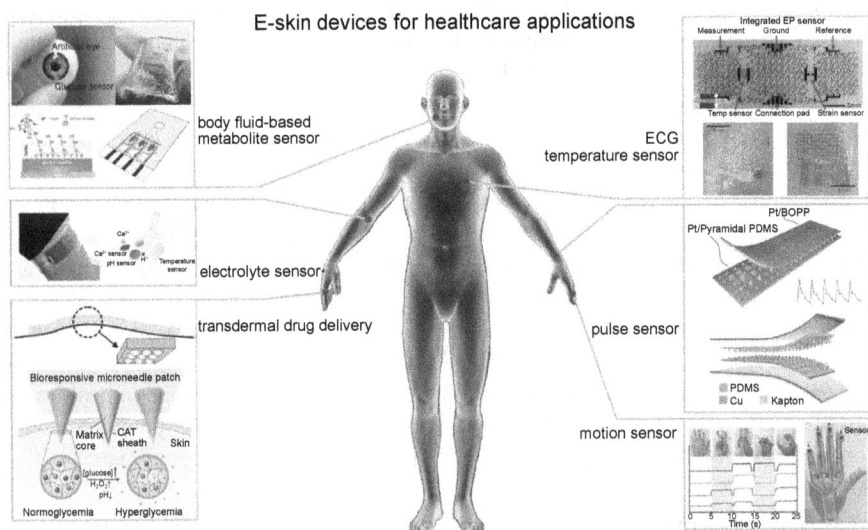

Fig. 1. Electronic skins for healthcare and sensor, showing the potential applications of e-skins and wearable electronic devices for various applications such as body fluid based sensors, drug delivery, electrolyte sensor, motion sensor, pulse sensor, temperature sensor, and ECG. Reproduced with permission from (Ma et al. 2019).

This chapter covers some of the fabrication approaches such as roll-to-roll technique (Yerushalmi et al. 2007), tattoo-based sensor development (Kabiri Ameri et al. 2017), and layer-by-layer printing processes (Fan et al. 2008).

Integration of power source in the device is important in designing a wearable energy harvester. The power source should deliver stable power to the device, which should also be helpful for reducing the overall size of the device (Purusothaman et al. 2018, Vivekananthan et al. 2019a, Sukumaran et al. 2020, Vivekananthan et al. 2020a). Here we are discussing about mechanical energy harvesting devices such as triboelectric nanogenerator, and piezoelectric nanogenerators (Vivekananthan et al. 2019b, 2020b), which generates electric power upon mechanical input from human body motions. By using the triboelectric and piezoelectric effect, the wearable devices can be made as an energy autonomous device. These energy autonomous devices can be further designed into a self-charging power cell by connecting to a storage device and made use as a single self-powered supercapacitor (Xue et al. 2014, Ramadoss et al. 2015). This can be used directly as a power source in the wearable sensor devices. The chapter also covers the integration of e-skin devices with external electronic circuits and be used for human machine interaction-based sensing. Some of the potential applications that are discussed in this chapter, such as physical (temperature, touch, and pressure), and chemical sensing (metal ions, glucose, sweat, and lactate) with high accuracy, pave the way towards commercializing in the near future.

2. Important Parameters in Designing an E-skin Device

There are several parameters which are considered to be important when designing an electronic skin-based device (Figure 2). The parameters such as stretchability, biocompatibility, biodegradability and self-healing capability should be considered deeply while we use it for biomedical and biosensor applications.

2.1 Stretchability

The stretchability of the device and the electronic materials are important, which increases the overall robustness of the device that protects from mechanical failure under various human body motions. To enhance the stretchable property of the device, strategies such as intermolecular interaction, strain engineering and hybrid materials to form composites films should be followed. Generally, stretchability is classified into two, that is intrinsic (Rogers et al. 2010) and extrinsic (Trung and Lee 2017) that depends on the materials that we have used for the fabrication of device.

Strain engineering is used for the materials that are rigid and brittle, which is used for making extrinsic type stretchability (Khang 2006). Under this strategy, different type of geometric structures were designed to release the strain applied (Kim et al. 2010). The widely used strain engineered structure for interconnection is the serpentine structure or wavy pattern. Figure 2a shows an example of serpentine structure by using gold interconnect, which has conformal contact with the human skin (Khang et al. 2006). This structure is highly helpful in connecting rigid island contacts with electronic components on elastomer-based film. This design allows biaxial strain on the device and eventually helps in designing high-performance

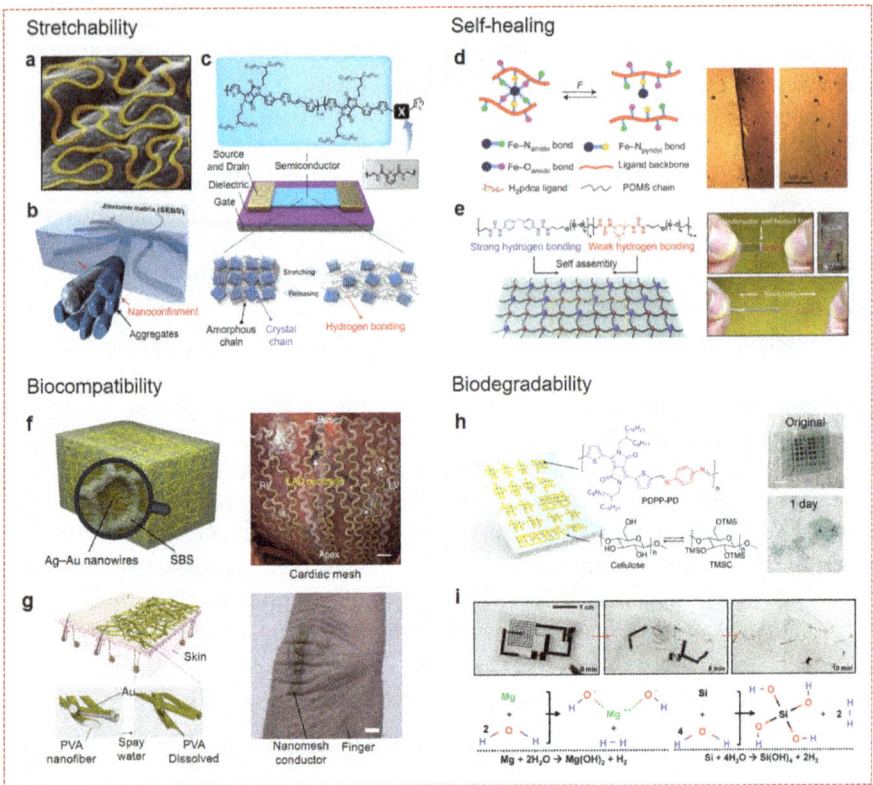

Fig. 2. Parameters in designing an e-skin device. Important parameters such as stretchability, self-healing ability, biocompatibility, and biodegradability are described in detail. (a) serpentine architecture showing strain engineering (b) nanoconfinement effect in elastomer using semiconducting polymer nanowire (c) dynamic bonding of molecular design (d) self-healing polymer using metal coordination (e) water-insensitive self-healing elastomeric polymers (f) biocompatible polymeric nanowire composite mesh (g) on skin nanomesh based skin electronics (h) degradable semiconducting polymer (i) bioresorbable electronic device based on silicone and magnesium. Reproduced with permission form (Oh and Bao 2019).

stretchable electronics. Recently, many reports suggested the use of percolation theory, by designing hybrid composites using elastomer with micro and nanomaterials. Nanostructured materials such as nanoparticles, nanowires, and nanoflakes are used as a filler material in the elastomers to form a stretchable conductor. The successful fabrication of this type of composite is still a challenging task and very few advances have been made so far in the area of stretchable semiconductors. The recent exploitation of nanoconfinement effect provided a breakthrough in the stretchable semiconductors by altering the mechanical properties of the polymer materials. This effect was investigated to form a 1D semiconducting polymer nanostructures in elastomeric matrix, which is shown in Figure 2b (Oh et al. 2016). The semiconducting film with nanoconfined polymer network can be stretched up to 100% strain while maintaining its μ to be > 1 cm^2 V^{-1} s^{-1}, a value comparable of amorphous silicon. The final strategy for better stretchability is dynamic intermolecular interaction, which provides intrinsic stretchability to the polymer. This type of intrinsic stretchable polymer can be obtained by incorporating functional molecular blocks providing the dynamic interaction between the molecules. This process can impart stretchability of the materials in insulating polymer and semiconducting polymer materials (Xu et al. 2017). Figure 1c shows the example of noncovalent dynamic crosslinking to improve the stretchability. This can be achieved by crosslinking weak hydrogen bonding between the polymers and organic moieties. Here the noncovalent dynamic crosslinking is achieved by attaching (2,6-pyridine dicarboxamide, PDCA) moieties to a 3,6-di(thiophen-2-yl)-2,5-dihydropyrrolo[3,4-c] pyrrole-1,4-dione conjugated polymer backbone. Incorporation of non-conjugated PDCA moieties at ≈ 10 mol% effectively reduced the elastic modulus of the polymer by almost ten times compared to the fully conjugated one. Furthermore, the resulting material became stretchable to above 100% strain (Xu et al. 2017).

2.2 Self-healing Ability

Self-healing ability of the materials helps in joining the ruptured layers of the e-skin device and also to get adhered on the other e-skin device or with human skin. Various strategies have been used so far to obtain a better self-healing ability (Tan et al. 2018). The self-healing ability of the materials are of two types such as intrinsic and extrinsic self-healing materials. Intrinsic self-healing systems are generally used due to their high repeatability and the simple formation procedures (Toohey et al. 2007, Cho et al. 2009), whereas extrinsic self-healing property of the materials incorporate capsules or vascular tubes filled with reactive chemical that releases under fractures at the damaged regions. This type of self-healing occurs only once in the same region upon rupture. Contrarily, the intrinsic type self-healing offers a strong intermolecular interaction such as electrostatic interaction, hydrogen bonding, π-π bonding, and metal coordination, which influences the self-healing and reversible properties (Cordier et al. 2008). For example, polydimethylsiloxane (PDMS) polymer chains were dynamically cross linked by a Fe (III)-2,6-PDCA coordination complex, in which the three dynamic bonding strengths in the metal coordination complex help to form intermolecular interactions, which is shown in Figure 2d (Herbst et al. 2013). This phenomenon helps in several ways to reform upon various types of ruptures during stretching. Similarly, Figure 2e shows a dual-hydrogen bonding system which has achieved the development of a self-healing elastomer with water resistant property (Li et al. 2016b).

2.3 Biocompatibility

To extend the applicability of e-skin devices as wearable or implantable, the device should be biocompatible in nature (Irimia-Vladu 2014). Various biocompatible and biodegradable materials have been developed so far especially towards biomedical applications, most of which are not conducting materials (Parker et al. 2009). To use it as an electronic device, the materials need to be conducting or semi conducting. Figure 2f shows a stretchable, highly conducting biocompatible material for implantable electronic device-based applications. The fabricated materials are composed of Ag-Au core-sheath nanowire and poly(styrene-butadiene-styrene) elastomer (Nawrocki et al. 2016). The material is highly conductive with the reported range of 41850 S cm^{-2} and the optimized stretchability reaches to 266% (Choi et al. 2018). The fabricated device is tested by implanting the device on human skin and swine heart, and successfully recording the electrophysiological signals (Miyamoto et al. 2017). Similarly, Figure 2g shows the Au nanomesh on human skin, in which Au was first thermally evaporated onto electro spun poly (vinyl alcohol) (PVA) nanofibers. The PVA nanomesh was then dissolved in water. The nanomesh was then found to get adhered on the skin easily and obtain a stretchability of 10% maintaining its electrical property. This type of devices provide an excellent opportunity to be used as an interface for measuring various physical quantities such as electromyogram, pressure, touch, and temperature.

2.4 Biodegradability

Biodegradability is an important aspect in the fast-growing electronic world, which helps to drastically reduce pollution due to electronic waste (Irimia-Vladu et al. 2012). It also helps during biomedical implants which can slowly degrade after it has served its purpose. The biodegradable materials mostly come from the organic and inorganic materials (Cheng and Vepachedu 2016). Materials such as silk fibers, cellulose, and collagen are naturally biodegradable polymers. Synthetic biodegradable polymers are also used in biomedical applications such as PVA, polylactic acid (PLA), polyethylene glycol, polyurethane, etc. But these materials are electrically insulating in nature, due to which most of the materials cannot be used directly, and can be used only on selected substrates (Yu et al. 2018). Figure 2h shows a biodegradable semi conducting polymer developed using reversible imine chemistry and the polymer was observed to disintegrate completely in 30 days. Similar to the organic materials, the inorganic materials are also used for degradation (Feig et al. 2018). The inorganic materials such as Si, SiO_2, Mg, MgO can be dissolved through water hydrolysis as shown in Figure 2h. The reaction kinetic shows that the degradation strongly depends on pH, and the transient time was controlled by the size of the encapsulation layer and the silicon membranes (Lei et al. 2017).

3. Materials for E-skin Devices for Physical and Chemical Sensing

Figure 3 shows the layer-by-layer configuration of the flexible electronics, which generally comprises of substrate, active layer and interface layer in-between them. The main function of the substrate is that it acts as a base to start the preparation of the device, while the materials or functional electrical circuits act as an active layer and interface has an important role of keeping the active layer on or inside the substrate (Fan et al. 2009). The designated targets are main role of the well-designed layers. It is hard to design the device on flexible substrate if the active layers consist of many electrical components or materials. The flexible device markets are highly dependent on advanced organic materials (Takei et al. 2014). The merits of the inorganic nanomaterials are physicochemical properties, high electron/hole mobility, chemical durability and mechanical strength. The standard lithography process is used to design flexible devices with certain active layers. The carbon nanotubes and

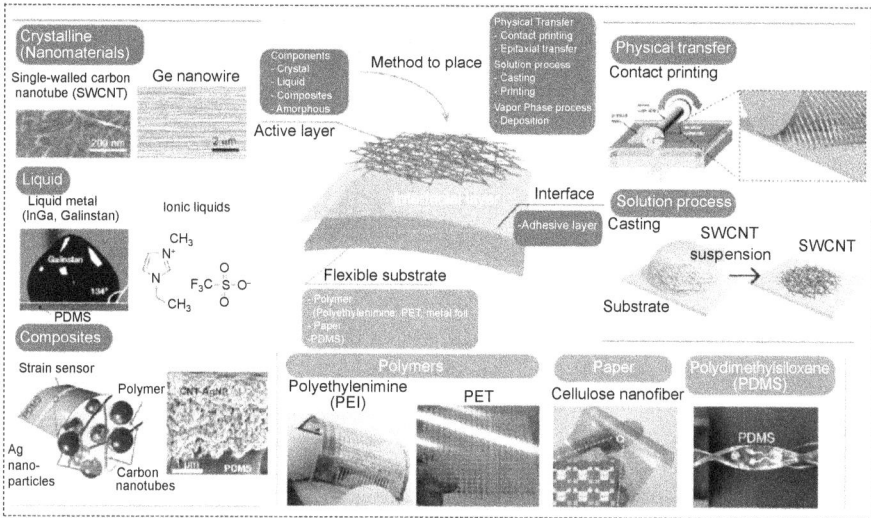

Fig. 3. Materials for e-skin devices for sensing. Device architecture and schematic showing the device made from the substrates, active layers, functionalization, and the adhesive layer keeping the active layers attached on the substrates. Reproduced with permission from (Gao et al. 2019).

graphene act as an active layer for the field effect transistors or sensors (Kiriya et al. 2014). The crystalline nanowire is highly semiconducting in nature leading to the high performance of FET (Ota et al. 2014).

Composite materials can act as active conductive layers on flexible substrates. To achieve optimized output or construction of active layer, the FET operation is highly influenced by interface layer between the active layer and substrate. The gate controllability is hard to obtain in the interface layer in case of the single walled carbon nanotube MOSFET. SWCNT are capable of control of electrostatic operation of MOSFET and design an interface layer. It has certain merits such as flexibility, stretchability, and conductivity (Fujisaki et al. 2014). The liquid metals can easily realize the wide range of applications such as wearables, robotics and prosthetics. Gallium metal alloys, particularly eutectic gallium indium (EGaIn) and Galinstan, are common liquid metals utilized for electrodes and sensors (Yeom et al. 2015). They can fabricate stretchable and flexible electronic skin. The self-healing property of the wire can be realized from good interconnects by liquid metal electrodes. It can withstand various extreme stretching, twisting, and bending (Wang et al. 2013). The choice for transferring process of the active layer onto the substrate is dependent on the synthesis for the materials of the active layers.

4. Silk Based E-skin for Temperature and Pressure Sensing

Since the several past years, large number of research groups have been interested in improving the performance of flexible sensors and monitoring several chemical as well as physical parameters like temperature, sweat, and heart pulse. Generally the sensors can detect only the major signal concerned to health issues, but unable to detect or monitor the full range health problem. Also, if the sensor is embedded to detect multiple signals, the sensitivity will be affected due to poor decoupling of signals. The information from the physiological data highly influences the flexible sensor. The sensing parameters obtained from the sweat such as glucose, calcium, potassium and metal ions are the main clinical data for the diagnosis. Various diseases can be selectively diagnosed by means of the multiple sensors integrated into one chip (Nakata et al. 2017). Keeping in view the many challenges, many research groups have taken step forward to combine at least two sensor types into a flexible device. Someya and co-workers pioneered in this context as they integrate the temperature and strain sensors (Someya et al. 2005). The main limitation of this work can be cross coupling so the individual signals could not be measured simultaneously. The stacking of double active layers as bimodal sensors generally helps to monitor the stimuli like pressure and temperature (Wang et al. 2017). Jeon et al. were able to decouple piezo resistive effect and thermoelectric effect by bimodal device configuration (Jung et al. 2017). The silk-nanofiber-derived carbon fiber membranes act as an active material in the e-skin that can detect temperature and pressure. The active materials exhibits a temperature sensitivity of 0.81% per centigrade and a high sensitivity gauge factor of ~8350 at a strain of 50%. The temperature sensor is not affected by any pressure signals and vice versa. Researchers have developed a multiple sensors system like temperature, light and pressure synchronously by means of a sandwich sensing system inspired from the skin layer of human being such as epidermis, dermis, and hypodermis, which is shown in Figure 4a and b. Zu et al. developed a dual nanoribbon as well as constant temperature feedback circuit, which is capable of measuring the artery temperature and pulse rate.

Much progress has been made in the area of integration of two or three sensors into one flexible device system and monitor the signals without cross coupling effect. The further integration of multiple sensors, signal processing, and power supply units bear thicker films (micrometers size) as supporting matrices due to ease of fabrication and handling. Skin irritation can be resolved by attaining a good adhesion between some flexible sensors and skin. Figure 4c and d shows developed healthcare patch by integrating a skin temperature sensor and a gel-less sticky ECG sensor (Gui et al. 2017). The thickness of film is responsible for accurately measuring the temperature of skin and the optimization of material concentration of CNT, PEIE, as well as PDMS that makes a gel-less ECG sensor (Yamamoto et al. 2017). Figure 4e and f shows that the integrated sensor patch is connected to a person's chest under various states of standing up, sitting down, standing up, and running. A 1°C difference is seen for the skin temperature between the flexible temperature sensor and commercial IR sensor. An optimization of the thickness of the film is still required to get a higher output. Figure 4g shows an ion-sensitive field-effect transistor (ISFET)-based pH sensor, which is incorporated with a printed temperature sensor, measuring the skin temperature and temperature effect compensation of the ISFET. The advantage of ISFET is attributed to its relatively simple direct current detection, while the conventional cyclic voltammetry (CV) method requires more measurements that are complex. The prediction and diagnosis of health problems can be monitored from chemical and physical parameters. When performing pH level and temperature recordings simultaneously based on the designed integrated flexible sensors, there

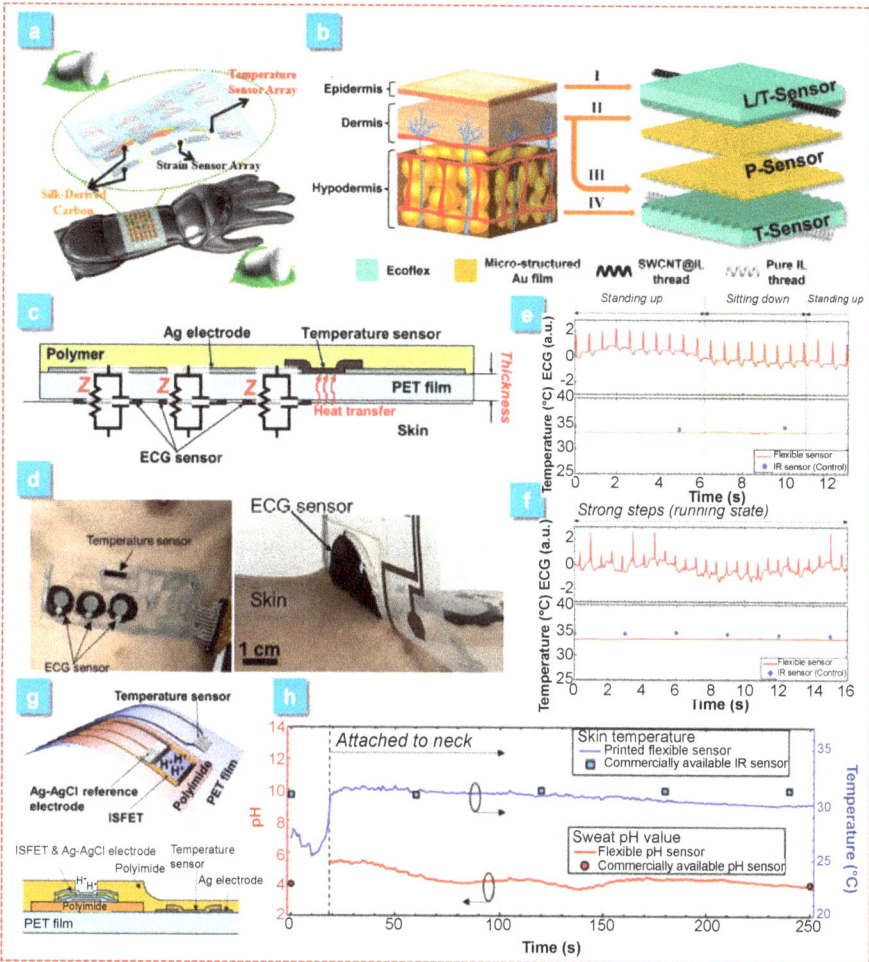

Fig. 4. Silk based e-skin for temperature and pressure sensing. (a) Schematic illustration of the e-skin device with sensor array arrangement (b) layer-by-layer schematic illustration from human skin to device layers (c) schematic image of the equivalent circuit of the ECG sensor and heat transfer through a PET film from skin to device (d) digital photo image of the fabricated device and device adhesion on the human skin (e) device output response upon standing up and sitting stages (f) strong moving and stepping stages (g) schematic of the device integrated with flexible pH and temperature sensors, with device cross-section (h) real-time analysis of the pH and temperature sensor results. Reproduced with permission from (Xu et al. 2019).

is a potential shift because of pH and temperature, leading to pH monitoring over a wide range of temperature. After attaching ISFET based pH sensor system on neck by a tape, it can record the sweat pH and skin temperature after exercising. The skin temperature is seen to be in between 30.0–31.7°C with the pH of 4.0, which matches with control experiments as depicted from Figure 6h. However, the ISFET is unable to detect urea or lactic acid, as they also vary with pH values. The complex fabrication routes as well as diverse materials' selection for various sensors are still changing to integrate many sensors unit into a flexible device/chip which may hinder the wearing comfort and reduce its sensing capabilities. Undoubtedly, the integrating plurality of sensors makes the e-skin comprising of intriguing functionalities (Xu et al. 2019).

5. Wearable Heart Rate Monitoring

In the present era, the flexible pressure sensors are capable of doing wide range of applications such as micro-robot, human-machine interface, wearable systems, health monitoring, etc. The strain of 200% can be easily sustained by liquid-metal based sensors without mechanical failure. However, this sensor is not capable of measuring heart rate monitoring as they cannot detect the minute changes in the pressure like kPa level. In this work, a microfluidic tactile pressure sensor has been studied. An equivalent circuit design was used like the Wheatstone bridge to replicate the diaphragm pressure sensor and fetch the tangential and radial strain fields, high sensitivity, linearity, resolution, and thermal stability. The pressure sensor is designed and fabricated based on embedded Galinstan microchannels with equal height and width of about 70 μm. The diaphragm pressure sensor is capable of fulfilling the potential applications. The health monitoring such as heart pulse rate investigation seems to be a feasible application. The authors designed a PDMS wristband with

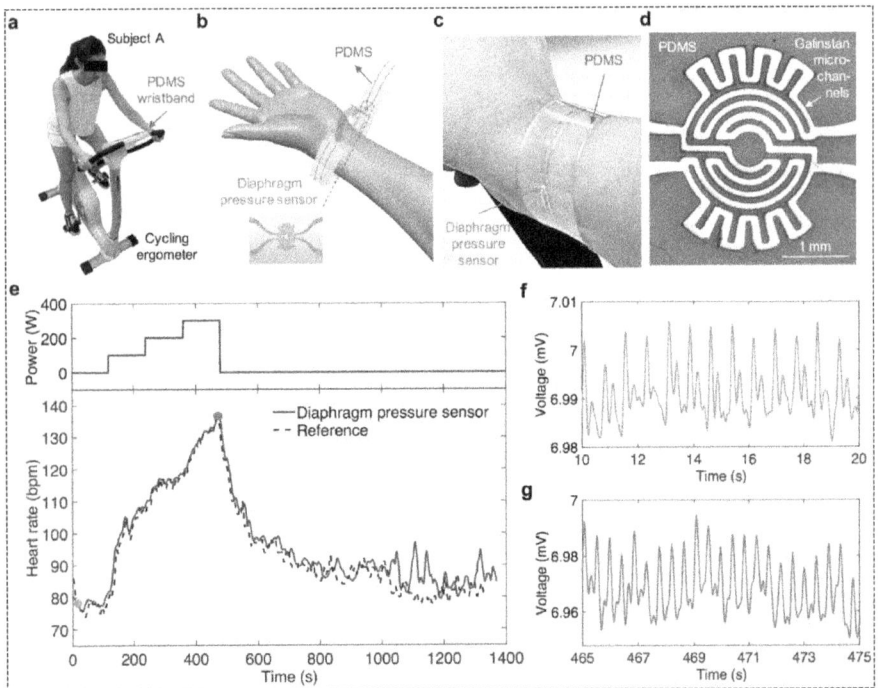

Fig. 5. Wearable heart-rate monitoring. (a) Digital image of the device worn on human hand (b) schematic of the PDMS device with sensor (c) digital photo of device worn on the human hand (d) optical image of the sensor device used on the PDMS wrist band (e) sensor output measured from the pulse rate compared with commercial device (f) and (g) real-time response of pulse rate and rest measuring data over 10s-time span. Reproduced with permission from (Gao et al. 2017).

embedded microfluidic diaphragm to support their concept. The fabricated pressure sensors traced the dynamic pulse. Figure 5a to c shows that the PDMS wrist band is attached to the subject's hand near the radial artery during cycling. The PDMS wrist band is quite flexible to attach with the hand. The pulse of the experimental subject was first traced for 2 mins before the cycling activity and further 2 mins of cycling r at 100, 200, and 300 W each.

The authors used a commercial TICKR heart rate monitor (Wahoo fitness) to act as a reference to the fabricated pressure sensor worn by person. Figure 5e showed the smoothed pulse result (in beat per min (bpm)) using time intervals between adjacent pulse peaks in comparison to the commercial sensor. A heart rate pulse of 75–80 bpm is recorded for subject before cycling and as the reproducibility of cycling increased, the heart rate pulse increased. It can be witnessed that the maximum heart pulse rate reaches to 137 bpm at 8 min rising with the intensity of cycling exercise, and the pulse rate comes back to normal with cycling stop. After a period of rest, the heart pulse rate decreased to initial starting stage. In Figure 5d, the diaphragm pressure sensor data looks equivalent to the reference commercial monitor. Figure 5f and g shows the heart pulse rate during rest and peak rate for over 10 seconds, respectively. It can be clearly seen that there is a rise in the heart beats strength that arises from the rapid exercises (Gao et al. 2017).

6. Tactile Sensing Glove for Pressure Sensing

The pressure sensors embedded in PDMS gloves were fabricated to test several experimental conditions. The aim is to realize the tactile pressure sensing having a wide pressure sensing range, high resolution and low detection limit. A smart glove was fabricated employing a 3D printed hand mold. As shown in Figure 6a and b, a sum of 17 diaphragm pressure sensors were embedded in the PDMS gloves.

The finger comprises of ten diaphragm pressure sensors and seven on the palm. The touching and holding objects were carried out to realize the comprehensive tactile sensing. The holding, gripping, grasping, squeezing, lifting, moving or touching objects are several types of hand motions done with the smart gloves that generated dynamic responses. Figure 6c and d shows the real time response from the index and thumb finger from the subject by gently dropping a grape. It can be clearly seen that the response enhances as the grape is grasped in the finger and the output went to the initial stage as the grape is released. Figure 6e shows the holding of bat by using the hand where the subject is wearing the PDMS gloves comprising of the multiple diaphragm pressure sensors. The gripping of the bat is depicted using a color mapping of relative change in the voltage ($\Delta V/V_0$) sent back from the corresponding sensors on the entire hand, showing the pressure distribution. The pressure sensitive glove can be actively used to collect the haptic feedback and find potential usage in human-machine interaction, virtual reality and extreme robotic applications. The extremely low detection limit and resolution, having a very fast response time of 90 ms, make the sensors capable of multiple applications (Gao et al. 2017).

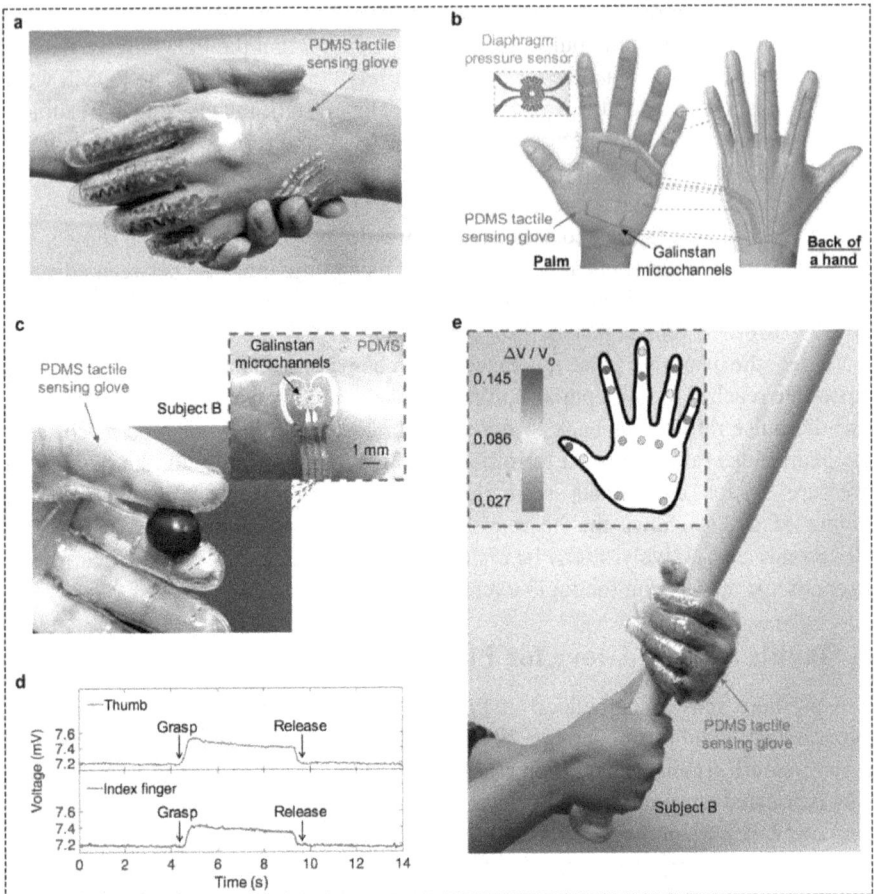

Fig. 6. Tactile sensing glove for pressure sensing. (a) digital photo of the tactile sensing glove worn on the hand showing a hand shaking gesture (b) schematic image of the tactile sensing glove (c) digital image of the tactile sensing glove holding grapes (d) real-time output response recorded while grasping and releasing the grapes (e) digital photographic image of the tactile sensing glove while holding a baseball bat and the corresponding simulated output response is given in the inset. Reproduced with permission from (Gao et al. 2017).

7. Wearable Sweat Sensor

Gao et al. reported *in-situ* perspiration analysis using wearable sensor array. Figure 7a shows the flexible integrated sensor array (FISA) used for the biomarker analysis in sweat. The flexible sensor was fabricated on plastic substrate and integrated circuit were arranged on flexible printed circuit board (FPCB) (Gao et al. 2016). They provide solutions for existing technological problems present in signal conditioning, signal transduction, signal processing, and wireless communication. The device consists of a sensor array which can detect various analytes. The sensor is capable of detecting glucose, lactate, sodium and potassium ions through electrochemical sensing approach. The developed FISA was capable of detecting

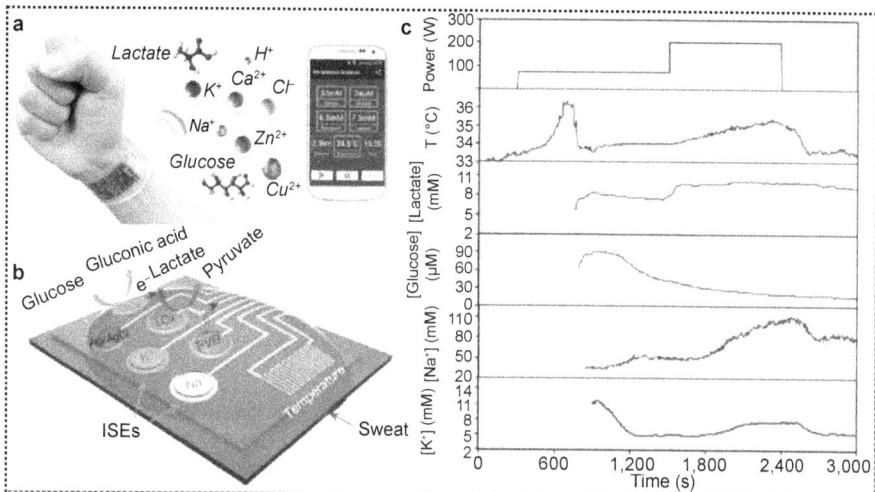

Fig. 7. Wearable sweat sensor. (a) integrated sweat sensor array worn on the human hand (b) schematic of the sweat sensor array with different metabolites and electrolytes (c) real-time sweat analysis data under exercise. Reproduced with permission from (Gao et al. 2017).

temperature as well as analysing different analytes in sweat. The FISA device was powered using a single rechargeable lithium-ion battery (Emaminejad et al. 2017).

Figure 7b schematically illustrates the sensor array with each electrode of 3 mm diameter. The enzyme lactate oxidase and glucose oxidase were immobilized for the amperometry sensing of glucose and lactate. The FISA works as autonomous sensor which doesn't need any external power. The ion-selective electrodes (ISE) were utilized for the sensing of potassium and sodium ion levels (Nyein et al. 2016). The Cr/Au microwires were used for the realization of temperature sensor. Figure 7c depicts the results of on-body real-time perspiration during cycling. The peak corresponds to rest, cycling at graded load and cool down condition. The sweat rate increases with high power leading to changes in glucose, lactate, sodium and potassium ion levels. Once the exercise stops, the response decreases and remains stable (Nyein et al. 2018).

8. Wearable Devices for Glucose Monitoring from Sweat

A fully integrated sweat analysis system was integrated as a wearable electronic device, through which glucose levels are identified from the sweat. The study involves diagnosis of the cystic fibrosis and blood/sweat glucose correlation (Heikenfeld 2016). The cystic fibrosis patients exhibit high electrolyte content in sweat. The extraction of sweat was carried out using iontophoresis interface. The inherent accessibility of sweat in sedentary patients in large quantities under *in situ* analysis has glucose levels. So, a wearable device with miniaturized iontophoresis interface is an excellent solution to overcome the above issue. But the challenge still remains in implementation of this process, which is that it can extract sufficient quantity of sweat for high accuracy sensing, without harming the patients and also to protect the device from corrosion. Emaminejad et al. had overcome this issue by introducing an electrochemically enhanced iontophoresis interface techniques integrated to wearable devices (Emaminejad et al. 2017). Figure 8a shows the autonomous sweat extraction and sensing system. The sensor can sense glucose, sodium and chloride ions in the sweat while maintaining high sensitivity. The sensing system comprises of electrode array (sweat induction and sensing) and wireless FPCB. The sweat rate profile with iontophoresis current of 1 mA is also shown in the Figure 8a. It is well known that the sweat composition and the sweat rate are inextricably linked. Considering this issue, sweat rate monitoring will be crucial for both fundamental and clinical investigations. To overcome this problem, Nyein et al. developed a microfluidic sweat sensor for sweat monitoring and analysis. This device is a sweat-sensing patch and can monitor the real time sweat rate monitoring and enhancing the sweat analysis (Nyein et al. 2018). The microfluidic sensor shown in Figure 8b consists of four layers: an insulation layer of Parylene C, sodium ion sensing electrodes, gold electrodes for sensing of sweat rate and spiral designed channel. The developed patch can be worn on the wrist and data can transfer wirelessly to the

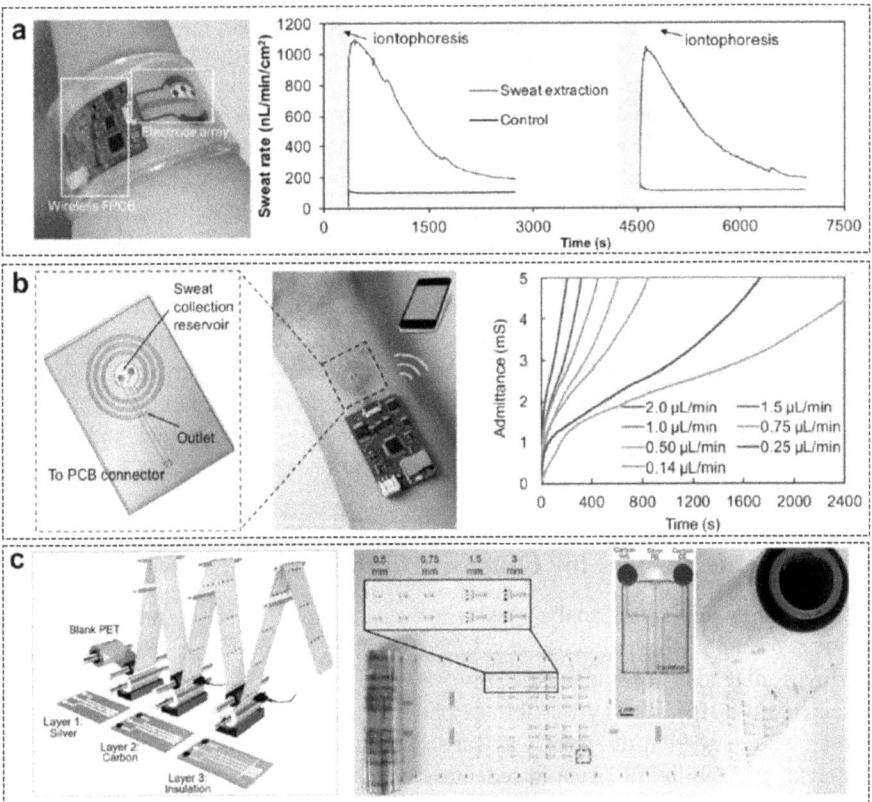

Fig. 8. Wearable devices for glucose monitoring from sweat. (a) wearable sensor array device with integrated iontophoresis module for sweat extraction (b) microfluidic based sensing system for sweat sampling (c) roll-to-roll gravure printing enabled mass production of sensors at low cost. Reproduced with permission from (Gao et al. 2017).

mobile phone. The results of admittance changes at different distance travelled by the fluid are also shown in Figure 8.

As the non-invasive techniques for biosensing are increasing, a cost effective, high throughput technique for fabricating sensing components is highly required. Moving towards achieve this goal, roll to roll (R2R) gravure printed electrodes that can be used for sensing applications were demonstrated by Bariya et al. shown in Figure 8c (Bariya et al. 2018). Figure shows R2R gravure printing on 150 m PET substrate. The figure clearly depicts the printed electrodes ranging from 0.5 mm to 3 mm. The technique involves ink rheology to combine printability requirements with mechanical and electrochemical robustness. The printed silver ink layer has a thickness at an average of 250 nm, the carbon layer is 1.3 μm thick, and the insulation layer is 1.5 μm thick under SEM analysis. The final printed device has a resistivity of 1.8×10^{-4} Ω.cm, comparable to the conductive silver inks used directly for gravure printing. The low-cost, size controllability and high throughput are the main attributes of the technique. The fabricated sensor from R2R technique is then secured conformally and a thin PDMS layer is used, surrounding the sensing electrodes, creating a well to accumulate the sweat. This also helps the sweat from evaporation and prevents the sensing layer from abrasion. The sweat sensing is then conducted by wearing it on arm and exercising for 51 minutes, including a 6 min warmup and 45 min cycling. The R2R printed electrodes can be used for sensing of analytes in sweat, sensing of ions and other analytes. The R2R technique will lead to the mass production of low-cost sensors (Bariya et al. 2018).

9. Energy Autonomous Electronic Skins and Wearable Devices

Energy autonomy is the key to next generation wearable and portable electronic systems and could be widely used for various applications. Among the various other systems, electronic skins were widely used due to their wide range of applications such as robotics, fashion, digital health and internet of things (IoT). The energy autonomous e-skin consists of multiple electronic components such as power source, actuators, electronic devices, and sensors (Núñez et al. 2017). The concept of energy autonomy comes from the limitations the batteries currently have, which is bulk in size, harmful to environment, disposal issues, over heating upon long time usage, and frequent recharging (Taube Navaraj et al. 2017). So, a promising power source is required which has to overcome the drawbacks present in the batteries. Incorporating energy sources with e-skin such as thermoelectric, photovoltaic, triboelectric, piezoelectric, and energy storage devices address the successful development of e-skin based devices. The e-skin device should work continuously for 24 h with high stability and reliability (Armand and Tarascon 2008). Low power conversion efficiency and discontinuous power supply are few of the drawbacks present in the energy harvesting devices. This can be overcome by reducing the size of the sensors and selection of highly efficient energy materials. Figure 9 shows the schematic diagram of energy autonomous electronic skin using the state-of-art techniques of energy harvesting and storage.

Few of the devices such as tactile skin powered by sunlight, pulse oximeter which was powered using thermoelectric generator, in which the thermal energy is

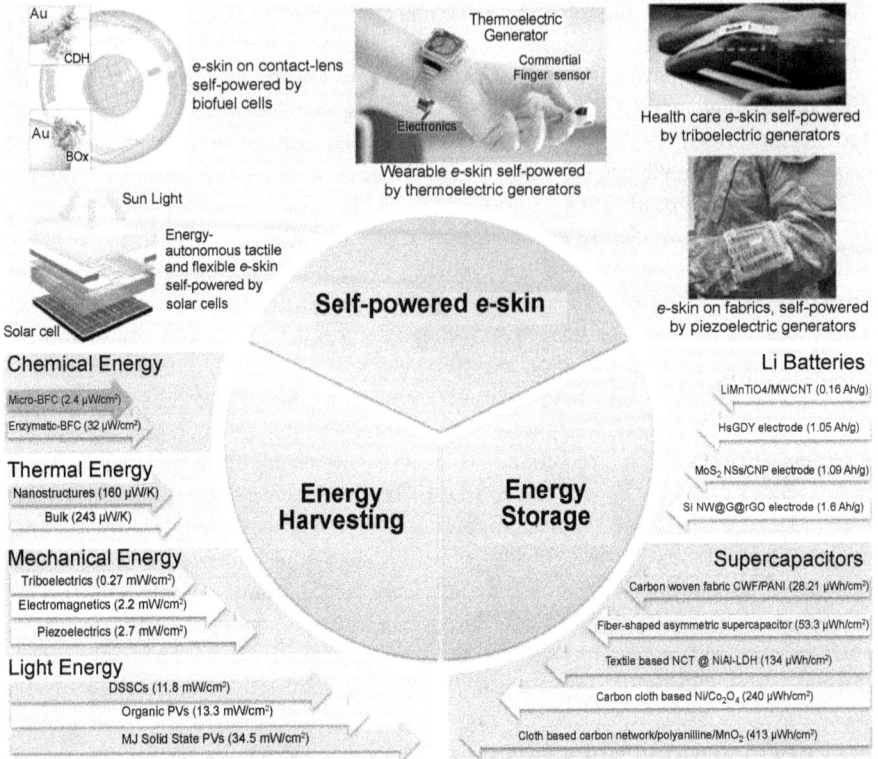

Fig. 9. Energy autonomous electronic skin and wearable devices. (a) schematic of the energy autonomous e-skin consisting of energy harvesting sources, energy storage device and examples of e-skin devices reported. Reproduced with permission from (García Núñez et al. 2019).

generated from the human body itself (Leonov and Vullers 2009). Similarly, health care electronic patches powered by the vibration energy from the human fingers (Yang et al. 2009), and multi-sensing energy autonomous electronic skin powered by human arm actions does not require any external power source (Yang et al. 2009). Also, Figure 9 shows some of the high-performance flexible energy storage and flexible harvesting devices from various energy sources such as chemical energy, thermal energy, mechanical, and light energy. The high-performance flexible energy storage devices such as Li batteries and supercapacitors (Li et al. 2016b) made using different materials are also depicted in Figure 9 (García Núñez et al. 2019).

10. Self-powered System for Wearable Health Monitoring

Energy system comprises of both energy harvesting and energy storage, which are two independent systems. Among all the existing energy harvesters, mechanical energy harvesting is an easy approach in which mechanical energy is ubiquitous. A compact mechanical energy harvester can be designed easily by utilizing the triboelectric effect and piezoelectric effects. The nanogenerators are easy to design, cost effective and have high performance under low frequency. Similarly, supercapacitors (SC) are considered to be one of the promising energy storage devices attributed to its high-power density, flexibility and simple fabrication process. By combining the nanogenerator and supercapacitor together as a single device, various self-powered systems or sensors can be designed. Therefore, many researchers are recently working on combining these two systems to perform various self-powered operations. Figure 10a depicts the energy harvesting TENG unit and energy storage

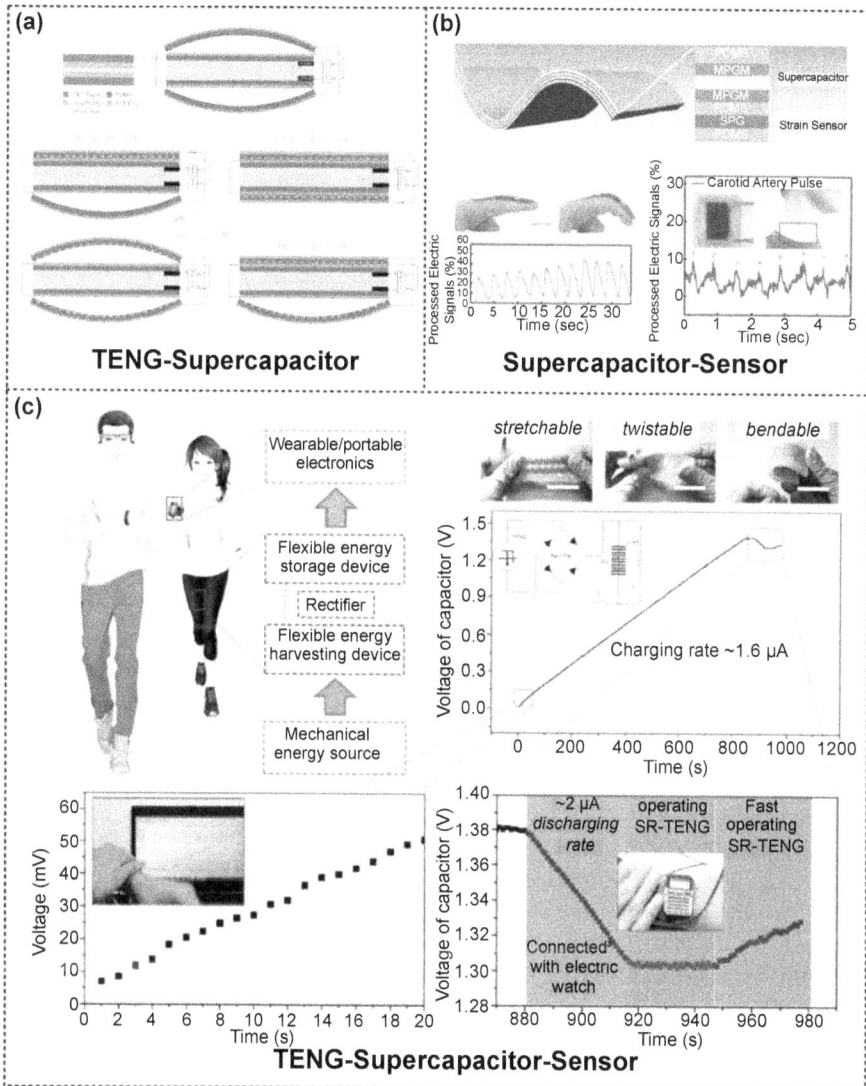

Fig. 10. Self-powered system for wearable health monitoring. (a) energy harvesting system with TENG and supercapacitor coupled for storing energy harvested from mechanical stimuli (b) energy utilization consisting of supercapacitor and sensor which can be worn and detect physical parameters (c) all in one integrated system comprising of harvesters, storage, and sensor for self-powered sensing. Reproduced with permission from (Chen et al. 2019).

unit (supercapacitor) for developing self-charging power unit (Song et al. 2016). The design consists of two TENG device and a supercapacitor in a TENG-SC-TENG architecture. The applied mechanical energy is directly converted into electrical energy and can be stored in the supercapacitor.

Another example is shown in Figure 10b, depicting an energy supply system reported by Fan et al. 2019 by the integration of the strain sensor and supercapacitor. The supercapacitor was binder free ultralight solid-state device. The strain sensor and supercapacitor are stacked together and attached on different body locations to detect the minor motions. The device is verified by attaching on the finger, and the device records the oscillating electrical signals synchronously with the folding and releasing of the finger (Li et al. 2017). Again, the device is attached near the carotid artery, periodical signals were collected and recorded a heartbeat of 96 min^{-1}. Wang and co-workers reported an all-in-one self-powered system shown in Figure 10c. The prototype involves energy harvesting, storage of the harvested energy and powering of wearable electronics. The stretchable TENG was fabricated by using silicone rubber and silver nanowires as electrode. An ultra-stretchable supercapacitor fulfils the purpose of energy storage. A kirigami paper-based supercapacitor was designed and it works as a flexible energy storage. The flexible device has a good stretchability up to 215%. For the energy harvesting device, an ultra-stretchable and shape-adaptive silicone rubber TENG was utilized, and the usage of silicone elastomer enables the device to be flexible. The combined flexible energy harvester and storage was integrated with a rectifier, and a highly flexible self-charging power system was achieved for powering up wearable electronic devices. Finally, a commercial watch was powered by converting the AC output of TENG via full wave rectifier to store the energy in supercapacitor (Guo et al. 2016).

11. Self-powered Tattoo Based Human Machine Interface (HMI) Towards Physiological Signals and Control

Human machine interface (HMI) with wearable capabilities has grown these days in the era of IoTs, and has gained a huge attention from both industries and academia. With the HMI, there are a lot of commercial products such as smart watches and smart glasses, which are gaining attention. But they have certain drawbacks such as their big size and rigidity, leading to inaccurate data collection and wearing discomfort. To overcome this problem, soft wearable devices are developed by mimicking the human skin and incorporating all its functionalities such as sensing to touch, temperature, pressure, sweat and moisture. Concerning this, An et al. (2020) fabricated a gold nanowire-based tattoo working with triboelectric effect for wearable HMI type device. The device is entirely self-powered and is able to operate without the usage of batteries. The device can also act as a tactile sensor which can sense physical quantities and can also control lights and vehicles wirelessly.

Figure 11a shows the schematic illustration of triboelectric pressure sensor made of Ecoflex/v-AuNWs. The v-AuNWs are sandwiched between two thin Ecoflex layers made by spin coating process. These nanowires are 200–300 nm in length and a diameter of 10 nm, resulting in a conductive percolation network. The device could sustain extreme stretching up to 500% without any rupture. Figure 11b shows the

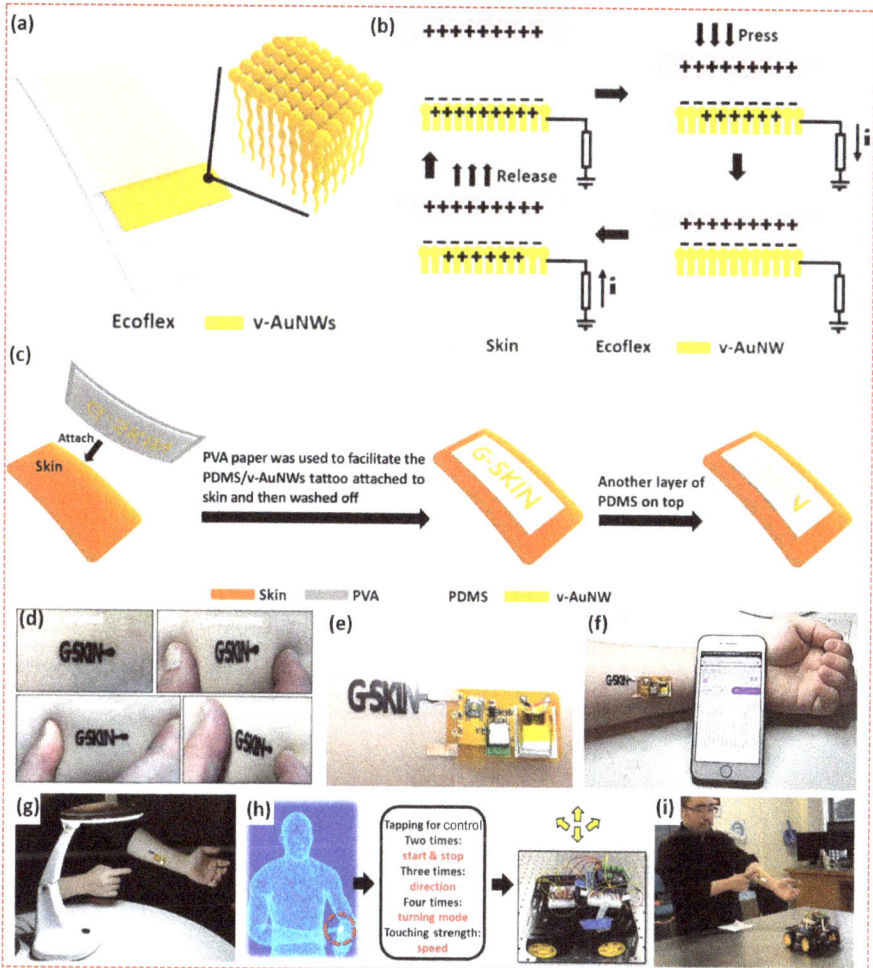

Fig. 11. Self-powered tattoo based human machine interface. (a) schematic illustration of v-AuNWs/ Ecoflex-based self-powered triboelectric pressure sensor (b) single electrode mode sensing mechanism (c) step-by-step schematic illustration showing the preparation steps on triboelectric tattoo (d) tattoo device on skin actuated under different directions (e) tattoo device integrated with a PCB module (f) tattoo device generates electrical signals and responses are received in smart phone through bluetooth communication (g) device acting as a human machine interface and used as a wireless switch for controlling light (h) and (i) wireless vehicle controlling using the HMI tattoo TENG and its practical photograph. Reproduced with permission from (An et al. 2020).

working mechanism of Ecoflex/v-AuNWs based triboelectric sensor with Ecoflex as negative and human skin as positive triboelectric layers. When the materials get in contact with the device, charges accumulate on the layers and electrons flow between the skin and elastomer layer, leading to generation of electrical signals. When the skin separates from the device, the electrons flow in the reverse direction creating opposite signal. Figure 11c shows the step-by-step fabrication process of tattoo based triboelectric sensor. v-AuNWs were first grown on PMMA substrates followed by spin coating of thin PDMS layer on the top of the NWs. After curing the PDMS layer with v-AuNWs (nanowires) are peeled from the PMMA substrate and transferred to PVA paper. The PVA paper can easily be used as a tattoo and comfortably able to transfer on the skin.

Figure 11d shows the tattoo on the human skin and it gets adapted to various deformations made on the skin without getting ruptured or peeled off. The prepared composite has a conductivity of 1000 S cm^{-1} at natural state and decreases when applied to strain. For processing the signals generated from the tattoo sensor and interfacing as HMI for practical application, a PCB board with wireless signal transmission module is connected as shown in Figure 11e. Upon touching the triboelectric tattoo sensor, the signals are generated and wirelessly monitored in smartphone using Bluetooth technique. Figure 11f shows the wireless lighting control using tattoo tribo-sensor with Bluetooth and microcontroller circuit and also can be used to control vehicle motion as shown in Figure 11h. Figure 11g shows the orders of movements with respect to touch. Tapping two times for start and stop moving, tapping three times for changing the directions such as forward and reverse, tapping four times for turning in different directions (An et al. 2020).

Further, a tactile sensing skin is developed by designing a cross-bar structure in a bi-layered layout. On the top of the pattern designed using v-AuNWs, a PDMS layer was added on the top as shown in Figure 12a. Those layers are then stacked

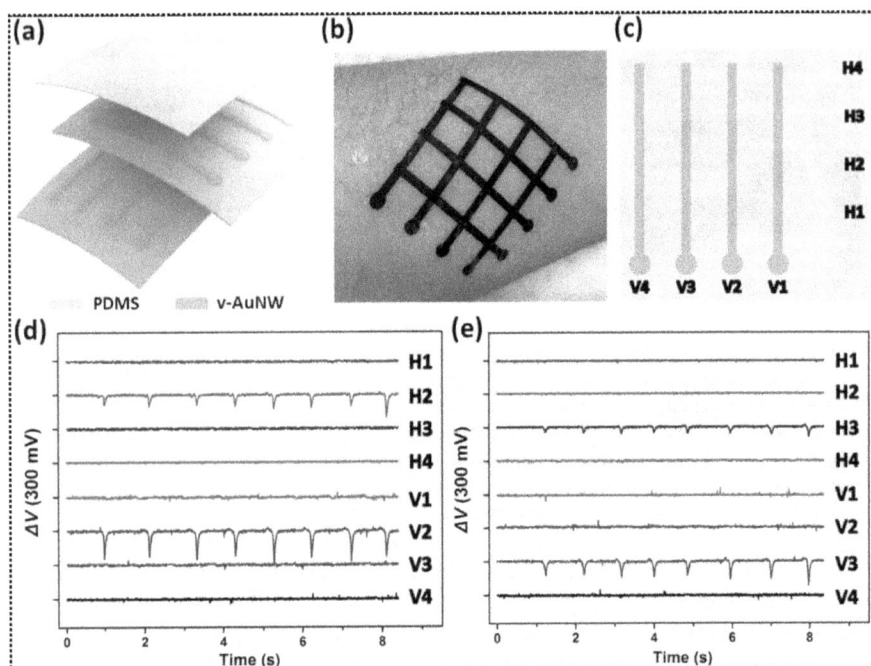

Fig. 12. Skin comfortable tactile sensor. (a) schematic illustration and structure of the tactile sensor device (b) digital photo of the developed tactile sensor attached on the human skin (c) 4 × 4 cross-bar structure of the tactile sensor with sixteen sensing points (d) and (e) output voltage and signal from each channel. Reproduced with permission from (An et al. 2020).

together firmly attached to the human skin in 4 × 4 pixels, with the PDMS layer at top that acts as triboelectric layer which is shown in Figure 12b.

The 4 × 4 structured tactile sensor provides 8 sensing points termed as V1 to V4 and H1 to H4 as shown in Figure 12c. Each point is connected to the oscilloscope via a signal conditioning unit for testing the fabricated device. A proof-of-concept demonstration of tactile sensing, the points V3H3 and V2H2 were tapped and the corresponding signals are shown in Figure 12d and e. Here it can be seen clearly that the points which are actuated show the signals and the remaining points remain silent with low/less interference. The design and the obtained results show that the v-AuNWs based tattoo triboelectric sensor paves way for the soft HMI technology for better interactions with human body and has potential in biomedical and sensing applications (An et al. 2020).

12. Conclusion

In summary, this chapter summarized the recent advances and futuristic approach of e-skin for sensing applications including device design, various physical sensors, tactile sensing, chemical sensors such as wearable glucose sensor, sweat sensor, metal ion sensor, and lactate sensors. In addition, the topic covers the energy autonomous e-skin for self-powered sensors and its advantages. It is clear from the above topics and the discussions that e-skin based devices are highly beneficial in the field of biomedical electronics and biosensors. Also, the chapter covers the major important parameters for a good e-skin device such as stretchability, self-healing ability, biodegradability and biocompatibility. The futuristic advancements in the e-skin device would be composed of nano sensors and nanoelectronics circuits on the soft materials with high conductivity and stretchability with light weight in nature. Along with medical and sensing applications, the e-skin devices can also be used for sensory system in robotics and prosthetics. The chapter also covers the integration of different energy technologies such as energy harvesting and storage in a single device to work as an energy autonomous e-skin to work towards self-powered sensing. The futuristic skin would overcome the drawbacks such as thermal expansion co-efficient, and effect of environments (temperature, humidity, and chemical stability) by using materials with multifunctionalities. So far, the fabrication problems had been rectified with various fabrication techniques such as layer-by-layer printing, flip chip bonding, and roll-to-roll printing. However, still there is a long way to go for the e-skin devices towards commercializing and use for real-time monitoring. But there is a great future and bright potential for e-skin devices for advanced healthcare and sensing systems.

Acknowledgements

This work was supported by the Basic Science Research Program through the National Research Foundation of Korea (NRF) grant funded by the Korea government (MSIT) (2019R1A2C3009747, 2021R1A4A2000934). The authors thank Mr. Sugato Hajra for his help in drafting the chapter.

References

An, T., D. V. Anaya, S. Gong, L. W. Yap, F. Lin, R. Wang, M. R. Yuce and W. Cheng. (2020). Self-powered gold nanowire tattoo triboelectric sensors for soft wearable human-machine interface. *Nano Energy* 77: 105295.

Armand, M. and J. M. Tarascon. (2008). Building better batteries. *Nature* 451(7179): 652–657.

Bariya, M., Z. Shahpar, H. Park, J. Sun, Y. Jung, W. Gao, H. Y. Y. Nyein, T. S. Liaw, L.-C. Tai, Q. P. Ngo, M. Chao, Y. Zhao, M. Hettick, G. Cho and A. Javey. (2018). Roll-to-roll gravure printed electrochemical sensors for wearable and medical devices. *ACS Nano* 12(7): 6978–6987.

Chen, H., Y. Song, X. Cheng and H. Zhang. (2019). Self-powered electronic skin based on the triboelectric generator. *Nano Energy* 56: 252–268.

Cheng, H. and V. Vepachedu. (2016). Recent development of transient electronics. *Theoretical and Applied Mechanics Letters* 6(1): 21–31.

Cho, S. H., S. R. White and P. V. Braun. (2009). Self-healing polymer coatings. *Advanced Materials* 21(6): 645–649.

Choi, S., S. I. Han, D. Jung, H. J. Hwang, C. Lim, S. Bae, O. K. Park, C. M. Tschabrunn, M. Lee, S. Y. Bae, J. W. Yu, J. H. Ryu, S.-W. Lee, K. Park, P. M. Kang, W. B. Lee, R. Nezafat, T. Hyeon and D.-H. Kim. (2018). Highly conductive, stretchable and biocompatible Ag–Au core–sheath nanowire composite for wearable and implantable bioelectronics. *Nature Nanotechnology* 13(11): 1048–1056.

Chortos, A., J. Liu and Z. Bao. (2016). Pursuing prosthetic electronic skin. *Nature Materials* 15(9): 937–950.

Cordier, P., F. Tournilhac, C. Soulié-Ziakovic and L. Leibler. (2008). Self-healing and thermoreversible rubber from supramolecular assembly. *Nature* 451(7181): 977–980.

Emaminejad, S., W. Gao, E. Wu, Z. A. Davies, H. Yin Yin Nyein, S. Challa, S. P. Ryan, H. M. Fahad, K. Chen, Z. Shahpar, S. Talebi, C. Milla, A. Javey and R. W. Davis. (2017). Autonomous sweat extraction and analysis applied to cystic fibrosis and glucose monitoring using a fully integrated wearable platform. *Proceedings of the National Academy of Sciences* 114(18): 4625–4630.

Fan, Z., J. C. Ho, Z. A. Jacobson, R. Yerushalmi, R. L. Alley, H. Razavi and A. Javey. (2008). Wafer-scale assembly of highly ordered semiconductor nanowire arrays by contact printing. *Nano Letters* 8(1): 20–25.

Fan, Z., J. C. Ho, T. Takahashi, R. Yerushalmi, K. Takei, A. C. Ford, Y.-L. Chueh and A. Javey. (2009). Toward the development of printable nanowire electronics and sensors. *Advanced Materials* 21(37): 3730–3743.

Feig, V. R., H. Tran and Z. Bao. (2018). Biodegradable polymeric materials in degradable electronic devices. *ACS Central Science* 4(3): 337–348.

Fujisaki, Y., H. Koga, Y. Nakajima, M. Nakata, H. Tsuji, T. Yamamoto, T. Kurita, M. Nogi and N. Shimidzu. (2014). Transparent nanopaper-based flexible organic thin-film transistor array. *Advanced Functional Materials* 24(12): 1657–1663.

Gao, W., S. Emaminejad, H. Y. Y. Nyein, S. Challa, K. Chen, A. Peck, H. M. Fahad, H. Ota, H. Shiraki, D. Kiriya, D.-H. Lien, G. A. Brooks, R. W. Davis and A. Javey. (2016). Fully integrated wearable sensor arrays for multiplexed in situ perspiration analysis. *Nature* 529(7587): 509–514.

Gao, W., H. Ota, D. Kiriya, K. Takei and A. Javey. (2019). Flexible electronics toward wearable sensing. *Accounts of Chemical Research* 52(3): 523–533.

Gao, Y., H. Ota, E. W. Schaler, K. Chen, A. Zhao, W. Gao, H. M. Fahad, Y. Leng, A. Zheng, F. Xiong, C. Zhang, L.-C. Tai, P. Zhao, R. S. Fearing and A. Javey. (2017). Wearable microfluidic diaphragm pressure sensor for health and tactile touch monitoring. *Advanced Materials* 29(39): 1701985.

García Núñez, C., L. Manjakkal and R. Dahiya. (2019). Energy autonomous electronic skin. *npj Flexible Electronics* 3(1): 1.

Gui, Q., Y. He, N. Gao, X. Tao and Y. Wang. (2017). A skin-inspired integrated sensor for synchronous monitoring of multiparameter signals. *Advanced Functional Materials* 27(36): 1702050.

Guo, H., M.-H. Yeh, Y.-C. Lai, Y. Zi, C. Wu, Z. Wen, C. Hu and Z. L. Wang. (2016). All-in-one shape-adaptive self-charging power package for wearable electronics. *ACS Nano* 10(11): 10580–10588.

Hammock, M. L., A. Chortos, B. C.-K. Tee, J. B.-H. Tok and Z. Bao. (2013). 25th anniversary article: the evolution of electronic skin (e-skin): a brief history, design considerations, and recent progress. *Advanced Materials* 25(42): 5997–6038.

Heikenfeld, J. (2016). Non-invasive analyte access and sensing through eccrine sweat: challenges and outlook circa 2016. *Electroanalysis* 28(6): 1242–1249.

Herbst, F., D. Döhler, P. Michael and W. H. Binder. (2013). Self-healing polymers via supramolecular forces. *Macromolecular Rapid Communications* 34(3): 203–220.

Irimia-Vladu, M., E. D. Głowacki, G. Voss, S. Bauer and N. S. Sariciftci. (2012). Green and biodegradable electronics. *Materials Today* 15(7): 340–346.

Irimia-Vladu, M. (2014). "Green" electronics: biodegradable and biocompatible materials and devices for sustainable future. *Chemical Society Reviews* 43(2): 588–610.

Jung, M., K. Kim, B. Kim, H. Cheong, K. Shin, O.-S. Kwon, J.-J. Park and S. Jeon. (2017). Paper-based bimodal sensor for electronic skin applications. *ACS Applied Materials & Interfaces* 9(32): 26974–26982.

Kabiri Ameri, S., R. Ho, H. Jang, L. Tao, Y. Wang, L. Wang, D. M. Schnyer, D. Akinwande and N. Lu. (2017). Graphene electronic tattoo sensors. *ACS Nano* 11(8): 7634–7641.

Khang, D.-Y., H. Jiang, Y. Huang and J. A. Rogers. (2006). A stretchable form of single-crystal silicon for high-performance electronics on rubber substrates. *Science* 311(5758): 208–212.

Kim, D.-H., J. Xiao, J. Song, Y. Huang and J. A. Rogers. (2010). Stretchable, curvilinear electronics based on inorganic materials. *Advanced Materials* 22(19): 2108–2124.

Kiriya, D., K. Chen, H. Ota, Y. Lin, P. Zhao, Z. Yu, T.-j. Ha and A. Javey. (2014). Design of surfactant–substrate interactions for roll-to-roll assembly of carbon nanotubes for thin-film transistors. *Journal of the American Chemical Society* 136(31): 11188–11194.

Lei, T., M. Guan, J. Liu, H.-C. Lin, R. Pfattner, L. Shaw, A. F. McGuire, T.-C. Huang, L. Shao, K.-T. Cheng, J. B.-H. Tok and Z. Bao. (2017). Biocompatible and totally disintegrable semiconducting polymer for ultrathin and ultralightweight transient electronics. *Proceedings of the National Academy of Sciences* 114(20): 5107–5112.

Leonov, V. and R. J. M. Vullers. (2009). Wearable electronics self-powered by using human body heat: The state of the art and the perspective. *Journal of Renewable and Sustainable Energy* 1(6): 062701.

Li, C.-H., C. Wang, C. Keplinger, J.-L. Zuo, L. Jin, Y. Sun, P. Zheng, Y. Cao, F. Lissel, C. Linder, X.-Z. You and Z. Bao. (2016). A highly stretchable autonomous self-healing elastomer. *Nature Chemistry* 8(6): 618–624.

Li, C., M. M. Islam, J. Moore, J. Sleppy, C. Morrison, K. Konstantinov, S. X. Dou, C. Renduchintala and J. Thomas. (2016). Wearable energy-smart ribbons for synchronous energy harvest and storage. *Nature Communications* 7(1): 13319.

Li, L., L. Pan, Z. Ma, K. Yan, W. Cheng, Y. Shi and G. Yu. (2018). All inkjet-printed amperometric multiplexed biosensors based on nanostructured conductive hydrogel electrodes. *Nano Letters* 18(6): 3322–3327.

Li, W., X. Xu, C. Liu, M. C. Tekell, J. Ning, J. Guo, J. Zhang and D. Fan. (2017). Ultralight and binder-free all-solid-state flexible supercapacitors for powering wearable strain sensors. *Advanced Functional Materials* 27(39): 1702738.

Luo, N., W. Dai, C. Li, Z. Zhou, L. Lu, C. C. Y. Poon, S.-C. Chen, Y. Zhang and N. Zhao. (2016). Flexible piezoresistive sensor patch enabling ultralow power cuffless blood pressure measurement. *Advanced Functional Materials* 26(8): 1178–1187.

Luo, N., Y. Huang, J. Liu, S.-C. Chen, C. P. Wong and N. Zhao. (2017). Hollow-structured graphene–silicone-composite-based piezoresistive sensors: decoupled property tuning and bending reliability. *Advanced Materials* 29(40): 1702675.

Ma, Z., S. Li, H. Wang, W. Cheng, Y. Li, L. Pan and Y. Shi. (2019). Advanced electronic skin devices for healthcare applications. *Journal of Materials Chemistry B* 7(2): 173–197.

Miyamoto, A., S. Lee, N. F. Cooray, S. Lee, M. Mori, N. Matsuhisa, H. Jin, L. Yoda, T. Yokota, A. Itoh, M. Sekino, H. Kawasaki, T. Ebihara, M. Amagai and T. Someya. (2017). Inflammation-free, gas-permeable, lightweight, stretchable on-skin electronics with nanomeshes. *Nature Nanotechnology* 12(9): 907–913.

Nakata, S., T. Arie, S. Akita and K. Takei. (2017). Wearable, flexible, and multifunctional healthcare device with an ISFET chemical sensor for simultaneous sweat pH and skin temperature monitoring. *ACS Sensors* 2(3): 443–448.

Nawrocki, R. A., N. Matsuhisa, T. Yokota and T. Someya. (2016). 300-nm imperceptible, ultraflexible, and biocompatible e-skin fit with tactile sensors and organic transistors. *Advanced Electronic Materials* 2(4): 1500452.

Núñez, C. G., W. T. Navaraj, E. O. Polat and R. Dahiya. (2017). Energy-autonomous, flexible, and transparent tactile skin. *Advanced Functional Materials* 27(18): 1606287.

Nyein, H. Y. Y., W. Gao, Z. Shahpar, S. Emaminejad, S. Challa, K. Chen, H. M. Fahad, L.-C. Tai, H. Ota, R. W. Davis and A. Javey. (2016). A wearable electrochemical platform for noninvasive simultaneous monitoring of Ca^{2+} and pH. *ACS Nano* 10(7): 7216–7224.

Nyein, H. Y. Y., L.-C. Tai, Q. P. Ngo, M. Chao, G. B. Zhang, W. Gao, M. Bariya, J. Bullock, H. Kim, H. M. Fahad and A. Javey. (2018). A wearable microfluidic sensing patch for dynamic sweat secretion analysis. *ACS Sensors* 3(5): 944–952.

Oh, J. Y., S. Rondeau-Gagné, Y. C. Chiu, A. Chortos, F. Lissel, G. N. Wang, B. C. Schroeder, T. Kurosawa, J. Lopez, T. Katsumata, J. Xu, C. Zhu, X. Gu, W. G. Bae, Y. Kim, L. Jin, J. W. Chung, J. B. Tok and Z. Bao. (2016). Intrinsically stretchable and healable semiconducting polymer for organic transistors. *Nature* 539(7629): 411–415.

Oh, J. Y. and Z. Bao. (2019). Second skin enabled by advanced electronics. *Advanced Science* 6(11): 1900186.

Ota, H., K. Chen, Y. Lin, D. Kiriya, H. Shiraki, Z. Yu, T.-J. Ha and A. Javey. (2014). Highly deformable liquid-state heterojunction sensors. *Nature Communications* 5(1): 5032.

Parker, S. T., P. Domachuk, J. Amsden, J. Bressner, J. A. Lewis, D. L. Kaplan and F. G. Omenetto. (2009). Biocompatible silk printed optical waveguides. *Advanced Materials* 21(23): 2411–2415.

Purusothaman, Y., N. R. Alluri, A. Chandrasekhar, V. Vivekananthan and S.-J. Kim. (2018). Direct *in situ* hybridized interfacial quantification to stimulate highly flexile self-powered photodetector. *The Journal of Physical Chemistry C* 122(23): 12177–12184.

Ramadoss, A., B. Saravanakumar, S. W. Lee, Y.-S. Kim, S. J. Kim and Z. L. Wang. (2015). Piezoelectric-driven self-charging supercapacitor power cell. *ACS Nano* 9(4): 4337–4345.

Rogers, J. A., T. Someya and Y. Huang. (2010). Materials and mechanics for stretchable electronics. *Science* 327(5973): 1603–1607.

Someya, T., Y. Kato, T. Sekitani, S. Iba, Y. Noguchi, Y. Murase, H. Kawaguchi and T. Sakurai. (2005). Conformable, flexible, large-area networks of pressure and thermal sensors with organic transistor active matrixes. *Proceedings of the National Academy of Sciences of the United States of America* 102(35): 12321–12325.

Song, Y., X. Cheng, H. Chen, J. Huang, X. Chen, M. Han, Z. Su, B. Meng, Z. Song and H. Zhang. (2016). Integrated self-charging power unit with flexible supercapacitor and triboelectric nanogenerator. *Journal of Materials Chemistry A* 4(37): 14298–14306.

Sukumaran, C., V. Vivekananthan, V. Mohan, Z. C. Alex, A. Chandrasekhar and S.-J. Kim. (2020). Triboelectric nanogenerators from reused plastic: An approach for vehicle security alarming and tire motion monitoring in rover. *Applied Materials Today* 19: 100625.

Takei, K., Z. Yu, M. Zheng, H. Ota, T. Takahashi and A. Javey. (2014). Highly sensitive electronic whiskers based on patterned carbon nanotube and silver nanoparticle composite films. *Proceedings of the National Academy of Sciences* 111(5): 1703–1707.

Tan, Y. J., J. Wu, H. Li and B. C. K. Tee. (2018). Self-healing electronic materials for a smart and sustainable future. *ACS Applied Materials & Interfaces* 10(18): 15331–15345.

Taube Navaraj, W., C. García Núñez, D. Shakthivel, V. Vinciguerra, F. Labeau, D. H. Gregory and R. Dahiya. (2017). Nanowire FET based neural element for robotic tactile sensing skin. *Frontiers in Neuroscience* 11(501).

Toohey, K. S., N. R. Sottos, J. A. Lewis, J. S. Moore and S. R. White. (2007). Self-healing materials with microvascular networks. *Nature Materials* 6(8): 581–585.

Trung, T. Q. and N.-E. Lee. (2016). Flexible and stretchable physical sensor integrated platforms for wearable human-activity monitoring and personal healthcare. *Advanced Materials* 28(22): 4338–4372.

Trung, T. Q. and N.-E. Lee. (2017). Recent progress on stretchable electronic devices with intrinsically stretchable components. *Advanced Materials* 29(3): 1603167.

Vivekananthan, V., N. R. Alluri, A. Chandrasekhar, Y. Purusothaman, A. Gupta and S.-J. Kim. (2019). Zero-power consuming intruder identification system by enhanced piezoelectricity of K0.5Na0.5NbO3 using substitutional doping of BTO NPs. *Journal of Materials Chemistry C* 7(25): 7563–7571.

Vivekananthan, V., A. Chandrasekhar, N. R. Alluri, Y. Purusothaman, G. Khandelwal, R. Pandey and S.-J. Kim. (2019). Fe$_2$O$_3$ magnetic particles derived triboelectric-electromagnetic hybrid generator for zero-power consuming seismic detection. *Nano Energy* 64: 103926.

Vivekananthan, V., A. Chandrasekhar, N. R. Alluri, Y. Purusothaman and S.-J. Kim. (2020). A highly reliable, impervious and sustainable triboelectric nanogenerator as a zero-power consuming active pressure sensor. *Nanoscale Advances* 2(2): 746–754.

Vivekananthan, V., N. P. Maria Joseph Raj, N. R. Alluri, Y. Purusothaman, A. Chandrasekhar and S.-J. Kim. (2020). Substantial improvement on electrical energy harvesting by chemically modified/sandpaper-based surface modification in micro-scale for hybrid nanogenerators. *Applied Surface Science* 514: 145904.

Wang, C., D. Hwang, Z. Yu, K. Takei, J. Park, T. Chen, B. Ma and A. Javey. (2013). User-interactive electronic skin for instantaneous pressure visualization. *Nature Materials* 12(10): 899–904.

Wang, C., K. Xia, M. Zhang, M. Jian and Y. Zhang. (2017). An all-silk-derived dual-mode e-skin for simultaneous temperature–pressure detection. *ACS Applied Materials & Interfaces* 9(45): 39484–39492.

Xu, J., S. Wang, G.-J. N. Wang, C. Zhu, S. Luo, L. Jin, X. Gu, S. Chen, V. R. Feig, J. W. F. To, S. Rondeau-Gagné, J. Park, B. C. Schroeder, C. Lu, J. Y. Oh, Y. Wang, Y.-H. Kim, H. Yan, R. Sinclair, D. Zhou, G. Xue, B. Murmann, C. Linder, W. Cai, J. B.-H. Tok, J. W. Chung and Z. Bao. (2017). Highly stretchable polymer semiconductor films through the nanoconfinement effect. *Science* 355(6320): 59–64.

Xu, K., Y. Lu and K. Takei. (2019). Multifunctional skin-inspired flexible sensor systems for wearable electronics. *Advanced Materials Technologies* 4(3): 1800628.

Xue, X., P. Deng, B. He, Y. Nie, L. Xing, Y. Zhang and Z. L. Wang. (2014). Flexible self-charging power cell for one-step energy conversion and storage. *Advanced Energy Materials* 4(5): 1301329.

Yamamoto, Y., D. Yamamoto, M. Takada, H. Naito, T. Arie, S. Akita and K. Takei. (2017). Efficient skin temperature sensor and stable gel-less sticky ECG sensor for a wearable flexible healthcare patch. *Advanced Healthcare Materials* 6(17): 1700495.

Yang, A., Y. Li, C. Yang, Y. Fu, N. Wang, L. Li and F. Yan. (2018). Fabric organic electrochemical transistors for biosensors. *Advanced Materials* 30(23): 1800051.

Yang, R., Y. Qin, C. Li, G. Zhu and Z. L. Wang. (2009). Converting biomechanical energy into electricity by a muscle-movement-driven nanogenerator. *Nano Letters* 9(3): 1201–1205.

Yeo, W.-H., Y.-S. Kim, J. Lee, A. Ameen, L. Shi, M. Li, S. Wang, R. Ma, S. H. Jin, Z. Kang, Y. Huang and J. A. Rogers. (2013). Multifunctional epidermal electronics printed directly onto the skin. *Advanced Materials* 25(20): 2773–2778.

Yeom, C., K. Chen, D. Kiriya, Z. Yu, G. Cho and A. Javey. (2015). Large-area compliant tactile sensors using printed carbon nanotube active-matrix backplanes. *Advanced Materials* 27(9): 1561–1566.

Yerushalmi, R., Z. A. Jacobson, J. C. Ho, Z. Fan and A. Javey. (2007). Large scale, highly ordered assembly of nanowire parallel arrays by differential roll printing. *Applied Physics Letters* 91(20): 203104.

Yu, X., W. Shou, B. K. Mahajan, X. Huang and H. Pan. (2018). Materials, processes, and facile manufacturing for bioresorbable electronics: a review. *Advanced Materials* 30(28): 1707624.

CHAPTER 9

Implantable and Non-Invasive Wearable and Dermal Nanosensors for Healthcare Applications

Joseph Sonia,[1,#] Kannan Sapna,[1,2,#] Ashaiba Asiamma,[1,2] Kodiadka Ayshathil Bushra,[1] Ananthapadmanabha Bhagwath Arun[2] and Kariate Sudhakara Prasad[1,3,]*

1. Introduction

In the research realm, non-invasive and implantable sensors have recently received much interest owing to their potential for real-time monitoring of human health, fitness, and a plethora of biomedical applications (Diamond et al. 2008, Buller et al. 2010, Turner 2013). The novel concept of smart skin opens up new possibilities for the continuous monitoring of the fitness of the user without interfering with their daily activities and also can be used for monitoring patients in clinical settings (Someya and Amagai 2019). The number of people suffering from lifestyle-related diseases such as hypertension, dyslipidemia, hyperuricemia and infectious diseases is also increasing, which propels the ceaseless advances for wearable, non-invasive, implantable, or generally personalized biosensing devices (Kaushik et al. 2014, Yamada et al. 2015, Sapna et al. 2020). The electrochemical detection modality is perhaps one of the best ways for wearable sensors due to its ease of miniaturization,

[1] Nanomaterial Research Laboratory (NMRL), Nano Division, Yenepoya Research Centre, Yenepoya (Deemed to be University), Deralakatte, Mangalore 575 018, India.
[2] Yenepoya Research Centre, Yenepoya (Deemed to be University), Deralakatte, Mangalore 575 018, India.
[3] Centre for Nutrition Studies, Yenepoya (Deemed to be University), Deralakatte, Mangalore 575 018, India.
* Corresponding author: ksprasadnair@yenepoya.edu.in
# equal contribution

high sensitivity, and relatively low power consumption. Just like the profound success of personal glucometers, electrochemical sensors have the potential for wide practical application. The wearable electronics typically comprise sensors, actuators, data communication systems, and power supplies (An et al. 2017, Sonia et al. 2020).

In the chapter, we provide an overview of the key advances in non-invasive nanosystems such as wearable devices and dermal implants for sensing and drug delivery from the past few years and discuss their potential as alternatives to invasive biomedical devices used in health care (Figure 1). Particularly, we confer how the

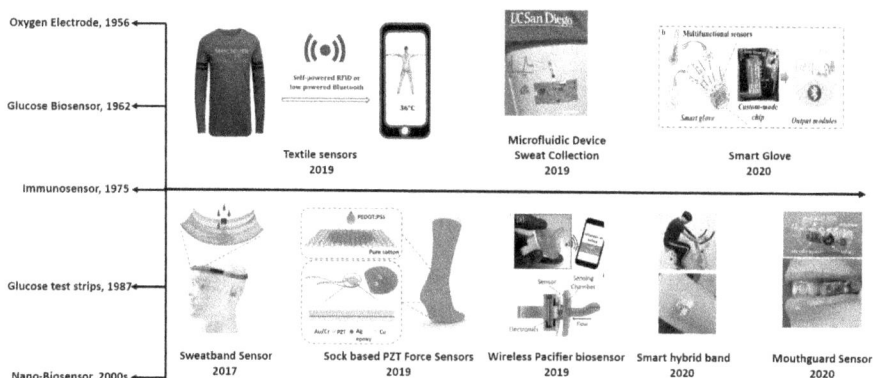

Fig. 1. Biosensor development into wearables. Reprinted with permission from Clark Jr. L.C. and C. Lyons. (1962). Electrode systems for continuous monitoring in cardiovascular surgery. *Annals of the New York Academy of Sciences* 102: 29–45. Janata, J. (1975). Immunoelectrode. *Journal of the American Chemical Society* 97: 2914–2916. Pokorski, J. K., J.-M. Nam, R. A. Vega, C. A. Mirkin and D. H. Appella. (2005). Cyclopentane-modified PNA improves the sensitivity of nanoparticle-based scanometric DNA detection. *Chemical Communications* (2005): 2101–2103. Wang, J. (2005). Electrochemical glucose biosensors. *Chemical Reviews* 108: 814–825. Martín, A., J. Kim, J. F. Kurniawan, J. R. Sempionatto, J. R. Moreto, G. Tang, A. S. Campbell, A. Shin, M. Y. Lee and X. Liu. (2017). Epidermal microfluidic electrochemical detection system: Enhanced sweat sampling and metabolite detection. *ACS Sensors* 2: 1860–1868. Bai, W., T. Kuang, C. Chitrakar, R. Yang, S. Li, D. Zhu and L. Chang. (2018). Patchable micro/nanodevices interacting with skin. *Biosensors and Bioelectronics* 122: 189–204. Kim, J., A. S. Campbell and J. Wang. (2018). Wearable non-invasive epidermal glucose sensors: A review. *Talanta* 177: 163–170. Afroj, S., N. Karim, Z. Wang, S. Tan, P. He, M. Holwill, D. Ghazaryan, A. Fernando and K. S. Novoselov. (2019). Engineering graphene flakes for wearable textile sensors via highly scalable and ultrafast yarn dyeing technique. *ACS Nano* 13: 3847–3857. García-Carmona, L., A. Martín, J. R. Sempionatto, J. R. Moreto, M. C. Gonzalez, J. Wang and A. Escarpa. (2019). Pacifier biosensor: toward noninvasive saliva biomarker monitoring. *Analytical Chemistry* 91: 13883–13891. Zhu, M., Q. Shi, T. He, Z. Yi, Y. Ma, B. Yang, T. Chen and C. Lee. (2019). Self-powered and self-functional cotton sock using piezoelectric and triboelectric hybrid mechanism for healthcare and sports monitoring. *ACS Nano* 13: 1940–1952. Arakawa, T., K. Tomoto, H. Nitta, K. Toma, S. Takeuchi, T. Sekita, S. Minakuchi and K. Mitsubayashi. (2020). A wearable cellulose acetate-coated mouthguard biosensor for *in vivo* salivary glucose measurement. *Analytical Chemistry* 92: 12201–12207. He, X., S. Yang, Q. Pei, Y. Song, C. Liu, T. Xu and X. Zhang. (2020). Integrated smart janus textile bands for self-pumping sweat sampling and analysis. *ACS Sensors*. Huang, J., J. Zeng, B. Liang, J. Wu, T. Li, Q. Li, F. Feng, Q. Feng, M. J. Rood and Z. Yan. (2020). Multi-arch-structured all-carbon aerogels with superelasticity and high fatigue resistance as wearable sensors. *ACS Applied Materials & Interfaces* 12: 16822–16830. (Clark Jr and Lyons 1962, Janata 1975, Pokorski et al. 2005, Wang 2008, Martín et al. 2017, Bai et al. 2018, Kim et al. 2018, Afroj et al. 2019, García-Carmona et al. 2019, Zhu et al. 2019, Arakawa et al. 2020, He et al. 2020, Huang et al. 2020).

fundamental principles of biosensor systems can be utilized for designing reliable biomedical devices with minimal mechanical and biological damages. Also, we have focused on the key challenges in the functioning of devices in specific non-invasive physiological biofluids and also the prospects of these devices for the biomedical field. We critically review pioneering studies that greatly influenced the sector of wearable and implantable biosensing and address future challenges to beat. Most of the research works addressed here comprise of biosensing devices based on electrochemical signal transduction and we highlight nanosystems directing at practical healthcare applications with promise for clinical translation in the near future. The following sections first focus mainly on biofluids such as sweat, tear and saliva, generally secreted by human body, which harbor diverse information relevant to health care providers.

1.1 Sweat

Human sweat contains abundant information about a person's health status and is considered as one of the best non-invasive body fluid for sensing the electrolytes and metabolites (Mohan et al. 2020). As a result of the increasing population and subsequent upsurge in the prevalence of diseases, better health care is needed. Recently, the wearable devices have grabbed more attention for non-invasive real-time, point of care disease analysis (Bandodkar and Wang 2014, Khan et al. 2016a, Trung and Lee 2016, Liu et al. 2017). Further, based on the analysis, the treatment can be given to the person at the earliest, so that the severity of the diseases can be reduced. The treatment mainly depends on the observed/monitored values, the accuracy, sensitivity, specificity, and overall maintenance of the quality of the system which is very crucial. The invasive methods with blood samples are inconvenient and painful. The body fluids like sweat can be used for continuous and easy disease diagnosis in many cases in a non-invasive manner (Huang et al. 2014, Mena-Bravo and De Castro 2014). Varieties of non-invasive, point-of-care diagnostic tools were designed and fabricated by scientists for sweat sample analysis. The sweat can be easily extracted and includes most of the biomarkers related to health issues, used for the disease identification, monitoring, and prevention of diseases (Rock et al. 2014, Anastasova et al. 2017). The quantitative/qualitative, labeled/non-labeled analysis of the sweat biomarkers can be done by developing non-invasive point of care devices with the integration of electrochemical/electronic biosensors with the use of modern digital techniques (Brasier and Eckstein 2019). The difficulties, challenges, technical issues, and problems related to sweat sampling, data collection, and processing are already reviewed by many experts in the field. But still detailed and deep studies are needed as a small variation may cause a drastic impact on human health management (Zeng et al. 2014, Grau et al. 2016, Khan et al. 2016b, Anastasova et al. 2017). The physiology of the sweat sensors was studied; accordingly, the challenges faced are the difference in human body nature, their habits, and activities (Kaya et al. 2019).

The sweat and blood compositions are mostly related; hence, the blood and sweat sensing can be correlated (Baker 2019). The presence as well as the concentration of metals, metal ions, minerals, metabolites, proteins, and hormones related to human health conditions can be identified in a non-invasive manner by sampling the sweat secretions. For each disease diagnosis, specific biomarker can be diagnosed by multiple procedures with an individual method having its pros and cons (Dervisevic et al. 2020). For sweat collection, absorbent pads, sweat-wicking pads, and guiding channels were placed on the soft polymer or silicon substrates (Cazalé et al. 2016, Dam et al. 2016, Gao et al. 2016a). The paper-based microfluidic pads can directly absorb and guide the sweat sample into the sensing area through microfluidic channels for real-time monitoring (Kumar et al. 2016, Hauke et al. 2018). On the other hand, sensing electrodes can be directly tattooed onto the surface for sweat sensing. Recently, several studies were done to make the physiological effect independent from individual body surface conditions and environment (Cramer and Jay 2014). Such sweat generation can be accomplished through exercise activity, thermal heating, stress, or iontophoresis stimulation.

The basic model of sweat based biosensors consists of a bioreceptor to sense the analyte of interest followed by a transducer, that produces a measurable signal for further boosting and processing. The sweat sensors were developed and validated so far with different substrates including fabrics, paper, silicon-based, soft polymers, and plastics (Rose et al. 2014, Zeng et al. 2014, Gao et al. 2016b, Koh et al. 2016, Emaminejad et al. 2017). For a particular application, the substrate can be modified with multiple nanomaterials or their combinations. The substrate may be biodegradable or non-biodegradable; usually, microfluidic substrates with different grades have good flow rate, are economic as well as easily available (Liu et al. 2018, Oh et al. 2018). Recently, researchers are working to develop cheaper devices in miniaturized size for sweat sensing by incorporating biodegradable cheaper substrate combined with nanotechnology (Han et al. 2017, Bhide et al. 2018, Mishra et al. 2018). Works have been carried out for developing sweat biosensing miniaturized devices along with simultaneously stimulated drug delivery systems (Lee et al. 2017). The sweat biosensor could detect multiple bioanalytes like Na^+, Cl^-, K^+, Ca^{2+}, Zn^{2+}, Cd^{2+}, Pb^{2+}, Cu^{2+}, Hg^+, glucose, lactate, uric acid, and ascorbic acid, ethanol, cortisol, DNA, proteins, peptides, urea, and pH values (Choi et al. 2016, Medi et al. 2017, Yilmaz et al. 2017, Bariya et al. 2018, Choi et al. 2018a). The developed device should be able to differentiate between normal and abnormal sweat secretion. The common electrochemical techniques such as chronoamperometry, potentiometric, voltametric, impedance spectroscopy along with transistor-based elctronic measurement techniques can be employed for sweat sample analysis (Jia et al. 2013, Choi et al. 2016, Zhao et al. 2019a, Ganguly et al. 2020, Nagamine et al. 2020). Selectivity, sensitivity, response time, noise elimination, accuracy, repeatability, and robustness are the key factors that need to be considered. The challenges like contamination of the sample, the effect of temperature, and false analysis should be taken care of. Most of the sweat sensors were wearable sensors, which may use enzymatic or non-enzymatic sensing for real-time monitoring, incorporating organic and inorganic nanomaterials, and noble metal nanoparticles (Qiao et al. 2020). The combination of biosensing, transducers, and signal processing unit results in a lab on chip technology. The printed circuit boards with all integrated units with self-generated or battery supply can be placed on the human body for continuous monitoring in health care systems (Reid and Mahbub 2020). The summary of sweat components and their analysis methods are given in Table 1.

Table 1. Summary of sweat components and their electrochemical method used.

Sweat components	Method used	Concentration	References
Electrolytes	Potentiometric	≈ 100 ng/ml–l0.5 mg/ml	(Bariya et al. 2020)
Metabolites	Amperometry/ Chronoamperometry/ Colorimetric	≈ 18 µg/ml–3 mg/ml	(Yokus et al. 2020)
Amino acids	Ion-exchange or liquid chromatography (IEC)	≈ 1 µg/ml–1 mg/ml	(Delgado-Povedano et al. 2016)
Proteins	Chromatography	≈ 3 pg/ml–1 mg/ml	(Zhang et al. 2018)
Hormone	ELISA/HPLC/Electrochemical	≈ 8 ng/ml–25 ng/ml	(Tu et al. 2020)

Source: Bariya, M., L. Li et al. (2020). Glove-based sensors for multimodal monitoring of natural sweat. *Science Advances* 6(35): eabb8308. Yokus, M. A., T. Songkakul et al. (2020). Wearable multiplexed biosensor system toward continuous monitoring of metabolites. *Biosensors and Bioelectronics* 153: 112038. Delgado-Povedano, M., M. Calderón-Santiago et al. (2016). Study of sample preparation for quantitative analysis of amino acids in human sweat by liquid chromatography–tandem mass spectrometry. *Talanta* 146: 310–317. Csősz, É., G. Emri et al. (2015). Highly abundant defense proteins in human sweat as revealed by targeted proteomics and label-free quantification mass spectrometry. *Journal of the European Academy of Dermatology and Venereology* 29(10): 2024–2031 and Tu, E., P. Pearlmutter et al. (2020). Comparison of colorimetric analyses to determine cortisol in human sweat. *ACS Omega* 5(14): 8211–8218.

The completely integrated, self-powered, wearable sweat glucose sensors were successfully demonstrated for continuous monitoring of sweat glucose (Zhao et al. 2019b). The microfluidic chip-based wearable colorimetric sensor for sweat glucose detection was investigated with pre-embedded glucose oxidase (GOD)-peroxidase-o-dianisidine reagents (Xiao et al. 2019). The multifunctional nickel phosphate nano/micro-flakes 3D electrode has been used for electrochemical energy storage, non-enzymatic glucose detection, and sweat pH sensing applications; hence, it could be used as a novel platform for non-enzymatic wearable sensor development (Padmanathan et al. 2018). Figure 2 highlights few of such wearable sweat sensors available for health care monitoring.

The quantitative analysis of the hormone named cortisol is done by sweat sensors, which is an important biomarker for sensing metabolites. The studies show that a large amount of cortisol in the human body leads to depression, anxiety, and Alzheimer's (Tu et al. 2020). Even though almost 300 proteins are found in the human body, which are biomarkers for various diseases, bone loss, cardiovascular diseases, and kidney-related diseases, their concentration is very less in sweat. So the peptides produced from hormones can also be used as biomarkers in medical practice (Peterson et al. 2016). The frequently used bioreceptors include enzymes, antibodies, and DNA/RNA (Qiao et al. 2020). The tracking of drug (levodopa) absorption used for the treatment of Parkinson's disease was done with sweatbands through the incorporation of electrochemical sensor units (Tai et al. 2019). Monitoring of methylxanthine drug, and caffeine was done with the electrochemical analysis of sweat. The sensor was developed by printing silver (Ag), carbon (C), and insulation layers on the substrate material (Tai et al. 2018). Sometimes detection of prohibited drugs such as marijuana, methamphetamine, and phencyclidine can be identified by sweat analysis (Jain and Bhattacharya 2020).

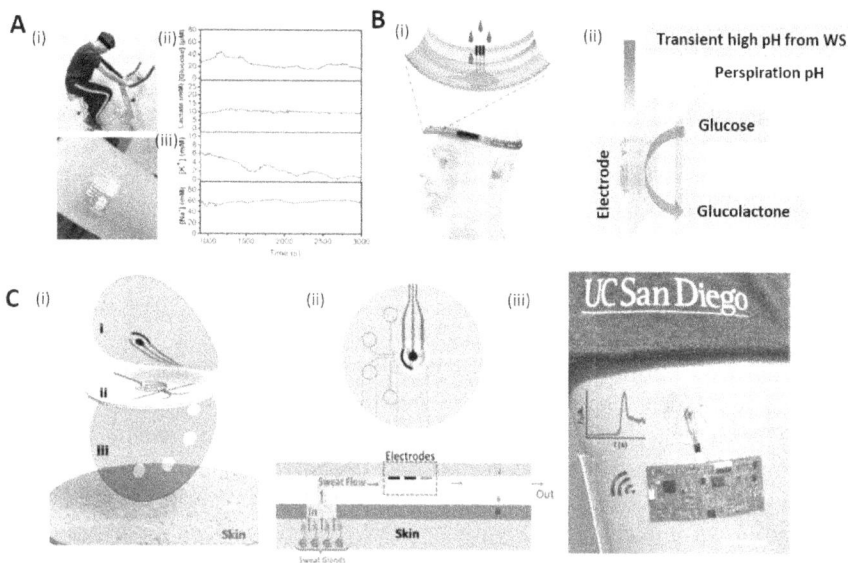

Fig. 2. (A) Self-pumping Janus textile bands for quantitatively tracking sweat markers. (a), (b) Smart hybrid band is tied on a volunteer's arm during stationary biking activity. (c) On-body, real-time, and *in situ* sweat monitoring. Signals acquisition for each sensor occurred upon an adequate sweat sample being collected (Reprinted with permission from Martin, A., J. Kim, J. F. Kurniawan, J. R. Sempionatto, J. R. Moreto, G. Tang, A. S. Campbell, A. Shin, M. Y. Lee and X. Liu (2017). Epidermal microfluidic electrochemical detection system: Enhanced sweat sampling and metabolite detection. *ACS Sensors* 2(12): 1860–1868) (B) (i) Sweatband with electrochemical sensor for perspiration glucose sensing. (ii) Nonenzymatic glucose sensing method with the WSE reaction on the MOF-based electrode (Reprinted with permission from Zhu, M., Q. Shi, T. He, Z. Yi, Y. Ma, B. Yang, T. Chen and C. Lee (2019). Self-powered and self-functional cotton sock using piezoelectric and triboelectric hybrid mechanism for healthcare and sports monitoring. *ACS Nano* 13(2): 1940–1952). (C) Microfluidic device design and operation. The soft epidermal microchip device conforms to the skin and routes the sampled sweat toward the electrochemical detector. (i) Schematic representation of layered microfluidic device configuration on skin composed of top PDMS layer with incorporated sensor electrodes, PDMS microfluidic device, an adhesive layer on the skin. (ii) Schematic representation of microfluidic device sweat collection and operation on the skin in top-down and cross-sectional views. (iii) Photograph of microfluidic device integrated with wireless conformal electronics on the skin with lithography-based gold current collectors and screen-printed silver–silver chloride (RE) and Prussian blue (WE and CE). Inset: Electrochemical temporal response to sweat metabolites. Scale bar, 5 mm (Reprinted with permission from He, X., S. Yang, Q. Pei, Y. Song, C. Liu, T. Xu and X. Zhang (2020). Integrated smart janus textile bands for self-pumping sweat sampling and analysis. *ACS Sensors*) (Martin et al. 2017, Zhu et al. 2019, He et al. 2020a).

Systematic studies of the electronics, microfluidics, and integration schemes establish the key design considerations and performance traits (Choi et al. 2018b). Patch-based multiple sensors were developed to measure lactate and glucose simultaneously along with the concentration of Na^+ and K^+. The resulted multisensory array could also measure body temperature (Sakata et al. 2020). The electrochemical based multiple sensors to detect glucose, Na^+, K^+, Ca^{2+}, and pH together by coating carbon nanotube (CNT) with recognition element are also reported (He et al. 2020a). The CdSSe (Cadmium Sulphoselenide) nanowire chip-based wearable sensor can read salt, and moisture content (Zhang et al. 2019). The identification of multiple analytes such as glucose, lactate, ascorbic acid, uric acid, Na^+, and K^+ are done by a patch sensor developed on the soft fiber substrate with carbon-based electrodes (He et al. 2019). Heavy metal ions such as Zn, Cd, Pb, Cu, and Hg detection in sweat can be done with electrochemical stripping voltammetry with Au and Bi as electrodes on a flexible substrate (Yu et al. 2020). The multisensory array that could sense pH (4.5–7.5), Cl^- (10–100 mM), K^+ (10–10 mM), and Na^+ (10–100 mM) by placing multi-walled CNT on flexible, stretchable material were fabricated with minimum sweat contamination (Parrilla et al. 2019).

A tattoo-based alcohol biosensing system for non-invasive alcohol monitoring in induced sweat has also been demonstrated. The skin-worn alcohol monitoring platform integrates an iontophoretic-biosensing temporary tattoo system along with flexible wireless electronics (Kim et al. 2016). Recent achievements in stretchable electronics and biocompatible tattoos are revolutionizing the design of wearable devices for sweat alcohol analysis (Kim et al. 2016). An electrochemical lactate sensor on a tattoo for real-time monitoring of human perspiration for the first time was developed in 2013 by the Wang group and successfully measured the lactate content in sweat through a skin-worn sensor by enzymatic reactions (Jia et al. 2013). Sweating can be induced by pilocarpine iontophoresis, removing some of the challenges for its sampling (Tricoli et al. 2017). The commercial silver/silver chloride reference electrode and platinum wire were used to form a complete three-electrode electrochemical system for sensing nicotine (Tai et al. 2020). Some of the sweat sensors can be incorporated into gloves for health and fitness monitoring (Bariya et al. 2020). The unique properties of two-dimensional materials are used for the development of multiple wearable and implantable biomedical biosensors such as humidity sensors and tactile sensors (Yang et al. 2017, 2020). Varieties of cheaper, simple colorimetric sensors were also designed, developed, and validated for human sweat analysis (Promphet et al. 2020, Tu et al. 2020). Wearable biosensors are used to measure vital signs like heart rate, respiratory rate, temperature, blood pressure, etc. These kinds of wearable biosensors help healthcare providers to track the vital traits of patients through smartphones. Post therapeutic treatment follow-ups are important for surgery and other high-risk illness such as cancer, cardiovascular, neurological, kidney diseases, and diabetes, etc. Researchers have developed a pair of wearables and ingestible sensors for monitoring the medicine intake by an individual. These biosensors automatically record the date and time of ingestion of antibiotics without any report from the patients (Belknap et al. 2013, Zhu et al. 2017).

The E4 wrist band and the embrace (Empatia) are the commercially available wrist band designed for detecting seizures (Regalia et al. 2019). The detection of cortisol in sweat by using a wearable biosensor will help in quantifying the sweat biomarkers like cortisol, further helping in monitoring stress levels in individuals. The monitoring of high and low levels of basal cortisol eventually paves way for diagnosis of Addison's and Cushing's diseases (Ragnarsson 2020, Saverino and Falorni 2020). Recently, researchers fabricated an integrated smart glove with multiwalled carbon nanotubes/poly (dimethyl siloxane) (MWNTs/PDMS) for finger dexterity, hand gesture, and temperature recognition. Lightweight, low cost, and easy to use are the advantages of this smart glove (Li et al. 2020). Modified graphene aerogel (MGA) based pressure sensor exhibited a fast response, wide sensing range,

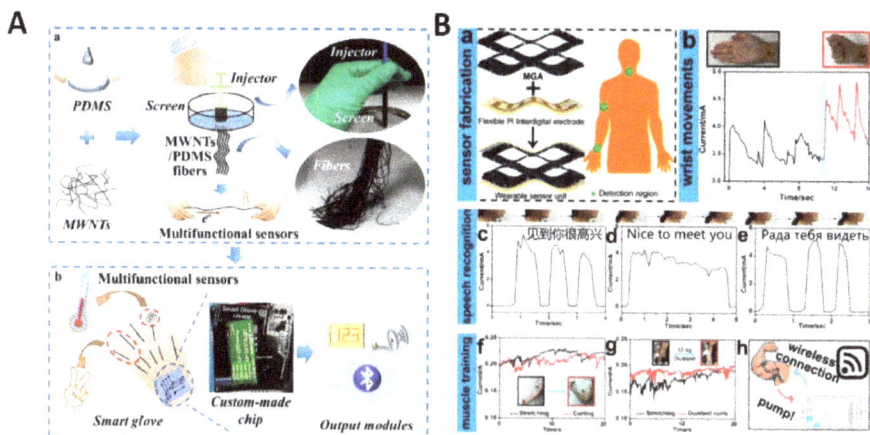

Fig. 3. (A) Schematic illustration of the fabrications of the MWNTs/PDMS fibers and a smart glove. (a) The fabrication process of the MWNTs/PDMS fibers as multifunctional sensors. The insets are photographs of the fabrication process. (b) Integration of the fibers with a smart glove for gesture recognition and temperature measurement (Reprinted with permission from Li, Y., C. Zheng, S. Liu, L. Huang, T. Fang, J. X. Li, F. Xu and F. Li. (2020). A smart glove integrated with tunable MWNTs/PDMS fibers made of one-step extrusion method for finger dexterity, gesture and temperature recognition. *ACS Applied Materials & Interfaces* (Li et al. 2020)). (B) Sensing applications of MGA on human movements, speech recognition, and muscle training. (a) Schematic illustration of the fabrication of the MGA-based sensor and detection regions of the human. The size of MGA in a wearable device was tailored to a cuboid shape ($1 \times 1 \times 0.5$ cm³). (b) Artery wrist pulse variation during an open and gripped hand. (c–e) Real-time current response of the fabricated sensor to different languages (Chinese, English, and Russian). (f) Motion detection of biceps brachii in training. (g) Motion detection with a loading of 10 kg dumbbell. (h) MGA wearable sensor is connected to a circuit implanted with a wireless system, which transmits the real-time current signal to a mobile phone (Reprinted with permission from Huang, J., J. Zeng, B. Liang, J. Wu, T. Li, Q. Li, F. Feng, Q. Feng, M. J. Rood and Z. Yan. (2020). Multi-arch-structured all-carbon aerogels with super elasticity and high fatigue resistance as wearable sensors. *ACS Applied Materials & Interfaces* 12(14): 16822–16830) (Huang et al. 2020).

and long-term sensing durability. The sensing properties of MGA-based pressure sensors have promising applications in wearable devices for detecting human biosignals and body motions. This aerogel-based wearable pressure sensor achieved high sensitivity toward subtle activities of a volunteer, such as a wrist pulse rating, throat vibration, and muscular motion, etc. (Figure 3) (Huang et al. 2020).

1.2 Tear

The implantable and noninvasive nano systems are important and, interestingly, the tear-based sensors open up better avenues to monitor clinical manifestations. A tear commonly contains proteins/peptides, enzymes, electrolytes, lipids, and metabolites from lacrimal glands, ocular surface epithelial cells, Meibomian glands, goblet cells, small organic molecules, and blood. In addition to this, dopamine and various other neurotransmitters have been found in the tears. The glucose level in the tear can be monitored by contact lenses (CL). The normal glucose level present in the tears has been estimated to be 0.2–0.6 mM, but in the case of diabetic conditions, the average concentration can be higher than 0.9 mM (Veli and Ozcan 2018, Sempionatto et al. 2019a). CLs integrated with an amperometric glucose sensor is fabricated by creating microstructures on the polymer substrate (titania sol-gel film), which is successively molded as a CL. This glucose sensor exhibited a fast response (20 s) with high sensitivity (240 Aμcm^{-2} mM^{-1}), good repeatability, and can detect a low concentration of glucose in the tear film (Yao et al. 2011). Another group fabricated a glucose sensor (NovioSense sensor) for the detection of glucose in tear fluid of patients with type 1 diabetes with polysaccharide coating and studied the clinical implications. When comparing with the other monitoring methods for diabetes which are invasive, painful, and expensive, the CL-based one overcomes the aforementioned bottlenecks. The NovioSense sensor is placed in the fornix of the lower eyelid and bathed in tear fluid. This device provides measurement during the basal tear flow and the results were well correlated with conventional glucose measurements in the blood (Figure 4A) (Kownacka et al. 2018).

For the non-invasive medical diagnostics, Park et al. developed a method to fabricate smart and soft CL that can indicate the diabetic condition by monitoring glucose levels in tears through the display with wireless operations. For achieving this, the smart lens and the electronic components including glucose sensor, light-emitting diode (LED) pixel, a rectifier circuit, and the stretchable transparent antenna were all integrated onto the mechanical stress-tunable hybrid substrate with refractive indices that allow attaining high optical transparency. The CL based on the hybrid substrate, in which the reinforced islands were embedded inside an elastic layer, can effectively distribute the mechanical strain and protect ordinary electronics from mechanical

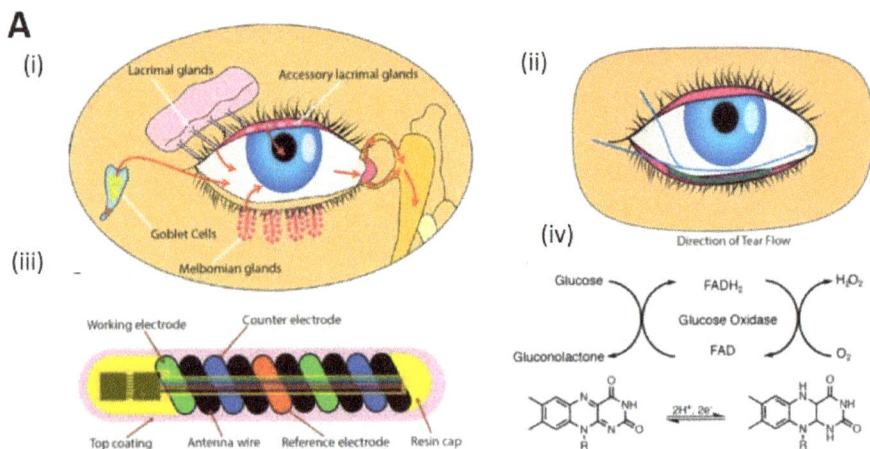

Fig. 4. (A) Schematic illustration and properties of NovioSense minimally-invasive tear glucose sensor. (i) Illustration showing the tear fluid production. (ii) The direction of the tear flow and anatomical position of the NovioSense device. (iii) Design and structure of the final product showing electronic components. (iv) Mechanism of glucose detection where glucose oxidase is used as the enzymatic sensing element (Reprinted with permission from Kownacka, A. E., D. Vegelyte, M. Joosse, N. Anton, B. J. Toebes, J. Lauko, I. Buzzacchera, K. Lipinska, D. A. Wilson and N. Geelhoed-Duijvestijn. (2018). Clinical evidence for use of a noninvasive biosensor for tear glucose as an alternative to painful finger-prick for diabetes management utilizing a biopolymer coating. *Biomacromolecules* 19(11): 4504–4511) (Kownacka et al. 2018).

deformations of the soft lens. The one-dimensional (1D), ultra-long metal nanofibers (mNFs) are used as transparent and stretchable electrodes and the charge transport is facilitated with the reduction of sheet resistance, by minimizing the number of junctions between the metal fibers. The glucose oxidase (GO), immobilized on the graphene sensor by π-π interaction, gets oxidized immediately when in contact with glucose. As soon as the graphene sensor detects glucose in tears, the transparent and stretchable antenna with a rectifier drives the pixels of the LED to display real-time sensing information wirelessly, thereby aiding in obtaining real-time glucose levels. In addition, no noticeable adverse effects were observed during *in vivo* tests for its stable operation in a live rabbit. Hence, the soft and smart CL provides a platform for wireless, continuous, and noninvasive monitoring of glucose concentration (Park et al. 2018). Alcohol is considered a major key target opportunity for wearable sensing devices as it is the most widely used substance of abuse worldwide. Sempionatto and group reported the development of a wearable eyeglasses-based platform with an alcohol biosensor flow detector mounted onto the nose-bridge pad for detection of alcohol in stimulated tears. The external miniaturized flow detector mounted on the eyeglasses pad helps in the collection of stimulated tears. The biosensor system consisted of a fluidic device for collecting the tears, and alcohol oxidase (AOx) modified electrochemical flow detector, wireless electronics, and the supporting eyeglasses platform. The developed CL can simultaneously detect the glucose, alcohol, and vitamin levels in a quantitative manner. The AOx catalyzes the oxidation of ethanol molecules present in the alcohol and transmits the reduction signal to the laptop using integrated wireless electronics. Due to the electroactive nature of vitamins, their concentration in tears can be monitored using highly sensitive and rapid square wave voltammetry (SWV) owing to their redox reactions. This is the first work demonstrating the non-invasive monitoring of vitamins toward potential personal nutrition applications. To address the drawbacks of current CL systems such as potential infections and impaired vision, the electrochemical detection system and its wireless electronic backbone are placed outside the eye area. The novel platform was also validated by continuously monitoring the tear alcohol levels of human subjects after alcohol consumption (Sempionatto et al. 2019b).

Lysozyme is a vital anti-microbial enzyme and isone of the common proteins found in tear fluid. One of the research groups demonstrated simple mobile sensing workflow, using commercially available CLs to collect lysozyme in tear-fluid for potential point-of-care diagnostics for diseases such as in case of dry eye disease (DED), Sjögren's syndrome, or viral conjunctivitis, and also used for detecting the presence of pathogens such as *Staphylococcus aureus* and *Acanthamoeba*. In this study, a CL-based mobile sensing approach was used to monitor the lysozyme levels of healthy participants and DED patients over a period of 2 weeks. The study consists of two groups of healthy participants, one is CL wearers and the other is the non-CL wearers. The authors employed a previously developed mobile phone-based well-plate reader for rapid analysis, wherein time-lapse imaging is used to capture an increased fluorescent signal produced by a commercial enzymatic assay. The calculated rate-of-change in the detected fluorescence is then used to indirectly quantify the concentration of the lysozyme adsorbed by the CLs by comparison to a standard curve. A significant increase for the CL wearers lysozyme concentration measurements was observed when they played a mobile-phone game during the wear-duration. This increase in concentration is due to digital eye-strain in the participants. While in the case of DED patients, a significantly smaller mean lysozyme level was found when compared to the non-DED patients (Ballard et al. 2020).

1.3 Saliva

The non-invasive detection of biomarkers or analytes of interest involved the use of biofluids such as tears, sweat, and saliva. Among these, saliva is regarded as a better option owing to its continuous and convenient availability and its good correlation with blood metabolites (Bandodkar and Wang 2014, Kim et al. 2014, Shanbhag and Prasad 2016). In addition, the oral cavity has a place where the sensor can be placed appropriately. Kim and friends, in 2014, fabricated a non-invasive mouthguard biosensor based on an immobilized lactate oxidase for the continuous in-mouth monitoring of salivary metabolites. The mouthguard was integrated with a printable amperometric enzymatic biosensor for sensing lactate. Compared to traditional in-built sensors, this miniaturized sensor is easily wearable and removable without any surgical interventions. In addition, since the level of salivary lactate corresponds to the blood lactate level, the sensor endowed accurate results. Besides, the continuous contact with the saliva also facilitated to get the readings without any interruption. In the sensors, the working electrode was modified with Prussian-Blue (PB) and coated with poly-orthophenylene diamine (PPD)/lactate-oxidase (LOx) reagent layer. The PPD is generally used for the electro polymeric entrapment of oxidases and to eliminate potential interferences and protection of the biosensor surface. The mouthguard biosensor was able to detect lactate in undiluted human salivary samples, which shows the applicability of the sensor for human health and fitness monitoring (Kim et al. 2014).

Detection of biomarkers in infants is also equally important for the rapid control of low birth weights and certain metabolic diseases such as diabetes which leads to long-term health problems. On the other hand, non-invasive sensors for newborns are highly recommended because of the difficulty in using skin-based wearable sensors. Notably, current techniques for glucose monitoring in newborns involve invasive and expensive techniques. Hence, there is a requirement for rapid, portable, and noninvasive monitoring in babies. Recently, a fully integrated wearable pacifier-based platform toward wireless noninvasive glucose sensing in saliva for monitoring newborn's health was reported. The biosensor is based on glucose-oxidase (GOx)-based enzymatic detection. The pacifier sensing platform incorporates a compatible wireless amperometric circuitry, combined with a Bluetooth communication system for miniaturization and low-power operation, an isolated electrochemical detector for preventing leakage of materials to the mouth, and a nontoxic polymeric nipple containing a safety rectifying channel for saliva sampling are fabricated (Figure 5B). The unidirectional flow of saliva combined with the pump-free systems helps it to be easily integrated into a baby's day to day life (García-Carmona et al. 2019).

The endocrine gland, which is the source of its discharge of saliva, is distributed throughout the body, especially in the hands, feet, lower back, and under the arms.

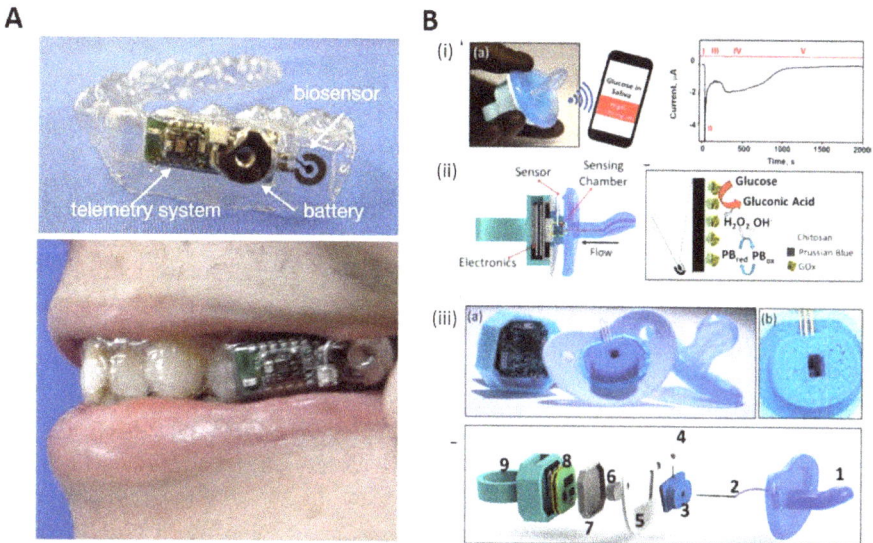

Fig. 5. (A) Photograph of the fabricated MG sensor integrated with a wireless module and battery and wearing of the mouthguard sensor. (B) (i). Glucose pacifier sensing concept. (ii) Schematic of the assembled wireless pacifier biosensor. (ii) Schematic of the glucose enzymatic biosensing approach on the PB electrode. (iii) Scheme of the assembling of the pacifier pieces: nipple of the pacifier (1), inlet for collecting saliva (2), electrochemical chamber (3), PB-GOx electrode (4), central piece (5), outlet (6), insulator pacifier cap (7), integrated wireless potentiostat (8), and back cap of the pacifier (9) (Reprinted with permission from García-Carmona, L., A. Martin, J. R. Sempionatto, J. R. Moreto, M. C. Gonzalez, J. Wang and A. Escarpa. (2019). Pacifier biosensor: toward noninvasive saliva biomarker monitoring. *Analytical Chemistry* 91(21): 13883–13891. Arakawa, T., K. Tomoto, H. Nitta, K. Toma, S. Takeuchi, T. Sekita, S. Minakuchi and K. Mitsubayashi. (2020). A wearable cellulose acetate-coated mouthguard biosensor for *in vivo* salivary glucose measurement. *Analytical Chemistry* 92(18): 12201–12207) (García-Carmona et al. 2019, Arakawa et al. 2020).

Human saliva contains glucose at a concentration of about 1/100 of the concentration of blood glucose, and glucose is secreted continuously at a rate of 0.56 mL/min. One of the major interferences in glucose monitoring in saliva using a mouthguard is cellulose acetate (CA). Glucose in the saliva is successfully measured *in vivo* without any pretreatment of human saliva. A mouthguard (MG) glucose sensor is developed by Arakawa and the group to monitor salivary glucose, which is reported to be associated with the blood glucose level and interference studies were also performed (Figure 5A). The fabricated MG sensor was able to detect the glucose concentration in the range of 1.75–10000 µmol/L, which includes a salivary sugar concentration of 20–200 µmol/L (Arakawa et al. 2020).

1.4 Dermal Patches

Implantable systems are extensively used in health care; however, most of them are known for their invasive characteristics. Nevertheless, non-invasive dermal implant showed better potential for drug delivery applications. Implantable drug delivery offers more advantages over conventional routes of drug delivery, and better patient compliances. The opportunity to deliver low dose of drug of interest and subsequent reduction in side-effects vouch for the development of non-invasive dermal implant-based drug delivery systems (Jung and Jin 2021). Recently, painless microneedle-array patch with glucose-responsive insulin delivery device for type 1 diabetics management has been reported. The transdermal route not only offers avoidance of first pass metabolism, and gastro-intestinal degradation but also non-invasiveness. The smart insulin microarray patch comprises glucose-responsive vesicles (GRVs), which are loaded with insulin and glucose oxidase (GOx) enzyme. Under hypoxic conditions, the hydrophobic component of GRVs that are self-assembled from hypoxia-sensitive hyaluronic acid (HS-HA) converts to hydrophilic 2-aminoimidazoles through bioreduction. During the enzymatic oxidation of glucose in the hyperglycemic state, a hypoxic condition is created which leads to the reduction of (HS-HA), triggering the release of insulin from vesicles. The synthetic glucose-responsive device in hypoxic conditions for the regulation of insulin release is the first demonstration done in a chemically induced type 1 diabetes mouse model. This promising therapy has the potential for translation into a human application for avoiding the risk of hypoglycemia (Yu et al. 2015). On a similar note, Lee et al. developed integrated feedback transdermal drug delivery module. Interestingly the device could be used as a wearable or disposable strip type for precise point-of-care therapy tool to deliver metformin in response to measured glucose levels with real time correction based on pH, temperature and humidity changes (Lee et al. 2017). Metformin is one of the first line drugs for treating type 2 diabetes; the delivery through transdermal route reduces the dosage of the drug than usual oral delivery and also prevents gastrointestinal side effects. Controlled release of formulations of Small molecules such as opioid fentanyl is used for controled release of drug delivery by enclosing in transdermal patch duragesic (Vargason et al. 2021). Recently, dermal microneedles' patches have been effectively developed for the release of contraceptives in a minimally invasive manner, where the authors reported biodegradable self-administrable microneedles with poly lactic acid and polylactic-co-glycolic acid for the controlled release of contraceptive hormone, levonorgestrel (Li et al. 2019). Furthermore, efforts are also made to develop ocular drug delivery platforms using eye patches with an array of detachable microneedles. The self-implantable microneedles with micro-reservoirs for drug of interest have been reported successfully for the delivery of DC101, anti-angiogenic monoclonal antibody and anti-inflammatory drug diclofenac (Than et al. 2018).

2. Conclusion and Future Perspectives

In the current chapter, we have outlined recent advances in non-invasive nano systems in the form of wearable devices and dermal implants for real-time monitoring of vital functions of the body along with detection of various biological signals, for monitoring and managing the health conditions and drug delivery. In the near future, wearable biosensors are expected to become more simplified, moving into the wearer's daily life through their applicability in textiles and fashion accessories. The wearable sensors market is thus expected to reach $2.86 billion by 2025 and continue to improve and transform people's lives. However, the realization of wearable non-invasive chemical sensors is a major challenge, which includes long-term stability, biocompatibility, and resilience of these devices such as in the case of wearable salivary sensors, where plaque develop within minutes in teeth and can possibly hinder their performance. Hence, more focus should be on efforts to increase the sensitivity of wearable sensors. The electronic components of the chemical sensors need to be separated from the aqueous environment but the chemical sensors can detect the presence of the analyte only in contact with the aqueous solution. This is an unsettling task to maintain the aesthetics of the miniaturized sensor device. Addressing these key challenges, wearable electrochemical sensors for continuously monitoring the human body for various biomedical and fitness applications can be introduced in the near future. These wearable sensors, which do not require any skilled person, can be used for bedside health monitoring of patients, especially in the case of elderly people for self-management of chronic diseases at home.

The need for multi-analyte sensing using a combinatorial of sensors is becoming a high priority for monitoring various metabolites in the body. Furthermore, real-time monitoring using Bluetooth integrated devices can aid in a broad assessment of the user's health-care status, including his or her stress and performance levels. Moreover, most of the implantable or non-invasive systems are integrated with the rapid growth of the internet of things (IoT); hence, social, ethical, and ecological issues need to be well addressed from end-user's perspectives. Steps need to be taken for the improvement in data gathering, privacy protection, data threats, and safety of the devices. Even though considerable amount of advancement has been achieved in the case of implants and wearables, the biocompatibility issues need to be scrutinized very well, since most of the substrates used are not approved by FDA, or the European Medicines Agency (EMA), or the China National Medical Product Administration (NMPA), or the Central Drugs Standard Control Organization (CDSCO); hence, clinical trial of the developed systems is difficult. As we consider the implants and wearables complex fabrication steps, delivery of therapeutic doses, regulatory/ethical issues and lack of clinical studies hamper the mass production and use in health care settings. In general, most of the non-invasive health care monitoring devices are considered as "general wellness" product with very low risk to users' safety; the original equipment manufacturers should obey and be liable to protect the public data and follow data protection by design and default. Moreover, designers and manufacturers should employ potential risk management that involves identifying, understanding, and preventing hazardous conditions to ensure utmost safety and effectiveness of the implants and non-invasive disease monitoring and drug delivery systems.

Acknowledgment

JS and KS are thankful to Senior Research Fellowship from Indian Council of Medical Research (ICMR). AB is thankful to Council of Scientific and Industrial Research (CSIR) for Senior Research Fellowship. ABA and AA are indebted to ICMR. KSP is grateful to have funding from the Department of Biotechnology (DBT), Govt. of India.

References

Afroj, S., N. Karim, Z. Wang, S. Tan, P. He, M. Holwill, D. Ghazaryan, A. Fernando and K. S. Novoselov. (2019). Engineering graphene flakes for wearable textile sensors via highly scalable and ultrafast yarn dyeing technique. *ACS Nano* 13(4): 3847–3857.

An, B. W., J. H. Shin, S.-Y. Kim, J. Kim, S. Ji, J. Park, Y. Lee, J. Jang, Y.-G. Park and E. Cho. (2017). Smart sensor systems for wearable electronic devices. *Polymers* 9(8): 303.

Anastasova, S., B. Crewther, P. Bembnowicz, V. Curto, H. M. Ip, B. Rosa and G.-Z. Yang. (2017). A wearable multisensing patch for continuous sweat monitoring. *Biosensors and Bioelectronics* 93: 139–145.

Arakawa, T., K. Tomoto, H. Nitta, K. Toma, S. Takeuchi, T. Sekita, S. Minakuchi and K. Mitsubayashi. (2020). A wearable cellulose acetate-coated mouthguard biosensor for *in vivo* salivary glucose measurement. *Analytical Chemistry* 92(18): 12201–12207.

Bai, W., T. Kuang, C. Chitrakar, R. Yang, S. Li, D. Zhu and L. Chang. (2018). Patchable micro/nanodevices interacting with skin. *Biosensors and Bioelectronics* 122: 189–204.

Baker, L. B. (2019). Physiology of sweat gland function: The roles of sweating and sweat composition in human health. *Temperature* 6(3): 211–259.

Ballard, Z., S. Bazargan, D. Jung, S. Sathianathan, A. Clemens, D. Shir, S. Al-Hashimi and A. Ozcan. (2020). Contact lens-based lysozyme detection in tear using a mobile sensor. *Lab on a Chip* 20(8): 1493–1502.

Bandodkar, A. J. and J. Wang. (2014). Non-invasive wearable electrochemical sensors: a review. *Trends in Biotechnology* 32(7): 363–371.

Bariya, M., H. Y. Y. Nyein and A. Javey. (2018). Wearable sweat sensors. *Nature Electronics* 1(3): 160–171.

Bariya, M., L. Li, R. Ghattamaneni, C. H. Ahn, H. Y. Y. Nyein, L.-C. Tai and A. Javey. (2020). Glove-based sensors for multimodal monitoring of natural sweat. *Science Advances* 6(35): eabb8308.

Belknap, R., S. Weis, A. Brookens, K. Y. Au-Yeung, G. Moon, L. DiCarlo and R. Reves. (2013). Feasibility of an ingestible sensor-based system for monitoring adherence to tuberculosis therapy. *PloS One* 8(1): e53373.

Bhide, A., S. Muthukumar, A. Saini and S. Prasad. (2018). Simultaneous lancet-free monitoring of alcohol and glucose from low-volumes of perspired human sweat. *Scientific Reports* 8(1): 1–11.

Brasier, N. and J. Eckstein. (2019). Sweat as a source of next-generation digital biomarkers. *Digital Biomarkers* 3(3): 155–165.

Buller, M., A. Welles, O. C. Jenkins and R. Hoyt.. (2010). Extreme health sensing: The challenges, technologies, and strategies for active health sustainment of military personnel during training and combat missions. Sensors, and Command, Control, Communications, and Intelligence (C3I) Technologies for Homeland Security and Homeland Defense IX, International Society for Optics and Photonics.

Cazalé, A., W. Sant, F. Ginot, J.-C. Launay, G. Savourey, F. Revol-Cavalier, J.-M. Lagarde, D. Heinry, J. Launay and P. Temple-Boyer. (2016). Physiological stress monitoring using sodium ion potentiometric microsensors for sweat analysis. *Sensors and Actuators B: Chemical* 225: 1–9.

Choi, C., Y. Lee, K. W. Cho, J. H. Koo and D.-H. Kim. (2018). Wearable and implantable soft bioelectronics using two-dimensional materials. *Accounts of Chemical Research* 52(1): 73–81.

Choi, D.-H., J. S. Kim, G. R. Cutting and P. C. Searson. (2016). Wearable potentiometric chloride sweat sensor: the critical role of the salt bridge. *Analytical Chemistry* 88(24): 12241–12247.

Choi, J., R. Ghaffari, L. B. Baker and J. A. Rogers. (2018). Skin-interfaced systems for sweat collection and analytics. *Science Advances* 4(2): eaar3921.

Clark Jr, L. C. and C. Lyons. (1962). Electrode systems for continuous monitoring in cardiovascular surgery. *Annals of the New York Academy of Sciences* 102(1): 29–45.

Cramer, M. N. and O. Jay. (2014). Selecting the correct exercise intensity for unbiased comparisons of thermoregulatory responses between groups of different mass and surface area. *Journal of Applied Physiology* 116(9): 1123–1132.

Dam, V., M. Zevenbergen and R. Van Schaijk. (2016). Toward wearable patch for sweat analysis. *Sensors and Actuators B: Chemical* 236: 834–838.

Delgado-Povedano, M., M. Calderón-Santiago, F. Priego-Capote and M. L. de Castro. (2016). Study of sample preparation for quantitative analysis of amino acids in human sweat by liquid chromatography–tandem mass spectrometry. *Talanta* 146: 310–317.

Dervisevic, M., M. Alba, B. Prieto-Simon and N. H. Voelcker. (2020). Skin in the diagnostics game: Wearable biosensor nano-and microsystems for medical diagnostics. *Nano Today* 30: 100828.

Diamond, D., S. Coyle, S. Scarmagnani and J. Hayes. (2008). Wireless sensor networks and chemo-/biosensing. *Chemical Reviews* 108(2): 652–679.

Emaminejad, S., W. Gao, E. Wu, Z. A. Davies, H. Y. Y. Nyein, S. Challa, S. P. Ryan, H. M. Fahad, K. Chen and Z. Shahpar. (2017). Autonomous sweat extraction and analysis applied to cystic fibrosis and glucose monitoring using a fully integrated wearable platform. *Proceedings of the National Academy of Sciences* 114(18): 4625–4630.

Ganguly, A., P. Rice, K.-C. Lin, S. Muthukumar and S. Prasad. (2020). A combinatorial electrochemical biosensor for sweat biomarker benchmarking. *SLAS Technology: Translating Life Sciences Innovation* 25(1): 25–32.

Gao, W., S. Emaminejad, H. Y. Y. Nyein, S. Challa, K. Chen, A. Peck, H. M. Fahad, H. Ota, H. Shiraki and D. Kiriya. (2016). Fully integrated wearable sensor arrays for multiplexed *in situ* perspiration analysis. *Nature* 529(7587): 509–514.

Gao, W., H. Y. Nyein, Z. Shahpar, L.-C. Tai, E. Wu, M. Bariya, H. Ota, H. M. Fahad, K. Chen and A. Javey. (2016). Wearable sweat biosensors. 2016 IEEE International Electron Devices Meeting (IEDM), IEEE.

García-Carmona, L., A. Martin, J. R. Sempionatto, J. R. Moreto, M. C. Gonzalez, J. Wang and A. Escarpa. (2019). Pacifier biosensor: toward noninvasive saliva biomarker monitoring. *Analytical Chemistry* 91(21): 13883–13891.

Grau, G., J. Cen, H. Kang, R. Kitsomboonloha, W. J. Scheideler and V. Subramanian. (2016). Gravure-printed electronics: recent progress in tooling development, understanding of printing physics, and realization of printed devices. *Flexible and Printed Electronics* 1(2): 023002.

Han, W., H. He, L. Zhang, C. Dong, H. Zeng, Y. Dai, L. Xing, Y. Zhang and X. Xue. (2017). A self-powered wearable noninvasive electronic-skin for perspiration analysis based on piezo-biosensing unit matrix of enzyme/ZnO nanoarrays. *ACS Applied Materials & Interfaces* 9(35): 29526–29537.

Hauke, A., P. Simmers, Y. Ojha, B. Cameron, R. Ballweg, T. Zhang, N. Twine, M. Brothers, E. Gomez and J. Heikenfeld. (2018). Complete validation of a continuous and blood-correlated sweat biosensing device with integrated sweat stimulation. *Lab on a Chip* 18(24): 3750–3759.

He, W., C. Wang, H. Wang, M. Jian, W. Lu, X. Liang, X. Zhang, F. Yang and Y. Zhang. (2019). Integrated textile sensor patch for real-time and multiplex sweat analysis. *Science Advances* 5(11): eaax0649.

He, X., S. Yang, Q. Pei, Y. Song, C. Liu, T. Xu and X. Zhang. (2020). Integrated smart janus textile bands for self-pumping sweat sampling and analysis. *ACS Sensors*.

He, Y., L. Zhao, J. Zhang, L. Liu, H. Liu and L. Liu. (2020). A breathable, sensitive and wearable piezoresistive sensor based on hierarchical micro-porous PU@ CNT films for long-term health monitoring. *Composites Science and Technology* 200: 108419.

Huang, J., J. Zeng, B. Liang, J. Wu, T. Li, Q. Li, F. Feng, Q. Feng, M. J. Rood and Z. Yan. (2020). Multi-arch-structured all-carbon aerogels with superelasticity and high fatigue resistance as wearable sensors. *ACS Applied Materials & Interfaces* 12(14): 16822–16830.

Huang, X., Y. Liu, K. Chen, W. J. Shin, C. J. Lu, G. W. Kong, D. Patnaik, S. H. Lee, J. F. Cortes and J. A. Rogers. (2014). Stretchable, wireless sensors and functional substrates for epidermal characterization of sweat. *Small* 10(15): 3083–3090.

Jain, M. and B. Bhattacharya. (2020). Examining the significance of saliva and sweat in forensic science. *Journal of Legal Studies and Criminal Justice* 1(1): 6–15.

Janata, J. (1975). Immunoelectrode. *Journal of the American Chemical Society* 97(10): 2914–2916.

Jia, W., A. J. Bandodkar, G. Valdes-Ramirez, J. R. Windmiller, Z. Yang, J. Ramírez, G. Chan and J. Wang. (2013). Electrochemical tattoo biosensors for real-time noninvasive lactate monitoring in human perspiration. *Analytical Chemistry* 85(14): 6553–6560.

Jung, J. H. and S. G. Jin. (2021). Microneedle for transdermal drug delivery: current trends and fabrication. *Journal of Pharmaceutical Investigation* 1–15.

Kaushik, A., A. Vasudev, S. K. Arya, S. K. Pasha and S. Bhansali. (2014). Recent advances in cortisol sensing technologies for point-of-care application. *Biosensors and Bioelectronics* 53: 499–512.

Kaya, T., G. Liu, J. Ho, K. Yelamarthi, K. Miller, J. Edwards and A. Stannard. (2019). Wearable sweat sensors: background and current trends. *Electroanalysis* 31(3): 411–421.

Khan, Y., M. Garg, Q. Gui, M. Schadt, A. Gaikwad, D. Han, N. A. Yamamoto, P. Hart, R. Welte and W. Wilson. (2016). Flexible hybrid electronics: Direct interfacing of soft and hard electronics for wearable health monitoring. *Advanced Functional Materials* 26(47): 8764–8775.

Khan, Y., A. E. Ostfeld, C. M. Lochner, A. Pierre and A. C. Arias. (2016). Monitoring of vital signs with flexible and wearable medical devices. *Advanced Materials* 28(22): 4373–4395.

Kim, J., G. Valdés-Ramírez, A. J. Bandodkar, W. Jia, A. G. Martinez, J. Ramírez, P. Mercier and J. Wang. (2014). Non-invasive mouthguard biosensor for continuous salivary monitoring of metabolites. *Analyst* 139(7): 1632–1636.

Kim, J., I. Jeerapan, S. Imani, T. N. Cho, A. Bandodkar, S. Cinti, P. P. Mercier and J. Wang. (2016). Noninvasive alcohol monitoring using a wearable tattoo-based iontophoretic-biosensing system. *ACS Sensors* 1(8): 1011–1019.

Kim, J., A. S. Campbell and J. Wang. (2018). Wearable non-invasive epidermal glucose sensors: A review. *Talanta* 177: 163–170.

Koh, A., D. Kang, Y. Xue, S. Lee, R. M. Pielak, J. Kim, T. Hwang, S. Min, A. Banks and P. Bastien. (2016). A soft, wearable microfluidic device for the capture, storage, and colorimetric sensing of sweat. *Science Translational Medicine* 8(366): 366ra165–366ra165.

Kownacka, A. E., D. Vegelyte, M. Joosse, N. Anton, B. J. Toebes, J. Lauko, I. Buzzacchera, K. Lipinska, D. A. Wilson and N. Geelhoed-Duijvestijn. (2018). Clinical evidence for use of a noninvasive biosensor for tear glucose as an alternative to painful finger-prick for diabetes management utilizing a biopolymer coating. *Biomacromolecules* 19(11): 4504–4511.

Kumar, L. S., X. Wang, J. Hagen, R. Naik, I. Papautsky and J. Heikenfeld. (2016). Label free nano-aptasensor for interleukin-6 in protein-dilute biofluids such as sweat. *Analytical Methods* 8(17): 3440–3444.

Lee, H., C. Song, Y. S. Hong, M. S. Kim, H. R. Cho, T. Kang, K. Shin, S. H. Choi, T. Hyeon and D.-H. Kim. (2017). Wearable/disposable sweat-based glucose monitoring device with multistage transdermal drug delivery module. *Science Advances* 3(3): e1601314.

Li, W., R. N. Terry, J. Tang, M. R. Feng, S. P. Schwendeman and M. R. Prausnitz. (2019). Rapidly separable microneedle patch for the sustained release of a contraceptive. *Nature Biomedical Engineering* 3(3): 220–229.

Li, Y., C. Zheng, S. Liu, L. Huang, T. Fang, J. X. Li, F. Xu and F. Li. (2020). A smart glove integrated with tunable MWNTs/PDMS fibers made of one-step extrusion method for finger dexterity, gesture and temperature recognition. *ACS Applied Materials & Interfaces*.

Liu, Q., Y. Liu, F. Wu, X. Cao, Z. Li, M. Alharbi, A. N. Abbas, M. R. Amer and C. Zhou. (2018). Highly sensitive and wearable In_2O_3 nanoribbon transistor biosensors with integrated on-chip gate for glucose monitoring in body fluids. *ACS Nano* 12(2): 1170–1178.

Liu, Y., M. Pharr and G. A. Salvatore. (2017). Lab-on-skin: a review of flexible and stretchable electronics for wearable health monitoring. *ACS Nano* 11(10): 9614–9635.

Martín, A., J. Kim, J. F. Kurniawan, J. R. Sempionatto, J. R. Moreto, G. Tang, A. S. Campbell, A. Shin, M. Y. Lee and X. Liu. (2017). Epidermal microfluidic electrochemical detection system: Enhanced sweat sampling and metabolite detection. *ACS Sensors* 2(12): 1860–1868.

Medi, B. M., B. Layek and J. Singh. (2017). Electroporation for dermal and transdermal drug delivery. Percutaneous Penetration Enhancers Physical Methods in Penetration Enhancement, Springer: 105–122.

Mena-Bravo, A. and M. L. De Castro. (2014). Sweat: a sample with limited present applications and promising future in metabolomics. *Journal of Pharmaceutical and Biomedical Analysis* 90: 139–147.

Mishra, R. K., A. Barfidokht, A. Karajic, J. R. Sempionatto, J. Wang and J. Wang. (2018). Wearable potentiometric tattoo biosensor for on-body detection of G-type nerve agents simulants. *Sensors and Actuators B: Chemical* 273: 966–972.

Mohan, A. V., V. Rajendran, R. K. Mishra and M. Jayaraman. (2020). Recent advances and perspectives in sweat based wearable electrochemical sensors. *TrAC Trends in Analytical Chemistry*: 116024.

Nagamine, K., A. Nomura, Y. Ichimura, R. Izawa, S. Sasaki, H. Furusawa, H. Matsui and S. Tokito. (2020). Printed organic transistor-based biosensors for non-invasive sweat analysis. *Analytical Sciences*: 19R007.

Oh, S. Y., S. Y. Hong, Y. R. Jeong, J. Yun, H. Park, S. W. Jin, G. Lee, J. H. Oh, H. Lee and S.-S. Lee. (2018). Skin-attachable, stretchable electrochemical sweat sensor for glucose and pH detection. *ACS Applied Materials & Interfaces* 10(16): 13729–13740.

Padmanathan, N., H. Shao and K. M. Razeeb. (2018). Multifunctional nickel phosphate nano/microflakes 3D electrode for electrochemical energy storage, nonenzymatic glucose, and sweat pH sensors. *ACS Applied Materials & Interfaces* 10(10): 8599–8610.

Park, J., J. Kim, S.-Y. Kim, W. H. Cheong, J. Jang, Y.-G. Park, K. Na, Y.-T. Kim, J. H. Heo and C. Y. Lee. (2018). Soft, smart contact lenses with integrations of wireless circuits, glucose sensors, and displays. *Science Advances* 4(1): eaap9841.

Parrilla, M., I. Ortiz-Gomez, R. Canovas, A. Salinas-Castillo, M. Cuartero and G. A. Crespo. (2019). Wearable potentiometric ion patch for on-body electrolyte monitoring in sweat: toward a validation strategy to ensure physiological relevance. *Analytical Chemistry* 91(13): 8644–8651.

Peterson, R. A., A. Gueniche, S. Adam de Beaumais, L. Breton, M. Dalko-Csiba and N. H. Packer. (2016). Sweating the small stuff: Glycoproteins in human sweat and their unexplored potential for microbial adhesion. *Glycobiology* 26(3): 218–229.

Pokorski, J. K., J.-M. Nam, R. A. Vega, C. A. Mirkin and D. H. Appella. (2005). Cyclopentane-modified PNA improves the sensitivity of nanoparticle-based scanometric DNA detection. *Chemical Communications* (16): 2101–2103.

Promphet, N., J. P. Hinestroza, P. Rattanawaleedirojn, N. Soatthiyanon, K. Siralertmukul, P. Potiyaraj and N. Rodthongkum. (2020). Cotton thread-based wearable sensor for non-invasive simultaneous diagnosis of diabetes and kidney failure. *Sensors and Actuators B: Chemical* 321: 128549.

Qiao, L., M. R. Benzigar, J. A. Subramony, N. H. Lovell and G. Liu. (2020). Advances in sweat wearables: Sample extraction, real-time biosensing, and flexible platforms. *ACS Applied Materials & Interfaces* 12(30): 34337–34361.

Ragnarsson, O. (2020). Cushing's syndrome–Disease monitoring: Recurrence, surveillance with biomarkers or imaging studies. *Best Practice & Research Clinical Endocrinology & Metabolism* 34(2): 101382.

Regalia, G., F. Onorati, M. Lai, C. Caborni and R. W. Picard. (2019). Multimodal wrist-worn devices for seizure detection and advancing research: focus on the Empatica wristbands. *Epilepsy Research* 153: 79–82.

Reid, R. C. and I. Mahbub. (2020). Wearable self-powered biosensors. *Current Opinion in Electrochemistry* 19: 55–62.

Rock, M. J., L. Makholm and J. Eickhoff. (2014). A new method of sweat testing: the CF Quantum® sweat test. *Journal of Cystic Fibrosis* 13(5): 520–527.

Rose, D. P., M. E. Ratterman, D. K. Griffin, L. Hou, N. Kelley-Loughnane, R. R. Naik, J. A. Hagen, I. Papautsky and J. C. Heikenfeld. (2014). Adhesive RFID sensor patch for monitoring of sweat electrolytes. *IEEE Transactions on Biomedical Engineering* 62(6): 1457–1465.

Sakata, T., M. Hagio, A. Saito, Y. Mori, M. Nakao and K. Nishi. (2020). Biocompatible and flexible paper-based metal electrode for potentiometric wearable wireless biosensing. *Science and Technology of Advanced Materials*.

Sapna, K., M. Tarique, A. Asiamma, T. N. R. Kumar, V. Shashidhar, A. B. Arun and K. S. Prasad. (2020). Early detection of leptospirosis using Anti-LipL32 carbon nanotube immunofluorescence probe. *Journal of Bioscience and Bioengineering* 130(4): 424–430.

Saverino, S. and A. Falorni. (2020). Autoimmune Addison's disease. *Best Practice & Research Clinical Endocrinology & Metabolism* 34(1): 101379.

Sempionatto, J. R., L. C. Brazaca, L. García-Carmona, G. Bolat, A. S. Campbell, A. Martin, G. Tang, R. Shah, R. K. Mishra and J. Kim. (2019). Eyeglasses-based tear biosensing system: Non-invasive detection of alcohol, vitamins and glucose. *Biosensors and Bioelectronics* 137: 161–170.

Sempionatto, J. R., I. Jeerapan, S. Krishnan and J. Wang. (2019). Wearable chemical sensors: emerging systems for on-body analytical chemistry. *Analytical Chemistry* 92(1): 378–396.

Shanbhag, V. K. L. and K. Prasad. (2016). Graphene based sensors in the detection of glucose in saliva—a promising emerging modality to diagnose diabetes mellitus. *Analytical Methods* 8(33): 6255–6259.

Someya, T. and M. Amagai. (2019). Toward a new generation of smart skins. *Nature Biotechnology* 37(4): 382–388.

Sonia, J., G. M. Zanhal and K. S. Prasad. (2020). Low cost paper electrodes and the role of oxygen functionalities and edge-plane sites towards trolox sensing. *Microchemical Journal*: 105164.

Tai, L. C., W. Gao, M. Chao, M. Bariya, Q. P. Ngo, Z. Shahpar, H. Y. Nyein, H. Park, J. Sun and Y. Jung. (2018). Methylxanthine drug monitoring with wearable sweat sensors. *Advanced Materials* 30(23): 1707442.

Tai, L.-C., T. S. Liaw, Y. Lin, H. Y. Nyein, M. Bariya, W. Ji, M. Hettick, C. Zhao, J. Zhao and L. Hou. (2019). Wearable sweat band for noninvasive levodopa monitoring. *Nano Letters* 19(9): 6346–6351.

Tai, L.-C., C. H. Ahn, H. Y. Y. Nyein, W. Ji, M. Bariya, Y. Lin, L. Li and A. Javey. (2020). Nicotine monitoring with wearable sweat band. *ACS Sensors*.

Than, A., C. Liu, H. Chang, P. K. Duong, C. M. G. Cheung, C. Xu, X. Wang and P. Chen. (2018). Self-implantable double-layered micro-drug-reservoirs for efficient and controlled ocular drug delivery. *Nature Communications* 9(1): 1–12.

Tricoli, A., N. Nasiri and S. De. (2017). Wearable and miniaturized sensor technologies for personalized and preventive medicine. *Advanced Functional Materials* 27(15): 1605271.

Trung, T. Q. and N. E. Lee. (2016). Flexible and stretchable physical sensor integrated platforms for wearable human-activity monitoring and personal healthcare. *Advanced Materials* 28(22): 4338–4372.

Tu, E., P. Pearlmutter, M. Tiangco, G. Derose, L. Begdache and A. Koh. (2020). Comparison of colorimetric analyses to determine cortisol in human sweat. *ACS Omega* 5(14): 8211–8218.

Turner, A. (2013). Biosensors: then and now. *Trends in Biotechnology* 31(3): 119.

Vargason, A. M., A. C. Anselmo and S. Mitragotri. (2021). The evolution of commercial drug delivery technologies. *Nature Biomedical Engineering*: 1–17.

Veli, M. and A. Ozcan. (2018). Computational sensing of *Staphylococcus aureus* on contact lenses using 3D imaging of curved surfaces and machine learning. *ACS Nano* 12(3): 2554–2559.

Wang, J. (2008). Electrochemical glucose biosensors. *Chemical Reviews* 108(2): 814–825.

Xiao, J., Y. Liu, L. Su, D. Zhao, L. Zhao and X. Zhang. (2019). Microfluidic chip-based wearable colorimetric sensor for simple and facile detection of sweat glucose. *Analytical Chemistry* 91(23): 14803–14807.

Yamada, Y., S. Hiyama, T. Toyooka, S. Takeuchi, K. Itabashi, T. Okubo and H. Tabata. (2015). Ultratrace measurement of acetone from skin using Zeolite: toward development of a wearable monitor of fat metabolism. *Analytical Chemistry* 87(15): 7588–7594.

Yang, C., A. Abodurexiti and X. Maimaitiyiming. (2020). Flexible humidity and pressure sensors realized by molding and inkjet printing processes with sandwich structure. *Macromolecular Materials and Engineering* 305(8): 2000287.

Yang, T., D. Xie, Z. Li and H. Zhu. (2017). Recent advances in wearable tactile sensors: Materials, sensing mechanisms, and device performance. *Materials Science and Engineering: R: Reports* 115: 1–37.

Yao, H., A. J. Shum, M. Cowan, I. Lähdesmäki and B. A. Parviz. (2011). A contact lens with embedded sensor for monitoring tear glucose level. *Biosensors and Bioelectronics* 26(7): 3290–3296.

Yilmaz, T., T. Ozturk and S. Joof. (2017). A comparative study for development of microwave glucose sensors. Proceedings of the 32nd URSI GASS, Montreal, QC, Canada: 19–26.

Yokus, M. A., T. Songkakul, V. A. Pozdin, A. Bozkurt and M. A. Daniele. (2020). Wearable multiplexed biosensor system toward continuous monitoring of metabolites. *Biosensors and Bioelectronics* 153: 112038.

Yu, J., Y. Zhang, Y. Ye, R. DiSanto, W. Sun, D. Ranson, F. S. Ligler, J. B. Buse and Z. Gu. (2015). Microneedle-array patches loaded with hypoxia-sensitive vesicles provide fast glucose-responsive insulin delivery. *Proceedings of the National Academy of Sciences* 112(27): 8260–8265.

Yu, Y., H. Y. Y. Nyein, W. Gao and A. Javey. (2020). Flexible electrochemical bioelectronics: the rise of *in situ* bioanalysis. *Advanced Materials* 32(15): 1902083.

Zeng, W., L. Shu, Q. Li, S. Chen, F. Wang and X. M. Tao. (2014). Fiber-based wearable electronics: a review of materials, fabrication, devices, and applications. *Advanced Materials* 26(31): 5310–5336.

Zhang, K., G. Liu and E. M. Goldys. (2018). Robust immunosensing system based on biotin-streptavidin coupling for spatially localized femtogram mL^{-1} level detection of interleukin-6. *Biosensors and Bioelectronics* 102: 80–86.

Zhang, M., S. Guo, D. Weller, Y. Hao, X. Wang, C. Ding, K. Chai, B. Zou and R. Liu. (2019). CdSSe nanowire-chip based wearable sweat sensor. *Journal of Nanobiotechnology* 17(1): 1–10.

Zhao, J., Y. Lin, J. Wu, H. Y. Y. Nyein, M. Bariya, L.-C. Tai, M. Chao, W. Ji, G. Zhang and Z. Fan. (2019). A fully integrated and self-powered smartwatch for continuous sweat glucose monitoring. *ACS Sensors* 4(7): 1925–1933.

Zhao, Y., Q. Zhai, D. Dong, T. An, S. Gong, Q. Shi and W. Cheng. (2019). Highly stretchable and strain-insensitive fiber-based wearable electrochemical biosensor to monitor glucose in the sweat. *Analytical Chemistry* 91(10): 6569–6576.

Zhu, M., Q. Shi, T. He, Z. Yi, Y. Ma, B. Yang, T. Chen and C. Lee. (2019). Self-powered and self-functional cotton sock using piezoelectric and triboelectric hybrid mechanism for healthcare and sports monitoring. *ACS Nano* 13(2): 1940–1952.

Zhu, X., W. Liu, S. Shuang, M. Nair and C.-Z. Li. (2017). Intelligent tattoos, patches, and other wearable biosensors. Medical Biosensors for Point of Care (POC) Applications, Elsevier: 133–150.

Section II

Multiplexed Sensing and Wireless Tools for Monitoring Real-time Health Status

CHAPTER 10

Monitoring Human Health in Real-Time using Nanogenerator-Based Self-Powered Sensors

Ammu Anna Mathew,[1] Charanya Sukumaran,[2] S. Vivekanandan[1] and Arunkumar Chandrasekhar[2],*

1. Introduction

The real-time health analysis of humans is made possible by biomedical and healthcare devices through continuous monitoring, evaluation and recording of physiological signals from the body. The recent developments in Nanogenerators (NG) and nanotechnology have contributed more to the medical field. Features like miniaturization, flexibility, user-friendliness, simplicity in design, energy harvesting, durability, low power consumption, etc., have aided the process in addition to the materials used in the fabrication. The miniaturized biocompatible and biodegradable portable sensors were fabricated, which finds its application as cardiac sensors, pacemakers, cell stimulators, biomechanical motion sensors, neuro-signal sensors, drug and gene delivery devices, etc. These devices based on nanogenerators with uninterrupted power supply have upgraded the lifestyle of humans across the world. The devices are powered by the biomechanical energy from the body, thereby developing a self-powered device (Khandelwal et al. 2020). The device operated devoid of an external power source is called self-powered device. The devices based on the nanogenerator approach have proved to exhibit higher sensitivity and efficiency with respect to the available sensors. Small-scale and large-scale energy harvesting is possible based on the nanogenerator approach and has almost replaced the lithium-ion battery-based devices. The prominent challenges related to batteries

[1] School of Electrical Engineering, Technology Tower, Vellore Institute of Technology, Vellore, Tamil Nadu – 632 014, India.
[2] Nanosensors and Nanoenergy Lab, Department of Sensor and Biomedical Technology, School of Electronics Engineering, Technology Tower, Vellore Institute of Technology, Vellore-632014, India.
* Corresponding author: arunkumar.c@vit.ac.in

such as periodic replacement and recycling are resolved by the nanogenerator approach (Yoon and Kim 2020).

The Piezoelectric Nanogenerators (PENG) and Triboelectric Nanogenerators (TENG) are current technology used to convert the biomechanical energy harvested from the body to electricity for powering the devices. Despite the challenges existing in this field, these approaches are being utilized widely due to ease of fabrication and device design, in addition to the availability of suitable materials. Both these nanogenerator approaches have several applications in the biomedical fields as shown in Figure 1. They are used for *in vivo* and *in vitro* treatments. This chapter discusses some of the applications based on these two nanogenerator approaches.

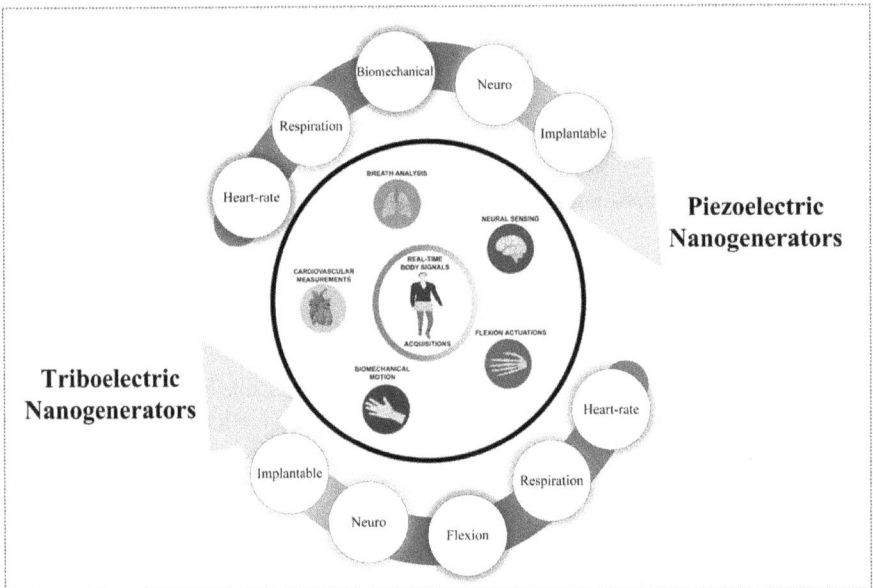

Fig. 1. Biomedical applications of nanogenerators.

2. Energy Conversion Mechanism

The technology of generating electricity by converting mechanical/thermal energy obtained from small-scale physical deformations is called Nanogenerator (NG). The NG is considered as a technology where the displacement current is the reason for the energy conversion irrespective of the materials used in fabrication. There are three different NG energy conversion approaches: piezoelectricity, triboelectricity and pyroelectricity. Mechanical to electrical energy conversion occurs in both piezoelectric and triboelectric nanogenerators whereas thermal energy to electrical energy conversion occurs in pyroelectric nanogenerators. The energy conversion mechanism is explained in detail below.

2.1 Piezoelectricity

The property where the electric charge is accumulated on certain solid materials as a result of the applied mechanical stress or pressure or latent heat is called piezoelectricity.

This technology has been widely accepted for its various integrated applications. This is due to the application of a variety of designs for the direct conversion of mechanical to electrical energy. The electric dipole moment phenomenon in solids has a close resemblance to the piezoelectric effect characteristics. Either the direct effect of the molecular groups or the ions due to asymmetric charge distribution on the crystal lattice induces the electric dipole moment (Zheng et al. 2017).

A Piezoelectric Nanogenerator (PENG) is a kinetic energy harvesting device that is able to transform the external kinetic energy to electrical power with the help of nano-structured piezoelectric materials. The invention of PENG has resolved the concerns regarding the source of energy to an extent and has led to easy integration of the device with other energy harvesters. PENG fabrication mainly involves flexible substrates and piezoelectric materials. The structure of design, encapsulation and material selection are the prime factors to be taken care of in PENG development. Two different working principles are considered based on the direction of the force exerted to the axis of the nanowires: (a) perpendicular and (b) parallel.

There are mainly three geometrical configurations for PENG based on the piezoelectric nanostructure configuration: (i) Vertical nanowire Integrated Nanogenerator (VING): A three-dimensional configuration with a three-layer stack consisting of base and counter electrode along with a piezoelectric nanostructure grown from the base electrode (vertically grown) and incorporated with counter electrode (ii) Lateral nanowire Integrated Nanogenerator (LING): A configuration with a base electrode, a Schottky contact with the metal electrode and a piezoelectric nanostructure (laterally grown) in two-dimensional pattern and (iii) Nano-composite Electrical Generators (NEG): A 3-D configuration with metal plate electrodes, piezoelectric nanostructure which is vertically grown along with a polymer matrix to fill the nanostructure.

2.2 Triboelectricity and Electrostatic Induction

The process of generating electricity by contact electrification is called triboelectricity. When there is a frictional contact between two different materials (thin organic/ inorganic films), one of the materials becomes electrically charged thereby generating the electric power. This happens due to the chemical bond formation on the surface at certain parts between the materials in contact and due to the movement of charges to gain or lose electrons to attain stability. The process of gaining and losing electrons will result in the generation of opposite charges on the materials in frictional contact leading to the generation of triboelectric potential. The drop in electric potential is balanced by the triboelectric potential. Though the triboelectric concept has been established earlier for mechanical energy harvesting in many fields including wasted mechanical energy, the full utilization of this concept is still in the study.

The Triboelectric Nanogenerator (TENG) is the combination of triboelectricity and electrostatic induction where external mechanical energy gets converted to electrical energy. The electrical charge redistribution in an object due to the effect of nearby charges is called electrostatic induction. The TENG is used due to the specific features like simplistic design, miniaturization, high efficiency and sensitivity, better accuracy, wide selection for materials which are mostly flexible, portable and cheap.

Based on the principle, there are four different configurations or modes of operation for TENG which exhibits different characteristics suitable for various applications, which are given in Table 1: (i) Vertical Contact—Separation (CS) Mode: Two vertically contacting frictional layers are formed (top and bottom) with dielectric films of different materials along with an electrode on the outer surface of each film (ii) Lateral Sliding (LS) Mode: Structure is similar to the CS mode with an exemption by creating a parallel relative sliding between the layers for electrification (iii) Single Electrode (SE) Mode: Here mechanical energy is harvested with the help of single electrode. Since the bottom electrode is grounded, the charge distribution is dependent on the movement of the top object and (iv) Freestanding Triboelectric—Layer (FT) Mode: Symmetric electrode pair is designed with same order in width and gap as that of the moving object under a dielectric film for generating asymmetric charge distribution on the material surface (Mathew et al. 2020).

Table 1. Operational Modes of TENG.

S. No.	Operational Modes	Positives	Negatives	References
1	CS Mode	Steady and better response, air breakdown situation is achieved with complexity	Complexity in design and fabrication process	(Mathew et al. 2020)
2	LS Mode	Electric charges are caused due to the to and fro motion	Problems due to the effect of use	(Mathew et al. 2020)
3	SE Mode	Design and fabrication process is simpler, has only one free layer	Electrostatic shield effect has negative impact	(Mathew et al. 2020)
4	FT Mode	Figure of Merits with high performance, single freely moving layer	Complex configuration and fabrication procedure	(Mathew et al. 2020)

Source: Data from Mathew et al. 2020

2.3 Pyroelectricity

The property exhibited by certain inherent electrically polarized crystals with huge electric fields that develops a temporary voltage on either heating or cooling that particular crystal is called pyroelectricity. The polarization characteristics of the material are varied as a result of the position change happening for the atoms due to the temperature variation in the crystal structure. The change in polarization develops a voltage across the crystal on opposite faces. The pyroelectric voltage is dependent on the temperature value. Once the temperature becomes constant, there is a reduction in the pyroelectric voltage as a result of the leakage current. The electron movement within the crystal, movement of ions in the air, etc., causes the leakage current. Almost all pyroelectric materials exhibit piezoelectric properties as well with few exceptions like Boron Gallium Nitride (BGaN) and Boron Aluminium Nitride (BAlN). The other pyroelectric effect exhibiting materials include gallium nitride, cobalt phthalocyanine, lithium tantalite, caesium nitrate, hafnium oxide and polyvinyl fluorides and so on.

The Pyroelectric Nanogenerator (PYNG) converts the external thermal energy to electrical energy with the help of nano-structured materials/nanowires with a composite structure having pyroelectric properties. In the view of the Seebeck effect, the charge carriers are diffused based on the difference in temperature between the two device ends. The spatial uniformity maintained by the temperature in the absence of a gradient hinders the application of the Seebeck effect for collecting the thermal energy from the fluctuating temperature which is time-varying. In such cases, the pyroelectric effect takes the role due to the polarization effect occurring in anisotropic solids due to temperature variation. Primary and secondary pyroelectric effects are the two operational modes of PYNG. The potential applications evolved from harvesting the wasted heat energy include wireless sensors, medical diagnostics, wearable electronics, thermal imaging and so on. The PYNG finds its application in various fields where temperature fluctuation with varying time occurs such as active sensors, temperature sensors, heat sensors, power generation, and nuclear fusion. This classification of NG generates an output with high voltage and small current (Zhao et al. 2019).

3. Real-time Health Monitoring Applications of Nanogenerators

A real-time biomedical healthcare system enables reciprocal and instant communication of physiological signals which has led to an evolution in the medical field. The development of automatic medical systems with wearable and implantable nanosensors based on nanogenerator approaches helped to detect bio-signals, even the weak ones, with greater accuracy and sensitivity. The nanogenerator-based self-powered sensors are capable of monitoring vital body signals. The time-to-time analysis helps to analyze the variation of signals from their normal value to avoid the risk of life (Novak et al. 2018). This chapter discusses the nanosensors based on the piezoelectric and triboelectric approaches.

3.1 PENG based Sensors in Healthcare

The problems in wearable healthcare devices related to power consumption are answered by the invention of self-powered sensors. The energy conversion using flexible harvesters with increased efficiency in output current is the highlighted feature of the piezoelectric effect. This is made possible by the suitable selection of materials to be used in fabrication which delivers a high piezoelectric charge coefficient. A summary of the PENG-based sensors for various applications are given in Table 2. This section explains the piezoelectric nanogenerator-based nanosensors for various applications (Park et al. 2017).

3.1.1 Implantable Sensors

A thin film piezoelectric implantable multi-frequency acoustic transducer was developed for Fully-Implantable Cochlear Implant (FICI) in the eardrum, mimicking the hair cell functions as shown in Figure 2 (A1–A3). An 8 cantilever beam multichannel structure was proposed, which resonated at a frequency range within 250 to 5000 Hz and the stimulation circuitry required output voltage was generated by sensing the eardrum vibration when resonant frequency matches the

Table 2. Summary of the PENG-based self-powered sensors.

S. No.	Device	Materials	Location	Electrical Output/ Sensitivity	References
1.	Respiration Sensor	P(VDF-TrFE)/ MWCNT	Under Nose	Peak output voltage = 0.1 V	(Zhao et al. 2019)
2.	Heartbeat Sensor	ZnO Nanowire	Rat Heart	Peak output voltage = 3 mV	(Zhao et al. 2019)
3.	Pulse Sensor	P(VDF-TrFE)/ MWCNT	Wrist and neck	Peak output voltage = 12 mV	(Zhao et al. 2019)
4.	Pulse Sensor	PZT	Wrist and neck	Peak output voltage = 65 mV	(Zhao et al. 2019)
5.	Blood Pressure Sensor	Al, PVDF	Aorta porcine	Maximum sensitivity = 14.3 mV/mmHg	(Zhao et al. 2019)

Source: Data from Zhao et al. 2019

excitation frequency thus initiating the neural stimulation. The proposed transducer made of Pulsed Laser Deposited (PLD) Lead ZirconateTitanate (PZT) with $5 \times 5 \times 0.2$ mm^3 volume and 12.2 mg mass produced a better sensitivity of 391.9 mV/Pa at 900 Hz frequency. The PLD-PZT is preferred for acoustic sensing due to its ferroelectric and piezoelectric characteristics. Figure 2 A1 shows the six mask micro fabrication details of the proposed transducer. The lift-off process is used in patterning the upper electrode and is made of chromium (30 nm)/gold (400 nm). The cantilevers are arranged in a particular pattern to obtain a smaller footprint. In the initial process, a prototype single-channel cantilever was designed, and assembled on a flexible carrier which mimics the eardrum operation when placed on the Parylene C membrane. Laser Doppler Vibrometer (LDV) was used to obtain the acceleration characterization of the prototype and other characterizations with a standard shaker table for acceleration levels ranging from 0.006 g to 0.6 g. Figure 2 A2 shows the acoustic characterization setup using the prototype. An output peak to peak voltage of 114 mV was delivered by this proposed device for a Sound Pressure Level (SPL) of 110 dB at a frequency of 1325 Hz as shown in Figure 2 A3. This device has overcome the cochlear implant disadvantages by considering the factors like stimulation signal, volume and mass. An interface circuit processed the piezoelectric sensor output which was converted to stimulation pulses. The stimulation frequency resolution is found to be in a linear relation with the number of channels, thereby improving the sound perception quality by limiting the factor of power consumption (İlik et al. 2018).

A real-time *in vivo* blood glucose detector was proposed based on the piezoelectric principle for diabetes prophylaxis treatment which is nontoxic and biocompatible. The implantable self-powered skin-like glucometer working is based on piezo-enzymatic-reaction coupling effect of GOx@ZnO nanowires as in Figure 2 (B1–B3) where GOx represents glucose oxidase. The zinc oxide (ZnO) nanowires were vertically aligned on titanium (Ti) substrate using the seed-assisted hydrothermal technique as shown in Figure 2 B1. The ZnO crystal characterization was done using X-ray powder diffraction (XRD) followed by Scanning Electron Microscope (SEM) and Transmission Electron Microscopy (TEM). Piezoelectric output signals,

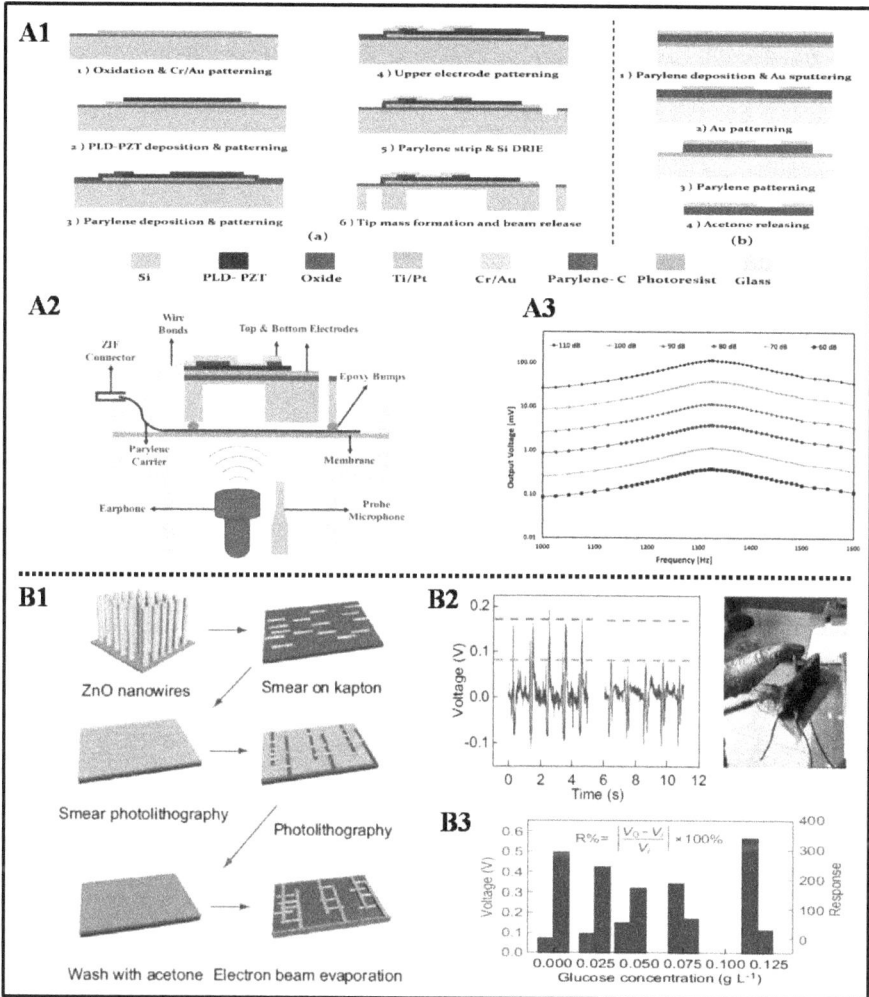

Fig. 2. (A1) Sensor fabrication technique of Cantilever thin film acoustic sensor (A2) Schematic view of characterization setup (A3) Acoustic results measured from the sensor for a sound pressure level ranging from 60 to 110 decibels SPL (Reprinted with permission from B. İlik, A. Koyuncuoğlu, Ö. Şardan-Sukas, and H. Külah, *Thin film piezoelectric acoustic transducer for fully implantable cochlear implants* (Sensors Actuators, A Phys. 2018), pp. 38–46. Copyright Elsevier). (B1) Fabrication procedure of implantable skin-like glucometer (B2) Change in piezoelectric output voltage prior to and after glucose injection (B3) Device response and piezoelectric output voltage versus glucose solution concentration (Reprinted with permission from W. Zhang et al., *Self-Powered Implantable Skin-Like Glucometer for Real-Time Detection of Blood Glucose Level In Vivo* (Nano-Micro Lett. 2018), pp. 1–11. Copyright Springer Nature).

measured using a low-noise preamplifier, are generated by applied deformations and give details of glucose detected inside the body. The experiment was done on a mouse abdomen underneath the skin for blood glucose detection as shown in Figure 2 B2. A stepper motor provides the force to run the device, thereby providing equal strain before and after glucose injection. The biosensing performance of the proposed device is shown in Figure 2 B3 for glucose concentrations

0.024 g/L, 0.045 g/L, 0.076 g/L and 0.119 g/L with a response of 16, 53, 188 and 340, respectively. The piezoelectric output decreases with an increase in glucose concentration. The output voltage acts as both biosensing signal and source signal for device operation. The device is also capable of harvesting tiny mechanical movements such as finger pressing, arm flexing and so on (Zhang et al. 2018).

3.1.2 Neuro-signal Monitoring Sensors

Electrical pulses are used to stimulate specific areas of the brain through a neurosurgical procedure called Deep Brain Stimulation (DBS), which is an effectual treatment for most of the neurologic and psychiatric disorders. A self-powered flexible DBS has been reported based on the piezoelectric effect to induce movements in the body as shown in Figure 3 (A1–A3). The high-performance thin-film single crystal $Pb(In_{1/2}Nb_{1/2})O_3$–$Pb(Mg_{1/3}Nb_{2/3})O_3$–$PbTiO_3$ (PIMNT) energy harvester with dimension 1.7×1.7 cm is placed on a Polyethylene Terephthalate (PET) substrate (125 μm) that enables the DBS in the test specimen (mice). The maximum threshold current for operating the real-time DBS to enable the forearm motion in mice is generated from the bending movement (mechanical movement), which generates an increasing current with the maximum limit at 0.57 mA and for biomechanical motion 11 V voltage is obtained, in addition to the instantaneous power of 0.7 mW. Figure 3 A1 shows the proposed PIMNT energy harvester which has implanted a stimulation electrode. The output measurement is done using a linear bending stage (periodic bending and unbending movements) along with the source-meter and Faraday cage. A slight horizontal displacement along with strain change is applied on the bending stage for analysis with a straining rate of 2.3% per second at a frequency of 0.32 Hz. The ferroelectric property in the thin PIMNT crystal is the reason for the high piezoelectric constant on the PET substrate. Figure 3 A2 shows the COMSOL simulation model of the piezo-potential distribution inside PIMNT energy harvester based on Finite Element Analysis (FEA) for estimating the output voltage and delivered a piezoelectric potential difference (ΔV) of 24.5 V. The generated electrical signals are transmitted to stimulation electrodes by means of metal wires. The harvested electrical energy is directly applied to specific brain parts to induce body movements; for example, when M1 cortex in mice is activated through an electrical pulse from the flexible energy harvester by which the forelimb muscles are contracted. The flexible PIMNT harvester is capable of supplying electric power and recharging the internal battery. Further applications of this device can be extended to areas like a motion-feedback neuronal stimulator, tremor intensity sensor and so on depending on the generated current. The external load resistance was varied from 120 ohms to 220 Mega-ohms for characterizing the instantaneous output power (maximum 0.7 mW for load resistance 70 kΩ). In addition to instantaneous power, the output voltage and the output current are also determined as shown in Figure 3 A3 (Hwang et al. 2015).

The C-reactive protein (CRP) was detected by an ultra-sensitive QCM immunosensor from the serum based on piezoelectric effect with Horseradish Peroxidase (HRP) labeled gold nanoparticles (AuNPs) signal tag as shown in Figure 3 (B1–B3). The multiplexed signal amplified immunosensor is fabricated by initially creating sandwich-type immunoreactions between the capture probes

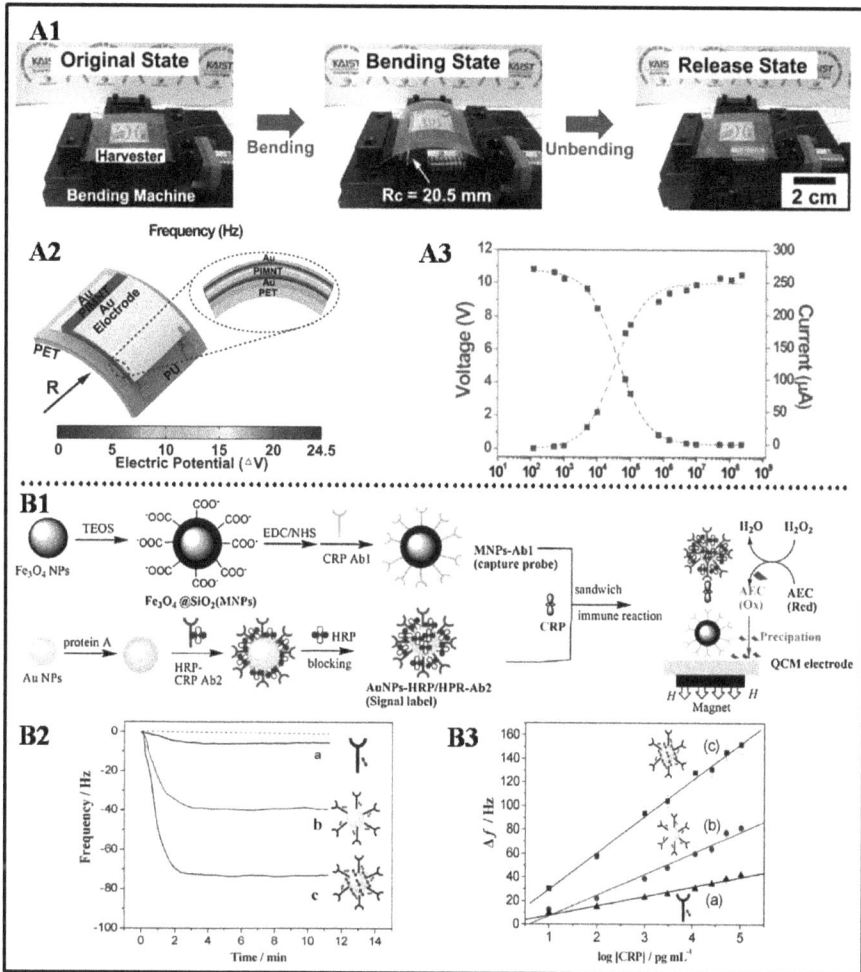

Fig. 3. (A1) Electrical output characteristics of flexible thin film PIMNT harvester in three states: original, bend and release positions (A2) The piezo-potential distribution MIM-type PIMNT energy harvester for a tensile strain of 0.3% (A3). The resultant voltage and current of flexible PIMNT harvester under various load resistance varied from 120 ohms to 220 Megaohms (Reprinted with permission from G. T. Hwang et al., *Self-powered deep brain stimulation via a flexible PIMNT energy harvester* (Energy Environ. Sci. 2015), pp. 2677–2684. Copyright Royal Society of Chemistry). (B1) The experimentation and detection principle of Fe_3O_4 on SiO_2-Ab1 and AuNPs-HRP/HRP-Ab2 amalgamation (B2) Immunosensor frequency traces with various signal tags (B3) Immunosensor calibration curve with different signal tags (Reprinted with permission from J. Zhou et al., *Ultratrace detection of C-reactive protein by a piezoelectric immunosensor based on Fe3O4@SiO2 magnetic capture nanoprobes and HRP-antibody co-immobilized nano gold as signal tags* (Sensors Actuators, B Chem. 2013), pp. 494–500. Copyright Elsevier).

MNPs-CRP Ab1 (SiO_2-coated magnetic Fe_3O_4 Nanoparticles (MNPs) marked with the primary CRP antibody), CRP (multiple concentrations) and signal tag AuNPs-HRP/HRP-CRP Ab2. Hydrogen peroxide and 3-amino- 9- ethylcarbazole (AEC) exposure was given to the immunocomplex. Figure 3 B1 shows the procedure for the two capture probes with its detection principle. The increase in the enhancement of CRP through magnetic separation for signal amplification and immobilization on the electrode for sensor preparation with an external magnet is done with the help of a MNPs-CRP Ab1 capture probe. Catalyzing and yielding more AEC's indissoluble oxidation product on the surface of a piezoelectric crystal thereby providing a change in frequency response is done by the signal tag, thereby obtaining multiplex signal amplification. The change in frequency is proportional to the CRP concentration with a low detection limit of 0.3 pg/mL for a concentration range from 0.001 to 100 ng/mL. Figure 3 B2 shows the frequency trace made by capture probe MNPs-CRP Ab1 with the following signal tags HRP-CRP Ab2, AuNPs-HRP-CRP Ab2 and AuNPs-HRP/HRP-CRP Ab2 on mixing with H_2O_2 and AEC, on a real-time basis. Figure 3 B3 shows the calibration curve of immunosensor to verify the amplification effect using different signal tags HRP-CRP Ab2, AuNPs-HRP-CRP Ab2 and AuNPs-HRP/HRP-CRP Ab2. The proposed system finds its application in the clinical investigation of ultra-trace CRP in heart diseases (Zhou et al. 2013).

3.2 TENG-based Sensors in Healthcare

Energy conversion based on the triboelectric nanogenerator approach is advancing technology in the field of customized and healthcare electronics (Fan et al. 2012). The biomechanical motion detected by the self-powered active nanosensors is activated by the bio-signals generated by the TENG output (Zheng et al. 2016). The electrical signals generated by the sensor are utilized by the self-powered device. A brief description of the TENG-based sensors for various applications is given in Table 3. This section focuses on the triboelectric-based sensors for various applications.

Table 3. Brief details of the TENG-based self-powered sensors.

S. No.	Device	Materials	Location	Electrical Output/ Sensitivity	References
1.	Power Pacemaker	Al, PDMS	Left chest	Peak average output voltage = 3.73 V	(Liu et al. 2019)
2.	Pulse Sensor	Copper, PDMS, FEP	Chest and wrist	Maximum sensitivity S_{max} = 150 mV/Pa	(Mathew et al. 2020)
3.	Obesity Sensor	Au, PDMS, PTFE, Ecofle	Stomach	Peak average output voltage = 60 mV	(Mathew et al. 2020)
4.	Heartbeat Sensor	Al, Parylene C, PTFE, PDMS	Heart	Peak average output voltage = 10 V and Peak average output current = 4 μA	(Liu et al. 2019)
5.	Endocardial Pressure Sensor	PTFE, Al, PDMS	Ventricle and Atrium	Peak average output voltage = 6.2 V	(Liu et al. 2019)

3.2.1 Implantable Sensors

The triboelectric effect-based applications are the recent advancement in the biomedical field. One of the milestone inventions among them is the Drug Delivery (DD) system, a miniaturized flexible self-powered system. Figure 4 A1 shows the schematic diagram of an implantable Drug Delivery system (iDDS) for ocular drug delivery with the help of an electrochemical micro-fluidic pump having a rate of flow from 5.3 μL/min to 40 μL/min powered by the TENG operating various rotating speeds. TENG is powered by human hand motion. The TENG is fabricated as two layers with metal copper in radial array pattern with an electrification layer made of Polytetrafluoroethylene (PTFE) film in between the copper layers encapsulated by glass epoxy. The copper (Cu) layer acts as a stator and rotator. The drug delivery system contains a reservoir (10 × 10 × 2 mm), PDMS (Polydimethylsiloxane) microtube (100 mm long with 0.5 mm and 0.3 mm diameters outside and inside, respectively) and gold as electrodes (5 × 5 × 0.5 mm) on a silicon substrate with Parylene C encapsulation. An *ex vivo* trans-sclera DD was performed in porcine eyes by implanting either in the subcutaneous tissues or under the eye's sclera for demonstration of the concept and the TENG was operated by the biokinetic energy from the human hand as shown in Figure 4 A2. The drug refilling ability of the reservoir is an advantage of the invention. The TENG with electrochemical pressure pump produced an output current of 1.5 mA for an output voltage of 15 V with 500 rpm rotating speed. Figure 4 A3 shows that the flow rate is measured for various rotating speeds of TENG till 600 rpm. Higher rotation speeds can damage the TENG; hence, lower speeds were utilized to characterize the TENG performance (Song et al. 2017).

Bone diseases are serious health issues affecting one's daily life which needs immediate treatment. Usually, the treatments available are long-term processes. To overcome this difficulty, an innovative implantable and portable technique based on triboelectric principle and infrared (IR) laser irradiation has been introduced as in Figure 4 (B1–B3). A self-powered low-level laser cure (SPLC) system has been proposed for treating osteogenesis. The SPLC system includes a flexible arch-shaped TENG and an IR laser excitation unit (wavelength: 850 nm, power: 10 mW, and power density: 35.4 mW/cm^2). The proposed contact-separation mode TENG (1.5 cm × 1.0 cm) is made of a pyramid array patterned (5 μm long) Polydimethylsiloxane (PDMS) acting as one triboelectric layer and indium tin oxide (ITO) acts as another triboelectric layer as shown in Figure 4 B1. This device was found effective for the mouse embryonic osteoblasts' proliferation (MC3T3-E1–15% enhancement in 2 days) and differentiation (16.9% acceleration in 5 days), thereby treating tooth and bone diseases giving an output: 70 nC charge transfer, 30 μA SC current and 115 V OC voltage as in Figure 4 B2 and B3. MTT assay was performed to evaluate the proliferation. The laser unit irradiation rate is 1 pulse/min and the implantable TENG (iTENG) output is 0.06 nA SC current and 0.2 V OC voltage. This system can harvest even energy from human walking, mouse breathing and so on, thereby ensuring its capability to utilize as a portable device. The experiment was performed on a human arm to harvest the swing movement during walking (Tang et al. 2015).

3.2.2 Neuro-signal monitoring sensors

The human sensory tissues are mimicked by the artificial sensory-substitution system based on the triboelectric effect as shown in Figure 5 (A1–A3). The self-powered multi-perception e-skin receptor contributes to the sensory-handicap patient perception. The

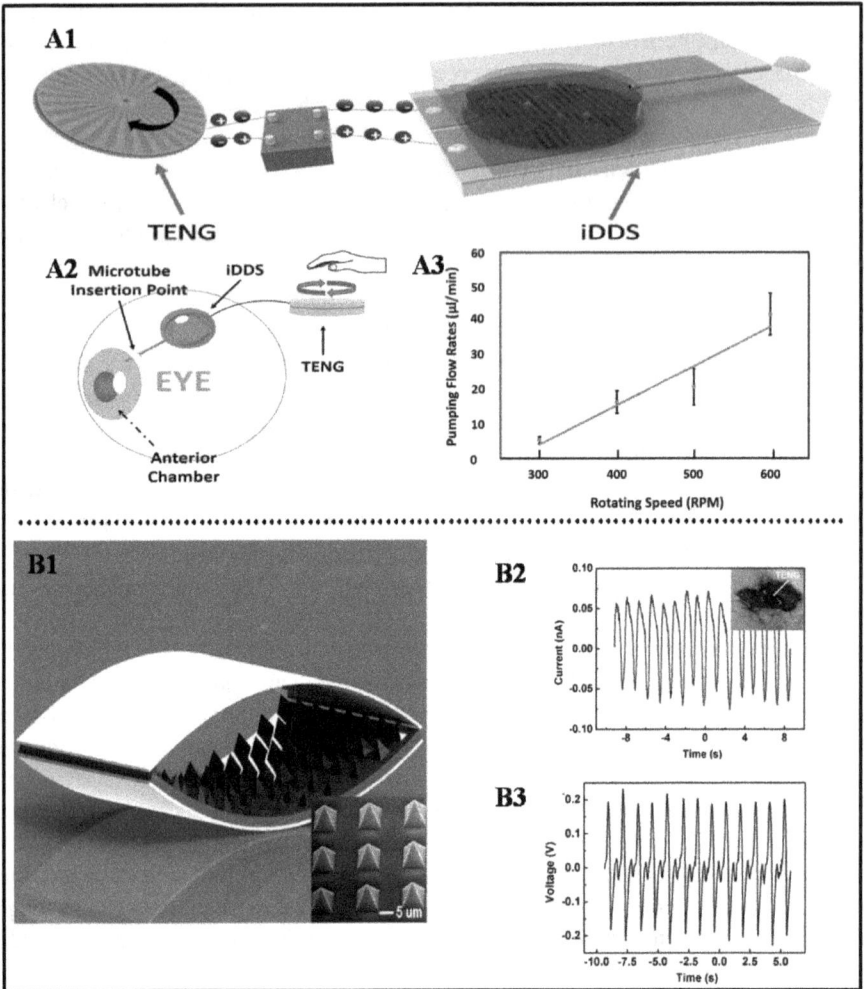

Fig. 4. (A1) Pictorial representation of self-powered iDDS (A2) Schematic representation of *ex vivo* drug delivery device implantation in the anterior chamber (A3) Pumping performance of iDDS: Flow rate versus rotating speed (Reprinted with permission from P. Song et al., *A Self-Powered Implantable Drug-Delivery System Using Biokinetic Energy* (Adv. Mater. 203), pp. 1–7. Copyright John Wiley & Sons). (B1) Schematic diagram of flexible arch-shaped TENG (B2) Output current of TENG implanted on rat (B3) Output voltage of TENG implanted between liver and diaphragm of a rat (Reprinted with permission from W. Tang et al., *Implantable Self-Powered Low-Level Laser Cure System for Mouse Embryonic Osteoblasts' Proliferation and Differentiation* (ACS Nano 2015), pp. 7867–7873. Copyright American Chemical Society).

proposed system substitutes tactile, visual, auditory, olfactory and gustatory sensations. The triboelectric sensory signal output from the device is directly given to the brain without an external power supply. The device supply is obtained from human motion energy through the triboelectric effect. The sensing unit arrays function as both the brain sensory receptors and power source for the device. The sensor unit arrays imitate various perceptions, thereby transmitting code signals to the brain. Figure 5 A1 shows the fabrication procedure of the multi-perception receptor on a 4 cm × 4 cm × 10 μm lithography patterned copper network. The copper foil attached to SiO_2 acts as source material with a 0.8 μm thick photoresist in positive charge. PDMS is spin-coated on the patterned Cu network (back electrode) with polypyrrole (Ppy-positive material) as the upper electrode. The groove-patterned flexible PDMS of size 4 cm × 4 cm × 0.5 mm

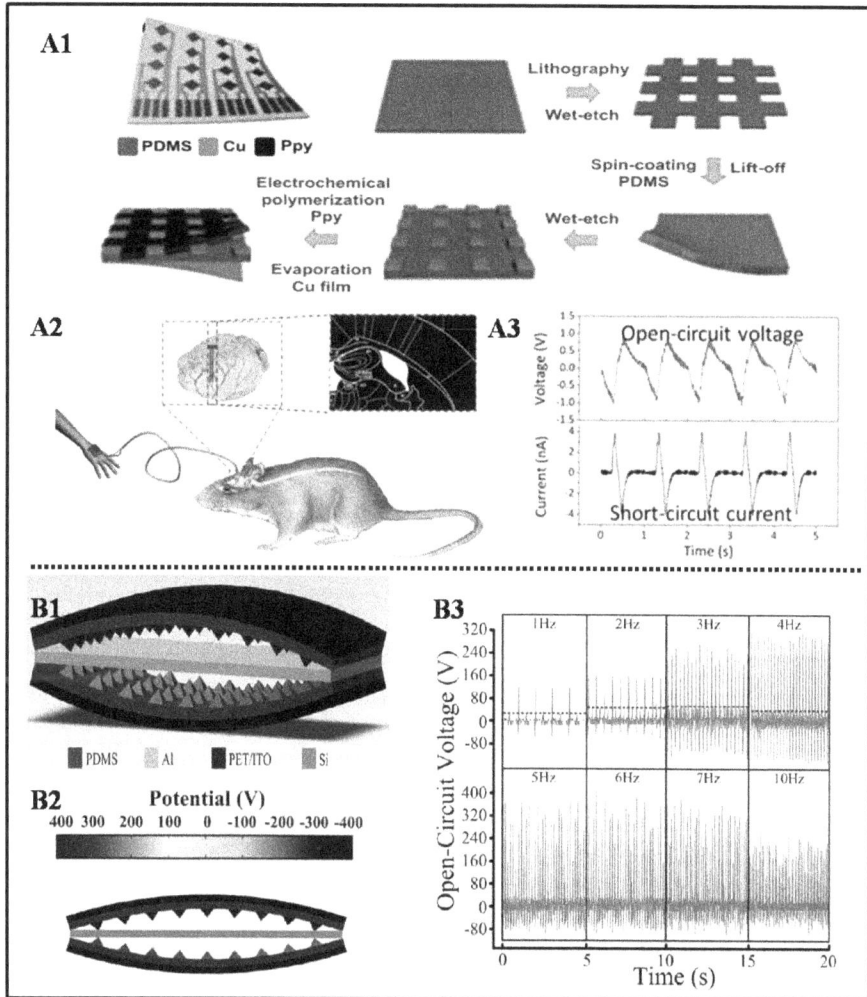

Fig. 5. (A1) Fabrication technique of self-powered multi-perception e-skin (A2) Schematic experiment details of sensory tribo-electrical signal transmission into mouse brain (A3) Output SC current and OC voltage of the individual sensor in e-skin for applied deformation (Reprinted with permission from Y. Fu et al., *A self-powered brain multi-perception receptor for sensory-substitution application* (Nano Energy 2018), pp. 43–52. Copyright Elsevier). (B1) Schematic view of the sandwich-shaped TENG—structural design (B2) Graphical diagram of the electrical potential generated (B3) The output performance characterization of the sandwich shaped TENG with nanostructured PDMS film for a frequency range of 1–10 Hz under external forces (Reprinted with permission from X. S. Zhang et al., *Frequency-multiplication high-output triboelectric nanogenerator for sustainably powering biomedical Microsystems* (Nano Lett. 2013), pp. 1168–132. Copyright American Chemical Society).

is the negative frictional material. The structure has three electrodes: the working electrode (platinum (Pt) wafer), the counter electrode (Pt wire) and the reference electrode (Ag/AgCl). The device characterization was done by SEM. The brain sensory receptors track force trace of tactility, taste, image recognition, smell analysis and audibility of words. There is a tracking force trace of tactile effect on the integrated sensor array unit. The trace is a result of the finger movement over the sensor unit, thus differentiating and distinguishing the static and dynamic press made in tactile sensing. The experimentation was done on a mouse brain at the primary somatosensory barrel cortex (S1BF) as shown in the schematic diagram Figure 5 A2. The S1BF on activation can resemble the perception of the mouse and will be able to control the mouse activities. On bending deformation of the device at an angle of $90°$ with a force of 22 N at 1 Hz frequency, a single sensor unit generates an output with SC current of ~ 4 nA and OC voltage of ~ 1 V as shown in Figure 5 A3. There is a linear relationship among the triboelectric output current, bending angle and force magnitude (Fu et al. 2018).

A device which gives electrical energy with double frequency from a low frequency mechanical energy is made possible with the invention of sandwich-shaped TENG as shown in Figure 5 (B1–B3), which has the capability of powering a 3-D microelectrode array (implantable) used for neural prosthesis. The frequency multiplication was achieved by introducing a structure where contact electrification occurs twice within a single external force cycle. The device (2 cm × 4 cm) is fabricated by placing an aluminium (Al) film (20 μm) between the two PDMS membranes (450 μm) in the sandwich model as shown in Figure 5 B1. The PDMS surface is designed with micro/nano dual-scale structures along with a PET thin film (125 μm) bend in arch-shape to improve the device output performance and an ITO thin film (125 μm) on the top of the PDMS membrane acts as an electrode. Initially, there is no electric potential between the Al and PDMS layers but during the single periodic compressive force these layers come in contact twice thereby generating a double frequency output. The electrical potential distribution based on the periodic compressive force is shown in Figure 5 B2. The device output is as follows: Peak voltage = 465 V, Current density = 13.4 μA/cm² and Energy volume density = 53.4 mW/cm³. The device performance for various frequencies under the compressive force is shown in Figure 5 B3. The open circuit voltage is increased from 120 to 320 V for a frequency increase from 1 Hz to 5 Hz but for a frequency increase from 5 to 7 Hz, the OC voltage is constant. Again, the open circuit voltage decreases to 218 V for frequency ranging from 7 Hz to 10 Hz. This is due to the contact and release of the sandwich structure under the effect of external force. This device is capable of powering biomedical micro-systems, thus extending its potential application in the biomedical fields (Zhang et al. 2013).

4. Conclusion

In summary, the evolution of technology has brought immense changes in the human lifestyle, which has led to the increased demand for customizable electronics. This increase in demand is due to certain specific features like miniaturization, flexibility, user-friendliness, simplicity in design, energy harvesting, durability, low power consumption and so on. This technological evolution by incorporating intelligent algorithms and the Internet of Things (IoT) had a great impact on the biomedical and healthcare field, which implemented a self-powered healthcare long-term wearable medical device. The challenges involved in obtaining the crucial body signals for real-time analysis were resolved to an extent with the introduction of nanotechnology in the biomedical and healthcare industry. The sensors fabricated by using nanomaterials have proved to produce higher sensitivity and reliability in a cost-effective way compared to the existing sensors in harvesting the biomechanical energy. Nanotechnology has paved the way for the increased demands by the introduction of cost-effective portable devices. The development of self-powered nanosensors for real-time applications has become a milestone in the healthcare industry, giving importance to output optimization and power management in design during device integration. The nanogenerator-based inventions have replaced most of the battery-based devices in almost all fields. Further studies in the field of nanogenerators incorporated with nanotechnology can result in an ultimate real-time health monitoring solution.

Conflict of interest

The authors declare no conflict of interest.

Acknowledgement

This work was supported by Vellore Institute of Technology, Vellore, Tamil Nadu, India.

References

Fan, F., Z. Tian and Z. Lin. (2012). Flexible triboelectric generator. *Nano Energy* 1(2): 328–334.

Fu, Y., M. Zhang, Y. Dai, H. Zeng, C. Sun, Y. Han, L. Xing, S. Wang, X. Xue, Y. Zhan and Y. Zhang. (2018). A self-powered brain multi-perception receptor for sensory-substitution application. *Nano Energy* 44(December 203): 43–52.

Hwang, G. T., Y. Kim, J. H. Lee, S. Oh, C. K. Jeong, D. Y. Park, J. Ryu, H. Kwon, S. G. Lee, B. Joung, D. Kim and K. J. Lee. (2015). Self-powered deep brain stimulation via a flexible PIMNT energy harvester. *Energy and Environmental Science* 8(9): 2677–2684.

İlik, B., A. Koyuncuoğlu, Ö. Şardan-Sukas and H. Külah. (2018). Thin film piezoelectric acoustic transducer for fully implantable cochlear implants. *Sensors and Actuators, A: Physical* 280: 38–46.

Khandelwal, G., N. P. Maria Joseph Raj and S. J. Kim. (2020). Triboelectric nanogenerator for healthcare and biomedical applications. *Nano Today* 33: 100882 (1–29).

Liu, Z., H. Li, B. Shi, Y. Fan, Z. L. Wang and Z. Li. (2019). Wearable and implantable triboelectric nanogenerators. *Advanced Functional Materials* 1808820: 1–19.

Mathew, A. A., A. Chandrasekhar and S. Vivekanandan. (2020). A review on real-time implantable and wearable health monitoring sensors based on triboelectric nanogenerator approach. *Nano Energy* 80(November 2020): 105566.

Novak, V., L. E. Blanc, R. L. E. Blanc and R. Martin. (2018). Influence of respiration on heart rate and blood pressure fluctuations. *Journal of Applied Physiology* 74(2): 63–626.

Park, D. Y., D. J. Joe, D. H. Kim, H. Park, J. H. Han, C. K. Jeong, H. Park, J. G. Park, B. Joung and K. J. Lee. (2017). Self-powered real-time arterial pulse monitoring using ultrathin epidermal piezoelectric sensors. *Advanced Materials* 29(37): 1–9.

Song, P., S. Kuang, N. Panwar, G. Yang, D. J. H. Tng, S. C. Tjin, W. J. Ng, M. B. A. Majid, G. Zhu, K. T. Yong and Z. L. Wang. (2017). A self-powered implantable drug-delivery system using biokinetic energy. *Advanced Materials* 29(11): 1–7.

Tang, W., J. Tian, Q. Zheng, L. Yan, J. Wang, Z. Li, Z. L. Wang, C. Academy, B. Science, M. Engineering, M. Science and U. States. (2015). Implantable self-powered low-level laser cure system for mouse embryonic osteoblasts' proliferation and differentiation. *ACS Nano* 9(8): 7867–7873.

Yoon, H. and S. Kim. (2020). Perspective nanogenerators to power implantable medical systems. *Joule* 4(7): 1–10.

Zhang, W., L. Zhang, H. Gao, W. Yang, S. Wang, L. Xing and X. Xue. (2018). Self-powered implantable skin-like glucometer for real-time detection of blood glucose level *in vivo*. *Nano-Micro Letters* 10(2): 1–11.

Zhang, X. S., M. Di Han, R. X. Wang, F. Y. Zhu, Z. H. Li, W. Wang and H. X. Zhang. (2013). Frequency-multiplication high-output triboelectric nanogenerator for sustainably powering biomedical microsystems. *Nano Letters* 13(3): 1168–132.

Zhao, L., H. Li, J. Meng and Z. Li. (2019). The recent advances in self-powered medical information sensors. *InfoMat* 2(1): 212–234 (12064).

Zheng, Q., B. Shi, Z. Li and Z. L. Wang. (2017). Recent progress on piezoelectric and triboelectric energy harvesters in biomedical systems. *Advanced Science* 4(7): 1–23 (1700029).

Zheng, Q., H. Zhang, B. Shi, X. Xue, Z. Liu, Y. Jin, Y. Ma, Y. Zou, X. Wang, Z. An, W. Tang, W. Zhang, F. Yang, Y. Liu, X. Lang, Z. Xu, Z. Li and Z. L. Wang. (2016). *In vivo* self-powered wireless cardiac monitoring via implantable triboelectric nanogenerator. *ACS Nano* 10(7): 6510–6518.

Zhou, J., N. Gan, T. Li, H. Zhou, X. Li, Y. Cao, L. Wang, W. Sang and F. Hu. (2013). Ultratrace detection of C-reactive protein by a piezoelectric immunosensor based on Fe_3O_4@SiO_2 magnetic capture nanoprobes and HRP-antibody co-immobilized nano gold as signal tags. *Sensors and Actuators, B: Chemical* 38: 494–500.

CHAPTER 11

Nanogenerator Based Self-Powered Sensors for Healthcare Applications

Gaurav Khandelwal,[1] *Pandey Rajagopalan,*[2] *Nirmal Prashanth Maria Joseph Raj,*[1] *Xiaozhi Wang*[2] and *Sang-Jae Kim*[1,*]

1. Introduction

The past decade was dominated by flexible and wearable sensors, the Internet of things (IoT), and smart electronics for many applications. The key challenge is to fulfill the power requirement of the sensors. Generally, Li-ion batteries are used to power up smart devices and sensors. Batteries suffer from short-life use, use of hazardous materials and are difficult to recycle. The market requires a portable, sustainable and eco-friendly alternative to batteries to power the low-rated electronics.

Numerous renewable energy sources like wind, solar, thermoelectric, etc., are utilized for various purposes. The energy harvesting devices convert different energy sources into electricity by utilizing mechanisms like contact-electrification, photovoltaic effect, piezoelectric effect, electromagnetic and electrostatic induction, etc. Among energy harvesting devices, nanogenerators (NGs), especially piezoelectric (PENG) and triboelectric (TENG), are of great interest due to their attractive attributes. The TENG and PENG are light-weight, can be designed as eco-friendly devices, and can be used for hybrid devices. The TENG has an added advantage of wide material choice, numerous device designs and high voltage. The advantages of NGs make them ideal for broad applications like physical, biological and chemical sensors, drug delivery, cell stimulation, implantable devices, water filtration, etc., as

---

[1] Nanomaterials and System Lab, Major of Mechatronics Engineering, Faculty of Applied Energy System, Jeju National University, Jeju 632-43, South Korea.
[2] Key Laboratory of Micro-nano Electronic Devices and Smart Systems of Zhejiang Province, College of Information Science & Electronic Engineering, Zhejiang University, Hangzhou, China.
* Corresponding author: kimsangj@jejunu.ac.kr

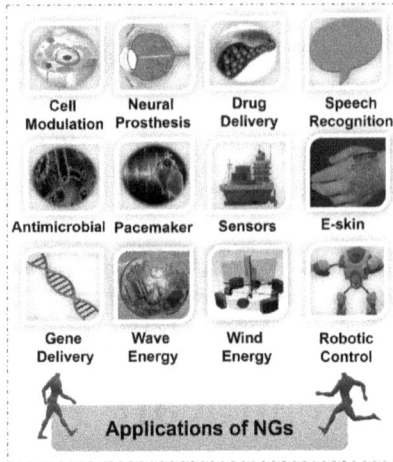

Fig. 1. Applications of TENG and PENG.

summarized in Figure 1. The application of NGs for healthcare sensors is paramount. The NG-based sensors can improve the patient's quality of life while maintaining the durability and never-ending uninterrupted power supply.

This chapter is focused on the use of TENG and PENG for healthcare sensors. The fundamentals of TENG and PENG are summarized in brief. The recent progress and achievement of NGs for various healthcare sensors like heartbeat sensors, sleep monitoring sensors, and chemical sensors is comprehensively summarized. Finally, the chapter is concluded with future perspectives and challenges.

2. Fundamentals of Nanogenerator

The triboelectric and piezoelectric nanogenerator (TENG and PENG) are two types of NG based on their different electricity generation mechanism. Maxwell's displacement current theory was considered the NG origin base.

2.1 Piezoelectric Nanogenerator (PENG)

Piezoelectric was discovered in the late 19th century by the Curie brothers upon hitting a quartz crystal (Kim et al. 2009, Rajagopalan et al. 2018, Richter et al. 2009). Piezoelectricity is the materials' ability to produce charge (or voltage) when stimulated by mechanical stress. This is due to non-inversion symmetry in the crystal lattice, which disrupts on account of strain, thus generating a dipole. The piezoelectric effect can be clarified utilizing a molecular model where the lattice is in an equilibrium state. Once an external force is applied upon the material, the lattice distorts, displacing the charge center. This results in the formation of a small dipole.

Mutually, all small dipoles contribute to a net dipole in the material. This phenomenon in materials is known as piezoelectricity. Figure 2a shows the schematic of the lattice shifting on account of strain. The experimental and theoretical investigations reveal that the piezoelectric material possesses direct and indirect

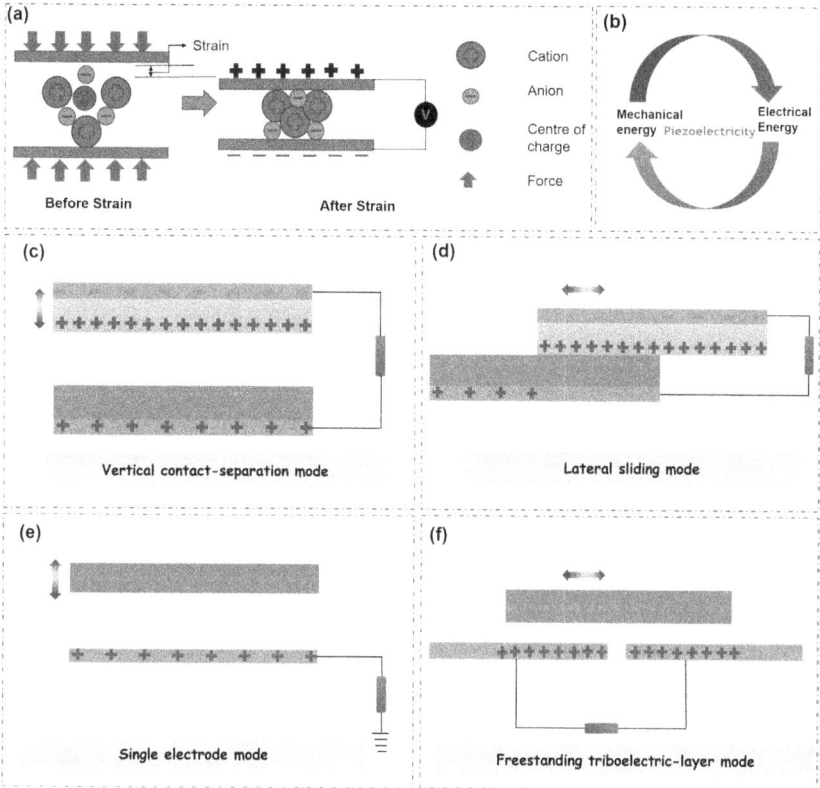

Fig. 2. (a,b) The principle and working mechanism of PENG. The fundamental working modes of TENG. (c) Contact-separation mode. (d) Lateral sliding mode. (e) Single electrode mode and (f) Freestanding triboelectric-layer mode. Reprinted with permission from (Khandelwal et al. 2020). Copyright Elsevier.

(converse) effects. In the direct effect, the strain can be converted into electricity. However, in the indirect effect, the electricity can be converted into motion, as shown in Figure 2b. These effects led to huge developments in the area of sensors, communications, transducers, health care, etc. Due to their dynamic nature, the earliest crystals were used in the field of communication and submarine detection demonstrated by Paul Langevin in the year 1917. Since then, the device has been used in a variety of applications. These materials have a long history of influencing the healthcare sectors in the most positive ways, like X-Rays and MRIs. These highly efficient crystals like lead zirconia titanate have been employed in non-invasive ultrasound diagnostic applications, which remains the fundamental hallmark of these materials in healthcare applications. This has led to some serious development in healthcare diagnostics and sensing. The direct and indirect principle was used for several other applications like inkjet printing, ultrasonic microbalances and non-destructive testing of materials. Figure 3 shows several applications of direct and converse piezoelectric effect.

In piezoelectric materials, elastic fields are coupled with the electric field by the following equations due to anisotropy (Rajagopalan et al. 2016).

$$T_{ij} = c_{ijkl}^{E} S_{kl} - e_{kij} E_{k} \tag{1}$$

$$D_{i} = e_{kij} S_{kl} + \varepsilon_{ij}^{S} E_{k} \tag{2}$$

where, T_{ij}, S_{kl}, E_{k}, D_{i}, c_{ijkl}^{E}, e_{kij} and ε_{ij}^{S} are the stress component, strain component, electric field component, electric displacement component, elastic constant, piezoelectric constant, and dielectric constants.

The changes in the atoms/molecules' position due to applied stress lead to the formation of net dipole moments that causes polarization and an electric

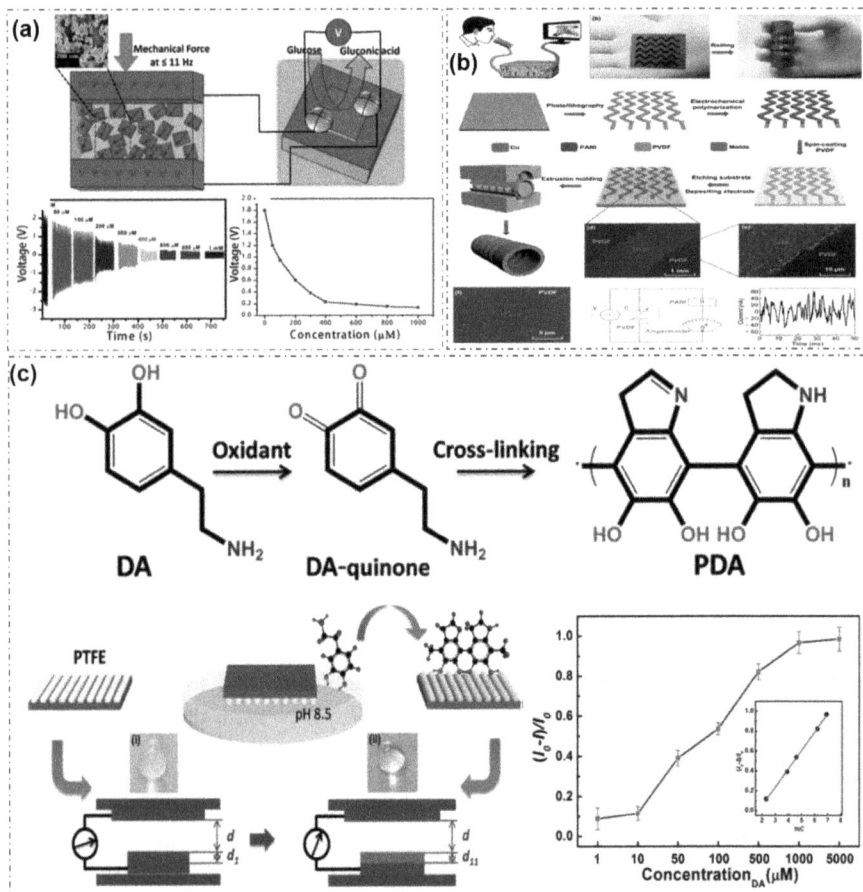

Fig. 3. (a) The working mechanism of self-powered biological sensing of glucose using a barium titanate film, the device's output under various concentrations of glucose. Reprinted with permission (Selvarajan et al. 2016). Copyright Elsevier. (b) Self-powered breath analyzer using PVDF/PANI. Figure shows the fabrication methodology along with the lumped model and current output of the data. Open Access. (c) The schematic illustration for the working mechanism of dopamine sensor and change in the current output at different dopamine concentrations. Reprinted with permission (Jie et al. 2015). Copyright American Chemical Society.

field, respectively. Many materials have been researched in the past century, and potentially, perovskites have been the clear winner in almost all the figures of merits (FOM). Arguably, lead zirconia titanate is one such versatile material that has worked pretty well in direct and indirect effects. However, because of lead, the governments worldwide want to replace this material with somewhat more biocompatible materials. This led to some serious research in piezoelectric materials in the last few decades. The other important materials are quartz, barium titanate, zinc oxide and PVDF, etc. Here, in this chapter, we have briefly discussed the self-powered characteristics of piezoelectric in the field of healthcare devices.

2.2 Triboelectric Nanogenerator (TENG)

The triboelectric effect is ubiquitous and happens when materials with distinct surface charges come in contact. The damage caused by triboelectricity makes it an adverse effect in many industries. Later, the triboelectric effect was successfully utilized in the friction machine and Van de Graff generator (Furfari 2005, Wu et al. 2019). Recently, in 2012, Wang et al. reported the TENG, which works based on the coupling effect of contact electrification (triboelectrification) and electrostatic induction (Fan et al. 2012). Since the day of its invention, the TENG has been utilized for a wide range of applications like physical, biological and chemical sensors, drug delivery, cell stimulation, implantable devices, blue and wind-wave energy harvesting, etc. The wide applications are ascribed to the advantages offered by TENG. Some of the TENG advantages are low-cost, light-weight, wide choice of designs and materials, flexibility, portability and high output, etc. (Feng et al. 2018, Wu et al. 2019). The charge transfer happens when two materials with opposite surface charges come in frictional contact. In TENG, either one or both materials are backed by an electrode, depending on the device model. Thus, opposite charges develop on the electrode via electrostatic induction. This phenomenon creates a potential difference during the dynamic process of contact and separation. The electrons will flow from one electrode to another (or ground), generating an A.C. output (Niu et al. 2013b, Yang et al. 2016a). The TENG operates in four different modes, as shown in Figure 2c–f.

2.2.1 Vertical Contact-separation (c-s) mode

The c-s mode is conventional TENG mode with a simple and straightforward fabrication process (Figure 2c). In the presence of force, when the opposite materials come close to each other, the opposite charges developed depending on their tendency to lose or gain electrons. When force is removed, the two layers try to regain the initial position due to the spacer, thus creating a potential difference. The electron flow from one electrode balances the potential difference to the other. This flow of electrons will generate a half cycle of the TENG output. On applying force, the top layer moves towards the bottom layer, leading to the flow of electrons in the opposite direction, thus completing a full cycle of the output. The process repeats in the presence and absence of force generating the output of TENG (Fan et al. 2012, Niu et al. 2013b, Wang et al. 2016c, Yang et al. 2016a).

2.2.2 Lateral Sliding (LS) mode

The LS mode works by the relative sliding of one layer over another (Figure 2d). The underlying working mechanism is pretty much similar to the vertical c-s mode. When the two layers come in contact with each other, they develop equal and opposite charges. The outward relative sliding of the top layer leads to a decrease in the contact area with the bottom layer, thus creating the charge separation. The electron will flow towards the top electrode until it slides completely out. The inward sliding motion of the layer will cause electrons to flow in the reverse direction, thus completing the LS TENG's full A.C. cycle. The LS mode suffers from wear and tear (Niu et al. 2013a, Wang 2014, Wang et al. 2016b).

2.2.3 Single Electrode (SE) mode

The use of two electrodes sometime may restrict the device applications. The SE mode offers the advantage of using one electrode so that the opposite layer can move freely (Figure 2e). In SE mode, the periodic contact and separation between the layers cause electrons to flow between the ground and electrode (Niu et al. 2014, Wang et al. 2016a). The SE mode can also work in the LS configuration (Wang et al. 2016a).

2.2.4 Freestanding Triboelectric Layer (FT) mode

The FT mode offers a high performance figure of merit (FOM_p). In FT mode, the freestanding layer is not attached to the electrode or any wires. The FT mode shown in Figure 2f has a dielectric layer with two symmetrical electrodes with a small gap placed underneath at a certain gap. The freestanding layer's movement concerning the electrodes creates an uneven charge. The electrons will flow from one electrode to another in order to balance the potential difference. The flow of electrons between the electrodes leads to the generation of the TENG output (Jiang et al. 2015).

The device fabrication materials can be selected by considering triboelectric series as a reference (Chen and Wang 2017, Liu et al. 2019c). The TENG performance depends on the materials' surface charge density and environmental factors (Pan and Zhang 2019).

3. NGs for Self-powered Biological and Chemical Sensors

The self-powered sensors are in burgeoning demand as they can work as a stand-alone system. Several researchers have worked in this area to sense bio analytes such as glucose, protein drugs, cysteine, etc., which are vital for the proper human upkeeping (Khandelwal et al. 2019b, Selvarajan et al. 2017a,b). The TENG and PENG are widely reported for self-powered sensing of heavy metal ions, glucose, tetracycline, catechin, benzene and phenol.

The first self-powered glucose sensor based on barium titanate (BTO) was reported by Sophia et al. (Selvarajan et al. 2016, 2017b). The BTO nanoparticles mixed in aqueous PVA were dropped, cast on the glass slide and heat-treated to remove PVA. Then, the device was connected across the composite piezoelectric nanogenerator. Hence, this device acts as a load resistance to the nanogenerator device. The load resistance value changes when glucose concentration varied in the range of 50 µM – 1 mM. Thus, the device's voltage changes making the sensing simple and easy compared with the electrochemical sensor. The authors cross verified the glucose-sensing with electrochemical sensing platforms also. The device schematic and device data are shown in Figure 3a. The device serves a dual purpose, i.e., energy harvesting and sensing. A similar approach was used to detect cysteine (Selvarajan et al. 2017a). Cysteine is an amino acid that plays an important role in homeostasis. It's an important biomarker whose uneven concertation can be related to important chronic ailments like rheumatoid arthritis, Parkinson's disease, Alzheimer's disease (Jung et al. 2012, Yang et al. 2016b, Zhang et al. 2007). In this work, the BTO particles were functionalized with (3-aminopropyl) triethoxysilane. The functionalized film was drop-casted upon the 3D agarose-based matrix. The I-V characterization was performed for the concentration range (10 µM – 1000 µM) with a limit of detection of 10 µM. In the configuration explained before, a self-powered device was also shown for rational understanding. Here, the nanogenerator was fabricated from pure agarose film. The electrical performance was governed by the hydroxyl and ether functional groups releasing the inductive charges on the respective electrodes. The produced potential difference between the respective electrodes directs the charge carriers' mobility, thus producing the current. The same group recently published a similar approach for self-powered cysteamine detection by functionalizing the BTO particles with casein (Selvarajan et al. 2020). This anticipated research can work out to be an introductory apparatus for investigating the possible interactions and application in theranostics before performing an invasive real-time experiment. Abisegapriyan et al. in year 2020 fabricated the hollow cylindrical bismuth ferrite-based sensor and piezoelectric energy harvester (K S et al. 2020). The device determines the concentration of catechol in the water. The device exhibits a wide concentration range of 25 µM – 250 mM concentration with a large linear range. The sensor showed excellent detection limits down to 10.2 µM. Fu et al. fabricated a gas sensor based on hollow PVDF/PANI (polyaniline) (Fu et al. 2018). The gases like acetone, methane, carbon monoxide, ethanol, etc. were sensed at different concentrations (0–600 ppm). Figure 3b shows the device fabrication schematic, lumped structure and data output.

In 2013, Zhang et al. reported a vertical c-s mode TENG fabricated between clothes for human motion energy harvesting and glucose sensing. The TENG consists of patterned Polydimethylsiloxane (PDMS) backed by copper (Cu) electrode and aluminum (Al) foil as the two layers. The TENG produced an output voltage of 17 V and a current density of 0.02 µA cm^{-2} during walking. A lithium-ion battery was charged by clapping clothes used to drive a glucose biosensor (Zhang et al. 2013).

Mercury is highly toxic to humans, while Hg^{+2} ions are highly stable and can bioaccumulate (Clarkson et al. 2003). In the same year, Lin et al. proposed mercury ion detection using TENG. Gold (Au) and PDMS were the active materials of vertical c-s mode TENG (Lin et al. 2013). Further, Au nanoparticles were assembled onto the metal plate. The Au nanoparticles enhance the TENG output by increasing contact between the layers. The sensitive and selective detection of Hg^{+2} ions was achieved by modifying the Au nanoparticles with 3-mercaptopropionic acid (3-MPA). The TENG produced an output voltage of 105 V and a current of 63 µA (1 cm × 1 cm). The TENG sensor was selective for Hg^{+2} ions with a linear range and detection limit of 100 nM – 5 μM and 30 nM, respectively. Moreover, in the case of 5 μM concentration, the detection was simplified by a glowing light-emitting diode (LED) in the absence of Hg^{+2} ions while LED extinguishes in the presence of Hg^{+2} (Lin et al. 2013).

Dopamine plays a significant role in the central nervous system (CNS), and its abnormality can cause diseases like Schizophrenia, Parkinson's and Huntington's disease (Howe et al. 2013, Phillips et al. 2003). Jie et al., in 2015, reported a vertical c-s mode TENG for self-powered detection of dopamine (Jie et al. 2015). The triboelectric nanosensor (TENS) consists of a polytetrafluoroethylene (PTFE) nanoparticle array and Al as active materials, while spring serves the purpose of the spacer. A 2 cm × 2 cm TENS produced an output of 116 V and 33 μA. The TENS works in a linear range of 10 μM to 1 mM ($R^2 = 0.99$) with a detection limit of 0.5 μM. The mechanism for detection and high selectivity is shown in Figure 3c. The dopamine detection happens as polydopamine (PDA) absorbed on PTFE changes the surface electrification and the material's permittivity. The dopamine converts to dopamine-quinone due to the oxidation of catechol in dopamine. The oxidation is pH-induced process. The dopamine-quinone led to the formation of PDA. The formation of PDA occurs by the conversion of dopamine-quinone to leukodopamine-chrome which was further oxidized to 5,6-dihydroxyindole. Finally, 5,6-dihydroxyindole oxidized further to form PDA via intermolecular cross-linking (Lee et al. 2007). Thus, the output change of TENS in the presence of PDA is attributed to change in thickness and permittivity. The short-circuit current of TENS reduces as the pH changes. The alkaline pH promotes oxidation, as explained above. The effect of dopamine concentration on TENS output is shown in Figure 3c. The sensor's selectivity was tested by studying the interference of other analytes like uric acid (UA), and ascorbic acid (AA). The UA and AA interact poorly with PTFE, thus producing less response than dopamine (Jie et al. 2015).

Volatile organic compounds (VOCs) are highly dangerous for humans. The benzene vapors are one of the VOCs which are carcinogenic and may cause confusion, headache, dizziness and blurred vision, etc. (Nguyen and El-Safty 2011). Khandelwal et al. proposed a highly selective benzene monitoring system using TENG (Khandelwal et al. 2019b). The effect of a simple phase-inversion process and TiO_2 fillers on the output of the TENG was also demonstrated. The cellulose acetate (CA) was a positive triboelectric layer while PVDF/PVDF-TiO_2 or phase inversed PVDF-TiO_2 acted as the negative layer. The phase inversion of PVDF with 10 wt% TiO_2 filler exhibits 10- and 7-times enhancement in the current and voltage output, respectively. The phase inversion process improves the crystallinity of the PVDF as confirmed by the FT-IR. The authors also demonstrated the use of vertical c-s mode TENG for wind energy harvesting by developing a crank-shaft based setup. For the sensing of benzene, a lab designed setup was used (Figure $4a_1$) where nitrogen acted as dilution and bubbling gas. The flow rate of bubbling and carrier gas was regulated to vary the concentration of benzene vapors. The output of TENG decreases with an increase in the concentration of benzene. The benzene adsorption reduces the triboelectric capacity of the PVDF-TiO_2 layer. The sensor's response with concentration and the total flow rate is shown in Figure $4a_2$-a_3. The sensor's sensitivity was 0.29176 V/sccm and 0.0035 V/ppm regarding the total flow rate and

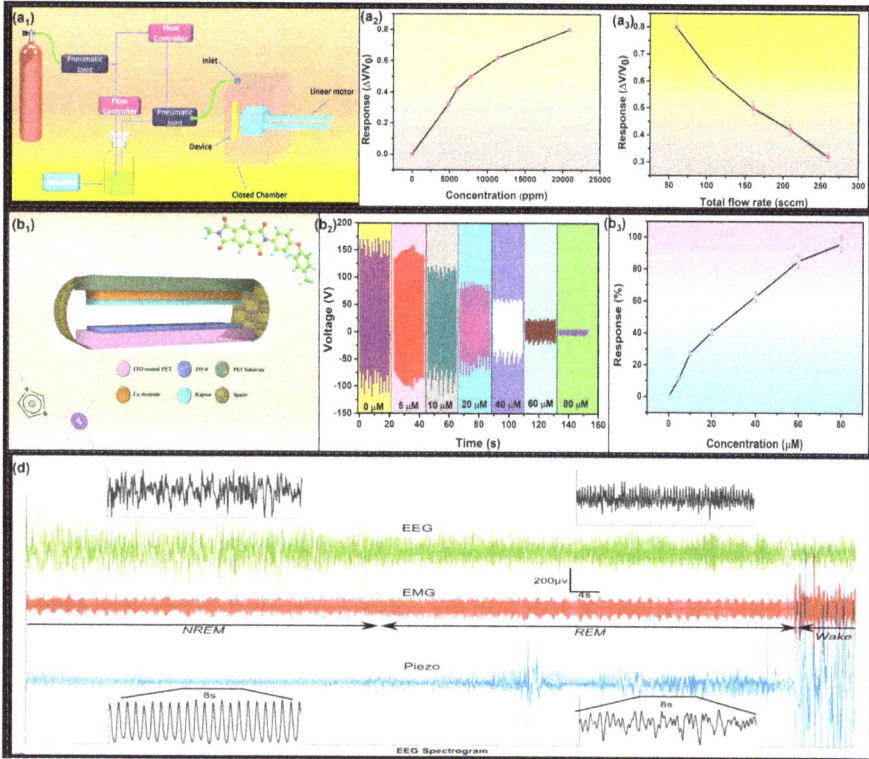

Fig. 4. The self-powered benzene sensing (a_1) setup and (a_2-a_3) response of the sensor concerning concentration and flow rate, respectively. Reprinted with permission (Khandelwal et al. 2019b). Copyright Elsevier. The self-powered tetracycline sensor. (b_1) 3D illustration of the device. (b_2) The voltage versus time graph corresponding to various tetracycline concentrations and (b_3) The % response versus concentration graph for the sensor. Reprinted with permission (Khandelwal et al. 2019a). Copyright John Wiley & Sons. (d) Comparing the respiratory pattern acquires by piezo device to the EMG and EEG signal during a typical NREM-REM-Wake transition in the mouse sleep cycle. Reprinted with permission (Yaghouby et al. 2016). Copyright Elsevier.

concentration, respectively. The device was also tested for the interference of other VOCs like acetone, toluene, etc. The benzene generates the highest response among all the VOCs confirming the selectivity of the sensor. The TENG sensor was also integrated with the Arduino UNO for triggering the alarm in the presence of benzene (Khandelwal et al. 2019b).

Khandelwal et al., for the first time, demonstrated the use of metal-organic framework (MOF) for TENG and self-powered sensors (Khandelwal et al. 2019a). The MOF offers numerous advantages like high surface area, high porosity and can be modified chemically without affecting the topology. The MOF-TENG was fabricated in traditional vertical c-s mode with zeolitic imidazole framework-8 (ZIF-8) as a positive triboelectric layer and Kapton as a negative layer. The ZIF-8 was grown on conducting ITO coated PET substrate at room temperature. Furthermore, the effect of different growth cycles on the output of TENG was also studied. The 20-cycle grown ZIF-8 produced the highest output of 164 V and 7 μA among all the cycles. The highest output of 20-cycle ZIF-8 was attributed to high surface roughness and sharp structures. The FE-SEM, 3D nanoprofiling and KPFM data support the high output of 20-cyc grown ZIF-8. Tetracycline antibiotic is effective against a broad range of bacteria (Li et al. 2018). The tetracycline works by inhibiting the protein synthesis in bacteria (Li et al. 2018). However, tetracycline causes several side effects and disorders. The release of tetracycline in water damages the aquatic environment. The MOF-TENG was successfully utilized to create a reusable tetracycline sensor (Figure $4b_1$). The tetracycline interacts with the ZIF-8 via π-π interactions. The voltage output of the device decreases with an increase in tetracycline concentrations (Figure $4b_2$). The device showed an excellent response of > 90% at the highest concentration, as shown in Figure $4b_3$. The MOF-TENG tetracycline sensor exhibits a sensitivity of 3.12 V μM^{-1}. The sensor is highly selective with no interference of methanol, ethanol, acetone and phenol. Moreover, the sensor can be reused by simple washing with methanol (Khandelwal et al. 2019a).

4. Heart rate, Respiratory Rate and Blood Pressure (B.P) Monitoring

Our human respiratory system is a combination of organs responsible for breathing. When we breathe in, we expand our diaphragm downwards, which allows us to inhale air. The normal inhalation in a healthy adult ranges from 12 to 20. The alteration in a heartbeat can be an early indication of several disorders in human health. Hence, a self-powered sensor can be of a very high advantage. Liu et al. fabricated a PVDF-based flexible respiratory monitoring setup on a silicone substrate (Liu et al. 2017). The device could power up to 1.5 V and 400 nA with a linear motor setup and, with the help of a bandage, reproducibly produced results with the diaphragm expansion and contraction. The device was also fabricated to observe delicate muscle movement and voice recognition. Nirmal et al. have fabricated the self-powered yarn type piezoelectric nanogenerator for breath sensing (Maria Joseph Raj et al. 2018). The novel device was very cost-effective and used on a simple cotton textile thread. Here, the authors used bismuth ferrite and PVDF composite and coated with a simple brush coating technique. The simple device was tested with a linear motor setup and could produce a high voltage output of 40 V and power 5 green LEDs. Finally, the device was demonstrated to monitor human breathing. Yaghouby et al. have tried to analyze the wakefulness and deep sleep parameters using the mouse's breath parameters (Yaghouby et al. 2016). The detection was non-invasive, and the signals were recorded from the cage floor of the mice. The device (with only Piezo signal) distinguished wakefulness with high sensitivity and specificity compared to gold standards of polysomnography, as shown in Figure 4d. The REM (rapid eye movement) state has moderate sensitivity compared to the Non-REM state which has poor sensitivity and high specificity. However, with a classifier, both sensitivity and specificity were increased for all the states.

Okano et al. have fabricated a multimodal cardiovascular system using a piezoelectric transducer and signal processing (Okano et al. 2017). The group proposed a concept of the device that can measure various inputs like heart rate variation, pulse wave propagation speed, and blood flow velocity. In this proposed device, an ultrasonic wave is projected towards the radial artery in the arms. The ultrasonic wave signals are reflected from the artery surface and absorbed by the sensing piezoelectric layer. The pulse of blood in the artery will alter the signal, and with the change in these alterations, the device could identify/calculate a variety of signals. Hence, the interval between the successive pulses of blood will reveal the heart rate using the Karvonen formula (Sornanathan and Khalil 2010). Moreover, a similar analysis can be used in the variation of the heart rate, which can be a sign of the underlining serious cardiovascular problem. This can also identify the pulse wave velocity in the patients, which can also be linked to a variety of diseases and psychological disorders like depression. Finally, the same signal can also identify the blood velocity motion. The device prototype was coupled with signal analysis and filtering devices.

In 2014, Bai et al. demonstrated respiratory and heartbeat sensing by utilizing a membrane-based triboelectric sensor (M-TES) (Bai et al. 2014). The M-TES is a multilayer structure composed of copper and FEP active layers. The M-TENG at 0.3 Hz frequency produced an output of 14.5 V. The M-TES works due to the pressure variation that occurs through the 2 mm air-conditioning channel in the center. The airbag with air tubes is connected to M-TES for mimicking a simple respiratory system. The expansion and contraction of the abdomen generates pressure changes, which M-TES can detect. The average breathing rate was analyzed using the voltage output produced by M-TES. The M-TES was also used to measure the heartbeat. The M-TES produced a signal of 0.06 V, which corresponds to a pulse of 72 beats per min (Bai et al. 2014).

A highly durable textile-based TENG (t-TENG) was demonstrated by Zhao et al. in 2016 (Zhao et al. 2016). The t-TENG comprises polyimide coated Cu-PET and Cu-PET as weft and warp yarns. The tapping of 10 cm s^{-1} generated an output of 4.98 V. The t-TENG was utilized for sensitive human respiratory monitoring by attaching it to a chest strap. The t-TENG produced its signal corresponding to the inhalation and exhalation of the chest cavity. The generated signal was processed. One complete breath cycle consists of inhalation and exhalation, which can be represented by $(T_{n-1}^{max}, T_n^{max})$. The tidal volume ($V_T$) was also measured as it directly related to the output voltage.

Zheng et al., in the same year, demonstrated self-powered cardiac monitoring via implantable TENG (i-TENG) (Zheng et al. 2016). The i-TENG consists of nanostructure PTFE as a negative layer and Al foil as the opposite electrode and a positive triboelectric layer (Figure 5a_1). The stability and resistance of i-TENG to work in *in vivo* conditions were enhanced by PDMS and parylene encapsulation. In phosphate buffer saline (PBS) solution, the voltage and current output generated by encapsulated i-TENG are 45 V and 7.5 μA, respectively. Moreover, the i-TENG exhibit no cytotoxicity on L929 cells. The *in vivo* cardiac monitoring was achieved by placing the i-TENG in between the pericardium and heart of the porcine. The electric signal produced was in good synchronization with the heartbeat. The effect of implantation site on the i-TENG output was also studied. Additionally, the effect of physiological parameters on the i-TENG was also considered in the study. The implantable wireless transmitter (iWT) and power management unit (PMU) were used to design a self-powered wireless transmission system. The wireless transmitted

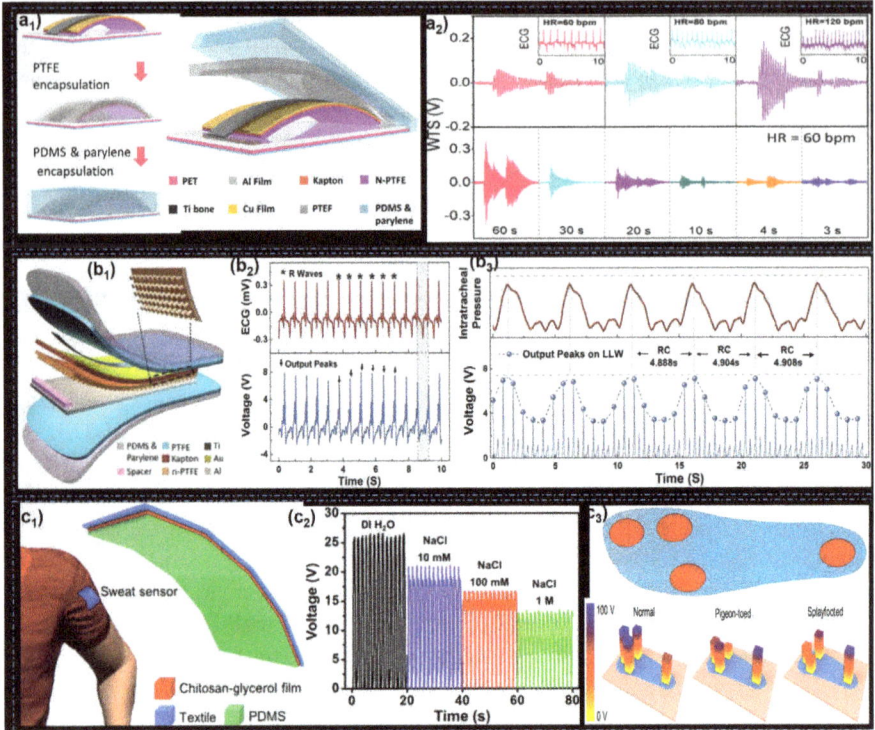

Fig. 5. (a₁) The fabrication process and device structure of i-TENG. (a₂) The wireless transmitted signal at different charging times (bottom) and at different heart rates (top). Reprinted with permission (Zheng et al. 2016). Copyright American Chemical Society. (b₁) The exploded view of multilayer iTEAS. (b₂) The ECG and iTEAS measured signals; asterisks represent the R wave. (b₃) The periodic fluctuation and stable output voltage when iTEAS was placed over the LLW. Reprinted with permission (Ma et al. 2016). Copyright American Chemical Society. (c₁) The illustration of the self-powered textile sweat sensor. (c₂) The effect of NaCl solution on the self-powered sensor and (c₃) The device location for gait sensing and the voltage distribution from the TENG units under different conditions. Reprinted with permission (Jao et al. 2018). Copyright Elsevier.

signal (WTS) corresponding to 60, 80 and 120 bpm is shown in Figure 5a₂. The developed i-TENG was stable even after 72 h of implantation.

Ma et al. reported a multilayer implantable triboelectric active sensor (i-TEAS) made up of biocompatible materials for health monitoring in real-time (Figure 5b₁) (Ma et al. 2016). The i-TEAS was implanted in the pericardial sac, which produced an output of 4 μA and ~ 10 V in response to the motion of organs. The output signal of implanted i-TEAS carried enough information for healthcare monitoring. The i-TEAS was first demonstrated for heart rate (H.R) monitoring with the electrocardiogram (ECG) as a standard to confirm the accuracy of the i-TEAS (Figure 5b₂). The i-TEAS signal corresponds to a heartbeat of 88.66 bpm, which was in good agreement with the ECG measured H.R of 89.46 bpm. The i-TEAS can also be used for blood pressure (B.P) monitoring as the peak voltage of the device is directly related to the B.P. The voltage increases with the rise in B.P and vice versa. A low R^2 value of 0.78

was achieved due to the dependency of B.P on other parameters. The i-TEAS was implanted at three different sites posterior wall (PW), right lateral wall (RLW) and left lateral wall (LLW) of the heart to determine the respiratory rate (R.R). As suggested in Figure 5b$_3$, the time interval of 5 s corresponds to a respiratory rate of 12 cpm (cycles per minute). The obtained results are consistent with the R.R controlled using artificial respiration. The i-TEAS is highly durable under *in vivo* conditions for up to 72 h while maintaining an accuracy of ~ 94% for H.R monitoring.

In 2018, a self-powered pulse sensor based on stretchable and waterproof TENG was reported by Chen and co-workers (Chen et al. 2018). The fully encapsulated single electrode TENG comprises polyurethane (PU) nanofibers as a positive and micro-structured PDMS as a negative layer. The wave information generated by placing the device on the wrist was used for healthcare monitoring. Cui et al. reported that TENG consists of trenched PDMS on the ITO-PET for pulse sensing (Cui et al. 2018). The TENG works in a single-electrode mode in which trench depth influences the pulse sensing. Therefore, the device performance was tested with different trench dimensions. The generated current pulse (0.7 s peak-peak interval) corresponds to 86 min^{-1} pulse rate. Moreover, the measured radial arterial augmentation index and the time difference between two peaks (ΔT_{DVP}) match well with the featured values.

Liu et al., in 2019,reported a microsphere-based TENG consisting of FEP and PDMS active layers for pulse monitoring (Liu et al. 2019b). A 0.8 cm × 0.8 cm device was used to generate a signal by placing it on the radial artery. The generated signal suggests the D-wave and P-wave wrist pulse characteristics with a pulse rate of 100 bpm.

5. Endocardial Pressure Sensor

In 2019, Liu et al. reported a minimal invasive endocardial pressure sensor by integrating the catheter with TENG based self-powered cardiac pressure sensor (SEPS) (Liu et al. 2019a). The SEPS (1 cm × 1.5 cm × 0.1 cm) was encapsulated in PDMS with nano-PTFE backed by Au electrode as a negative and Al foil (100 μm) as a positive triboelectric layer. A corona discharge was used to enhance the performance of SEPS by increasing the surface charge density on the nano-PTFE layer. The coagulation and hemolysis studies confirmed the hemocompatibility of the PDMS layer with an average hemolysis rate of 1.08%. The least invasive implantation was achieved with a heparin-coated PVC catheter. The left ventricle of Yorkshire pig was selected as the SEPS implantation site. The SEPS generates electric output in response to change in ventricular pressure occurred by contraction and relaxation. The signal during resting, arousal and functional condition by ECG, SEPS and femoral arterial pressure (FAP) was recorded. The left ventricular pressure (LVP) and FAP are related to left ventricle contractions; thus, commercial FAP sensor was taken as standard. The SEPS measured LVP signal exhibits linearity (R^2) of 0.974 and is consistent with FAP. The SEPS was also checked for detection of arrhythmia, and abnormal left atrial pressure (LAP).

6. Sweat and Gait Phase Detection

The textile-based TENG can be used for various wearable healthcare sensors. Sweat can provide information on many vital signs for human healthcare (Lee et al. 2017). The gait phase analysis can be useful for the sports person. In 2018, Jao et al. reported a sweat and gait phase sensing using a biodegradable TENG (C-TENG) based on chitosan film (Jao et al. 2018). The C-TENG comprises PTFE as a negative layer and nanostructured chitosan-glycerol as a positive triboelectric layer (Figure $5c_1$). The analysis was carried out using a fabric dipped in different concentrations of NaCl solution and deionized water (D.I.). Figure $5c_2$ shows the variation in the C-TENG output with respect to NaCl concentration. The change in output was due to the alteration in the fabric polarity. The gait phase monitoring was carried out by with the help of four C-TENG sensing units fabricated by coating chitosan on socks and PTFE inside the sole (Figure $5c_3$). The sensing units are capable of detecting different walking positions attributed to different force distribution used during walking. Figure $5c_3$ shows the voltage map corresponding to normal, pigeon-toed and splay footed conditions. The splay footed and pigeon-toed created unequal force distribution on four C-TENG units compared to equal force distribution in normal conditions resulting in the output shown in Figure $5c_3$.

Sarkar et al. reported a thermoplastic starch-based TENG (b-TENG) for self-powered gait sensing (Sarkar et al. 2019). The b-TENG produced an output voltage and current density of ~ 560 V and 120 mA m^{-2}, respectively. The authors used b-TENG as a gait analysis in the stance phase. The device generated a positive peak when the shoe pressed the b-TENG and vice versa when the shoes were removed. The signal magnitude, and time duration can be used to analyze the different movements and weight distribution. Lin et al. demonstrated real-time gait monitoring using triboelectric sensor-based smart insole (Lin et al. 2019). The two triboelectric sensors were placed at the rear and front sides of the smart insole. The triboelectric sensor comprises a concave TENG with an elastic air chamber (EAC) placed underneath. The triboelectric sensor was used to recognize several gait patterns like running, stepping and walking. Moreover, the triboelectric sensor was also demonstrated for gait monitoring of triple jump, gait rehabilitation and warning of fall down (Lin et al. 2019).

7. Sleep Monitoring

Sleep disorders are health-threatening and exist in one-third of the world population (McAlpine et al. 2019, Roenneberg 2013). Sleep behavior can be used to monitor or evaluate the quality of sleep, disease detection and prevention. Polysomnography (PSG) is considered a standard technique for sleep monitoring. PSG provides accurate information, but the placement of 22 signal cables is undesirable and restricts nocturnal movements (Zhang et al. 2020). Sleep monitoring using a smartwatch doesn't include much important information. Lin et al. developed a TENG array-based pressure-sensitive textile for sleep monitoring (Lin et al. 2018). The design of TENG based textile consists of wave-shaped PET placed in between the conductive fiber arrays laminated between the top and bottom fabrics. An excellent pressure sensitivity of 0.77 V Pa^{-1} in the lower pressure region was observed. The sensitivity decreases in the pressure region above 5.2 kPa. The smart textile was used for signal collection and integrated with a circuit for signal processing for self-powered sleep monitoring. The results chromatically depict the live behavior of a person lying on the smart mattress. The mattress can provide information on pressure distribution, body posture and position. Moreover, the active hours for the time from 11:00 p.m. to 8:00 a.m. were monitored and used to generate the sleep quality report. The mattress was also used for a self-powered warning system by triggering alert signals in emergencies (Lin et al. 2018).

Ding et al. developed a TENG-based tactile sensor using a CNT-doped porous PDMS and Al as the contact layers (Figure 6a_1) (Ding et al. 2018). The CNT-doped PDMS showed enhanced output compared to pure PDMS. A sleep monitoring belt was developed to capture heartbeat and breathing signals during sleep (Figure 6a_2). The real-time signal was used for the heartbeat measurement. The person has a heartbeat of nearly 70 bpm. The raw signal was processed for breath testing. Point A to B in Figure 6a_3 corresponds to the inhalation, and B to C represents the exhalation process (Ding et al. 2018). Recently in 2020, Zhang et al. developed a self-powered wireless sleep monitoring pillow based on triboelectric sensors (Figure 6b_1) (Zhang et al. 2020). The pillow contains feather-like triboelectric body sensors offering the advantage of good breathability, softness and zero connections. The triboelectric sensor in the pillow consists of two electrodes. The triboelectric sensor was fabricated on flexible and light substrates. The pillow distinguished numerous activities like snoring, breathing, teeth grinding and turning over, etc., as shown in Figure 6b_2. The sensor provides excellent results consistent with polysomnography (PSG) (Zhang et al. 2020).

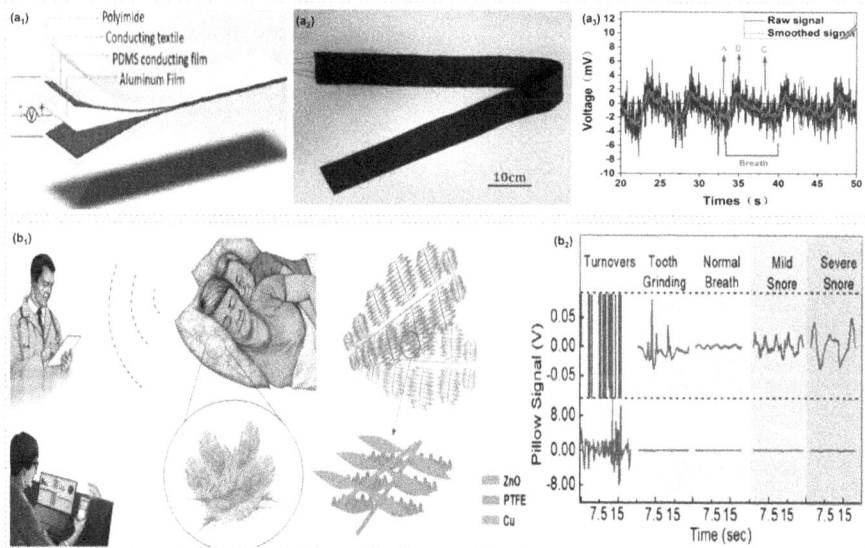

Fig. 6. (a$_1$) The 3D depiction of the TENG design and structure. (a$_2$) The as-designed sleep monitoring belt and (a$_3$) voltage output in real-time for breath and heartbeat. Open Access. (b$_1$) The schematic illustration of a self-powered sleep monitoring pillow filled with triboelectric sensors. (b$_2$) The signal is generated by pillow under different events. Reprinted with permission (Zhang et al. 2020). Copyright Elsevier.

8. Conclusion and Challenges

In this chapter, the recent advancements in healthcare sensors based on PENG and TENG are summarized. Most of the developments in the area have occurred in the last decade. This holds a high hope that this area will find even diverse and interesting research interventions in healthcare devices. Moreover, with better technologies and scientific advancements, huge developments are awaited to happen in this area, vital for a high-quality life. The humungous advancement in the area will soon lead to self-powered sensors' commercialization. Several challenges that need to be addressed for future commercialization are optimization of device size, effective encapsulation for long-term *in vivo* application, detailed studies on body immune response and miniaturization of the electrical circuits etc. Ongoing research will lead to the commercialization of self-powered devices for numerous applications shortly.

Acknowledgement

This work was supported by the Basic Science Research Program through the National Research Foundation of Korea (NRF) grant funded by the Korean government (MSIT) (2019R1A2C3009747, 2021R1A4A2000934).

References

Bai, P., G. Zhu, Q. Jing, J. Yang, J. Chen, Y. Su, J. Ma, G. Zhang and Z. L. Wang. (2014). Membrane-based self-powered triboelectric sensors for pressure change detection and its uses in security surveillance and healthcare monitoring. *Advanced Functional Materials* 24: 5807–5813.

Chen, J. and Z. L. Wang. (2017). Reviving vibration energy harvesting and self-powered sensing by a triboelectric nanogenerator. *Joule* 1: 480–521.

Chen, X., L. Miao, H. Guo, H. Chen, Y. Song, Z. Su and H. Zhang. (2018). Waterproof and stretchable triboelectric nanogenerator for biomechanical energy harvesting and self-powered sensing. *Applied Physics Letters* 112: 203902.

Clarkson, T. W., L. Magos and G. J. Myers. (2003). The toxicology of mercury—current exposures and clinical manifestations. *New England Journal of Medicine* 349: 1731–1737.

Cui, X., C. Zhang, W. Liu, Y. Zhang, J. Zhang, X. Li, L. Geng and X. Wang. (2018). Pulse sensor based on single-electrode triboelectric nanogenerator. *Sensors and Actuators A: Physical* 280: 326–331.

Ding, X., H. Cao, X. Zhang, M. Li and Y. Liu. (2018). Large scale triboelectric nanogenerator and self-powered flexible sensor for human sleep monitoring. *Sensors* 18: 1713.

Fan, F.-R., Z.-Q. Tian and Z. Lin Wang. (2012). Flexible triboelectric generator. *Nano Energy* 1: 328–334.

Feng, H., C. Zhao, P. Tan, R. Liu, X. Chen and Z. Li. (2018). Nanogenerator for biomedical applications. *Advanced Healthcare Materials* 7: 1701298.

Fu, Y., H. He, T. Zhao, Y. Dai, W. Han, J. Ma, L. Xing, Y. Zhang and X. Xue. (2018). A self-powered breath analyzer based on pani/pvdf piezo-gas-sensing arrays for potential diagnostics application. *Nano-Micro Letters* 10: 76.

Furfari, F. A. T. (2005). A history of the van de graaff generator. *IEEE Industry Applications Magazine* 11: 10–14.

Howe, M. W., P. L. Tierney, S. G. Sandberg, P. E. M. Phillips and A. M. Graybiel. (2013). Prolonged dopamine signalling in striatum signals proximity and value of distant rewards. *Nature* 500: 575–579.

Jao, Y.-T., P.-K. Yang, C.-M. Chiu, Y.-J. Lin, S.-W. Chen, D. Choi and Z.-H. Lin. (2018). A textile-based triboelectric nanogenerator with humidity-resistant output characteristic and its applications in self-powered healthcare sensors. *Nano Energy* 50: 513–520.

Jiang, T., X. Chen, C. B. Han, W. Tang and Z. L. Wang. (2015). Theoretical study of rotary freestanding triboelectric nanogenerators. *Advanced Functional Materials* 25: 2928–2938.

Jie, Y., N. Wang, X. Cao, Y. Xu, T. Li, X. Zhang and Z. L. Wang. (2015). Self-powered triboelectric nanosensor with poly(tetrafluoroethylene) nanoparticle arrays for dopamine detection. *ACS Nano* 9: 8376–8383.

Jung, H. S., T. Pradhan, J. H. Han, K. J. Heo, J. H. Lee, C. Kang and J. S. Kim. (2012). Molecular modulated cysteine-selective fluorescent probe. *Biomaterials* 33: 8495–8502.

K S, A., N. P. Maria Joseph Raj, N. R. Alluri, A. Chandrasekhar and S.-J. Kim. (2020). All in one transitional flow-based integrated self-powered catechol sensor using bifeo3 nanoparticles. *Sensors and Actuators B: Chemical* 320: 128417.

Khandelwal, G., A. Chandrasekhar, N. P. Maria Joseph Raj and S.-J. Kim. (2019a). Metal-organic framework: A novel material for triboelectric nanogenerator-based self-powered sensors and systems. *Advanced Energy Materials* 9: 1803581.

Khandelwal, G., A. Chandrasekhar, R. Pandey, N. P. Maria Joseph Raj and S.-J. Kim. (2019b). Phase inversion enabled energy scavenger: A multifunctional triboelectric nanogenerator as benzene monitoring system. *Sensors and Actuators B: Chemical* 282: 590–598.

Khandelwal, G., N. P. Maria Joseph Raj and S.-J. Kim. (2020). Triboelectric nanogenerator for healthcare and biomedical applications. *Nano Today* 33: 100882.

Kim, H., Y. Tadesse and S. Priya. (2009). Piezoelectric energy harvesting. pp. 3–39. *In*: Priya, S. and D. J. Inman (eds.). Energy Harvesting Technologies. Boston, MA: Springer US.

Lee, H., S. M. Dellatore, W. M. Miller and P. B. Messersmith. (2007). Mussel-inspired surface chemistry for multifunctional coatings. *Science* 318: 426–430.

Lee, H., C. Song, Y. S. Hong, M. S. Kim, H. R. Cho, T. Kang, K. Shin, S. H. Choi, T. Hyeon and D.-H. Kim. (2017). Wearable/disposable sweat-based glucose monitoring device with multistage transdermal drug delivery module. *Science Advances* 3: e1601314.

Li, P., S. Kumar, K. S. Park and H. G. Park. (2018). Development of a rapid and simple tetracycline detection system based on metal-enhanced fluorescence by europium-doped agnp@sio2 core–shell nanoparticles. *RSC Advances* 8: 24322–24327.

Lin, Z.-H., G. Zhu, Y. S. Zhou, Y. Yang, P. Bai, J. Chen and Z. L. Wang. (2013). A self-powered triboelectric nanosensor for mercury ion detection. *Angewandte Chemie International Edition* 52: 5065–5069.

Lin, Z., J. Yang, X. Li, Y. Wu, W. Wei, J. Liu, J. Chen and J. Yang. (2018). Large-scale and washable smart textiles based on triboelectric nanogenerator arrays for self-powered sleeping monitoring. *Advanced Functional Materials* 28: 1704112.

Lin, Z., Z. Wu, B. Zhang, Y.-C. Wang, H. Guo, G. Liu, C. Chen, Y. Chen, J. Yang and Z. L. Wang. (2019). A triboelectric nanogenerator-based smart insole for multifunctional gait monitoring. *Advanced Materials Technologies* 4: 1800360.

Liu, Z., S. Zhang, Y. M. Jin, H. Ouyang, Y. Zou, X. X. Wang, L. X. Xie and Z. Li. (2017). Flexible piezoelectric nanogenerator in wearable self-powered active sensor for respiration and healthcare monitoring. *Semiconductor Science and Technology* 32: 064004.

Liu, Z., Y. Ma, H. Ouyang, B. Shi, N. Li, D. Jiang, F. Xie, D. Qu, Y. Zou, Y. Huang, H. Li, C. Zhao, P. Tan, M. Yu, Y. Fan, H. Zhang, Z. L. Wang and Z. Li. (2019a). Transcatheter self-powered ultrasensitive endocardial pressure sensor. *Advanced Functional Materials* 29: 1807560.

Liu, Z., Z. Zhao, X. Zeng, X. Fu and Y. Hu. (2019b). Expandable microsphere-based triboelectric nanogenerators as ultrasensitive pressure sensors for respiratory and pulse monitoring. *Nano Energy* 59: 295–301.

Liu, Z., H. Li, B. Shi, Y. Fan, Z. L. Wang and Z. Li. (2019c). Wearable and implantable triboelectric nanogenerators. *Advanced Functional Materials* 29: 1808820.

Ma, Y., Q. Zheng, Y. Liu, B. Shi, X. Xue, W. Ji, Z. Liu, Y. Jin, Y. Zou, Z. An, W. Zhang, X. Wang, W. Jiang, Z. Xu, Z. L. Wang, Z. Li and H. Zhang. (2016). Self-powered, one-stop, and multifunctional implantable triboelectric active sensor for real-time biomedical monitoring. *Nano Letters* 16: 6042–6051.

Maria Joseph Raj, N. P., N. R. Alluri, V. Vivekananthan, A. Chandrasekhar, G. Khandelwal and S.-J. Kim. (2018). Sustainable yarn type-piezoelectric energy harvester as an eco-friendly, cost-effective battery-free breath sensor. *Applied Energy* 228: 1767–1776.

McAlpine, C. S., M. G. Kiss, S. Rattik, S. He, A. Vassalli, C. Valet, A. Anzai, C. T. Chan, J. E. Mindur, F. Kahles, W. C. Poller, V. Frodermann, A. M. Fenn, A. F. Gregory, L. Halle, Y. Iwamoto, F. F. Hoyer,

C. J. Binder, P. Libby, M. Tafti, T. E. Scammell, M. Nahrendorf and F. K. Swirski. (2019). Sleep modulates haematopoiesis and protects against atherosclerosis. *Nature* 566: 383–387.

Nguyen, H. and S. A. El-Safty. (2011). Meso- and macroporous co3o4 nanorods for effective voc gas sensors. *The Journal of Physical Chemistry C* 115: 8466–8474.

Niu, S., Y. Liu, S. Wang, L. Lin, Y. S. Zhou, Y. Hu and Z. L. Wang. (2013a). Theory of sliding-mode triboelectric nanogenerators. *Advanced Materials* 25: 6184–6193.

Niu, S., S. Wang, L. Lin, Y. Liu, Y. S. Zhou, Y. Hu and Z. L. Wang. (2013b). Theoretical study of contact-mode triboelectric nanogenerators as an effective power source. *Energy & Environmental Science* 6: 3576–3583.

Niu, S., Y. Liu, S. Wang, L. Lin, Y. S. Zhou, Y. Hu and Z. L. Wang. (2014). Theoretical investigation and structural optimization of single-electrode triboelectric nanogenerators. *Advanced Functional Materials* 24: 3332–3340.

Okano, T., S. Izumi, T. Katsuura, H. Kawaguchi and M. Yoshimoto. (2017). Multimodal cardiovascular information monitor using piezoelectric transducers for wearable healthcare, in 2017 IEEE International Workshop on Signal Processing Systems (SiPS), 1–6.

Pan, S. and Z. Zhang. (2019). Fundamental theories and basic principles of triboelectric effect: A review. *Friction* 7: 2–17.

Phillips, P. E. M., G. D. Stuber, M. L. A. V. Heien, R. M. Wightman and R. M. Carelli. (2003). Subsecond dopamine release promotes cocaine seeking. *Nature* 422: 614–618.

Rajagopalan, P., V. Singh and I. A. Palani. (2016). Investigations on the influence of substrate temperature in developing enhanced response zno nano generators on flexible polyimide using spray pyrolysis technique. *Materials Research Bulletin* 84: 340–345.

Rajagopalan, P., V. Singh and I. A. Palani. (2018). Enhancement of zno-based flexible nano generators via a sol–gel technique for sensing and energy harvesting applications. *Nanotechnology* 29: 105406.

Richter, B., J. Twiefel and J. Wallaschek. (2009). Piezoelectric equivalent circuit models. pp. 107–128. *In*: Priya, S. and D. J. Inman (eds.). Energy Harvesting Technologies. Boston, MA: Springer US.

Roenneberg, T. (2013). The human sleep project. *Nature* 498: 427–428.

Sarkar, P. K., T. Kamilya and S. Acharya. (2019). Introduction of triboelectric positive bioplastic for powering portable electronics and self-powered gait sensor. *ACS Applied Energy Materials* 2: 5507–5514.

Selvarajan, S., N. R. Alluri, A. Chandrasekhar, S.-J. Kim. (2016). Batio3 nanoparticles as biomaterial film for self-powered glucose sensor application. *Sensors and Actuators B: Chemical* 234: 395–403.

Selvarajan, S., N. R. Alluri, A. Chandrasekhar and S.-J. Kim. (2017a). Direct detection of cysteine using functionalized batio3 nanoparticles film based self-powered biosensor. *Biosensors and Bioelectronics* 91: 203–210.

Selvarajan, S., N. R. Alluri, A. Chandrasekhar and S.-J. Kim. (2017b). Unconventional active biosensor made of piezoelectric batio3 nanoparticles for biomolecule detection. *Sensors and Actuators B: Chemical* 253: 1180–1187.

Selvarajan, S., N. R. Alluri, A. Chandrasekhar and S.-J. Kim. (2020). Biocompatible electronic platform for monitoring protein-drug interactions with potential in future theranostics. *Sensors and Actuators B: Chemical* 305: 127497.

Sornanathan, L. and I. Khalil. (2010). Fitness monitoring system based on heart rate and spo2 level, in Proceedings of the 10th IEEE International Conference on Information Technology and Applications in Biomedicine, 1–5.

Wang, Z. L. (2014). Triboelectric nanogenerators as new energy technology and self-powered sensors—principles, problems and perspectives. *Faraday Discussions* 176: 447–458.

Wang, Z. L., L. Lin, J. Chen, S. Niu and Y. Zi. (2016a). Triboelectric nanogenerator: Single-electrode mode, in Triboelectric nanogenerators, 91–107. Cham: Springer International Publishing.

Wang, Z. L., L. Lin, J. Chen, S. Niu and Y. Zi. (2016b). Triboelectric nanogenerator: Lateral sliding mode, in Triboelectric nanogenerators, 49–90. Cham: Springer International Publishing.

Wang, Z. L., L. Lin, J. Chen, S. Niu and Y. Zi. (2016c). Triboelectric nanogenerator: Vertical contact-separation mode, in Triboelectric nanogenerators, 23–47. Cham: Springer International Publishing.

Wu, C., A. C. Wang, W. Ding, H. Guo and Z. L. Wang. (2019). Triboelectric nanogenerator: A foundation of the energy for the new era. *Advanced Energy Materials* 9: 1802906.

Yaghouby, F., K. D. Donohue, B. F. O'Hara and S. Sunderam. (2016). Noninvasive dissection of mouse sleep using a piezoelectric motion sensor. *Journal of Neuroscience Methods* 259: 90–100.

Yang, B., W. Zeng, Z.-H. Peng, S.-R. Liu, K. Chen and X.-M. Tao. (2016a). A fully verified theoretical analysis of contact-mode triboelectric nanogenerators as a wearable power source. *Advanced Energy Materials* 6: 1600505.

Yang, C., X. Wang, L. Shen, W. Deng, H. Liu, S. Ge, M. Yan and X. Song. (2016b). An aldehyde group-based p-acid probe for selective fluorescence turn-on sensing of cysteine and homocysteine. *Biosensors and Bioelectronics* 80: 17–23.

Zhang, H., Y. Yang, T.-C. Hou, Y. Su, C. Hu and Z. L. Wang. (2013). Triboelectric nanogenerator built inside clothes for self-powered glucose biosensors. *Nano Energy* 2: 1019–1024.

Zhang, M., M. Yu, F. Li, M. Zhu, M. Li, Y. Gao, L. Li, Z. Liu, J. Zhang, D. Zhang, T. Yi and C. Huang. (2007). A highly selective fluorescence turn-on sensor for cysteine/homocysteine and its application in bioimaging. *Journal of the American Chemical Society* 129: 10322–10323.

Zhang, N., Y. Li, S. Xiang, W. Guo, H. Zhang, C. Tao, S. Yang and X. Fan. (2020). Imperceptible sleep monitoring bedding for remote sleep healthcare and early disease diagnosis. *Nano Energy* 72: 104664.

Zhao, Z., C. Yan, Z. Liu, X. Fu, L.-M. Peng, Y. Hu and Z. Zheng. (2016). Machine-washable textile triboelectric nanogenerators for effective human respiratory monitoring through loom weaving of metallic yarns. *Advanced Materials* 28: 10267–10274.

Zheng, Q., H. Zhang, B. Shi, X. Xue, Z. Liu, Y. Jin, Y. Ma, Y. Zou, X. Wang, Z. An, W. Tang, W. Zhang, F. Yang, Y. Liu, X. Lang, Z. Xu, Z. Li and Z. L. Wang. (2016). *In vivo* self-powered wireless cardiac monitoring via implantable triboelectric nanogenerator. *ACS Nano* 10: 6510–6518.

CHAPTER 12

Minimally Invasive Microneedle Sensors
Developments in Wearable Healthcare Devices

Akshay Krishnakumar,[1,2] *Ganesh Kumar Mani,*[2]
Raghavv Raghavender Suresh,[3] *Arockia Jayalatha Kulandaisamy,*[3]
Kazuyoshi Tsuchiya[2,4] and *John Bosco Balaguru Rayappan*[3,*]

1. Introduction

Invasiveness is a process involving the entry of a diagnostic tool into a living tissue, that contributes to a painful post-treatment therapy. Studies show that a significant percentage (20%) of the global population is suffering from Trypanophobia, the fear of needles. Such preposterous or unnecessary fear of needles intends people to avoid any medical treatment which leads to various other health complications. Moreover, such invasive treatments are contagious and can lead to various unwanted infections and health complications. Though various non-invasive techniques were developed to accurately measure the disease biomarkers, it is inevitably true that the subcutaneous way of detection yields the maximum accuracy. To counteract such a conundrum, it is cardinally crucial to develop minimally invasive sensors for the detection of biomarkers. The prime objective of such sensors is to reduce

[1] Department of Electrical and Electronics Engineering, Tokai University, 4 Chome-1-1 Kitakaname, Hiratsuka, Kanagawa, Japan 259-1207.
[2] Micro Nano Technology Center, Tokai University, 4 Chome-1-1 Kitakaname, Hiratsuka, Kanagawa, Japan 259-1207.
[3] Centre for Nano Technology & Advanced Biomaterials (CeNTAB) and SASTRA Deemed University, Thanjavur, India 613401.
[4] Department of Precision Engineering, Tokai University, 4 Chome-1-1 Kitakaname, Hiratsuka, Kanagawa, Japan 259-1207.
* Corresponding author: rjbosco@ece.sastra.edu

the discomfort, infection rate, and pain recovery time in patients. Thus, this section deals with a collection of minimally invasive sensors and their sensing strategies. Point-of-care devices and the need for minimally invasive wearable electronics were discussed along with their contribution to detection, diagnosis, and treatment.

Human skin consists of three layers namely: stratum corneum, epidermal and dermal layer, in which dermal layer involves pain receptors (Cahill and Cearbhaill 2015, Waghule et al. 2019) (Figure 1). The stratum corneum is the outermost layer of the skin consisting of keratinized cells which allows lipophiles and low molecular weight drugs (10–40 microns). The epidermal layer consists of flat, scale-like squamous cells that act as a protective medium to the dermal layer of the skin (Xie et al. 2020).

The dermal layer is the actual connecting tissue to the skin containing fibroblasts, collagen, proteoglycans, and is rich in blood and lymph vessels (Madden et al. 2020). Hypodermic needles, transdermal patches, and topical creams are used for surface analyte detection and drug delivery applications. Transdermal patches pass through the stratum corneum to have an appreciable therapeutic and sensing effect, similar to that of non-invasive sensors. Though this technique is less contagious and is a painless way of detection, the action of the sensor or the drug delivery system is extremely slow with low bioavailability. On the other hand, hypodermic needles penetrate the dermal layer of the skin directly attacking the sensory neurons responsible for causing pain. Thus, minimally invasive sensors such as microneedles bypass the stratum corneum and collect samples near the epidermal-dermal junction

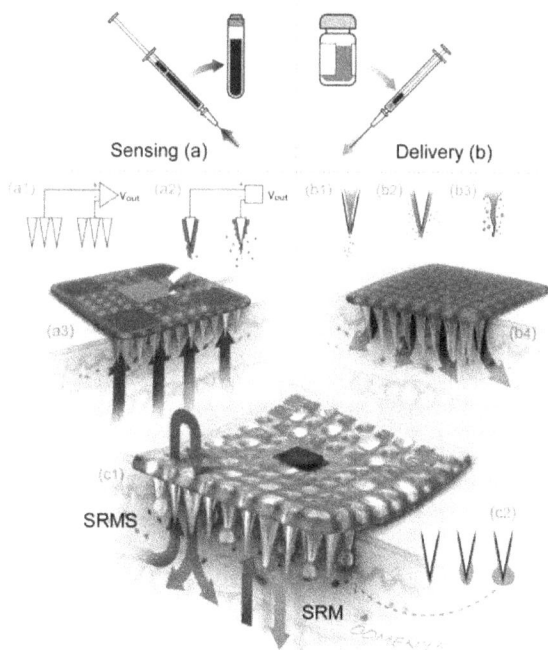

Fig. 1. Microneedle functionalities towards (a) biosensing and (b) drug delivery applications (Reproduced with permission from (Cahill and Cearbhaill 2015) Copyright 2015 ACS).

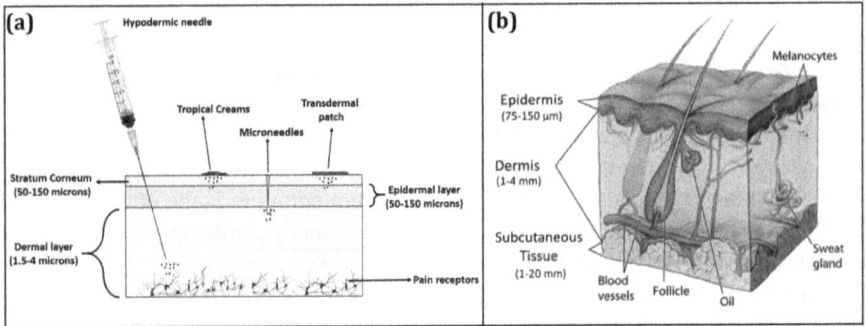

Fig. 2. (a) Comparison of conventional and wearable sensors (Inspired from (Waghule et al. 2019)) (b) Skin anatomy (Reproduced with permission from (Madden et al. 2020) Copyright Elsevier 2020).

(Figure 2(a)). The minimally invasive sensors have sufficient bioavailability, and faster onset action with better compliance from the patient. Thus, such painless detection of bio analytes could be achieved by microneedles and microelectrode patches.

Minimally invasive sensing devices such as microneedles are particularly designed to analyze the dermal interstitial fluids (fluids contained within the cells). Though disease analysis is performed in various body fluids such as urine, blood (Tsuchiya et al. 2005), saliva (Tsuchiya et al. 2014), cerebrospinal fluid (Mani et al. 2017, 2021) and sweat, the intestinal fluid-based detection could yield better physiological significance with minimal pain and contagion using MNs (Figure 2(b)) (Samant and Prausnitz 2018, Madden et al. 2020).

2. Wearable Minimally Invasive Wearable Sensors

The following section deals with various invasive and non-invasive types of wearable sensors, with their sensing strategies and materials used. A short review of the previously developed sensors is also highlighted in these sections.

2.1 Microneedles

Microneedles (MNs) are pharmaceutical devices consisting of several small needle-like structures, designed to create micro-conduits on the surface of the skin. Such small structures can easily penetrate the dermal tissue layer of the skin without touching the pain receptors (Xie et al. 2005). An ideal MN features a depth of 150 μm ~1 mm and a width of a few hundred microns. Thus, MNs can also be defined as a miniaturized transdermal needle that does not require a precarious insertion procedure. MNs are classified as solid, coated, hollow, dissolvable, and many more according to their application and the formation process. Thus, this subsection deals with the classification of MNs and materials utilized to prepare such MNs (Figure 3(a)).

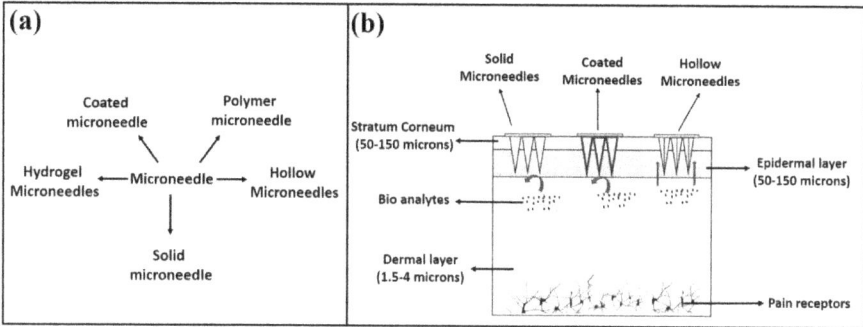

Fig. 3. (a) Classification of MNs (b) Action mechanism of MNs.

2.2 Types of Microneedles

Solid Microneedles

Solid MNs are used to create a microchannel pore on the surface of the skin. Solid MNs provide sufficient mechanical strength required for the material to pierce into the stratum corneum layer of the skin (Figure 3(b)). Such MNs are utilized to prepare biosensing electrodes or channels for the drug to enter the skin. Solid MNs can also be utilized for skin pretreatment in drug delivery applications (Waghule et al. 2019). Various materials used for the preparation of such solid MNs and their synthesis procedures are elucidated in detail in the next section. Solid MNs can provide a direct measurement of the bio analytes from the intestinal fluids. Though solid MNs are highly reliable for direct label detection, the inability to extract the intestinal fluids and skin perspiration poses a challenging limitation (Zhou et al. 2010). Besides, the long contact time of the MNs adds to the risk of pathogenic infection in the case of solid microneedles (Gupta et al. 2011).

Coated Microneedles

Coated MNs are similar to that of solid MNs, which are used to enhance the surface interaction (Figure 3(b)). Coated MNs are not an efficient way of drug delivery since the drug dose level is restricted to 1 mg. On the other hand, coated MNs coated with electroactive material can act as a highly sensitive biosensing medium. These MNs can also be coated with enzymes that could enhance the selectivity and sensitivity of the biosensor (Figure 4). Different architectural types of coated MNs can be utilized to yield an enhanced surface interaction area (Zhang et al. 2012b). Additionally, mechanical strength and biocompatibility of the MNs can be easily modified with the functional layer coated. The active area of such MNs layer is to be protected from external disturbances employing polymer-based material (Liu et al. 2019).

Hollow Microneedles

Hollow MNs are conduits to extract the intestinal fluids or to deliver the drug into the tissues. Rapid continuous analysis can be derived from such MNs by extracting blood and intestinal fluids. By this method, more precise discrimination of bio analytes can be easily obtained (Figure 3(b)). The primary disadvantage of such MN is that they

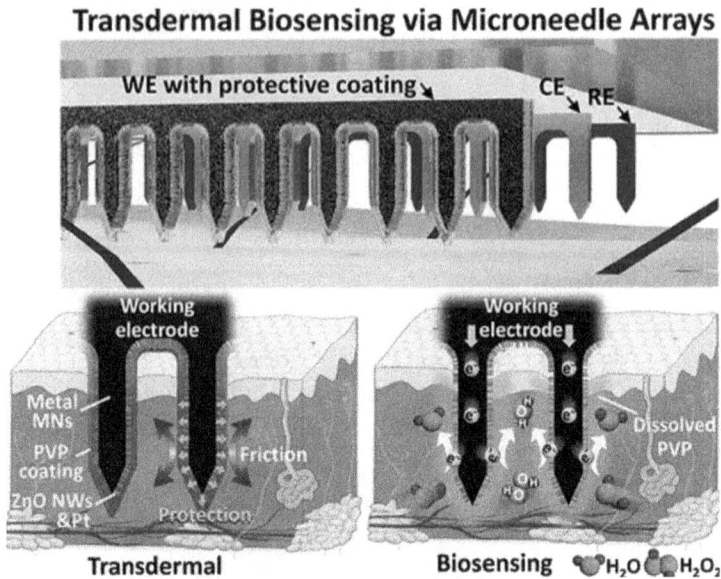

Fig. 4. Coated MNs mechanism of action on the surface of the human skin (Reproduced with permission from (Liu et al. 2019) Copyright 2019 ACS).

are highly prone to clogging which might oppose the fluid flow (Strambini et al. 2015).

Porous Microneedles

Porous MNs are similar to that of hollow MNs used to extract the interstitial liquids. The extraction of the system depends on the pores and geometry of the MN structure. As mentioned, porous MNs also have the possibility of clogging and thus oppose its liquid flow.

Swellable Microneedles

Swellable based MNs are constructed using swellable hydrogels. This MN is an alternative to hollow MNs, which are used to extract the body fluids by swelling (Andar et al. 2017). Construction of such MNs is relatively simple by mold casting and can extract a large number of liquid samples (Madden et al. 2020). In drug delivery, the drug is exchanged by the microcirculation process. These MNs suffer a primary demerit of low mechanical strength (Kim et al. 2012).

Dissolvable Microneedles

Unlike coated MNs, dissolvable MNs are completely dissolving into the skin to identify the bio analytes and to deliver drugs (Mcgrath et al. 2013). Such MNs provide a path of biosensing after which it dissolves into the skin. They are relatively safe and biodegradable than hollow and solid MNs (Lahiji et al. 2015). Degradability of the MNs avoids the risk of infections and biohazardous wastes hence follow the patient. Dissolvable MN chemicals can cause skin irritation and allergies to the patient's skin. Moreover, such MNs require a relatively long time to dissolve and will have to be completely dissolved before removing the patches (Kim et al. 2012).

2.3 Materials and Fabrication Techniques

MNs must possess essential properties like toughness, flexibility, and permeability depending upon their specific application. Thus, various materials used for the fabrication of the MNs biosensors are discussed below (Figure 5(a and b)).

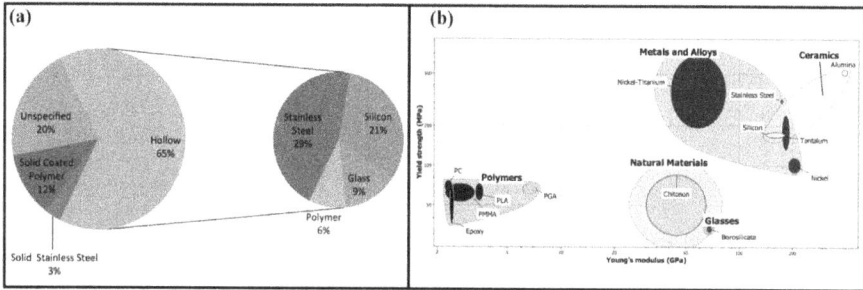

Fig. 5. (a) MN constituents and (b) Yield strength vs. Young's modulus of different materials (Reproduced with permission from (Cahill and Cearbhaill 2015) Copyright 2015 ACS).

Metal-based MNs

Pure metal such as tungsten (Francisco 1972), titanium (Li et al. 2017), nickel, palladium, and stainless-steel (Rajabi et al. 2016) based MNs was utilized for the fabrication of the biosensing element. Also, various noble metals such as gold (Bollella et al. 2019a, Senel et al. 2019), silver (Goud et al. 2019, Rawson et al. 2019), and platinum (Yoon et al. 2013) were also used due to their enhanced sensitivity towards the bio analytes. Besides, micro-nano structured metals could be employed for MN fabrication to enhance its electrochemical interaction (Xie et al. 2020). Metal-based MNs sensors might be allergic to sensitive skins and the choice of the material might increase its cost (Andar et al. 2017). Such metal-based MNs can be utilized for the fabrication of solid, hollow, and coated MNs. Metal MNs can be prepared using laser ablation (Lutton et al. 2015, Nejad et al. 2018), electrode position (Mansoor et al. 2013, Simfukwe et al. 2017), laser cutting and wet etching techniques (Ji et al. 2006, Shikida et al. 2006).

Silicon-based MNs

Similar to that of metal-based MNs, silicon MNs are hard and biocompatible. With the advent of semiconductors, silicon-based MNs have standard fabrication procedures, thus can be manufactured in various modalities (Pradeep Narayanan and Raghavan 2017). Similar to that of the glass-based MNs, these are highly brittle and with waste disposal concerns (Bodhale et al. 2010). Additionally, fabrication of silicon-based MNs need cleanroom processing and are extremely prone to fracture during transportation. Such silicon-based MNs are used for the fabrication of solid, coated, and hollow MNs (Figure 5(b)).

Ceramic based MNs

Composition of metallic and non-metallic elements of specific ratio with dominant ionic bonding results in the formation of ceramic compounds (Ita 2018). Bio-ceramics such as alumina, zirconia, and hydroxyapatiteare used to replace teeth, bones, and other hard tissues in our body. Such ceramic-based MNs are naturally porous with high stability towards temperature and humidity. Its inherent property avoids particle inflammation, and thus is extremely biodegradable and biocompatible. Similar to glass and silicon MNs, ceramics are considerably brittle and also require a detailed synthesis procedure. Such ceramic-based MNs are prepared using soft lithography (Hartmann et al. 2015), pulsed laser deposition (Gittard et al. 2009), plain (Cai et al. 2015), two-photon polymerization (Cai et al. 2014), and sacrificial micro-molding (Ita 2018). Ceramics are used for the preparation of hollow and dissolving MNs (Figure 5(b)).

Glass based MNs

Glass-based MNs are transparent electrodes that are inherently cheap and chemically inert (Hosu et al. 2019). Such glass-based materials can be used for the fabrication of solid and hollow MNs (Wang et al. 2005, Hu et al. 2020b). The primary disadvantage of such glass-based MNs is that they are highly brittle, thus are extremely dangerous. Glass based MNs can be fabricated for thermal pullers (Wang et al. 2005), micropipette pullers (Ayittey et al. 2009), and thermoplastic drawing (Hu et al. 2020b).

Polymer-based MNs

Degradable and non-degradable MNs can be constructed using polymer-based materials. Degradable polymers dissolve into the human skin successively delivering a particular transdermal drug or a disposable biosensor. Degradable MNs are made up of polyvinyl alcohol (Nguyen et al. 2018), hyaluronic acid, polyvinyl pyrrolidine (Sun et al. 2013), poly(glycolic acid), polyethylene glycol (PEG (Liu et al. 2017)), and cellulose, whereas non-degradable MNs are designed using Poly Dimethyl Siloxane (PDMS (Nejad et al. 2018)), polyacrylic acid (PAA (Tian et al. 2019)) and polyvinyl methyl vinyl ether (PMVE). The preparation techniques of such MNs are very simple and robust. The primary disadvantage of such MNs is low mechanical rigidity and Young's modulus. Polymer-based materials can be employed to fabricate all kinds of MNs.

Carbohydrate and Sugar-based MNs

Carbohydrate and sugar-based materials can be used to fabricate MNs by micro-molding procedures and drawing lithography (Donnelly et al. 2009, Lutton et al. 2016). Carbohydrates are naturally occurring substances that are better alternatives to other metal or silicon-based MNs since they are extremely safe. Post solidification of such MNs possess high mechanical strength which can pierce into the stratum corneum of the skin. They are primarily utilized in the construction of dissolving MNs. Though such sugar and carbohydrate-based MNs are extremely biocompatible and biodegradable, they are extremely hygroscopic on cooling, require tedious processing, and have relatively low strength compared to other MNs (Miyano and Tobinaga 2005, Donnelly et al. 2009, Serrano-castañeda et al. 2018).

2.4 Biofluid Extraction and Sensing Strategies

Analyzing the biofluids under the epidermal layer reveals the biochemical and physiological characteristics of the diseases. Biofluid extraction from the MN act as a critical diagnostic tool for various disease biomarker detection and therapeutic effect (Xie et al. 2020). Hollow MNs and solid MNs are the two common types of MNs used for the biofluid interaction (Madden et al. 2020). Capillary rise from the MN collector pulls the fluid into the analyzing reservoir for further analysis (Yao et al. 2019). A maximum of 100 μL of the biofluid is ideally extracted from the MN and channeled in the reservoir. Analysis of such biofluids for disease biomarker discrimination is carried out by these techniques.

Raman Analysis

Surface-Enhanced Raman Scattering or Spectroscopy (SERS) is a surface analysis technique that responds to the adsorbed molecules on the surface of the metals and metallic compounds by means of scattering. SERS has an expanded range of applications from solid-state physics, chemical physics, analytical chemistry, and electrochemistry (Sun 2013). SERS is a combined effect of Raman and surface plasmon resonance spectroscopy, with localized plasmons enhancing the shift (Xie et al. 2020). The shift observed by the system is of an extremely high order, thus can even detect a single molecular compound and the nature of charge transfer rate (Heather and Metiu 1989). SERS based MNs are used to detect the presence of the disease biomarker on the surface of the MN. Cancer biomarkers are the most common diseases detected by this technique and do not require pretreatment of the samples before analysis. Henceforth, MN-based SERS sensors exhibit swift response for an optimal for a real-time environment. However, lasers from Raman spectroscopy cannot propagate beyond 200 μm, whereas the area of interest is below 700 μm ~1 mm (Krafft et al. 2009, Yuen and Liu 2014). Therefore, these sensors are to be tested outside the skin after the application of the MN into the specimen. Figure 6(a) describes a highly sensitive SERS surface for glucose biomolecular detection. The glucose-sensing mechanism of SERS interaction substrate functionalized with phenylboronic acid is depicted in the schematic. Change in the alkyl peaks (1996 cm^{-1}) and biologically silent region (1800 cm^{-1} –2800 cm^{-1}) was used to detect the glucose analyte presence on the surface. Figure 6(b) shows a SERS encoded particle with a plasmon core, protective core, SERS reporter, and the surface interaction elements. The images describe MCF7 breast cancer cells and their characteristic morphologies. The Raman shift observed from the SERS reporter layer was used for the detection and quantification of the cancer biomarker (Kong et al. 2014). Current acquisition times of SERS type biosensors are apparently slower than normal optical biosensors. The sensitivity of such biosensors is comparatively low, which poses a serious challenge when deployed in a real-time environment (Bruzas et al. 2018). Unlike other wearable sensors, SERS cannot provide results instantaneously, since these sensors are needed to be removed and analyzed. Such sensors can be used as glucometer type instruments, where the MN patch is inserted into the device. Moreover, designing SERS type glucometer or cancer detector has not been proven to have a cost advantage over the currently available technology.

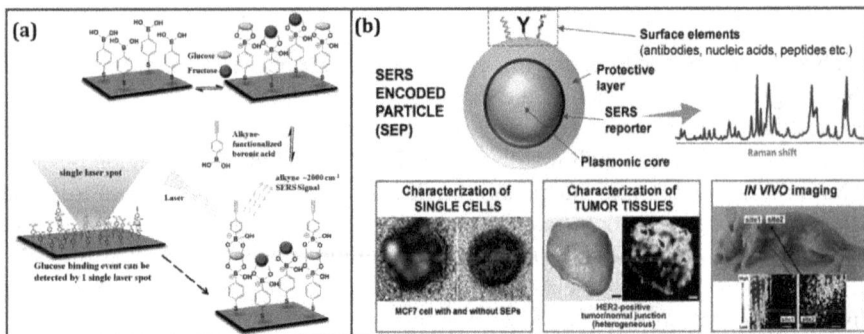

Fig. 6. (a) Schematic of glucose sensing and (b) Tumor cell detection by SERS (Reproduced with permission from (Guerrini et al. 2017) and (Kong et al. 2014) Elsevier).

Colorimetric Determination

Colorimetric sensors are visual sensors in which a particular analyte could be detected and quantified using the color change. Additionally, such sensors do not require any trained professionals and are of extremely low cost (Chen et al. 2020). With the increase in the concentration of the target analyte, the colorific characteristics of the sensing medium get altered. This type of wearable sensors can be used to fabricate "use and throw" type MN assay patches. Various materials are used for the fabrication of a calorimetric wearable sensor. A colorimetric-based HIV-1 sensor was prepared using quartz crystal microbalance that was employed in the detection of the HIV-1 virus (Figure 7(a)) (Saylan et al. 2019). An enzyme-linked immunosorbent was used for the detection of thrombin as shown in Figure 7(b) (Lu et al. 2012, Babamiri et al. 2018, Guo et al. 2020). The sensitivity of colorimetric sensors is extremely low, similar to that of SERS sensors. Colorimetric is mostly used for qualitative detection since it's necessary to amplify the bio signals achieved from this sensor for quantitative analysis (Babamiri et al. 2018).

Electrochemical Sensors

Electrochemical redox reactions generating electrical impulses with respect to the concentration of the target analyte is a popular way of biomarker detection. MN-based modified electrodes were used for such electrochemical detection and quantification of bio-analytes (Goud et al. 2019). Non-enzymatic and enzymatic detections are possible strategies for electrochemical recognition. Enzymatic detection of bio-analytes possesses enhanced figures of merits such as selectivity, sensitivity, and effective catalysis. On the other hand, non-enzymatic sensors are low cost, have high stability, long life, and are not affected by enzyme degeneration. This detection technique can be used in any MNs form and are the most efficient type of sensor. Electrochemical based MNs with medical applicability have gained their prominence in the diagnosis and classification of various pathogens. Though such sensors are extremely viable in a practical environment, the sensitivity of the MNs is an existing challenge (Dardano et al. 2019). A general stratagem to improve sensitivity is to enrich the electron transfer rate by increasing the electroactive surface of the electrode. Even with advanced high surface area materials such as

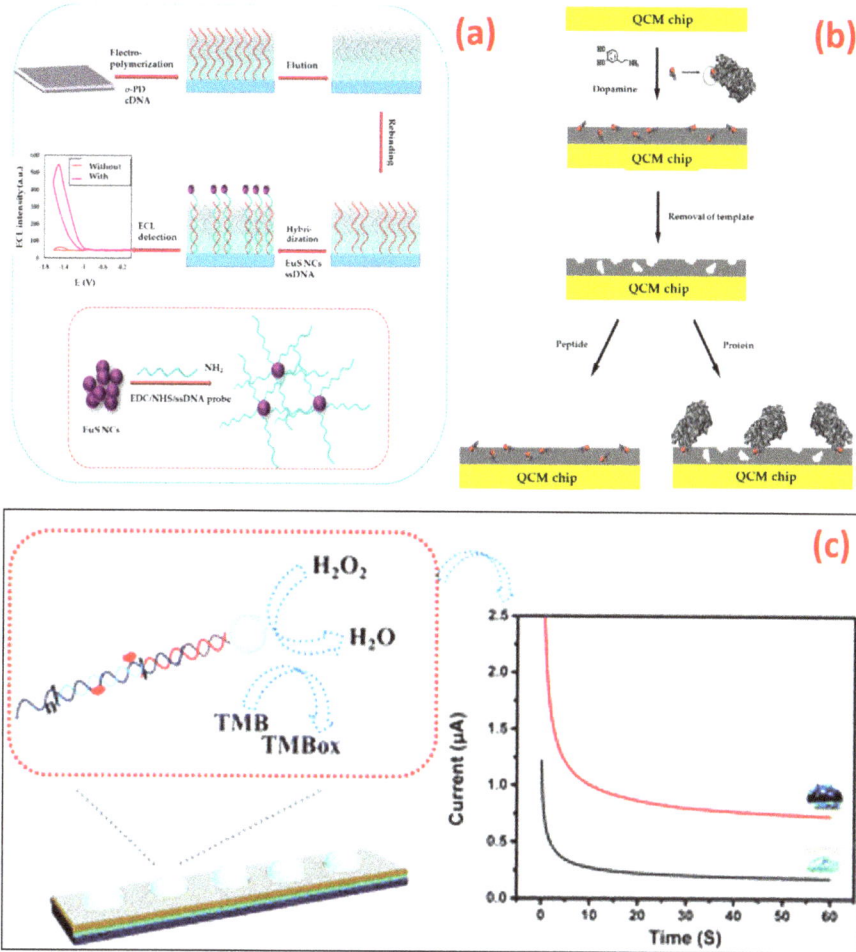

Fig. 7. (a&b) HIV-1 detection and (c) PDMS based colorimetric sensor with current changes for thrombin detection (Reproduced with permission from (Saylan et al. 2019) MDPI 2019 and (Guo et al. 2020) RSC 2020).

CNT and rGO, enhancing the sensitivity of the MNs is still a challenging one (Tasca et al. 2019).

2.5 Short Review of Microneedle/Microelectrode-based Biosensors

Microneedle Sensing

Developing biosensors non-invasively to monitor the concentration of the target analyte with minimal sample requirement has been the major point of interest (Hosu et al. 2019). Most of the MNs are used for glucose and cancer detection and monitoring. Herein, in this section, a short review of glucose, cancer, and other biomolecule detection strategies are elucidated.

Glucose Detection

Carbon-based materials were used as the working electrode for the electrochemical detection of glucose. Bollella et al. (Bollella et al. 2019b) devised an MN biosensor array for simultaneous detection of lactate and glucose in an artificial interstitial fluid environment. MWCNT/POLY Methylene blue modified MNs were used as the working electrode with glucose dehydrogenase enzyme. The linear range and detection limit were found to be (10–100 μM) and 3 μM towards lactate whereas (0.05–5 mM) and 7 μM towards glucose molecule, respectively. Glucose oxidase(GO) modified Au electrode was used for continuous glucose detection in a transdermal sensor setup (Hwa et al. 2015). The sensing unit consists of an array of needles immobilized with GO along with a self-assembled monolayer and 3-Mercaptopropionic acid (MPA). The linear range of the electrochemical unit was found to be in the range of $30-400\,\mathrm{mg\,dL^{-1}}$. Another group designed FAD-glucose oxidase modified terthiophene carboxylic acid MNs for the electrochemical determination of glucose. The group achieved a selectivity, detection limit, and linear range of $0.22\ \mu\mathrm{A\ mM^{-1}\ cm^{-1}}$, 19.4 μM, and 0.05 to 20.0 mM, respectively, towards glucose analyte (Kim et al. 2019). Lee et al. (Lee et al. 2016) developed a non-enzymatic 3D stainless steel biosensor with a Pt black active sensing layer. The fabricated 650 μm high and 150 μm wide MNs exhibited a selectivity, detection limit, and response time of $1.62\ \mu\mathrm{A\ mM^{-1}}$, 50 μM, and 13 s, respectively. Similarly, Chinnadayyala et al. (Chinnadayyala et al. 2018) used Pt black and nafion modified MN electrodes to achieve selectivity, detection limit, and response time of $175\ \mu\mathrm{A\ mM^{-1}}$, 23 μM and 2 s, respectively. Nafion acted as a protective layer to the Pt black modified layer resulting in the enhanced glucose-sensing characteristics of the MNs. Also, PVDF based nafion membranes coated MNs were used for transcutaneous implantable glucose sensing with a selectivity, detection limit, and linear range of $0.23\ \mu\mathrm{A\ mM^{-1}}$, 23 μM, and 0 to 20 mM, respectively (Chen et al. 2015).

Cancer Detection

Sulaiman et al. (Al Sulaiman et al. 2019) designed minimally invasive MNs to detect cell-free nuclei from liquid biopsies for early-stage cancer probing. This group designed MNs with alginate–peptide nucleic acid hybrid material towards the detection of nucleic acid from the interstitial fluid (Al Sulaiman et al. 2019). Hydrogel coated microneedles provided a channel of fast sampling kinetics with large surface capacity (~6.5 μL in 2 min). AgPO4 nanocomposite-based microneedles are used for early-stage cancer detection due to their good photocatalytic and antibacterial properties that could be exploited in bone regeneration and repair (Jiang et al. 2016, Panthi et al. 2017). Such silver-based thin film materials are used in tissue engineering and biotechnology due to their biodegradability and low toxicity (Zhang et al. 2012a). Chen et al. (Chen et al. 2020) designed such Ag3PO4 labeled carcinoembryonic antigens for early-stage cancer diagnosis. Colorimetric-based detection technique was used for the breast cancer diagnosis with the blue color response (Chen et al. 2020). In this way, disease propagation throughout the human body could be monitored and overlays a path for easier personal diagnosis.

Keum et al. (Keum et al. 2015) designed a real-time electrical impulse detection from nitic oxide for *in situ* cancer detection by the process of endomicroscopy.

Polycaprolactone (PCL) MNs coated with poly(3,4-ethylene dioxythiophene) (PEDOT) functionalized with hemin molecules on the surface were used for biosensing and endomicroscopy analysis. The sensor was targeted towards the electrochemical determination of NO for the detection of colon cancer (Keum et al. 2015).

Other Biomarkers and Smart Detection Techniques

Mishra et al. (Mishra et al. 2017) utilized a carbon paste array electrode transducer in hollow MNs for organophosphate presence detection using rapid square wave voltammetry measurements of p-nitrophenol (Figure 8(a)). The prepared biosensor showed a linear range of 20–180 µM with maximum selectivity in the presence of interferons (ascorbic acid, uric acid). The prepared MNs were tested *ex vivo* in mice skin samples using methyl paraoxon in presence of organophosphate hydrolase enzyme. This could be further analyzed for fast monitoring decontamination loop and an effective measure for protecting civilians. Solid microneedle with various coatings and polymeric membranes were used to enhance its selectivity. f-MWCNT based MNs assembled in polydimethylsiloxane (PDMS) substrates were used for potassium quantification (Parrilla et al. 2019). Bollella et al. (Bollella et al. 2019c) methylene blue (MB developed an MN-based biosensor towards continuous monitoring of lactate in dermal interstitial fluid. The developed biosensor utilized gold MNs as the working electrode using MWCNTs, polyethylene blue, and lactate oxidase immobilization. It was attributed that the optimized biosensor was used as a wearable device in clinical care instruments for continuous lactate monitoring. Esfandyarpour et al. (Esfandyarpour et al. 2013) designed an MNs method for label-free protein detection using microfluidic channels. This work utilized biotinylated bovine serum albumin for reception and streptavidin as the target molecule. This biosensor can directly detect the cancer cells biomarkers and integrated portable instrument in clinical care diagnosis.

Gold nanoparticles-horseradish peroxidase was modified in a tungsten microwire towards ultrasensitive hydrogen peroxide detection. They exhibited high electrochemical activity assisted by H_2O_2 reduction with a linear range, detection limit, and sensitivity as 5 nM to 5 µM, 800 pM, 490.59 mA.mM^{-1}.cm^{-2}. Such extreme electrocatalytic behavior was probably due to the strong synergetic effect between horseradish peroxidase and Au nanoparticles (Narayanan and Slaughter 2018).

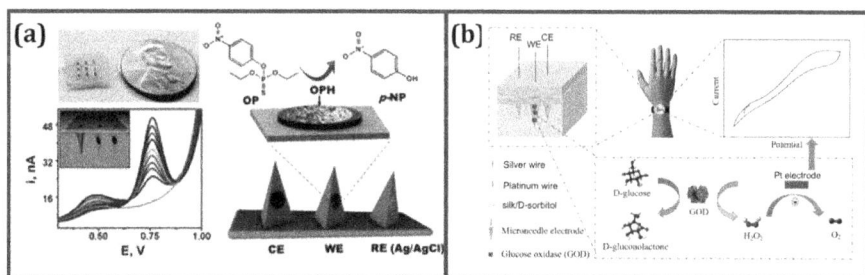

Fig. 8. (a) Organophosphate detection using MNs and (b) Silk/D-sorbitol MNs (Reproduced with permission from (Mishra et al. 2017) Copyright 2017 RSC and (Zhao et al. 2020) Copyright 2020 RSC).

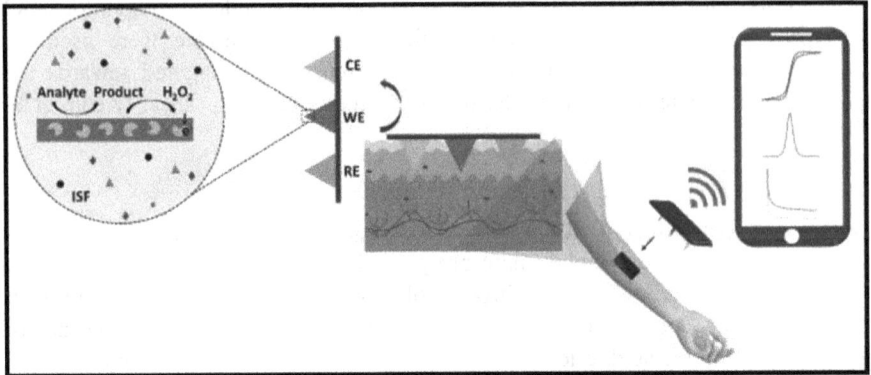

Fig. 9. Solid MN Schematic (Reproduced with permission from (Madden et al. 2020) Copyright Elsevier 2020).

Zaho et al. (Zhao et al. 2020) innovateda device consisting of microneedles, platinum (Pt) and silver wire electrodes with immobilized glucose oxidase. The silk worn MNs device was fabricated using an assembly of pyramidal needles using Pt and Ag wires with microcavity opening. They were found to be highly biocompatible (Figure 8(b)). The working schematic of a solid microneedle biosensor as a handheld device is depicted in Figure 9. The reaction schematic of the system is modified with an enzymatic layer for the analyte reaction. The data collected from the device is transmitted from the patch to the phone for continuous self-assessment (Madden et al. 2020). The skin penetration performance and the biosensing performance towards ethanol detection in the artificial interstitial fluid were elucidated (Mohan et al. 2017). Liu et al. (Liu et al. 2019) designed PVP protected ZnO nanowires stainless steel microneedle for transdermal detection of H_2O_2 in living organisms (Figure 10(a and b)). ZnO nanowires are hydrothermally synthesized on the MN surface in Pt thin layer surface. The PVP coating on the surface of the MNs promoted the sensing functionalities about 3 folds to that of the one without the polymer coating. Moreover, the depreciation and degradation of MN-based sensors are to be analyzed along with various pragmatic challenges. Further, such microneedles could be intergraded into various RF-ID techniques and IoT platforms to yield a continuous monitoring system.

Microelectrode Sensing

Microelectrode and microprobes are being used for measuring neural activity. The biosensor gives the neuron function, brain temperature, and intracranial pressure by using theoretical calculations. The biosensor uses three layers of metal and polymer films with the encapsulated brain temperature sensor. Estimated temperatures and simulators have been considered for estimating parameters of interest using micro electric arrays. Also, the electrode was used for neural activity distribution on any curved surface in the field of analysis (Panwar and Saxena 2019). Similarly, microelectrode arrays fabricated using aerosol jet micro additive was used with the spaces between Ag traces around 30–180 μm. Such electrodes were used for glucose and hydrogen peroxide detection. The 30 μm distance electrode had the highest

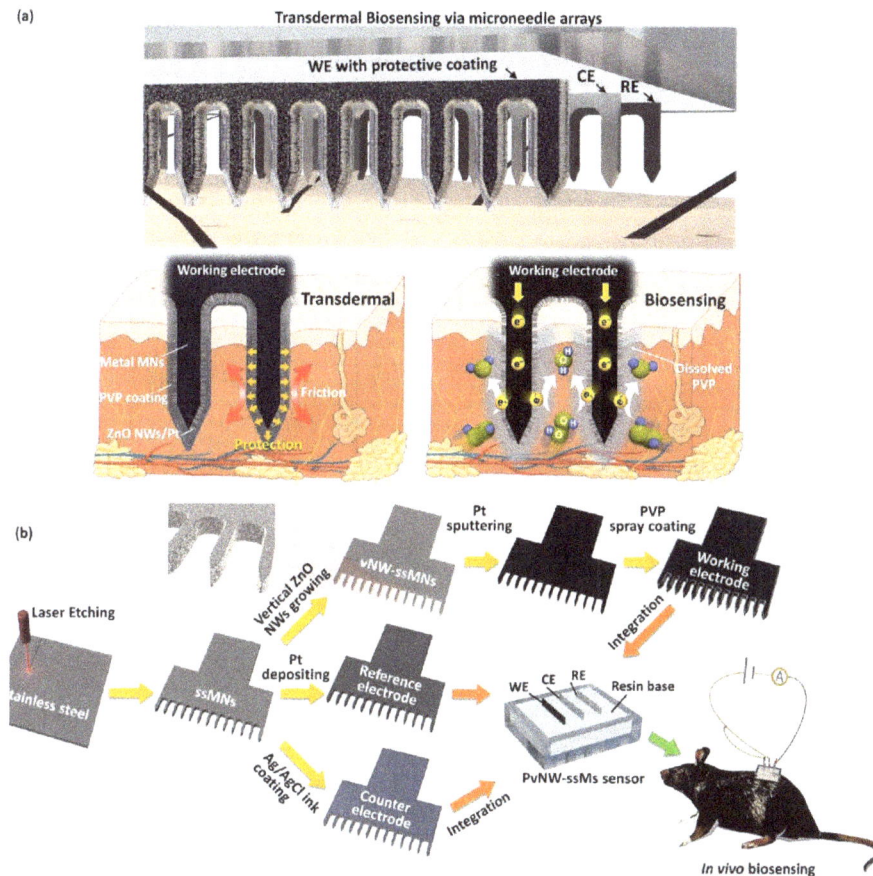

Fig. 10. PVP protected ZnO nanowires stainless steel microneedle for transdermal detection of H2O2 (a) reaction mechanism and (b) fabrication procedure (Reproduced with permission from (Liu et al. 2019) Copyright 2019 ACS).

sensitivity (13.3 µA mM^{-1}) and a low detection limit (2.7 µM) due to its MEA 30 level (12 nA), whereas the electrode with 100 µm distance had sensitivity, detection limit, and MEA 100 level as 9.9 µA mM^{-1} 0.45 µM, and 1.5 nA, respectively. This was probably due to the diffusion layer of neighboring electrodes in 30 µm (Yang et al. 2016) (Figure 11).

Hemanth et al. (Hemanth et al. 2018) developed an enzyme-based electrochemical 3D carbon microelectrode using biofunctionalized reduced graphene oxide. Glucose was used as the point of interest for the biosensor to compare 2D with 3D carbon electrodes. The 3D carbon electrode exhibited a sensitivity of 23.55 µA. mM^{-1}cm^{-2} compared to that of 2D electrodes (10.19 µA. mM^{-1}cm^{-2}). The microelectrode exhibited reproducibility in results over 7 days and high selectivity in the presence of various interferons. Nasr et al. (Nasr et al. 2018) designed self-organized nanostructure modified microelectrodes towards electrochemical detection of glutamate in stem cells of brain organoids. The sensor exhibited a catalytic reaction with a sensitivity, linear range, and limit of detection to be around 93 ± 9.5 nA·µM^{-1}·cm^{-2} (5 µM to 0.5 mM)

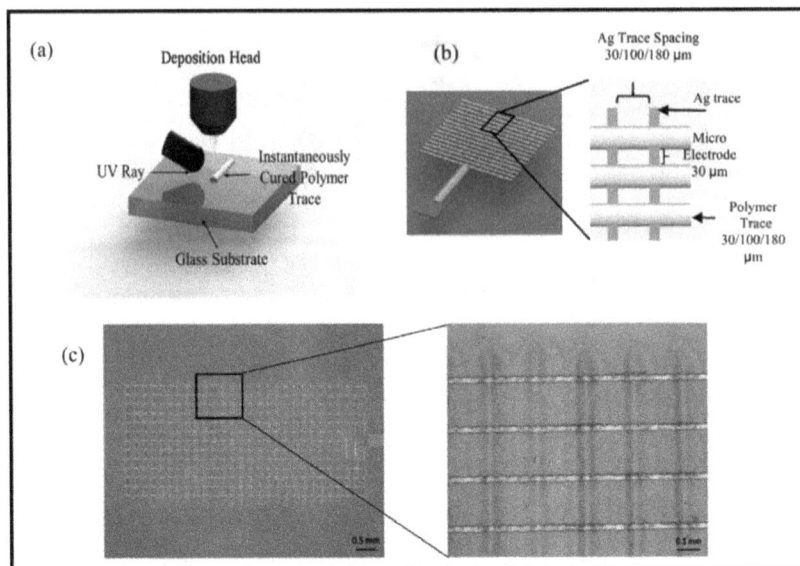

Fig. 11. Microelectrode arrays using aerosol jet spray printed microneedle array at different magnifications. (Reproduced with permission from (Yang et al. 2016) Copyright 2015 Elsevier).

and $5.6 \pm 0.2 \mu M$, respectively. Despite the ginormous success of MNs in the past few decades, minimal products are available in the market environment. This might be attributed to the rigorous approval process by various drug administration agencies before market availability. Also, such microneedles require engineering in material development and wirelessly controllable systems before it could be deployed in a practical application. Besides, innovative advances in the material used and synthesis procedures could accentuate the scope of minimally invasive sensors.

2.6 Invasive and Smart Electronics

Simultaneous Responsive Systems

As mentioned before, MNs are used for the electrochemical determination of disease biomarkers and bioanalyses. Transdermal therapeutic delivery of drugs via minimally invasive MNs is a recently developed strategy. The microchannels and micro conduits are used for the therapeutic effect whereas the outer surface of the MNs acts as an electroactive surface (Cahill and Cearbhaill 2015). Such combined MNs can be used for controlled and sustained drug delivery in accordance with the potential arising from the bio interactions. Swellable, coated,and dissolving MNs are usually preferred for therapeutic drug delivery applications (Wang et al. 2019). Huang et al. (Huang et al. 2007) designed a system to extract blood from the epidermal layer, evaluate the elevated glucose level, and deliver inulin into the body using a closed-loop control system. Similarly, Jayaneththi et al. (Jayaneththi et al. 2019) designed a magnetic polymer-driven handheld device for biosensing and transdermal delivery as shown in Figure 12. Such rapid response-delivery systems can be used to achieve simultaneous bio maker analysis and treatment. On the application side, such

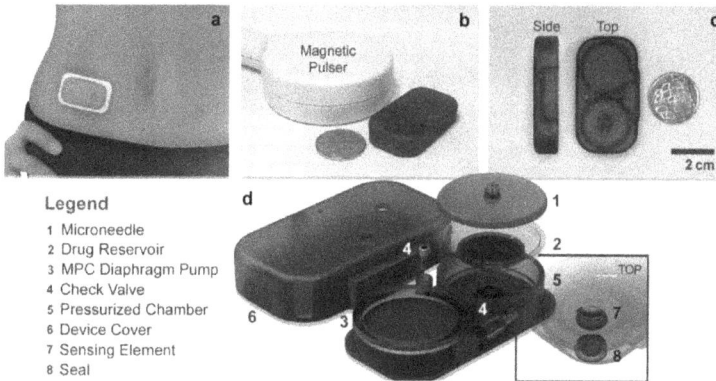

Fig. 12. Schematic photograph of the handheld simultaneous response and recovery systems designed by Jayaneththi et al. (Reproduced with permission from (Jayaneththi et al. 2019) Copyright 2019 Elsevier).

systems enhance low fabrication cost and long-term stability that could be integrated into clinical systems (Xie et al. 2020).

Edible Sensors

As this section deals with minimally invasive sensors, a completely new invasive type of sensors are introduced briefly. Edible sensors/electrodes are used in comprehensive devices using miniaturized flexible electronics to perform a particular action such as sensing, drug delivery, and imaging. Such sensors are biocompatible and are of the dissolvable type which do not pose any change in human physiology (Kalantar-zadeh et al. 2018). These sensors are encapsulated in placebo capsules which are usually digested within 2 days of intake. Such sensors are painless, bio erodible, environmentally sustainable, and are an alternative to invasive tests such as colonoscopy and endoscopy. Edible sensors establish a direct detection of biomarkers which is impractical with conventional techniques. The gas-sensing performance of the encapsulated flexible sensors was also demonstrated using this edible sensor. Fermentation of the microorganisms in the gastrointestinal tract (GI) induces fatty acids, CO_2, H_2, and CH_4 species that could be used to determine GI tract diseases (Ou et al. 2015). Specific abnormalities such as internal bleeding and swelling results in a rise in body temperature (Bettinger 2015). The thermistor-based sodium perturbation window was constructed using edible sensors to scavenge heat and to detect internal bleeding (Bettinger 2019). Thus, such sensors can also be used for self-temperature harvesting to detect bio analytes. Edible electrodes packed in food products such as a cookie, green bean, milky candy, cheese, and almonds were used for electrochemical characterization analysis. Bio-analytes such as ascorbic acid, salivary uric acid, and dopamine were used to discriminate disease biomarkers (Kim et al. 2017). Similarly, sodium-ion battery based current sources with activated carbon surface was used to detect the bio-analytes (Kim et al. 2013) (Figure 13). With such an actively powered battery-based electrode, heart rate, body temperature and metabolic activity of the human body could be administered. Though such sensors are extremely impressive with immense applications, such green materials have poor conductivity and their

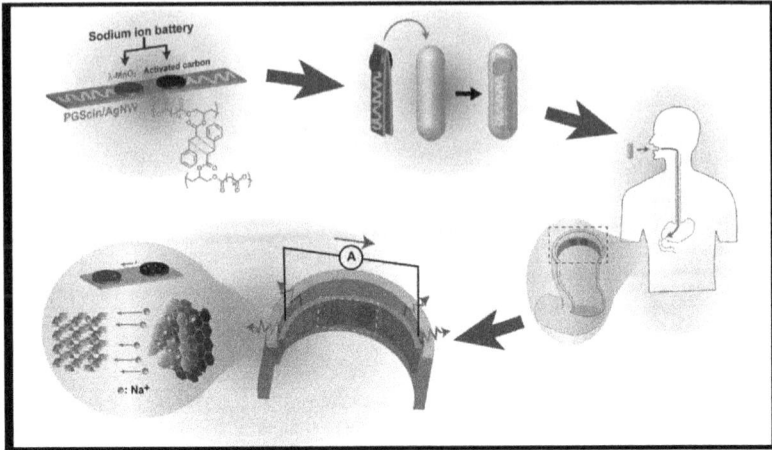

Fig. 13. Edible battery-based sensors (Reproduced with permission from (Kim et al. 2013) Copyright 2013 Elsevier).

performance degrades over time. In addition to poor longevity, material instability leads to limited performance capabilities.

The next section enunciates the various physio-chemical parameters of the human body which could be inferred from the various wearable sensors. The focus of this section is to enlighten the biomedical, mental, physical, and physiological parameters that could be deduced using wearable sensors. Most of the physiological parameters are still to be integrated with wearable MN-based sensors.

3. Wearable Sensors

3.1 Mental Acuity

In recent times, along with physical fitness, the role of mental well-being in determining the overall well-being of an individual has garnered significant attention. Thus, terms such as mental or cognitive acuity are being increasingly used in this context. The term mental acuity includes an assortment of highly interdependent brain functions such as information processing, memory, attention, and situational judgment. The overwhelming information from the environment in terms of sensory stimulus recognized *via* visual, tactile receptors and more, demands an efficient way to sort the information obtained so as to aid in its processing. Quickness in processing information ameliorates the brain's ability to discern and categorize information based on its relevance or importance and further aids in processing this information into memories. Along with improved information processing, "attention", which refers to the brain's ability to focus on a particular stimulus amongst many other signals in an environment, also aids in the memory storing activities of the brain (Harold Pashler 2005). The pre-mentioned functions cumulatively aid in the activity of "situational judgment", which indicates one's ability to take quick and ideal decisions in a fast-paced environment to which an individual is subjected on an everyday basis, thus attributing immense importance to this brain function. Hence, monitoring cognitive acuity aspects of an individual gives valuable information on one's attentiveness which may find immense use in smart driving alertness applications (Dhanusha et al. 2020, Pradesh 2020) and more. On the other hand, mental acuity based information may prove to be immensely useful towards highlighting factors/aspects affecting one's attention which may pave way for introspection and subsequently anchor the improvement on the identified factors by the individual, so as to improve his/her memory storing and decision-making ability. However, monitoring cognitive acuity *via* changes in biochemical factors proves to be complex. Thus, in this context, wearable devices have garnered significant attention and acceptance towards continual and efficient monitoring of cognitive parameters. Some of the recent works carried out in this domain are discussed in this sub-section.

Electroencephalography (EEG) bio-signal is a popular signature towards discerning the electrical activity and hence, potential abnormality (if any) in the brain. The human brain is a hard-wired network of billions of neurons, where each of these neurons responds to stimuli by varying ionic gradient across them by pumping/ taking in ions to/from the external environment and thus exhibits modulation of ionic potentials. An electrode material, generally placed in the scalp of a subject, appropriately converts these variations in ionic potentials to electrical potentials, which are then suitably recorded, processed, and analyzed. It must be noted that the output obtained as a result of the synchronous firing of neurons does not correspond to the firing of a particular one/few neurons. However, harnessing this bio-signal is immensely advantageous owing to its relatively simple hardware requirements and appreciable temporal resolution characteristics. Event-related potentials (ERP) from EEG signals may be used to monitor the fluctuations in brain activity, while performing a work/task, whereas frequency domain analysis of EEG involves its spectral component and aids the analysis of oscillations associated with the bio-signal. Correspondingly, waves of different frequency ranges are labialized and mapped to particular mental activity states of an individual;for example, waves, usually seen in babies, exhibit high amplitudes, and correspond to < 4 Hz frequency range; θ waves associated with drowsiness state lie between 4 to 7 Hz (Cahn and Polich 2006b); α waves, which exhibits variations during mental exertion, lie between 7 to 13 Hz (Gerrard and Malcolm 2007); μ waves also lie in the same frequency range, suppression of these waves are monitored towards mirror neuron based studies. Variations in ß waves, which lie between 13 to 30 Hz, are associated with thinking and activities involving physical movements (Pfurtscheller and Lopes 1999). Apart from these, γ waves of EEG lie between 30–100 Hz, the pre-mentioned waves as observed by (Pfurtscheller and Lopes 1999). This bio-signal finds potential applications towards studying sleep disorders (Griffiths and Gibson 1996), epileptic seizure detection applications (Sharmila and Geethanjali 2016), and more. Variations in Electrodermal Activity (EDA), also known as Galvanic Skin Resistance (GSR), may be correlated to the psychological or physiological changes in a subject. Arousal of the sympathetic nervous results in changes in sweat secretions and correspondingly modulates the skin conductivity. Owing to the non-uniform distribution of sweat glands in the human body, specific regions such as palms and legs are chosen for harnessing these bio-signals. However, environmental factors such as humidity and temperature and its delayed response of about 3 s prove to be bulwark towards its reliability towards commenting on the emotional states of an individual.

Two popular models namely, phenomenal and accessible consciousness models, have been widely discussed towards pondering the enigmatic concept of 'consciousness' (Block 2007). While the former considers only conscious experiences of individuals, the latter gives insight on the relationship between awareness and unawareness-based responses and is seen to be more optimal to ponder the effects of induced awareness through meditation practices. The way in which information obtained from the environment is perceived inturn regulates our emotions and this is achieved *via* a complex interaction between conscious and unconscious mechanisms. Apart from coordinating trivial senses associated with motion, touch, and more, unconscious mechanisms may also bring in past memories that may be strongly or sparsely related to the environment, which is an individual experience and thus tends to divert his/her attention from activities being performed. Such digressions affect an individual's performance and may also add to negative or stressful feelings upsetting his/her mental well-being. Mindfulness meditation, a Buddhist mediation practice, emphasizes paying attention and the need to control implicit feelings and accepting such mind states, thereby fostering one towards handling stressful life events in a constructive manner, hence contributing to one's psychophysical well-being (Tang et al. 2015). Evidence of this meditation technique to ameliorate several neuronal functions have been reported. Electrophysiological studies involving EEG measurements showed an increase in α and ß waves in frontal regions of the brain which implied improvements in focus and attention associated with the individual (Cahn and Polich 2006a). The realization of such improvements by an individual *via* frequently monitoring the progress may bolster one's intention to continue these practices and promote an individual towards the mediation's regular practice. In this context, Balconi et al. (Balconi et al. 2017) utilized commercially available wearable sensor Muse™ and reported the effects of periodic intervention during mindfulness meditation practice sessions on stress regulation in the subjects. As shown in Figure 14(a), Muse™ wearable device involves non-invasive measurement of EEG signals and is allied to a smartphone application, which provides an interactive user interface and aids in the process of practicing meditation techniques. Oscillatory profiles and information processing indicators were obtained via frequency and time-domain analysis of EEG, respectively. Changes in the ratio of α to ß waves indicated the effect of intervention by the device on the relaxation and activation states of the brain during meditation, which is shown in Figure 14(c) Stress levels reported in terms of perceived stress scale indicated a greater reduction in their levels on subjects exposed to periodic intervention in contrast to those that were not exposed to such periodic interventions, as seen in Figure 14(d) (Balconi et al. 2017).

Cognitive load provides information on mental power spent in performing a work/task and has been widely assessed through exosomatic parameters such as those obtained through GSR. However, it is difficult to obtain appreciable skin conductance values using this technique for individuals with dry skin conditions. On the other hand, endosomatic parameters, which harnesses bio-potentials in absence of an external electrical stimulus,are reported to be more suitable for providing significant information on the psychological and mental load of subjects.

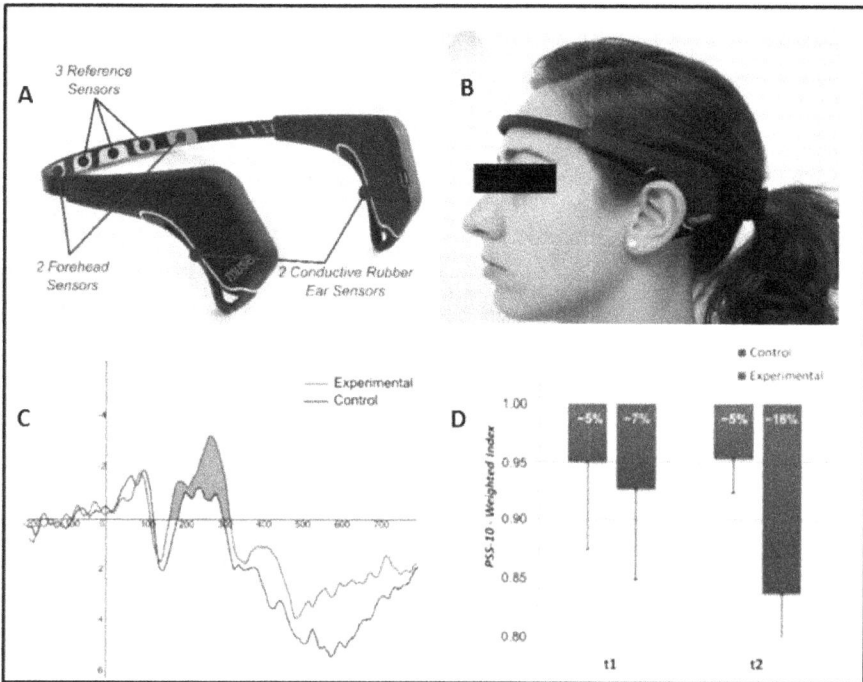

Fig. 14. (a) MuseTM wearable device, (b) Subject monitoring, (c) EEG Signal and (d) Stress levels (Reproduced with permission from (Balconi et al. 2017) Copyright 2017 Elsevier).

In this regard, Jaiswal et al. (Jaiswal et al. 2020) reported a reliable assessment of cognitive load during numerical addition task, by coupling both endosomatic and exosmotic bio-signals obtained through a 4 channel acquisition system and Shimmer GSR device, respectively. 4 channel data acquisition system was connected to index and middle finger on both hands and middle finger of both legs, while GSR measurements were harnessed from subject's palm. The random forest classifier algorithm was initially used for estimating accuracy from GSR and endosomatic parameters separately. Mutual features between signals obtained from 4 channel acquisition device and GSR signals obtained via combining results from feature discovery platform and maximal information coefficient technique was subjected to random forest classifier algorithm. This method resulted in greater accuracy and thus supports the concept of utilizing both endosomatic and exosmotic bio-signals towards the efficient assessment of cognitive load (Jaiswal et al. 2020).

Du et al. (Du et al. 2013) studied the role of *"cognitive reappraisal"*, one's ability to critically analyze and monitor events occurring, and thus exert control on his/her emotional changes, using a wearable GSR device. The emotional state of an individual and transition between different emotions was analyzed graphically *via* Arousal, Valence & Stance (AVS) space. The pre-mentioned parameters occupy three separate axes of the 3d space and they indicate excitement, pleasure and one's tendency to assess a situation critically. Distance between points in the 3d space may be used to unravel the extent of transition between different emotional states.

Emotional transitions are more plausible when an individual is subjected to external stimuli and the inherent relationship between such transitions was understood *via* Hidden Markov Model in this work. GSR wearable sensor consists of a multi-channel data acquisition device, which is used to harness other physiological parameters, "*medical silver chloride electrode*" was used for acquiring GSR signals, and the data obtained were transmitted using a Bluetooth device. GSR bio-signal is seen as a reliable signature towards different emotional states of an individual. Measurements of bio-signals using the pre-mentioned wearable device setup were carried out on subjects between the age of 21–25; all of them were asked to watch 14 different videos in random sequence, with/without self-assessing their emotions and, GSR bio-signals were monitored during each such case; this was repeated five times per video. Arousal measurements were seen to be correlated to GSR measurements over valence state measurements (Du et al. 2013).

Instances of micro-sleep or lapse in awareness have been inextricably associated with lesser productivity in work and motor accidents. These subsequently leave an indelible economic impact annually. Such instances are more often manifested in narcoleptic or sleep-deprived individuals and this has, in turn, resulted in an explosion of drug overdose/misuse cases by such individuals (Greene et al. 2008, Berman et al. 2008). Thus, the need to continuously monitor individuals for micro-sleep detection has attracted immense attention. Existing camera-based and polysomnography (PSG) techniques have certain limitations such as the inability to monitor other physiological signals other than eye movements and complex setup, respectively. In this context, Pham et al. (Pham et al. 2020) reported the fabrication of a wearable system for reliable micro-sleep detection. The Orexin system regulates brain activity, modulates eye-lid movements, and activates the sympathetic nervous system, whose effects are manifested as sweat gland activation and facial muscle movement in an individual. Subsequently, the activity of this system may be correlated to the awareness state of an individual and is assessed *via* monitoring EEG, Electrooculogram (EOG), Electromyogram (EMG), and EDA bio-signals of the individual. Consequently, micro-sleep may result in evident changes to the above signals, such as a shift to θ from ß and α wave in EEG, and in eyelids' gradual closing. Owing to the ear's anatomy, sufficient regions were chosen in the region and the obtained signals are referred to as '*behind the ear*' (BTE) signals. Since the magnitude of the signals were very low and were significantly affected by noise, they were appropriately amplified along with removing noise and then subjected to classification algorithms according to the methodology as reported. Silicon, owing to its flexibility and skin safety, was used in ear-piece and a wire within this silicon arrangement offered grip for its wearability. Gold is inert and exhibits minimal erosion on exposure to body sweat and oil secretions; thus, gold-coated copper electrodes, which exhibit very low resistance and good contact impedance,were chosen for the intended application. Motion artifacts and noise were removed while activities such as walking, and driving were being performed and these BTE signals hence obtained showed promise towards micro-sleep detection. BTE and PSG signals were recorded in an individual who was monitored in a sleep-inducing environment. '*Leave-one-subject-one-cross-validation*' classification revealed precision and recall of 76% and 85%, respectively (Pham et al. 2020).

Despite enhanced connectivity *via* social networking platforms, the Covid-19 pandemic may contribute to an increased sense of seclusion amongst individuals owing to a reduction in in-house visits, and social gatherings. Such feelings of loneliness may pave way for depression and other mental ailments (Killgore et al. 2020). Hence, the practice of meditation techniques is highly recommended in order to boost the morale of individuals. Thus, in this context, wearable sensors that monitor and provide real-time feedback and suggestions to users, depending on their brain activity during the practice of mindfulness meditation techniques, may garner greater attention and immense market value. As mentioned above, monitoring of awareness states of an individual is of immense importance to various fields, although EEG based bio-signals are used in this context, their subsequent processing and analysis offer certain difficulties owing to these bio-signals' smaller amplitudes and biological and motion artifacts. These factors highlight the importance of advanced signal processing and filtering techniques, which may correspondingly ameliorate subsequent classification and decision-making steps and lead to an overall improvement of the product.

3.2 Physical Performance

The key attribute of '*physical fitness*' proves to be a hub, connecting and regulating several pivotal health aspects such as mental well-being, blood pressure, weight control, immune response, and more. Regular physical activity is reported to improve the stress response and social skills of individuals (The ETO 2006) and boost their cardiovascular performance. Menopause in women leads to an increase in fat content, a decrease in muscle mass, a reduction in bone strength, which subsequently increases the chances of bone fracture and osteoporosis. Thus, regular engagement in physical activities tends to address the pre-mentioned effects associated with menopause and hence improve their overall health (Barbara Sternfeld 2012). In this context, the practice of less demanding exercises such as walking, jogging and swimming and strenuous exercises such as weightlifting and running is seen in a population of all ethnicity and age groups to bolster their physical fitness. In order to reap the best outcomes from physical activities, it is important to neither over-strain nor under-exert oneself during the practice and hence it is imperial to carry out these activities at/for an optimum '*intensity, duration and frequency*' which is personalized to an individual (Barbara Sternfeld 2012). Ill-effects associated with over/under-exertion of physical activities may be suitably addressed through one's regular monitoring of their physical fitness defining parameters (Haskell et al. 2012), such as aerobic capacity, muscle strength/exertion, stamina, flexibility and mobility of joints, shoulders and more. Subsequently, wearable sensors have capitalized on this niche environment towards continuous monitoring of one's physical performance which has attracted significant welcoming in the domain of sports and rehabilitation. On the other hand, neurodegenerative disorders such as Alzheimer's (AD) and Parkinson's disease (PD), which results in impaired motor functions in an individual, makes it exceedingly difficult for them to exhibit simple locomotion and other volitional activities (Kluger et al. 1997, Grabli et al. 2012, Chen et al. 2013). This subsequently renders them to greater risks of falling or to some other kind of physical injury, while walking or performing other simple tasks. In this context, the use of wearable sensors towards fall detection or movement monitoring in such disease afflicted patients and the corresponding development of fall prevention technologies have garnered significant attention. Wearable sensors designed for the pre-mentioned purposes are reported in this sub-section.

Accelerometers are electromechanical devices that are used in a plethora of applications, ranging from mobile phones to automobiles. The force associated with gravity exhibits changes on inclination; such changes monitored *via* accelerometers give information on the angle of inclination which, if exceeds a threshold limit, may be conclusively used to comment on the balanced state of an individual and hence can be utilized towards fall detection applications. These accelerometers work on piezoelectric or capacitive based principles, where changes associated with stress or capacitance between layers of microcrystal present in the device is correspondingly relayed as changes in acceleration values. Apart from the pre-mentioned application, these devices can be used towards tremor detection *via* monitoring the dynamic forces associated with involuntary vibrations exhibited by an individual. Tri-axial accelerometers, which monitor acceleration values along all three axes, are preferred for the applications mentioned above (Goodrich et al. 2019).

Tamura et al. (Tamura 2005) reported the usage of a wearable accelerometer that was leveraged towards the analysis of ambulatory patterns of individuals through changes in their acceleration profile. It was seen that acceleration patterns were periodic for normal, undisturbed walking, whereas abrupt/irregular walking movements that just preceded falling indicated the possibility of falling before the event takes place. Thus, this continual monitoring wearable device on integration with smart feedback strategy holds promise towards real-time fall detection/prevention applications (Tamura 2005). Lima et al. (Silva de Lima et al. 2020) performed a study to statistically highlight greater incidences of falling in PD patients over non-PD individuals via a study that involved over 2000 PD patients and control subjects, whose ambulatory profile were monitored for a span of over 2 years using commercially available necklace type wearable sensor. 'Age, comorbidity, living conditions, and gender' were common parameters between the groups. This sensor consisted of a barometer and accelerometer and initiated an automatic response signal which was relayed to a central unit on the occurrence of a fall event. Such relayed signals were immediately followed up by a phone call to the subject to both confirm occurrence of such events and provide appropriate medical assistance if required. Apart from overcoming the potential problem of unreliability associated with personally documenting such fall events by individual patients, this study highlights the suitability of wearable sensors towards long-term and effective monitoring of individuals (Silva de Lima et al. 2020).

Multiple Sclerosis (MS) is an autoimmune disorder that initiates inflammatory reactions and affects the myelin sheath covering of neurons in the brain (Compston and Coles 2008). This reaction subsequently leads to motor, sensory problems, and tremors. The manifestations of upper limb tremors are inturn used by clinicians to comment on the degree of MS of patients. However, such observations are affected by subjectivity and are more prone to errors. In this context, Teufl et al. (Teufl et al. 2017) reported monitoring of upper limb tremors using wrist-worn tri-axial accelerometers and hence comment on the degree of MS associated with patients. The group performed analysis on 12 individuals with self-reported MS and confirmed upper limb tremors by analyzing frequency components (if any) between 3 to 15 Hz *via* Fast Fourier Transform (FFT) technique and changes in acceleration in two out of three directions. The results obtained from the method adopted were correlated with a therapist and high sensitivity and specificity characteristics were reported (Teufl et al. 2017).

Ischemic stroke resulting due to impaired blood infusion into neuronal tissues of the brain is manifested as blood clots in that organ and results in exponential disruption of neurons, synapses, and myelin sheath in the surrounding region in a very short time (Saver 2006). Subsequently, this results in "*movement dysfunction*", which demands immediate addressing through relearning, and repeated practice of necessary movements/actions. Owing to the inherent plasticity of the human brain, which aids in reforming and re-establishing neuronal contacts and synapse formation, respectively, an early intervention following the unfortunate event of stroke is necessary. The extent of stroke and one's inherent ability to respond to therapy based interventions differs between patients and hence, personalized therapy has gained more attention. Thus, robotics-based wearable devices for rehabilitation have gained popularity. Problems associated with permanent cramps in fingers, whose long-term persistence may hinder day to day activities, is a cause of concern. In this context, Korzeniewska et al. (Korzeniewska et al. 2020) formulated a study catered towards the rehabilitation of fingers, whose movements are highly strained in stroke patients afflicted with upper limb dysfunction. Although commercially available TimStip© gloves (Figure 15(a and b)) are intended for such rehabilitation purposes, these are ineffective in patients with severe spasticity. Thus, the work involved deposition of commercially obtained highly conductive silver by physical vapor deposition (PVD) on a Band-Aid as seen in Figure 15(e), which was then integrated with TimStip© gloves. Although polyurethane in Band-Aid was randomly aligned (Figure 15(d)), a thin continuous electro-conductive layer of silver was obtained after the PVD process. However, the surface of the deposited layer was not uniform as seen in Figure 15(c). On the other hand, PVD proves to be more advantageous over conventional printing techniques such as inkjet printing, especially towards thin layer deposition of metals because the latter process results in soaking of flexible substrates such as cloth with metal inks, which subsequently results in greater deposition of metals on the substrate and hence limits flexibility. Rehabilitation *via* the proposed method carried out on a 36-year-old patient indicated promising results, which are reported in (Korzeniewska et al. 2020).

Although works involving the usage of wearable tri-axial accelerometers or wearable gyroscopes towards gait or motion analysis have rendered greater versatility towards carrying out ambulatory studies, such devices are limited by the need for firm surfaces for their integration with the subject. On the other hand, motion capture and camera-based technology along with pressure mat aids in an excellent spatial and temporal analysis of a subject's motion *via* providing a visual representation of dynamics exhibited; however, these are confined to lab atmosphere and is not suitable for continuous monitoring of subjects (Giansanti et al. 2003, Simon 2004, Schepers et al. 2007). Addressing these limitations, Low et al. (Low et al. 2020) fabricated

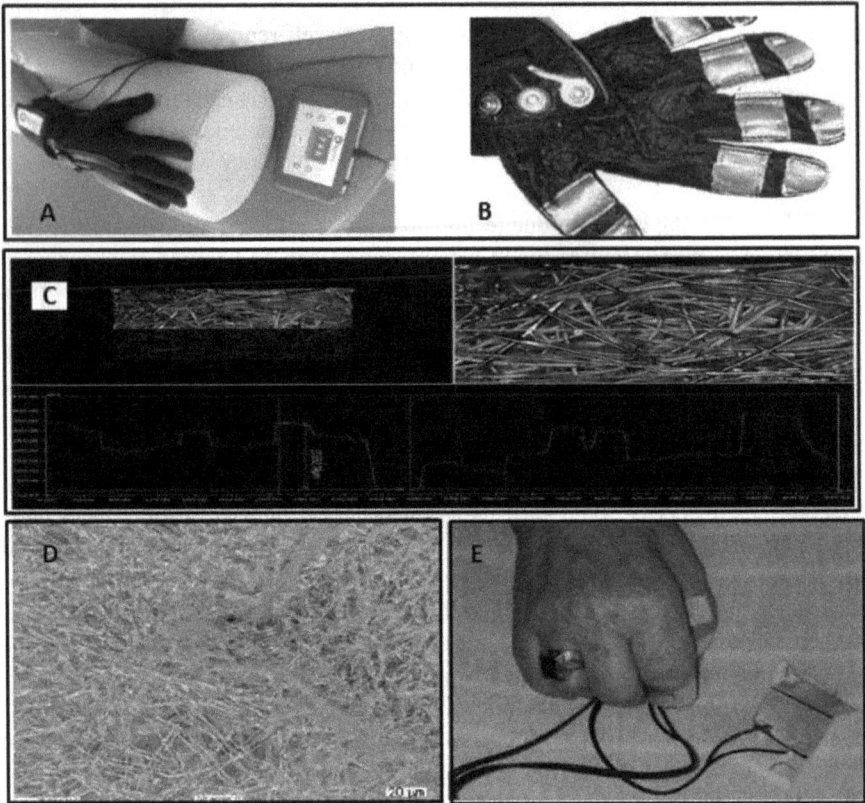

Fig. 15. TimStip© glove (a) on the medical stand, (b) internal view, (c) Structure and surface profile of PVD processed textile, Band-Aid, (d) Microscopic image, and (e) Silver deposited (Reproduced with permission from (Korzeniewska) Copyright 2020 MDPI).

a highly flexible spiral-shaped resistive sensor, integrated it into the sole portion of the shoe, and utilized it towards discerning the gait patterns of an individual. Liquid Eutectic Gallium-Indium, which is non-toxic, was slowly injected into the microfluidic patterns formulated on an elastomer. It was ensured that no air bubbles were formed during the liquid's injection, the injection was followed by attaching a copper film onto the ends of the fabricated flexible device. Owing to greater stress experienced by the metatarsal and heel regions in the foot during walking, the fabricated sensor was appropriately placed in these regions in the sole of a shoe. The stress produced due to walking initiates deformation in the microfluidic pattern changes the cross-sectional area and subsequently changes the resistance of the sensor which is relayed as voltage signals. The non-linear and linear resistance change characteristics of the fabricated sensor towards normal and axial stretching, respectively, were highlighted. Also, hysteresis exhibited by the sensor may be attributed to elastomer and is well within tolerating limits towards sense. The sensor exhibited negligible changes in resistance on the application of cyclic stress and periodic strain (30%) for 100 cycles, thus indicating its robust nature. For both the walking speeds of 2 and 5 kmh^{-1}, which was simulated using a treadmill, the metatarsal region makes a greater contribution to output, subsequently indicating greater stress experienced by this region over heels. The output signals obtained are relayed *via* Bluetooth to smartphone devices and hence can be effectively used towards real-time gait monitoring (Low et al. 2020).

Cricket, a sport played by 105 countries around the world, has gained immense viewership and tremendous market value in the past two decades. The increasing involvement of technology is being seen towards ameliorating the game experience to the viewers and to ease the decision-making process of umpires. Decision Review System (DRS), hawk-eye, and camera technologies augment the latter aspects *via* aiding decision making in Leg Before Wicket (LBW) and no ball declarations. However, not much technological interventions are seen towards commenting on the bowling action of bowlers. It must be noted that claim and subsequent confirmation of improper bowling action may result in serious repercussions and may even lead to the ban of a player. Hence, bowlers who are issued such notice on their bowling actions must have to appear for a test before an official panel within a stipulated time; the panel members will comment on the validity of the bowler's action and produce reports to the International Cricket Council (ICC). Thus, monitoring of angle between forearm and upper-arm may give useful insights on the validity of a bowler's action (For et al.). Such monitoring procedures carried out through conventional markers-based method prove to be cumbersome owing to the requirement of infrastructure, and specialist panel for decision making. Apart from these, the required infrastructure is only available in few regions around the globe and hence incurs travel charges to either player or their sponsors. Hence, in this context, Ahmed et al. (Ahmed et al. 2015) reported the fabrication of a real-time usable wearable sensor for bowling action monitoring; the device consisted of a wearable band, force sensor, and a Bluetooth setup to relay the output to a smartphone. The wearable band from Spectra Symbols consisted of an internal layer coated with resistive ink, which experiences a change in resistance on physical deformation of unidirectional bending of about 1°. Since this band is in contact with the subject's arm as shown in Figure 15 (a), the resistance changes may be extrapolated as the angle between forearm and upper arm and are subsequently used to comment on the validity of the action. Force sensor from Interlink Electronics was attached to a flexible ring-shaped material in the finger as shown in Figure 15 (e) and, on the release of the ball from the bowler's hand, the force experienced by the sensor changes. At this moment, the as obtained force and angle values obtained from force and flex sensor, respectively,are relayed to a microcontroller, which takes a decision. The data thus obtained are relayed to the smartphone *via* Bluetooth, which works on low power mode and thus ensures long-term utilization of the device towards the intended purpose (Ahmed et al. 2015). Apart from the pre-mentioned works, readers can refer to excellent articles (Spain et al. 2012, Tao et al. 2012, Chander et al. 2020, Zampogna et al. 2020) to garner more ideas on the works pertaining to wearable sensors in the field of fall detection and gait analysis.

Sectors such as gyms, therapists, and rehabilitation revolve and contribute towards several aspects of physical performance. Just like other sectors, these have also been worst affected by the Covid-19 pandemic. Owing to the impossibility in terms of maintaining sufficient social distancing and coming infrequent contacts with various surfaces and equipment in gyms, the latter have been forced to remain indefinitely closed. This has inturn promoted the practice of physical exercise at home and has subsequently opened up the market for wearable devices towards monitoring the progress of an individual (practicing exercise) *via* measuring pivotal physical parameters. Covid-19 has also worsened the situation for adults with motor impairments by reducing the other-wise frequent therapy sessions, thus subjecting adults living alone to greater risks. In this context, fall and gait monitoring sensors and subsequently fall prevention wearable technology, which obviates the requirement of skilled individuals or therapists for the pre-mentioned purposes, may garner greater demand. However, as mentioned elsewhere in the chapter, the requirement of rigid surfaces for placing wearable devices such as tri-axial accelerometers and goniometers limits their usage. This implies more work has to be done towards fabricating a fully flexible device as reported in (Low et al. 2020), which can be used effectively irrespective of the nature of work/activity performed by the individual.

According to Le Chatlier's principle, any system responds to an external perturbation in a way that opposes the effects of the latter. Living systems are an impeccable example to the previous statement owing to their exceptional ability to maintain minimally fluctuating steady-state conditions of physical and chemical parameters, which subsequently anchor the optimum performance and maintenance of physiological factors. The pre-mentioned phenomenon of "Homeostasis" is achieved *via* a complex interplay of contributing factors that can be broadly classified as biochemical and physiological parameters. Significant disturbances in these parameters may disrupt the health of an individual whose manifestations may or may not be immediate. This calls for a need to continuously monitor and efficiently convey adequate information pertaining to one's physiological and biochemical parameters to an individual so as to create awareness and aid him/her in early diagnosis or prognosis of ailment, if any, and improve the quality of living. Hence, the subsequent sub-sections cloister a discussion on wearable sensors that are catered towards monitoring above mentioned parameters along with including different works/reports performed in these domains.

3.3 Physiological Parameters

The living system, at every organizational level from organs to individual cells, may be effectively analyzed as a control system, where input received by a receptor is monitored by a control center, which *via* efficient feedback mechanisms control the response of effectors and facilitates homeostatic balance. Physiological variables associated with these processes are body temperature, arterial blood pressure, blood pH, fluidic balance, blood oxygen content, neurotransmission characteristics, and heart rate. It must be noted that physiological and biochemical parameters are interdependent, and hence an inextricable association lies between them, and subsequently, the effect of changes in one is manifested in the other. Thus, a significant deviation from steady-state values of the physiological variables may also trigger changes to biochemical counterparts, and these variations act as an indicator of dysfunction associated with organs, and hence sensors that facilitate continuous monitoring of these parameters have garnered attention. On the other hand, monitoring functions of organs in a living system may also be cloistered within the domain of physiological monitoring. Analysis of a complex system such as the entire human body may become more intuitive and easier *via* deciphering changes in a sub-system, i.e., at an organ level, and thus works on physiological monitoring of sub-systems such as respiratory, auditory,and urinary systems to name a few are being increasingly pursued. Some of the pertinent works are included in this sub-section.

Heart rate is more preferred over pulse rate owing to the latter's complex nature associated with the subject's vascular sclerosis characteristics. Electrocardiogram (ECG) signals possess p wave, QRS complex, and a T wave, which are representative of atrial depolarization, ventricular depolarization, and repolarization, respectively. Owing to the p wave's smaller amplitude, due to the smaller muscle mass of atria, heart rate may be obtained from calculating the RR interval of an ECG signal. On the other hand, physiological body temperature is monitored using thermistors, which undergo changes in resistance to changes in temperature. Thermistors with a negative temperature coefficient possess greater sensitivity and exhibit changes in resistance values for wide temperature ranges and are subsequently preferred (Hu et al. 2020a).

The quintessential requirement to achieve continual measurement of physiological parameters in newborns is a wearable device that is soft and extremely comfortable to the baby. Hu et al. (Hu et al. 2020a) reported designing such a sensor and utilized it to monitor the neo-natal's blood pressure, blood oxygen levels, and heart rate. The wearable device, shown in Figure 16(b), consists of a headband, which houses a blood oxygen monitoring module consisting of an optical frequency converter and bi-color Light Emitting Diode (LED) and conductive electrodes of silicon patch for measuring ECG signals. Light reflected from LED, which is in contact with the head of the baby as shown in Figure 16(b), is appropriately converted into frequency values, which are then passed through a timer (shown in Figure 16(a)), to obtain pulse rate and blood oxygen concentration of the baby. ECG signal measurement was performed in a single bipolar lead configuration using an ECG acquisition module, whose inherent architecture houses an amplifier and an analog to digital converter (ADC). The latter allows the direct transfer of obtained signal to a Bluetooth device *via* aserial peripheral interface (SPI), shown in Figure 16(a). Measurements of physiological parameters using the designed device were found to be on par with that obtained using a commercially available pulse rate, and heart rate monitoring device. However, errors associated with motion artifacts were not addressed in this work, which could hinder ECG measurements obtained (Hu et al. 2020a).

In an effort to aid the sensory characteristics of patients afflicted with peripheral nerve damage, Li et al. (Li et al. 2020) fabricated conducting flexible smart gloves that had the potential for temperature monitoring. The wearable device could also be used for gesture monitoring. Owing to excellent mechanical flexibility, and thermal and chemical stability, polydimethylsiloxane (PDMS) was chosen as the primary matrix material; to this variable, amounts of multi-walled carbon nanotubes (MWCNTs) were added to impart conductivity to the resultant composite (Wu 2019, Shi et al. 2020). Albeit, wet or electro-spinning based techniques are utilized for obtaining fibrous structures, which can be readily integrated into clothes

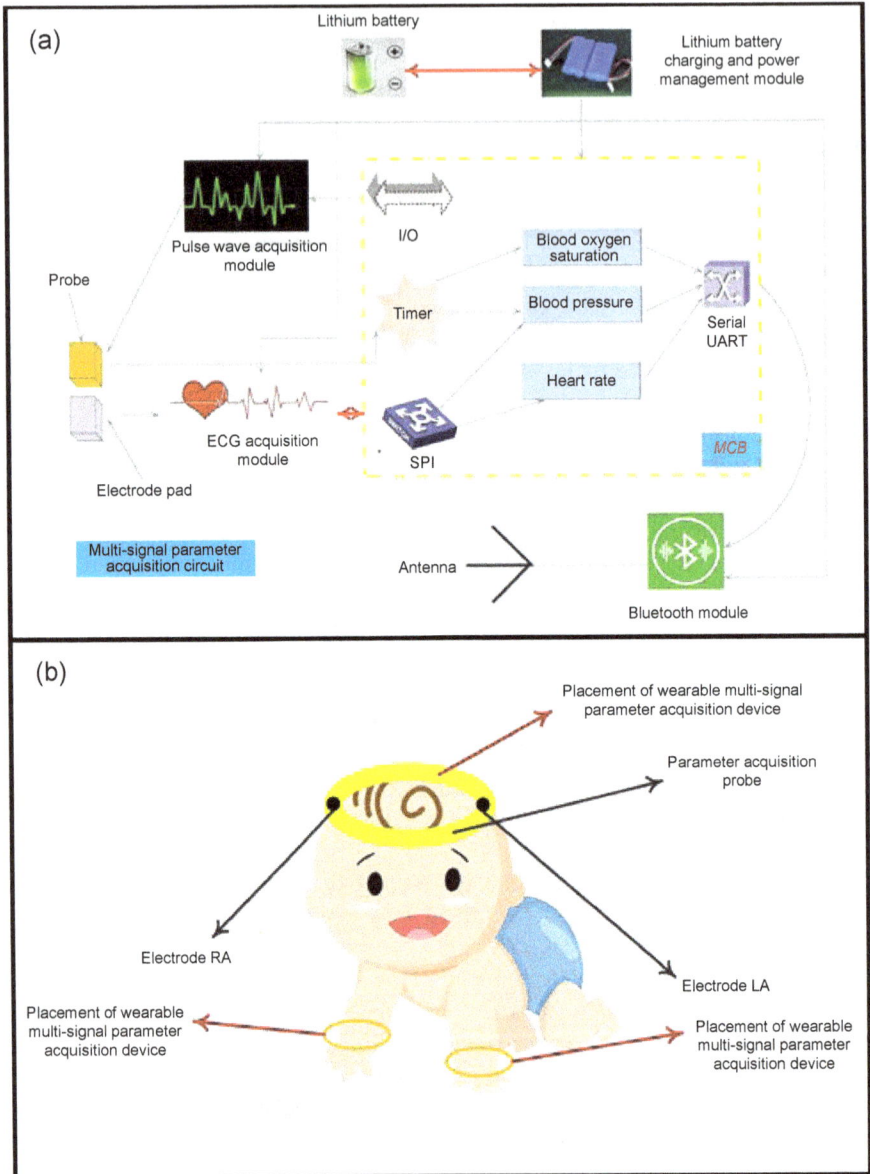

Fig. 16. Schematic of (a) Hardware, (b) Overall design (Reproduced with permission from (Hu et al. 2020) Copyright 2020 Elsevier).

or textiles, these techniques are limited by requirements of the bulk experimental setup, time, and skilled professionals for their operations. In this context, the group resorted to the one-step extrusion method, which involves squeezing pre-crosslinked composites having different filler composition into a screen having pores in micrometer range *via* a syringe. The obtained structures were further cross-linked to obtain final fibrous structures. Scanning electron microscopy (SEM) images revealed firm embedding of MWCNTs in the PDMS matrix, the obtained fibers were knitted in a sunflower like pattern as a testimony to its flexibility characteristics. A decrease in conductivity of the fibers was observed on stretching and this was attributed to an increased parallel distribution of MWCNTs and decreased point of contact between these filler structures. Also, stretching of fibers increases the length and reduces the cross-sectional area which leads to an increase in resistance of these structures. It was reported that fibers obtained with composites made of 0.2 g of MWCNTs in 1 g of PDMS exhibited ideal characteristics of both flexibility and conductivity. Stress-strain analysis of these fibers exhibited hysteresis and thus indicated a delay in recovery of original micro-structure, implying these fibers could be used only for certain loading frequencies in practical applications. The resistance changes of these structures exhibited linear relationship with the tensile elongation; also, the fibers exhibited real-time usability owing to their ability to withstand 20,000 stretching-releasing cycles for 50% strain values. Such structural robustness was attributed to MWCNTs' ability to undergo changes in tandem with deformations in PDMS. On the other hand, these fibers exhibited a linear decrease in changes associated with relative resistance on an increase in temperature, which may be due to increased charge hopping characteristics between filler particles. The decrease in the energy barrier of the transfer characteristics due to an increase in thermal energy at increasing temperatures may also be a reason. Owing to the fiber's ability to appropriately respond to both mechanical and thermal changes, they were suitably fixed onto a flexible glove and connected to a microcontroller. Fibers printed onto the tip of the index finger served as a temperature sensor, which can be utilized towards body temperature monitoring. On the other hand, finger movements, which subsequently initiate changes to relative resistance values, can be harnessed towards converting sign language into appropriate information (Li et al. 2020).

Urinary incontinence, a condition associated with dysfunctions of sphincter muscles in the urinary tract, results in involuntary micturition and proves to be an immense problem to the patients. Continual monitoring of bladder volume and appropriate triggering through an alarm when volume exceeds a certain level may signal an otherwise well patient to micturition and hence prevent the involuntary leakage of urine and substantially improve the patient's lifestyle. In this context, Gaubert et al. (Gaubert et al. 2020) designed an experiment and reported proof of concept for the intended application of urine volume monitoring in the bladder. The tissue environment is a collection of cells, which consists of intracellular and extracellular conducting fluids (owing to the presence of ions) and plasma membrane, the latter is insulating owing to its constituents of fatty acids and lipids and this arrangement imparts capacitive nature to the tissue architecture. Urine, which is relatively more conductive, is housed in the urinary bladder whose tissues are non-conducting and this arrangement promotes the usage of the non-invasive technique of bio-electrical impedance analysis (BIA) towards monitoring of bladder volume. The technique involves monitoring changes in voltage and frequency as a function of alternating current applied across electrodes in contact with skin. Of three configurations of the electrode, BIA on printed textile setup reveals configuration three to exhibit greater variations in relative resistance and is chosen for further studies. Studies were carried out on a pelvic model, whose interior consists of a pig bladder, which is filled *via* burette with solutions of different conductivity. The exterior and interior of textile electrodes, also known as "textrodes", also consists of a moisture sensor. Textrodes were fabricated by industrial knitting "*silver-plated nylon yarns*"; silver proved to be a prudent material of choice owing to its biocompatibility, conductivity, and antibacterial characteristics (Liu et al. 2014). Relative resistance changes for ionic solutions representative of urine, which indicates textrodes, are responsive to changes in the conductivity of the solution. Although these textrodes, which are dry electrodes, exhibit appreciable response to changes in solution's conductivity, their performance is lower in comparison to gel electrodes owing to parasitic capacitance associated with these dry electrodes. Albeit the work stands out as a proof of concept for the intended application, real-time bladder volume monitoring is replete with limitations associated with interference of other body fluids like sweat and other analytes involved in biochemical interactions in the body; also, noise due to motion artifacts must be adequately addressed *via* advanced signal processing techniques to enable the lab to produce transcendence of this work (Gaubert et al. 2020).

Monitoring respiratory rates of individuals at different postures is clinically relevant, especially in patients undergoing physiotherapy or those who have undergone surgery on the spine. On the other hand, continual and comfortable monitoring of respiratory rates may be carried out during sleep to comment on the quality of sleep and provide early intervention in conditions such as sleep apnea. In this purview, Al-Halhouli et al. (Al-Halhouli et al. 2020) fabricated a wearable, inkjet-printed respiratory rate monitoring strain sensor by inkjet printing silver ink on a PDMS substrate (Figure 17(a)). The sensor was attached to the xiphoid process as seen in Figure 17(b), whose resistance value changes owing to volume changes in abdomen/rib-cage. The sensor is part of a Wheatstone circuit with three other fixed resistance values, the changes in the sensor's resistance values cause a change in voltage, which is then fed to a microcontroller. The signals obtained are subsequently processed *via* a bandpass filter and frequency components obtained through Fast Fourier Transform are converted into the respiratory rate. Homogeneity of variance test revealed statistical non-significance and this minimized Type I error using ANOVA and hence the latter was performed using SPSS (statistical package for social sciences) package. ANOVA was performed on respiratory rates for different trials, postures (Figure 17(d)), and inkjet-printed and reference sensors. As seen in Figure 17(c), not any difference is seen between respiratory rates measurements for two different trials; in the figure, the mean is indicated by a horizontal line within the box plot, first and third quartiles are represented by ends of the box and maximum and minimum values are represented by whiskers. The mask-based sensors (e-Health) are reported to be less comfortable by users and are hence not preferred for continual monitoring of respiratory rates; in this aspect, the pre-mentioned fabricated sensor addresses this limitation and is found to be more comfortable for the intended purpose. However, it must be noted that the study was carried out only on healthy individuals, and measurements were carried out on subjects at rest and hence in this context, much work has to be carried out to comment on the effects of motion artifacts and other activities on the accuracy of this device (Al-Halhouli et al. 2020).

Owing to bulkiness and rigidity of existing physiological signal monitoring devices and the compulsion for the subject involved to remain stationary or being

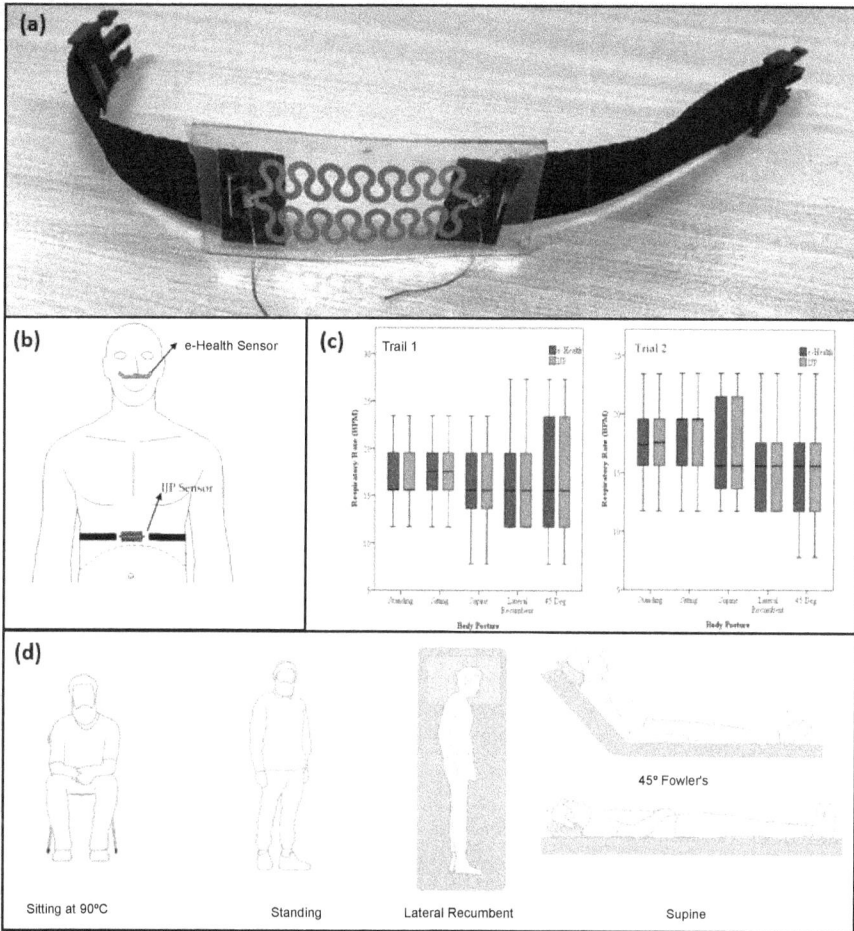

Fig. 17. (a) Silver ink printed on PDMS substrate, (b) inkjet-printed sensor on xiphoid process and e-Health sensor as mask, (c) Respiratory rates and statistical parameters for 2 different trails and (d) Different postures (i) sitting (90°) (ii) standing (iii) lateral recumbent (iv) 45° Fowler's and (v) supine (Reproduced with permission from (Al-Halhouli et al. 2020) Copyright 2019 MDPI).

allowed to only minimally move while carrying out studies on them have subsequently impeded the progress of performing real-time studies in babies or neo-natal. Hence, scientists or groups engaged in studying activities of babies with greater clarity to unravel more information on many of the intriguing aspects such as biomechanics associated with babies' crawling, heartbeat variations while performing activities and more may leverage ameliorations in the field of wearable devices. Owing to these devices' or sensors' ability to be appropriately integrated with babies' clothing and carry out measurements causing minimal discomfort to the babies has paved way for continual monitoring of certain physiological parameters in them. As mentioned earlier in the article, usage of wearable sensors may prove to be highly useful to geriatric patients, thereby allowing continual monitoring of their physiological parameters and alerting the appropriate authority or care-takers during critical times;

this subsequently lessens the patient's wholesome dependence on a caretaker and may lessen the tedious or vigilance the work generally demands from the latter. Apart from the pre-mentioned fields, wearable sensors have gained immense popularity towards continuous monitoring of physiological parameters and are being widely used in sports and rehabilitation domains. Data of each specific individual may be appropriately accessed by a physician or professional trainer and the personalized data hence obtained augments his/her understanding of an individual's requirement or physical condition and aid in providing appropriate interventions as required by an individual. Albeit information obtained is appropriately relayed at a short distance to necessary recipients *via* Bluetooth and Radio Frequency Identification Devices (RFID), more work in terms of information security and data transfer must be carried out to achieve reliable and safe long-distance transmission of information.

3.4 Biochemical Composition

A living system may be analogized to a machine, whose sub-systems or organs are exceptionally propelled to carry out day to day activities by electrochemical potential gradients generated by biochemical composition in their respective physiological micro-environment. Thus, significant variations in the concentration of such biochemical constituents may lead to immediate or a delayed dysfunction of organs and may proceed to affect physiological systems such as respiratory systems or affect urodynamics and more. Some of the commonly analyzed biochemical constituents are sodium, potassium and calcium ions, urea, cofactors such as magnesium ions, zinc ions, metabolites such as glucose, lactate, and more. Other biochemical constituents or parameters include biomolecules such as lipids, fatty acids and proteins, differential expression of small molecules such as micro-RNA (miRNA), and variations in the composition of cerebrospinal fluid, whose composition regulates brain activities. The blood glucose levels are tightly regulated whose concentration, when increases, propels the beta cells produced by islets of the pancreas to secrete insulin into the bloodstream. On the other hand, low blood glucose levels trigger the production of alpha cells which then increases glucagon and fatty acids levels (Koeslag et al. 2003). Diabetes mellitus type I is a disease closely associated with the disruption in pre-mentioned homeostatic control, where sufficient beta cells are not available to regulate the blood glucose levels and lead to hyperglycemia. An increase in the concentration of plasma calcium ions concentration is linked to parathyroid (Akizawa and Fukagawa 1999), which promotes bone resorption and renders one more prone to fracture. On the other hand, sodium ions are associated with regulating blood pressure (Preston et al. 2016) and potassium ions regulate aldosterone production (Büssemaker et al. 2010). Studies on small molecules of miRNA have been performed and differential expression of a panel of biomarkers have been utilized towards early diagnosis and prognosis of cancers (Rai et al. 2001, Timotheadou et al. 2007, Costa et al. 2011, Zhu et al. 2011, Yahalom et al. 2013). Apart from these, monitoring of pH in biological fluids such as saliva and urine may be performed towards commenting on the plausibility of gingivitis and renal disease, respectively (Hee et al. 2020). Data obtained from continual monitoring of these biochemical constituents may not only aid in the early diagnosis and prognosis of a disease but also in unraveling some of the many unknown mechanisms associated with the manifestation of resultant ailments. This subsequently demands minimally invasive or non-invasive monitoring of these constituents. Hence, selective monitoring of metabolites and ions from bio-fluids such as sweat, tears, saliva has been highly preferred (Heikenfeld 2016, Ferrero et al. 2018, Lin et al. 2018). Since it is not possible to encompass much of the works that have been carried out in this field, we have included some of the most recent works that promise to achieve non-invasive and continual monitoring of some of the very commonly monitored and physiologically relevant metabolites and ions in this sub-section. Apart from these, readers can refer to some of the excellent articles (Heikenfeld et al. 2019, Mccaul et al. 2017, Gao et al. 2018, Kim et al. 2018, Chen et al. 2019, Seshadri et al. 2019) in this field.

The ionic composition in sweat gives information about the hydration levels of an individual and subsequently, its continual monitoring has garnered attention in sports, where dehydration may be directly correlated to a player's performance and fitness. Moreover, analysis of body fluids like sweat is non-invasive and thus proves to be advantageous. In this purview, Copped et al. (Copped et al. 2020) performed experiments to validate the ability of ion-selective wearable textile-based organic electrochemical transistor (OECT) for monitoring Na^+, K^+ and Ca^{2+} in electrolyte samples. Bi-functionalization of textile was achieved by immersing yarn in poly (3,4-ethylenedioxythiophene):poly (styrene sulfonate) (PEDOT:PSS) and potassium or calcium selective membrane solution according to the procedure given in (Schefer et al. 1986, Tarabella et al. 2012), respectively. Figure 18.I(a) shows the schematic of the OCET and Figure 18.I(b) shows the experimental setup with an Ag wire gate electrode and the space between PEDOT:PSS/ion-selective membrane solution coated textile and gate electrode is filled with a sample electrolyte solution. Studies were performed at a constant drain-source voltage of −0.05 V and relative changes in drain current as a function of the concentration of ions for different gate voltages were obtained. Maximum response was obtained for gain voltage of 1 V for both calcium and potassium selective membrane functionalized textile electrodes. The mechanism of ion detection was reported as follows, where initially applied fixed drain voltage resulted in the drift of holes along conducting polymer backbone; on the application of positive gain voltage, the ions from electrolyte occupy the holes and decrease the conductivity due to holes. On bringing back the gain voltage to 0, the ions collected were emitted back to the electrolyte highlighting the reversible nature of the electrode. Figure 18.II(a) and (b) shows the response curves of potassium and calcium ion-selective membrane functionalized textile electrodes, respectively, where the relative changes in current response are more significant to K^+ and Ca^{2+} ions in comparison to Na^+ ions for which changes are seen only above 10^{-2} M. Figure 18.II(c) shows the kinetics associated with the response of individual ions where the response time for K^+ ions is much lower than that of Na^+ ions. The selectivity towards Ca2+ and K+ were 10 and 5 times of Na+, respectively, for their respective ion-selective membrane functionalized electrodes. Concentrations of ions were varied from 10^{-5} to 1 M and detection limits of the fabricated OCETs were in accordance with the concentration of these ions found in human sweat and prove to be promising towards practical sweat sensing applications (Copped et al. 2020).

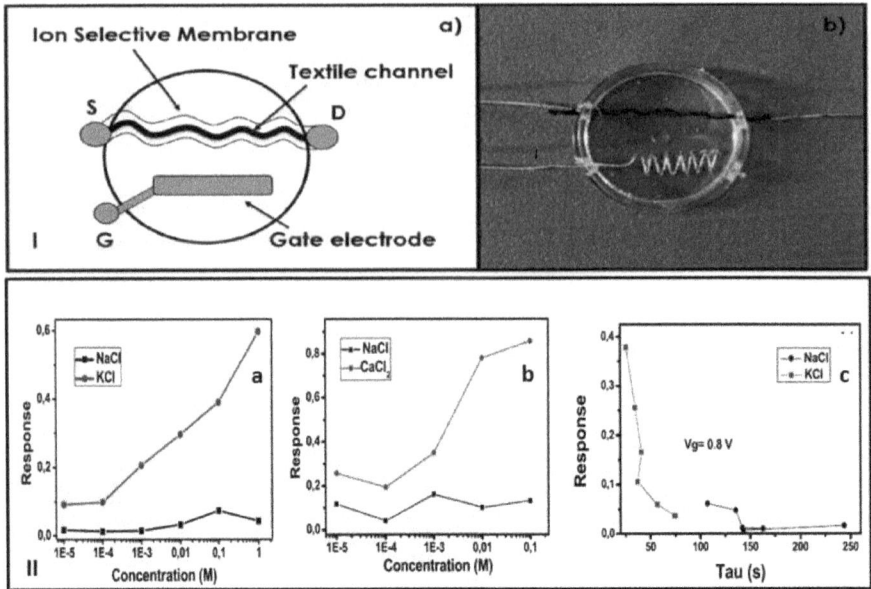

Fig. 18. I (a) Schematic of textile OCET. (b) Plastic vial containing electrolyte sample solution and fabricated OCET. II. Response curves for (a) potassium (b) calcium ion selective functionalized OCET and (c) kinetic studies on potassium ion selective functionalized OCET (Reproduced with permission from (Copped et al. 2020) Copyright 2020 Elsevier).

Diabetes, a lifestyle disorder, is expected to affect about 700 million people by the year 2045. Diabetes is also known to trigger dysfunction of other organs like eyes (Ellis et al. 1999), cause foot ulcers (Hinchliffe et al. 2020), and more; hence, coupled and continual monitoring of both blood sugar levels and impairment of organs may aid in unraveling unknown relationships between the progression of the disease and its effects on the organs. There are reports suggesting greater impairments in kidney functions and its correlation with diabetes. In this context, a proof of concept experiment was designed by Promphet et al. (Promphet et al. 2020) towards achieving a wearable, simultaneous monitoring of blood sugar levels indicative of diabetes and levels of impairments of kidneys. Glucose and urea are some of the many metabolites and small molecules that can be obtained from human sweat and may act as a bio-signature to comment on the blood sugar levels and kidney functioning of an individual. The group designed a sensor using a cotton thread dipped in cellulose nanofiber (CNF) solution, followed by dipping it in a graphene oxide/chitosan (GO/Chi) solution. One region in CNF-GO/Chi cotton fiber was appropriately exposed and coated with glucose oxidase and chromogenic solution for detection of glucose and the other region was coated with a solution of urease and phenol red for detection of urea. The flaking regions shown in Figure 19.I are coated with wax (Promphet et al. 2020). Cotton threads possess small gaps and this provides capillary action, which facilitates analysis of very low volumes of samples (Reches et al. 2010). Cotton threads dipped in 0.7% CNF solution for 1 h promotes interaction of GO/Chi with the cotton threads. GO enhances the effective interaction between enzymes and cellulose fibers through van der Waals and hydrogen bonding forces.

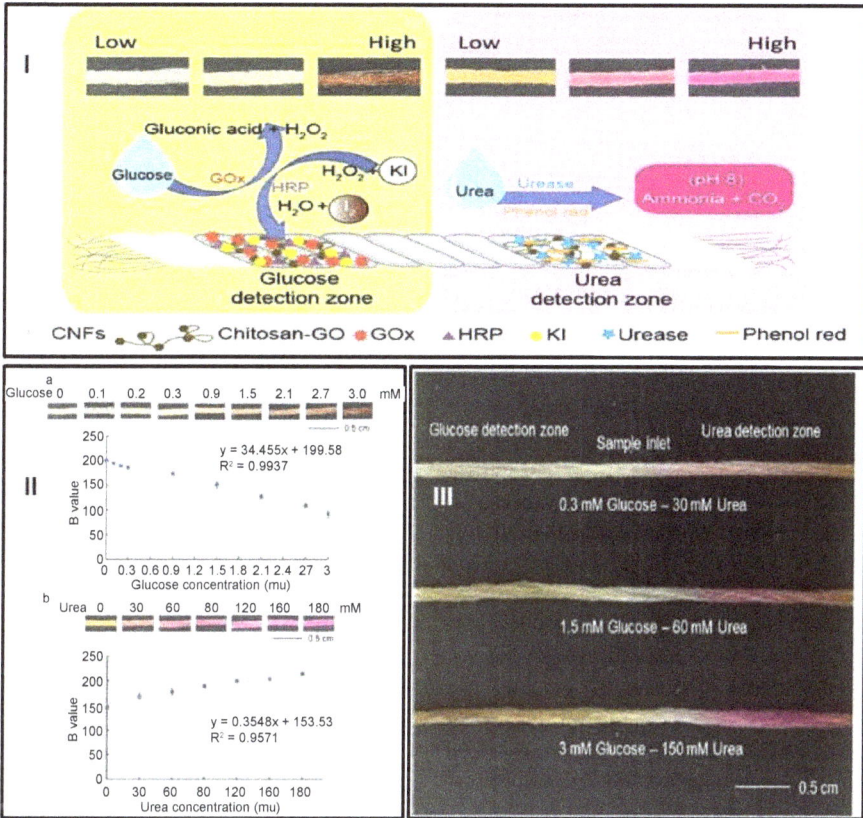

Fig. 19. I. Compartmentalization and mechanism for glucose and urea detection on cotton fiber sensor II. Blue color value (vs) (a) Glucose and (b) Urea concentration III. Color change was exhibited for different glucose and urea concentration in the sample (Reproduced with permission from (Promphet et al. 2020) Copyright 2020 Elsevier).

Chitosan owing to its positive charge electrostatically interacts with negatively charged cellulose fibers and provides the ideal micro-environment for sensing by aiding direct electron transfer between GO/Chi and enzyme surface. Figure 19.II(a) and (b) shows the response characteristics in terms of color change for varying concentrations of glucose and urea, respectively. Color change is achieved through mechanism depicted in schematic shown in Figure 19.I, where hydrogen peroxide liberated due to interaction of glucose with glucose oxidase, in the presence of Horse Radish Peroxidase (HRP), interacts with iodide ions from potassium iodide to give away iodine gas, which mediated color change. The limit of detection (LOD) was reported to be 0.1 – 0.3 mM, which is the level of glucose found in the sweat of normal healthy individuals. On the other hand, urea on interaction with urease liberates NH_3 and CO_2 which increases pH and results in a change of color in phenol red-coated regions as shown in Figure 19.III. LOD for urea was reported to be 30–180 mM, where urea levels above 65 mM are indicative of diabetes in an individual and hence this sensor finds real-time application towards colorimetric sensing of diabetic individuals using their sweat samples (Promphet et al. 2020).

As mentioned earlier, studies pertaining to differential expression of micro-RNAs (miRNAs) have garnered greater attention, especially towards early diagnosis or prognosis of cancer. miRNA-21 is one such biomarker, which on differential expression may hint plausibility of non-small cell lung cancer (Schwarzkopf and Pierce 2016). Conventional methods such as quantitative real-time polymerase chain reaction (qPCR) and other methods such as next-generation sequencing and Raman spectroscopy are used for estimation of mi-RNA concentration; however, these methods are hindered by the requirements of sophisticated equipment and skilled personnel for carrying out experiments and analyzing results. In this context, portable bio-sensors that can be further upgraded as a wearable device prove to be advantageous and hence attract more attention. Subsequently, Shin et al. (Shin et al. 2020) fabricated a screen-printed rGO/Au/ssDNA-based portable smart device that could selectively detect miRNA-21 from saliva. Saliva proves to be a safe and non-invasive option and is more preferred to harness information pertaining to biomarkers. Also, miRNAs are found insufficient concentration in saliva (Weber et al. 2010) and hence was chosen as the sample for this work. GO/HAuCl$_4$ solution was drop-casted onto the working electrode (WE) of the commercially available disc-type screen-printed electrode, followed by subjecting the setup to 10 cycles of electrochemical pre-treatment in NaOH solution. This latter process facilitated the reduction of both GO and HAuCl$_4$ to rGO and Au, respectively, which was confirmed *via* increasing current responses for each cycle in cyclic voltammetry (CV) and Energy dispersive spectra (EDS) analysis. A solution of ssDNA modified with thiol at its ends was added to the above electrode setup and incubated for 1 h, strong interaction between thiol and Au facilitated the immobilization of ssDNA onto WE. Following this, miRNA-21 was added and incubated for an hour in this electrode setup. CV and Differential Pulse Voltammetry (DPV) experiments were performed after each step on bare electrodes; rGO/Au electrode, rGO/Au/ssDNA and rGO/Au/ssDNA/miRNA21 immobilized WE using K$_4$Fe(CN)$_6$ as the analyte and corresponding results were obtained. Owing to the enhanced surface area and synergistic electro-catalytic properties of rGO/Au, maximum peak current response was obtained for this rGO/Au immobilized electrode. On addition of ssDNA, due to the presence of phosphate groups that hinder the interaction of K$_4$Fe(CN)$_6$ with rGO/Au composite, the peak current response decreases. The decrease was much more in case of rGO/Au/ssDNA/miRNA21 owing to presence of greater number of phosphate groups on a double stranded DNA structure. Interference studies with dissimilar miRNA such as miRNA-141 and miRNA with single base-mismatch indicated higher selectivity of fabricated sensor to miRNA21. The as fabricated electrode was connected to a circuit board, which was also connected to a smart phone device. The latter sends command to initiate DPV on the electrode/circuit setup, after which the circuit sends the resulting DPV data to the smart phone via Bluetooth as shown in Figure 20(a and b), which can be appropriately visualized using user-interactive application (Shin et al. 2020).

Monitoring of biomarkers such as nitric oxide from saliva may be useful towards heart disease monitoring/detection whereas monitoring saliva pH may be used towards discerning the plausibility of periodontal diseases (Pataranutaporn et al. 2019). Thus, continual monitoring and data collection using the saliva sample opens up broad avenues in terms of obtaining information about the biochemical composition of an individual, and these sensors can be further coupled with sensors monitoring physiological signals and may aid in the continual overall assessment of the user's health. In this context, proof of concept towards continual saliva monitoring and data collection *via* smartphone was demonstrated by Pataranutaporn et al. (Pataranutaporn et al. 2019). A wearable device, that consists of a paper roll with a plastic film at its bottom such that it undergoes color changes on interaction with saliva, and owing to plastic film, the effects of such interactions are not felt by the underlying biochemical paper. The roll-to-roll motion of the paper roll is controlled by a stepper motor and the color changes are sensed by the RGB sensor. The obtained data is then transmitted to a smartphone, which uses the IMU data, i.e., filtered data omitting variations in saliva during speaking or other regular activities, and is then reported to the user (Pataranutaporn et al. 2019).

As highlighted in the previous sub-section, both physiological and biochemical parameters are highly related, and thus monitoring one gives either direct or indirect information about the other. In fact, monitoring of biochemical parameters gives a molecular-level understanding of changes associated with physiological sub-systems and its continual monitoring may aid in early detection of certain dysfunctions associated with an organ and hence provide early intervention towards a diagnosis of an ailment if any. However, owing to the presence of plenary of such biochemical markers in each biological sample such as blood, sweat, saliva, tears, and more, subsequent processes associated with selectively extracting information on a biomarker or metabolite of interest becomes tedious and demands highly selective and sensitive methodology. Although conventional chromatography techniques achieve the pre-mentioned requirement, they are hindered by requirements of a sophisticated lab environment to perform quantification and hence fail to satisfy the continuous monitoring criterion. In this aspect, the usage of enzymatic biosensors have garnered

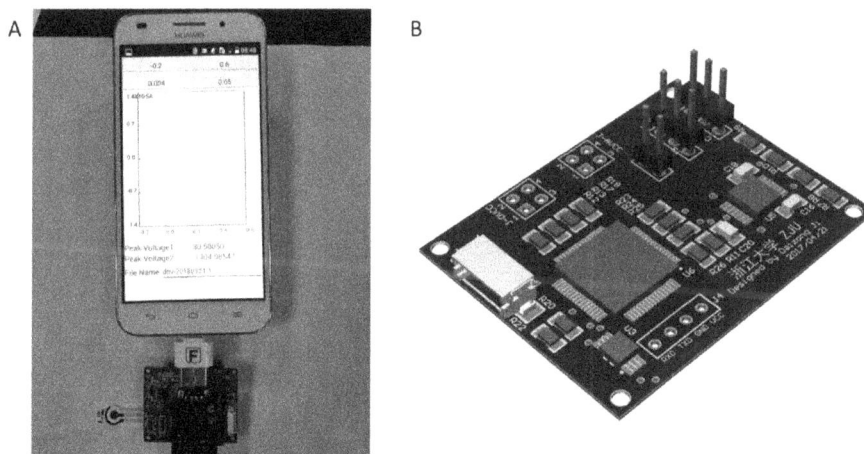

Fig. 20. Set-up of the smartphone-based electrochemical biosensing system for miRNA detection. (A) The rGO/Au composite-modified electrode was connected to the circuit board and the DPV result was transmitted and displayed on smartphone via Bluetooth. (B) Design of the printed circuit board used in this work (Reproduced with permission from (Shin et al. 2020) Copyright 2020 Elsevier).

immense attention; biosensors, owing to their miniature size, are portable and can be easily integrated into textile or other accessories can be used as a wearable device. On the other hand, enzymes are highly specific biomolecules and interact only with the analyte of interest and hence address aspects of selectivity. Electrochemical biosensors utilize redox mediators that augment the electron transfer reactions and subsequently improve the sensitivity of the sensor. Monitoring lactate levels gives credible information about heart failures (Lelyavina et al. 2015), and has greater relevance towards monitoring fatigue in individuals. Strenuous physical activities demand more energy which is obtained from glucose produced predominantly *via* glycolysis; however, owing to a decrease in oxygen amounts, lactate is produced due to an anaerobic process. Lactate detection from sweat has been advantageous owing to the latter's non-invasive nature; however, along with lactate sweat contains other metabolites such as uric acid (UA), glucose, ascorbic acid (AA), and more. Thus, Zhang et al. (Zhang et al. 2020) fabricated a carbon micro-band-type screen-printed electrode with Ag/AgCl as a reference electrode. A solution of osmium mediator, MWCNT, HRP, and PEDGE, which initiates crosslinking between osmium/MWCNT and HRP, were drop-casted on the WE. Following this, lactate oxidase (LOx) gel-entrapped in poly-(phenylenediamine) matrix was drop-casted on the WE area and the film was grown by subjecting electrode to electrochemical pre-treatment cycles from –0.5 to 1.2 V for 50 cycles. Lactate, on interaction with LOx, gives away pyruvate and hydrogen peroxide, which reduces HRP; the reduced HRP is inturn oxidized by osmium complex, which then attracts an electron from WE and causes a change in current. This response is utilized to quantitatively estimate lactate concentration in the sample. The prepared sensor was appropriately attached to an eyeglass and ensured that it is in proper contact with the skin of the person. Owing to appreciable current response at 0 V, as indicated from CV studies, this voltage was chosen for carrying out further amperometry analysis towards the quantitative estimation of lactate. Such low voltage requirements accorded further advantages such as not causing any skin irritation. The shelf-life of the sensor was appreciable and reproducible results were obtained. Measurements in the presence of physiologically relevant amounts of other metabolites such as UA, AA and glucose were performed and high selectivity towards lactate was confirmed. Real-time studies on healthy subjects carrying out static cycling activity were performed, and the current response indicative of lactate concentration was obtained. Trends from electrochemical sensing carried out on forehead sweat samples of individuals was similar to that obtained from the sides (in contact with eyeglasses) and hence revealed the aptness of the measurements. Though the trends in current responses were similar, measurements from lactate sensor as epidermal patch attached to the deltoid region had noise associated with body movements while performing the activity and in this context, placement of the sensor on the eyeglasses proved to be appropriate and thus holds promise towards effective and continual lactate monitoring in an individual (Zhang et al. 2020).

Although usage of insulin injection (Strauss et al. 2002) is popular amongst diabetic patients, which is generally infused following intake of food, after prior monitoring of blood glucose levels, it must be kept in mind that the measurements thus obtained are only instantaneous values. Owing to the inherent dynamics associated with biochemical parameters such as glucose, whose values temporally change based on the activity performed, it is imperative to consider the history of glucose concentration to provide the result. This demands a continual and non-invasive monitoring method that improves the reliability of the result and can be achieved *via* a wearable sensor. These continuous and reliable measurements open up avenues for simultaneous detection and treatment of ailments such as reported in (Lee et al. 2017), where depending on the variations in blood glucose levels, insulin can be infused into the individual through minimally invasive micro-needles attached to the epidermal surface. Thus, use of microneedles obliterates an external and possible error prone monitoring of glucose levels by individuals and promises maintenance of glucose levels near to one's homeostatic requirements. Apart from glucose, monitoring of other metabolites such as lactate and more as indicated in the article has gained attention. However, most of the metabolites' effective monitoring is achieved using enzymes; while the latter renders augmented selectivity towards sensing analyte of interest, it demands laborious immobilization procedures as indicated in (Zhang et al. 2020) which proves to be a challenge. Other than these, sensing of sodium and potassium ions has opened avenues towards monitoring hydration levels in individuals, which proves to be of high relevance to the sports sector. Advancements in genomics and data analysis have inturn fueled studies on differential expression of biomarkers such as miRNA which, as mentioned previously in this article, holds immense prospects towards early diagnosis and prognosis of diseases such as cancer; hence, this calls for an integrative approach between the field of genomics, data analysis, and sensors to initially identify a panel of biomarkers that are differentially expressed, whose clinical significance is analyzed through statistical tests followed by formulating appropriate sensing strategy towards sensing of clinically relevant biomarkers. The latter aspect of the fabrication of sensors is also benefitted by immense advancements in the *de novo* synthesis of DNA nucleotide sequences. Albeit, this sub-section covered certain aspects of popularly monitored biomarkers and clinically relevant metabolites and ions, these prove to be a speckle in the vast expanse of biochemical constituents regulating many specific aspects of homeostasis living system and hence, day to day developments and findings in this field of biosensors is inextricably associated to health care monitoring and appropriate integration of these sensors with wearable devices or textile ultimately ameliorates the quality of living of an individual.

4. Conclusions and Future Perspectives

Though the pandemic has caused global havoc leaving an indelible economic impact, it has indirectly opened broad avenues for wearable sensors and smart healthcare monitoring devices to cash in, owing to much preferred "social distancing", which has become the newnorm (triggered by Covid-19) and hence technologies that promote health monitoring by self, thereby reducing the frequency of clinic visits, are welcoming. Preparation protocols, materials used, and sensing modalities of the sensors are guided in this chapter. Also, the elaborately discussed parameters in sub-section 4 were biomedical, mental, physical, and physiological parameters with each one of them by themselves having an enormous market for wearable sensors,thus rendering the field to be extremely promising in the very near future.

Technological innovations in MNs based sensors would propel the discovery of new materials and swift fabrication procedures. Biocompatibility, sensitivity, selectivity, and long-term toxicity analysis of the microneedles are the most crucial parameters to be focused on during the material development and interface modifications. In par, the flexibility of the wearable sensors is one of the major challenges in recent times, since they highly contribute to the comfort of patients. In addition, flexibility can potentially alter the surface electrical parameters in the sensor system which are also to be addressed. Power consumption demand and flexible electronic circuitry of the sensor is an additional discrepancy that is to be addressed. Minimally invasive sensors have ameliorated biosensing application and yet is limited to very few technological innovations. To the best of our knowledge, minimally invasive sensors used for the measurement of the physiological parameters is still an unexplored field of interest. The assistance of a multi-disciplinary field of interest is quintessential to alleviate a minimally invasive continuous monitoring system.

Acknowledgements

The authors would like to thank SASTRA Deemed University, Thanjavur, India and Tokai University, Hiratsuka, Japan for the infrastructural and research support. GKM and KT would like to thank Japan Society for the Promotion of Science (P19076) and Micro/Nano Technology Centre (MNTC), Tokai University (Shonan Campus), Japan for providing infrastructure.

References

Ahmed, A., M. Asawal, M. J. Khan and H. M. Cheema. (2015). A wearable wireless sensor for real time validation of bowling action in cricket. 2015 IEEE 12th Int Conf Wearable Implant Body Sens Networks, BSN 1–5.

Akizawa, T. and M. Fukagawa. (1999). Modulation of parathyroid cell function by calcium ion in health and uremia. *Am J Med Sci* 317: 358–362.

Al-Halhouli, A., L. Al-Ghussain, S. El Bouri, F. Habash, H. Liu and D. Zheng. (2020). Clinical evaluation of stretchable and wearable inkjet-printed strain gauge sensor for respiratory rate monitoring at different body postures. *Appl Sci* 10: 480.

Al Sulaiman, D., J. Y. H. Chang, N. R. Bennett, H. Topouzi, C. A. Higgins, D. J. Irvine and S. Ladame. (2019). Hydrogel-coated microneedle arrays for minimally invasive sampling and sensing of specific circulating nucleic acids from skin interstitial fluid. *ACS Nano* 13: 9620–9628.

Andar, A. U., R. Karan, W. T. Pecher, P. DasSarma, W. D. Hedrich, A. L. Stinchcomb and S. DasSarma. (2017). Microneedle-assisted skin permeation by nontoxic bioengineerable gas vesicle nanoparticles. *Mol Pharm* 14: 953–958.

Ayittey, P. N., J. S. Walker, J. J. Rice and P. P. De Tombe. (2009). Glass microneedles for force measurements: A finite-element analysis model. *Pflugers Arch Eur J Physiol* 457: 1415–1422.

Babamiri, B., A. Salimi and R. Hallaj. (2018). A molecularly imprinted electrochemiluminescence sensor for ultrasensitive HIV-1 gene detection using EuS nanocrystals as luminophore. *Biosens Bioelectron* 117: 332–339.

Balconi, M., G. Fronda, I. Venturella and D. Crivelli. (2017). Conscious, pre-conscious and unconscious mechanisms in emotional behaviour. Some applications to the mindfulness approach with wearable devices. *Appl Sci* 7: 1–14.

Barbara Sternfeld, S. D. (2012). Physical activity and health during the menopausal transition. *Obsetrics & Gynecology Clinics* 38: 1–28.

Berman, S., J. O'Neill, S. Fears, G. Bartzokis and E. D. London. (2008). Abuse of amphetamines and structural abnormalities in the brain. *Annals of the New York Academy of Sciences* 1141: 195–220.

Bettinger, C. J. (2015). Materials advances for next-generation ingestible electronic medical devices. *Trends Biotechnol* 33(10): 575–585, ISSN 0167-7799, https://doi.org/10.1016/j.tibtech.2015.07.008.

Bettinger, C. J. (2019). Edible hybrid microbial-electronic sensors for bleeding detection and beyond. *Hepatobiliary Surgery and Nutrition* 8(2): 157–160. doi: 10.21037/hbsn.2018.11.14.

Block, N. (2007). Consciousness, accessibility, and the mesh between psychology and neuroscience. *Behavioral Brain Science* 30: 481–548.

Bodhale D. W., A. Nisar and N. Afzulpurkar. (2010). Design, fabrication and analysis of silicon microneedles for transdermal drug delivery applications. *IFMBE Proc* 27: 84–89.

Bollella, P., S. Sharma, A. E. G. Cass, F. Tasca and R. Antiochia. (2019a). Minimally invasive glucose monitoring using a highly porous gold microneedles-based biosensor: Characterization and application in artificial interstitial fluid. *Catalysts* 9(7): 580. https://doi.org/10.3390/catal9070580.

Bollella, P., S. Sharma, A. E. G. Cass and R. Antiochia. (2019b). Minimally-invasive Microneedle-based biosensor array for simultaneous lactate and glucose monitoring in artificial interstitial fluid. *Electroanalysis* 31: 374–382.

Bollella, P., S. Sharma, A. E. G. Cass and R. Antiochia. (2019c). Microneedle-based biosensor for minimally-invasive lactate detection. *Biosens Bioelectron* 123: 152–159.

Bruzas, I., W. Lum, Z. Gorunmez and L. Sagle. (2018). Advances in surface-enhanced Raman spectroscopy (SERS) substrates for lipid and protein characterization: Sensing and beyond. *Analyst.* 143: 3990–4008.

Büssemaker, E., U. Hillebrand, M. Hausberg, H. Pavenstädt and H. Oberleithner. (2010). Pathogenesis of hypertension: Interactions among sodium, potassium, and aldosterone. *American Journal of Kidney Diseases* 55(6): 1111–1120, ISSN 0272-6386, https://doi.org/10.1053/j.ajkd.2009.12.022.

Cahill, E. M. and E. D. O Cearbhaill. (2015). Toward biofunctional microneedles for stimulus responsive drug delivery. *Bioconjugate Chemistry, ACS.*

Cahn, B. R. and J. Polich. (2006a). Meditation states and traits: EEG, ERP, and neuroimaging studies. *Psychol Bull* 132: 180–211.

Cai, B., W. Xia, S. Bredenberg and H. Engqvist. (2014). Self-setting bioceramic microscopic protrusions for transdermal drug delivery. *J Mater Chem B* 2: 5992–5998.

Cai, B., W. Xia, S. Bredenberg, H. Li and H. Engqvist. (2015). Bioceramic microneedles with flexible and self-swelling substrate. *European Journal of Pharmaceutics and Biopharmaceutics* 94: 404–410, ISSN 0939-6411, https://doi.org/10.1016/j.ejpb.2015.06.016.

Chander, H., R. F. Burch, P. Talegaonkar, D. Saucier, T. Luczak, J.E. Ball et al. (2020). Wearable stretch sensors for human movement monitoring and fall detection in ergonomics. *Int J Environ Res Public Health* 17: 1–18.

Chen, D., C. Wang, W. Chen, Y. Chen and John X. J. Zhang. (2015). PVDF-Nafion nanomembranes coated microneedles for *in vivo* transcutaneous implantable glucose sensing. *Biosensors and Bioelectronics* 74: 1047–1052. ISSN 0956-5663, https://doi.org/10.1016/j.bios.2015.07.036.

Chen, L., C. Zhang, J. Xiao, J. You, W. Zhang, Y. Liu, L. Xu, A. Liu, H. Xin and X. Wang. (2020). Local extraction and detection of early stage breast cancers through a microneedle and nano-Ag/MBL film based painless and blood-free strategy. *Materials Science and Engineering: C* 109: 110402. ISSN 0928-4931, https://doi.org/10.1016/j.msec.2019.110402.

Chen, P., R. Wang, D. Liou and J. Shaw. (2013). Gait disorders in parkinson's disease : Assessment and management q. *International Journal for Gerontology* 7: 189–193.

Chen, Y., W. Ji, K. Yan, J. Gao and J. Zhang. (2019). Fuel cell-based self-powered electrochemical sensors for biochemical detection. *Nano Energy* 61: 173–193. ISSN 2211-2855, https://doi.org/10.1016/j.nanoen.2019.04.056.

Chinnadayyala, S. R., I. Park and S. Cho. (2018). Nonenzymatic determination of glucose at near neutral pH values based on the use of nafion and platinum black coated microneedle electrode array. *Microchim Acta* 185: 250. https://doi.org/10.1007/s00604-018-2770-1.

Compston, A. and A. Coles. (2008). Multiple sclerosis. *Lancet* 372: 1502–1517.

Copped, N., M. Giannetto, M. Villani, V. Lucchini, E. Battista, M. Careri and A. Zappettini. (2020). Ion selective textile organic electrochemical transistor for wearable sweat monitoring. *Organic Electronics* 78: 105579, ISSN 1566-1199, https://doi.org/10.1016/j.orgel.2019.105579.

Costa, V. L., R. Henrique, S. A. Danielsen, M. Eknaes, P. Patrício, A. Morais et al. (2011). TCF21 and PCDH17 methylation: An innovative panel of biomarkers for a simultaneous detection of urological cancers. *Epigenetics* 6: 1120–1130.

Dardano, P., I. Rea and L. De Stefano. (2019). Microneedles-based electrochemical sensors: New tools for advanced biosensing. *Curr Opin Electrochem* 17: 121–127.

Dhanusha, P. B., A. Lakshmi and K. Saravanan. (2020). Smart driving system with automatic driver alert and braking mechanism. *3C Tecnología. Glosas de innovación aplicadas a la pyme. Edición Especial, Marzo*, 287–299. http://doi.org/10.17993/3ctecno.2020.specialissue4.287-299.

Donnelly, R. F., D. I. J. Morrow, T. R. R Singh, K. Migalska, P. A. McCarron, C. O'Mahony and A. D. Woolfson. (2009). Processing difficulties and instability of carbohydrate microneedle arrays. *Drug Development and Industrial Pharmacy* 35: 10, 1242–1254, DOI: 10.1080/03639040902882280.

Du, R., Y. Zhu, L. Lu, W. Wenhua, D. Cheng and Y. Kuilan. (2013). The SVM-based algorithm for chinese micro-blog opinion sentence identification. *Journal of Hunan University & Technology* 27: 89–93.

Ellis, J. D., C. J. Macewen and A. D. Morris. (1999). Should diabetic patients be screened for glaucoma? *British Journal of Ophthalmology* 83: 369–372.

Esfandyarpour, R., H. Esfandyarpour, M. Javanmard, James S. Harris and Ronald W. Davis. (2013). Microneedle biosensor: A method for direct label-free real time protein detection. *Sensors and Actuators B: Chemical* 177: 848–855, ISSN 0925-4005, https://doi.org/10.1016/j.snb.2012.11.064.

Ferrero, G., F. Cordero, S. Tarallo, M. Arigoni, F. Riccardo, G. Gallo, G. Ronco, M. Allasia, N. Kulkarni, G. Matullo, P. Vineis, R. A. Calogero, B. Pardini and A. Naccarati. (2018). Small non-coding RNA profiling in human biofluids and surrogate tissues from healthy individuals: description of the diverse and most represented species. *Oncotarget* 9(3): 3097–3111. https://doi.org/10.18632/oncotarget.23203.

Francisco, S. (1972). Technical contribution tungsten microneedles" a simple method of production, 425–426.

Gao, B. B., A. Elbaz, Z. Z. He, Z. Y. Xie, H. Xu, S. Q. Liu, E. B. Su, H. Liu and Z. Z. Gu. (2018). Bioinspired kirigami fish-based highly stretched wearable biosensor for human biochemical—physiological hybrid monitoring. *Advanced Material Technology* 3: 1700308. https://doi.org/10.1002/admt.201700308.

Gaubert, V., H. Gidik and K. Vladan. (2020). Smart underwear, incorporating textrodes, to estimate the bladder volume: Proof of concept on a test bench. *Smart Mater Struct* 29 085028.

Gerrard, P. and R. Malcolm. (2007). Mechanisms of moda fi nil : A Review of Current Research 3: 349–364.

Giansanti, D., V. Macellari, G. Maccioni and A. Cappozzo. (2003). Is it Feasible to Reconstruct Body Segment 3-D Position and Orientation Using Accelerometric Data? 50: 476–483.

Gittard, S. D., R. J. Narayan, C. Jin, N. A. Monteiro-Riviere, A. Ovsianikov, B. N. Chichkov, S. Stafslien and B. Chisholm. (2009). Pulsed laser deposition of antimicrobial silver coating on Ormocer (registered) microneedles. United Kingdom: N. p. Web. doi: 10.1088/1758-5082/1/4/041001.

Goodrich, R. Accelerometers_ What They Are & How They Work (Live Science - Blog).

Goud, K. Y., C. Moonla, R. K. Mishra, C. Yu, R. Narayan, I. Litvan and J. Wang. (2019). Wearable electrochemical microneedle sensor for continuous monitoring of levodopa: Toward parkinson management. *ACS Sensors* 4(8): 2196-2204. DOI: 10.1021/acssensors.9b01127.

Grabli, D., C. Karachi, M. Welter, B. Lau, E. C. Hirsch, M. Vidailhet and C. François. (2012). Normal and pathological gait: What we learn from Parkinsons Disease? *Journal of Neurology, Neurosurgery & Psychiatry* 83: 979–985.

Greene, S. L., F. Kerr and G. Braitberg. (2008). Review article : Amphetamines and related drugs of abuse. *Emergency Medicine Australasia*, 20: 391–402. https://doi.org/10.1111/j.1742-6723.2008.01114.x.

Griffiths, J. and J. Gibson. (1996). Automated recognition of eeg changes respiratory sleep disorders. *Sleep* 19(4): 296–303, https://doi.org/10.1093/sleep/19.4.296.

Guo, W. J., X. Y. Yang, Z. Wu and Z. L. Zhang. (2020). A colorimetric and electrochemical dual-mode biosensor for thrombin using a magnetic separation technique. *Journal of Materials Chemistry B* 8: 3574–3581.

Gupta, J., H. S. Gill, S. N. Andrews and M. R. Prausnitz. (2011). Kinetics of skin resealing after insertion of microneedles in human subjects. *Journal of Controlled Release* 154: 148–155.

Harold Pashler, J. C. J. (2005). Attentional limitations in dual-task performance. *Trends in Cognitive Science* 9: 126–135.

Hartmann, X. H. M., P. Van Der Linde, E. F. G. A Homburg, L. C. A. Van Breemen, A. M. De Jong and R. Luttge. (2015). Insertion process of ceramic nanoporous microneedles by means of a novel mechanical applicator design. *Pharmaceutics* 7: 503–522, https://doi.org/10.3390/pharmaceutics7040503.

Haskell, W. L., R. P. Troian, J. A. Hammond, M. J. Phillips, L. C. Strader, D. X. Marquez et al. (2012). Physical activity and physical fitness. *Amepre* 42: 486–492.

Heather, R. and H. Metiu. (1989). Time-dependent theory of Raman scattering for systems with several excited electronic states: Application to a H^{3+} model system. *Chemical Physics* 90: 6903–6915.

Hee, J., S. Kim, H. Jun, Y. K. Kim, D. X. Oh, Han-Won Cho, K. G. Lee, S. Y. Hwang, J. Park, B. G. Choi. (2020). Biosensors and Bioelectronics Highly Self-healable and Flexible Cable-type pH Sensors for Real-time Monitoring of Human Fluids. 150, https://doi.org/10.1016/j.bios.2019.111946.

Heikenfeld, J. (2016). Non-invasive Analyte Access and Sensing through Eccrine Sweat : Challenges and Outlook circa. *Electroanalysis*, 28: 1242–1249. https://doi.org/10.1002/elan.201600018.

Heikenfeld, J., A. Jajack, B. Feldman, Steve W. Granger, S. Gaitonde, G. Begtrup and Benjamin A. Katchman. (2019). Accessing analytes in biofluids for peripheral biochemical monitoring. *Nature Biotechnolgy* 37: 407–419. https://doi.org/10.1038/s41587-019-0040-3.

Hemanth, S., A. Halder, C. Caviglia, Q. Chi and S. S. Keller. (2018). 3D carbon microelectrodes with bio-functionalized graphene for electrochemical biosensing. *Biosensors* 19; 8(3): 70

Hinchliffe, R. J., S. Nikol, R. O. Forsythe and E. J. Boyko. (2020). Guidelines on diagnosis, prognosis, and management of peripheral artery disease in patients with foot ulcers and diabetes. *Diabetic Metabolism Research and Reviews* 36: 1–12.

Hosu, O., S. Mirel, R. Sǎndulescu and C. Cristea. (2019). Minireview: Smart tattoo, microneedle, point-of-care, and phone-based biosensors for medical screening, diagnosis, and monitoring. *Anal Lett* 52: 78–92.

Hu, X., J. Cao and H. Wu. (2020a). A wearable device for collecting multi-signal parameters of newborn. *Comput Commun* 154: 269–277.

Hu, Z., C. S. Meduri, R. S. J. Ingrole, H. S. Gill and G. Kumar. (2020b). Solid and hollow metallic glass microneedles for transdermal drug-delivery. *Appl Phys Lett* 116: 203703.

Huang, C. J., Y. H. Chen, C. H. Wang, T. C. Chou and G. B. Lee. (2007). Integrated microfluidic systems for automatic glucose sensing and insulin injection. *Sensors Actuators B Chem* 122: 461–468.

Hwa, K. Y., B. Subramani, P. W. Chang, M. Chien and Jung-Tang Huang. (2015). Transdermal microneedle array-based sensor for real time continuous glucose monitoring. *Int J Electrochem Sci* 10: 2455–2466.

Ita, K. (2018). Ceramic microneedles and hollow microneedles for transdermal drug delivery: Two decades of research. *J Drug Deliv Sci Technol* 44: 314–322.

Jaiswal, D., M. Moulick, D. Chatterjee, R. Ranjan, R. K. Ramakrishnan, A. Pal and R. Ghosh. (2020). Assessment of cognitive load from bio-potentials measured using wearable endosomatic device. WearSys 2020 - Proc 6th ACM Work Wearable Syst. *Appl Part MobiSys* 2020 13–18.

Jayaneththi, V. R., K. Aw, M. Sharma, J. Wenb, D. Svirskisb and A. J. McDaida. (2019). Controlled transdermal drug delivery using a wireless magnetic microneedle patch: Preclinical device development. *Sensors Actuators B Chem* 297: 126708.

Ji, J., F. E. H Tay, J. Miao and C. Iliescu. (2006). Microfabricated silicon microneedle array for transdermal drug delivery. *J Phys Conf Ser* 34: 1127–1131.

Jiang, J., L. Li, K. Li, G. Li, F. You, Y. Zuo et al. (2016). Antibacterial nanohydroxyapatite/polyurethane composite scaffolds with silver phosphate particles for bone regeneration. *J Biomater Sci Polym Ed* 27: 1584–1598.

Kalantar-zadeh, K., K. J. Berean, N. Ha, A. F. Chrimes, K. Xu, D. Grando, J. Z. Ou, N. Pillai, J. L. Campbell, R. Brkljača, K. M. Taylor, R. E. Burgell, C. K. Yao, S. A. Ward, C. S. McSweeney, J. G. Muir and P. R. Gibson. (2018). A human pilot trial of ingestible electronic capsules capable of sensing different gases in the gut. *Nat Elec* 1: 79–87.

Keum, D. H., H. S. Jung, T. Wang, M. H. Shin, Y.-E. Kim, K. H. Kim, G.-O. Ahn and S. K. Hahn. (2015). Microneedle biosensor for real-time electrical detection of nitric oxide for *in situ* cancer diagnosis during endomicroscopy. *Adv Healthc Mater* 4: 1153–1158.

Killgore, W. D. S., S. A. Cloonan, E. C. Taylor and N. S. Dailey. (2020). Loneliness: A signature mental health concern in the era of COVID-19. *Psychiatry Res* 290: 113117.

Kim, J., I. Jeerapan, B. Ciui, M. C. Hartel, A. Martin and J. Wang. (2017). Edible Electrochemistry: Food Materials Based Electrochemical Sensors 1700770: 1–9.

Kim, J., A. S. Campbell and J. Wang. (2018). Wearable non-invasive epidermal glucose sensors : A review. *Talanta* 177: 163–170.

Kim, K. B., W. C. Lee, C. H. Cho, D.-S. Park, S. JeCho and Y.-B. Shim. (2019). Continuous glucose monitoring using a microneedle array sensor coupled with a wireless signal transmitter. *Sensors Actuators B Chem* 281: 14–21.

Kim, Y. C., J. H. Park and M. R. Prausnitz. (2012). Microneedles for drug and vaccine delivery. *Adv Drug Deliv Rev* 64: 1547–1568.

Kim, Y. J., S. E. Chun, J. Whitacre and C. J. Bettinger. (2013). Self-deployable current sources fabricated from edible materials. *J Mater Chem B* 1: 3781–3788.

Kluger, A., J. G. Gianutsos, J. Golomb, S. H. Ferris, A. E. George, E. Franssen and B. Reisberg. (1997). Patterns of Motor Impairment in Normal Aging, Mild Cognitive Decline, and Early Alzheimer's Disease 52: 28–39.

Koeslag, J. H., P. T. Saunders and E. Terblanche. (2003). A reappraisal of the blood glucose homeostat which comprehensively explains the type 2 diabetes mellitus-syndrome X complex. *J Physiol* 549: 333–346.

Kong, K. V., C. J. H. Ho, T. Gong, W. K. O. Laub and M. Olivo. (2014). Sensitive SERS glucose sensing in biological media using alkyne functionalized boronic acid on planar substrates. *Biosens Bioelectron* 56: 186–191.

Korzeniewska, E., A. Krawczyk, J. Mróz, E. Wyszyńska and R. Zawiślak. (2020). Applications of smart textiles in post-stroke rehabilitation. *Sensors* 20: 2370, https://doi.org/10.3390/s20082370.

Krafft, C., B. Dietzek and J. Popp. (2009). Raman and CARS microspectroscopy of cells and tissues. *Analyst* 134: 1046–1057.

Lahiji, S. F., M. Dangol and H. Jung. (2015). A patchless dissolving microneedle delivery system enabling rapid and efficient transdermal drug delivery. *Sincentific Reports Nature*, 1–7.

Lee, H., C. Song, Y. S. Hong, M. S. Kim, H. R. Cho, T. Kang et al. (2017). Wearable/disposable sweat-based glucose monitoring device with multistage transdermal drug delivery module. *Science Adv.* 3: e1601314: 1–8.

Lee, S. J., H. S. Yoon, X. Xuan, J. Y. Park, S. J. Paik and M. G. Allen. (2016). A patch type non-enzymatic biosensor based on 3D sus micro-needle electrode array for minimally invasive continuous glucose monitoring. *Sensors Actuators B Chem* 222: 1144–1151.

Lelyavina, T., M. Sitnikova and E. Shlyakhto. (2015). Diagnostic and prognostic value of lactate threshold and ph - threshold determination during cardiopulmonary testing in patients with chronic heart failure. *Br J Med Med Res* 5: 289–296.

Li, J., B. Liu, Y. Zhou, Z. Chen, L. Jiang, W. Yuan and L. Lian. (2017). Fabrication of a Ti porous microneedle array by metal injection molding for transdermal drug delivery. *PLOS one Journal* 12(2): e0172043.

Li, Y., C. Zheng, S. Liu, L. Huang, T. Fang, J. Xinze Li, F. Xu and F. Li. (2020). Smart glove integrated with tunable MWNTs/PDMS fibers made of a one-step extrusion method for finger dexterity, gesture, and temperature recognition. ACS *Appl Mater Interfaces* 12: 23764–23773.

Lin, Z. Y., X. Y. Han, Z. H. Chen, G. Shi and M. Zhang. (2018). Label-free non-invasive fluorescent pattern discrimination of thiols and chiral recognition of cysteine enantiomers in biofluids using a bioinspired copolymer–Cu^{2+} hybrid sensor array regulated by pH. *J Mater Chem B* 6: 6877–6883.

Liu, F., Z. Lin, Q. Jin, Q. Wu, C. Yang, H.-J. Chen et al. (2019). Protection of nanostructures-integrated microneedle biosensor using dissolvable polymer coating. *ACS Appl Mater Interfaces* 11: 4809–4819.

Liu, H., M. Lv, B. Deng, J. Li, M. Yu, Q. Huang and C. Fan. (2014). Laundering durable antibacterial cotton fabrics grafted with pomegranate-shaped polymer wrapped in silver nanoparticle aggregations. *Scientific Report*, 1–9.

Liu, S., D. C. Yeo, C. Wiraja, H. L. Tey, M. Mrksich and C. Xu. (2017). Peptide delivery with poly (ethylene glycol) diacrylate microneedles through swelling effect, *Bioengineering and Translational Medicine* 1–10.

Low, J. H., P. S. Chee, E. H. Lim and V. Ganesan. (2020). Design of a wireless smart insole using stretchable microfluidic sensor for gait monitoring. *Smart Mater Struct*, 29.

Lu, C. H., Y. Zhang, S. F. Tang, Z.-B. Fang, H.-H. Yang, X. Chen and G.-N. Chen. (2012). Sensing HIV related protein using epitope imprinted hydrophilic polymer coated quartz crystal microbalance. *Biosens Bioelectron* 31: 439–444.

Lutton, R. E. M., E. Larrañeta, M. C. Kearney, P. Boyd, A. D. Woolfson and R. F. Donnelly. (2015). A novel scalable manufacturing process for the production of hydrogel-forming microneedle arrays. *Int J Pharm* 494: 417–429.

Lutton, R. E. M., A. D. Woolfson and R. F. Donnelly. (2016). Microneedle arrays as transdermal and intradermal drug delivery systems : Materials science, manufacture and commercial development. *Materials Science and Engineering: R: Reports* 104: 1–32.

Madden, J., C. O'Mahony, M. Thompson, AlanO'Riordan and P. Galvin. (2020). Biosensing in dermal interstitial fluid using microneedle based electrochemical devices. *Sens Bio-Sensing Res* 29: 100348.

Mani, G. K., K. Miyakoda, A. Saito, Y. Yasoda, K. Kajiwara, M. Kimura and K. Tsuchiya. Microneedle pH sensor: direct, label-free, real-time detection of cerebrospinal fluid and bladder pH. *ACS Sensors* 9(26): 21651–21659.

Mani, G. K., M. Kentaro and T. Kazuyoshi. (2021). Cleanroom and template free fabrication of single polygonal shaped microneedle. *J Nanos and Nanotech* 21(9): 4861–4864.

Mansoor, I., Y. Liu, U. O. Häfeli and B. Stoeber. (2013). Fabrication of hollow microneedle arrays using electrodeposition of metal onto solvent cast conductive polymer structures. pp. 373–376. *In*: 2013 Transducers & Eurosensors XXVII: The 17th International Conference on Solid-State Sensors, Actuators and Microsystems (Transducers & eurosensors XXVII).

Mccaul, M., T. Glennon and D. Diamond. (2017). Challenges and opportunities in wearable technology for biochemical analysis in sweat. *Curr Opin Electrochem* 3: 46–50.

Mcgrath, M. G., S. Vucen, A. Vrdoljak, A. Kelly, ConorO'Mahony, A. M. Creana and A. Mooreac. (2013). European journal of pharmaceutics and biopharmaceutics production of dissolvable microneedles using an atomised spray process : Effect of microneedle composition on skin penetration. *Eur J Pharm Biopharm* 2: 200–211.

Mishra, R. K., A. M. Vinu Mohan, F. Soto, R. Chrostowski and J. Wang. (2017). A microneedle biosensor for minimally-invasive transdermal detection of nerve agents. *Analyst* 142: 918–924.

Miyano, T. and Y. Tobinaga. (2005). Sugar Micro Needles as Transdermic Drug Delivery System, 185–188

Mohan, A. M. V., J. R. Windmiller, R. K. Mishra and J. Wang. (2017). Continuous minimally-invasive alcohol monitoring using microneedle sensor arrays. *Biosens Bioelectron* 91: 574–579.

Narayanan, J. S. and G. Slaughter. (2018). AuNPs-HRP microneedle biosensor for ultrasensitive detection of hydrogen peroxide for organ preservation. *Med Devices Sensors* 1: e10015.

Nasr, B., R. Chatterton, J. H. M. Yong, P. Jamshidi , G. M. D'Abaco, A. R. Bjorksten, O.Kavehei, G. Chana, M. Dottori and E. Skafidas. (2018). Self-organized nanostructure modified microelectrode for sensitive electrochemical glutamate detection in stem cells-derived brain organoids. *Biosensors* 8(1): 14.

Nejad, H. R., A. Sadeqi, G. Kiaee and S. Sonkusale. (2018). Low-cost and cleanroom-free fabrication of microneedles. *Microsystems Nanoeng*. 4: 1–7.

Nguyen, H. X., B. Dasht, Y. Kim, A. Wieber, G. Birk, D. Lubda and A. K. Banga. (2018). European journal of pharmaceutics and biopharmaceutics poly (vinyl alcohol) microneedles : Fabrication, characterization, and application for transdermal drug delivery of doxorubicin. *Eur J Pharm Biopharm* 129: 88–103.

Ou, J. Z., C. K. Yao, A. Rotbart, J. G. Muir, P. R. Gibson and K. Kalantar-zadeh . (2015). Human intestinal gas measurement systems : *In vitro* fermentation and gas capsules. *Trends Biotechnol*, 1–6.

Panthi, G., S. J. Park, S. H. Chae, Tae-Woo Kimd Hea-Jong Chung, Seong-Tshool, Honge Mira Parka and Hak-YongKima. (2017). Immobilization of Ag3PO4 nanoparticles on electrospun PAN nanofibers via surface oximation: Bifunctional composite membrane with enhanced photocatalytic and antimicrobial activities. *J Ind Eng Chem* 45: 277–286.

Panwar, A. and R. S. Saxena. (2019). Chapter 117 PSPICE Circuit Simulation of Microbolometer IRFPA Unit Cell Using Sub-circuit Model of Microbolometer. Springer International Publishing.

Parrilla, M., M. Cuartero, S. Padrell Sánchez, M. Rajabi, N. Roxhed, F. Niklaus and G. A. Crespo. (2019). Wearable all-solid-state potentiometric microneedle patch for intradermal potassium detection. *Anal Chem* 91: 1578–1586.

Pataranutaporn, P., A. Jain, C. M. Johnson, P. Shah and P. Maes. (2019). Wearable Lab on Body : Combining Sensing of Biochemical and Digital Markers in a Wearable Device, 3327–3332.

Pfurtscheller, G. and F. H. Lopes. (1999). Event-related EEG/MEG Synchronization and Desynchronization : Basic Principles. *Clinical Neurophysiology* 110: 1842–1857.

Pham, N., T. Dinh, Z. Raghebi, N. Bui, H. Truong, T. Nguyen, F. Banaei-Kashani, A. Halbower, T. Dinh, P. Nguyen and T. Vu. (2020). A Behind-the-ear Wearable System for Microsleep Detection, 404–418.

Pradeep Narayanan, S. and S. Raghavan. (2017). Solid silicon microneedles for drug delivery applications. Int. *J Adv Manuf Technol* 93: 407–422.

Pradesh, A. (2020). Smart driving with drowsiness detection and alert SYSTEM. XIII: 244–247.

Preston, R. A., B. J. Materson, D. J. Reda, D. W. Williams, R. J. Hamburger, W. C. Cushman and R. J. Anderson. (2016). Age-race subgroup compared with renin profile as predictors of blood pressure response to antihypertensive therapy. *Clinical Cardiology* 280: 1168–1172.

Promphet, N., J. P. Hinestroza, P. Rattanawaleedirojn, N. Soatthiyanon, K. Siralertmukul, P. Potiyaraj and N. Rodthongkuma. (2020). Sensors and Actuators B : Chemical Cotton thread-based wearable sensor for non-invasive simultaneous diagnosis of diabetes and kidney failure. 321.

Rai, A. J., Z. Zhang, J. Rosenzweig, I.-M. Shih, T. Pham, E. T. Fung, L. J. Sokoll and D. W. Chan. (2001). Proteomic approaches to tumor marker discovery identification of biomarkers for ovarian cancer. *Arch Pathol Lab Med* 126(12): 1518–26.

Rajabi, M., N. Roxhed, R. Z. Shafagh, T. Haraldson, A. C. Fischer, W. van der Wijngaart, G. Stemme and F. Niklaus. (2016). Flexible and stretchable microneedle patches with integrated rigid stainless steel microneedles for transdermal biointerfacing. *PLoS One* 11: 1–13.

Rawson, T. M., S. A. N Gowers, D. M. E Freeman, R. C. Wilson, S. Sharma, M. Gilchrist, A. MacGowan, A. Lovering, M. Bayliss, M. Kyriakides, P. Georgiou, A. E. G. Cass, D. O'Hare and A. H. Holmes. (2019). Microneedle biosensors for real-time, minimally invasive drug monitoring of henoxymethylpenicillin: A first-in-human evaluation in healthy volunteers. *Lancet Digit Heal* 1: e335–e343.

Reches, M., K. A. Mirica, R. Dasgupta, M. D. Dickey, M. J. Butte and G. M. Whitesides. (2010). Thread as a Matrix for Biomedical Assays 2(6): 1722–8.

Samant, P. P. and M. R. Prausnitz. (2018). Mechanisms of sampling interstitial fluid from skin using a microneedle patch. *Proc Natl Acad Sci U S A* 115: 4583–4588.

Saver, J. L. (2006). Time Is Brain—Quantified The Growth Function of an Ischemic Stroke, 263–266.

Saylan, Y., Ö. Erdem, S. Ünal and A. Denizli. (2019). An Alternative Medical Diagnosis Method: Biosensors for Virus Detection. *Biosensors* 9(2): 65.

Schefer, U., D. Ammann, E. Pretsch and U. Oesch. (1986). Neutral carrier based Ca+ -selective electrode with detection limit in the sub-nanomolar range. *The Journal of Phsyiology* 2285: 2282–2285.

Schepers, H. M., S. Member, H. F. J. M. Koopman and Peter H. Veltink, (2007). Ambulatory Assessment of Ankle and Foot Dynamics. *IEEE Transactions on Biomedical Engineering* 54: 895–902.

Schwarzkopf, M. and N. A. Pierce. (2016). Multiplexed miRNA northern blots via hybridization chain reaction. *Nucleic Acids Res* 44: e129.

Senel, M., M. Dervisevic and N. H. Voelcker. (2019). Gold microneedles fabricated by casting of gold ink used for urea sensing. *Mater Lett* 243: 50–53.

Serrano-castañeda, P., J. J. Escobar-chávez, I. M. Rodríguez-cruz and L. M. Melgoza-contreras. (2018). Microneedles as enhancer of drug absorption through the skin and applications in medicine and cosmetology. *Journal of Pharmacy & Pharmaceutical Science*, 73–93.

Seshadri, D. R., R. T. Li, J. E. Voos, J. R. Rowbottom, C. M. Alfes, C. A. Zorman and C. K. Drummond. (2019). Wearable sensors for monitoring the physiological and biochemical profile of the athlete. *NPJ Digit Med*. 2.72.

Sharmila, A. and P. Geethanjali. (2016). DWT based detection of epileptic seizure from EEG signals using naive bayes and k-NN classifiers. *IEEE Access* 4: 7716–7727.

Shi, J., S. Liu, L. Zhang, B. Yang, L. Shu, Y. Yang, M. Ren, Y. Wang, J. Chen, W. Chen, Y. Chai and X. Tao. (2020). Smart Textile-Integrated Microelectronic Systems for Wearable Applications. 32(5).

Shikida, M., T. Hasada and K. Sato. (2006). Fabrication of a hollow needle structure by dicing, wet etching and metal deposition. *J Micromechanics Microengineering* 16: 2230–2239.

Shin, S., Y. Pan, D. Ji, Y. Li, Y. Lu, Y. He, Q. Chen and Q. Liu. (2020). Sensors and Actuators B: Chemical Smartphone-based portable electrochemical biosensing system for detection of circulating microRNA-21 in saliva as a proof-of-concept. *Sensors and Actuators B:* 308.

Silva de Lima, A. L., T. Smits, S. K. L. Darweesh, G. Valenti, M. Milosevic, M. Pijl, H. Baldus, N. M. de Vries, M. J. Meinders and B. R. Bloem. (2020). Home-based monitoring of falls using wearable sensors in Parkinson's disease. *Mov Disord* 35: 109–115.

Simfukwe, J., R. E. Mapasha, A. Braun and M. Diale. (2017). Biopatterning of keratinocytes in aqueous two-phase systems as a potential tool for skin tissue engineering. *MRS Adv* 357: 1–8.

Simon, S.R. (2004). Quantification of human motion : Gait Analysis—Benefits and limitations to its application to clinical problems. *Journal of Biomechanics* 37: 1869–1880.

Spain, R. I., R. J. St. George, A. Salarian, M. Mancini, J. M. Wagner, F. B. Horak and D. Bourdette. (2012). Body-worn motion sensors detect balance and gait deficits in people with multiple sclerosis who have normal walking speed. *Gait Posture* 35: 573–578.

Strambini, L. M., A. Longo, S. Scarano, T. Prescimone, I. Palchetti, M. Minunni, D. Giannessi and G. Barillaro. (2015). Biosensors and Bioelectronics Self-powered microneedle-based biosensors for pain-free high-accuracy measurement of glycaemia in interstitial fluid. *Biosens Bioelectron* 66: 162–168.

Strauss, K., H. De. Gols, I. Hannet, T.-M. Partanen and A. Frid. (2002). A pan-european epidemiologic study of insulin injection technique in patients with diabetes. *Practical Diabetes International* 19: 71–76.

Sun, M. (2013). Surface-Enhanced raman scattering. *Handb Mol Plasmon* 27: 321–354.

Sun, W., Z. Arac, M. Inayathullah, S. Manickam, X. Zhang, M. A. Bruce, M. P. Marinkovich, A. T. Lane, C. Milla, J. Rajadas and M. J. Butte. (2013). Polyvinylpyrrolidone microneedles enable delivery of intact proteins for diagnostic and therapeutic applications. *ACTA Biomater* 9(8): 7767–74.

Tamura, T. (2005). Wearable accelerometer in clinical use. *Annu Int Conf IEEE Eng Med Biol - Proc* 7: 7165–7166.

Tang, Y., B. K. Hölzel and M. I. Posner. (2015). The neuroscience of mindfulness meditation. *Nature Reviews Neuroscience* 16: 213–225.

Tao, W., T. Liu, R. Zheng and H. Feng. (2012). Gait analysis using wearable sensors. *Sensors* 12: 2255–2283.

Tarabella, G., M. Villani, D. Calestani, R. Mosca, S. Iannotta, A. Zappettini et al. (2012). A single cotton fiber organic electrochemical transistor for liquid electrolyte saline sensing. *J Mater Chem* 22: 23830–23834.

Tasca, F., C. Tortolini, P. Bollella and R. Antiochia. (2019). Microneedle-based electrochemical devices for transdermal biosensing: A review. *Curr Opin Electrochem* 16: 42–49.

Teufl, S., J. Preston, F. V. Wijck and B. Stansfield. (2017). Objective Identification of Upper Limb Tremor in Multiple Sclerosis Using a Wrist-Worn Motion Sensor : Establishing Validity and Reliability 80: 10.

The ETO. (2006). Excercise for Mental Health 8: 2–3.

Tian, Z., J. Cheng, J. Liu and Y. Zhu. (2019). Dissolving Graphene/Poly (Acrylic Acid) microneedles for potential transdermal drug delivery and photothermal therapy. *J Nanoscience Nanotechnology* 19: 2453–2459.

Timotheadou, E., D. Skarlos, E. Samantas, S. Papadopoulos, S. Murray, J. Skrickova, C. Christodoulou, C. Papakostantinou, D. Pectasides, P. Papakostas, J. Kaplanova, E. Vrettou, M. Karina, P. Kosmidis and G. Fountzilas. (2007). Evaluation of the prognostic role of a panel of biomarkers in stage IB-IIIA Non-small cell lung cancer patients. *Anti Cancer Research* 4490: 4481–4489.

Waghule, T., G. Singhvi, S. K. Dubey, M. M. Pandey, G. Gupta, M. Singh and K. Duac. (2019). Microneedles: A smart approach and increasing potential for transdermal drug delivery system. *Biomed Pharmacother* 109: 1249–1258.

Wang, P. M., M. Cornwell and M. R. Prausnitz. (2005). Minimally invasive extraction of dermal interstitial fluid for glucose monitoring using microneedles. *Diabetes Technol Ther* 7: 131–141.

Wang, S., M. Zhu, L. Zhao, D. Kuang, S. C. Kundu and S. Lu. (2019). Insulin-loaded silk fibroin microneedles as sustained release system. *ACS Biomater Sci Eng* 5: 1887–1894.

Weber, J. A., D. H. Baxter, S. Zhang, D. Y. Huang, K. H. Huang, M. J. Lee, D. J. Galas and K. Wang. (2010). The microRNA spectrum in 12 body fluids. *Clin Chem* 56: 1733–1741.

Wu, R. (2019). Silk Composite Electronic Textile Sensor for High Space Precision 2D Combo Temperature–Pressure Sensing. *Nano Micro Small* 15(31).

Xie, L., H. Zeng, J. Sun and W. Qian. (2020). Engineering microneedles for therapy and diagnosis: A survey. *Micromachines* 11: 1–28.

Xie, Y., B. Xu and Y. Gao. (2005). Controlled transdermal delivery of model drug compounds by MEMS microneedle array. *Nanomedicine Nanotechnology, Biol Med* 1: 184–190.

Yahalom, G., D. Weiss, I. Novikov, T. B. Bevers, L. G. Radvanyi, M. Liu et al. (2013). An antibody-based blood test utilizing a panel of biomarkers as a new method for improved breast cancer diagnosis. *Biomark Cancer* 5: 71–80.

Yang, H., M. T. Rahman, D. Du, R. Panat and Y. Lin. (2016). 3-D printed adjustable microelectrode arrays for electrochemical sensing and biosensing. *Sensors Actuators B Chem* 230: 600–606.

Yao, W., C. Tao, J. Zou, H. Zheng, J. Zhu, Z. Zhu, J. Zhu, L. Liu, F. Li and X. Song. (2019). Flexible two-layer dissolving and safing microneedle transdermal of neurotoxin: A biocomfortable attempt to treat Rheumatoid Arthritis. *Int J Pharm* 563: 91–100.

Yoon, Y., G. S. Lee, K. Yoo and J. B. Lee. (2013). Fabrication of a microneedle/CNT hierarchical micro/ nano surface electrochemical sensor and its *In-vitro* glucose sensing characterization. *Sensors* (Switzerland) 13: 16672–16681.

Yuen, C. and Q. Liu. (2014). Towards *in vivo* intradermal surface enhanced Raman scattering (SERS) measurements: Silver coated microneedle based SERS probe. *J Biophotonics* 7: 683–689.

Zampogna. A., I. Mileti, E. Palermo, C. Celletti, M. Paoloni, A. Manoni, I. Mazzetta, G. D. Costa, C. Pérez-López, F. Camerota, L. Leocani, J. Cabestany, F. Irrera and A.Suppa. (2020). Fifteen years of wireless sensors for balance assessment in neurological disorders. *Sensors* (Switzerland) 20: 1–32.

Zhang, H. Y., E. A. Tehrany, C. J. Kahn, M. Poncot, M. Linder and F. Cleymand. (2012a). Effects of nanoliposomes based on soya, rapeseed and fish lecithins on chitosan thin films designed for tissue engineering. *Carbohydr Polym* 88: 618–627.

Zhang, L., J. Liu, Z. Fu and L. Qi. (2020). A wearable biosensor based on bienzyme gel-membrane for sweat lactate monitoring by mounting on eyeglasses. *Jounral of Nanoscience and Nanotechnology* 20: 1495–1503.

Zhang, Y., Z. Jiang, M. Xue, S. Zhang, Y. Wang and L. Zhang. (2012b). Toxicogenomic analysis of the gene expression changes in rat liver after a 28-day oral Tripterygium wilfordii multiglycoside exposure. *J Ethnopharmacol* 141: 170–177.

Zhao, L., Z. Wen, F. Jiang, Z. Zheng and S. Lu. (2020). Silk/polyols/GOD microneedle based electrochemical biosensor for continuous glucose monitoring. *RSC Adv* 10: 6163–6171.

Zhou, C. P., Y. L. Liu, H. L. Wang, P. K. Zhang and J. L. Zhang. (2010). Transdermal delivery of insulin using microneedle rollers *in vivo*. *Int J Pharm* 392: 127–133.

Zhu, C. S., P. F. Pinsky, D. W. Cramer, D. F. Ransohoff, P. Hartge, R. M. Pfeiffer, N. Urban, G. Mor, R. C. Bast Jr., L. E. Moore, A. E. Lokshin, M. W. McIntosh, S. J. Skates, A. Vitonis, Z. Zhang, D. C. Ward, J. T. Symanowski, A. Lomakin, E. T. Fung, P. M. Sluss, N. Scholler, K. H. Lu, A. M. Marrangoni, C. Patriotis, S. Srivastava, S. S. Buys, C. D. Berg and P. L. C. O. Project Team. (2011). A framework for evaluating biomarkers for early detection: validation of biomarker panels for ovarian Cancer. *Cancer Prev Res (Phila)* 375–384.

Human Information Processing Vision, Memory, and Attention.

V. S. Ramachandran - The Tell-Tale Brain_ A Neuroscientist's Quest for What Makes Us Human -W. W. Norton (2011).pdf.

CHAPTER 13

Smart Wireless Nanosensor Systems for Human Healthcare

*Rajesh Ahirwar** and *Nabab Khan*

1. Introduction

The primitive analytical techniques of quantitative or qualitative determination of chemical, biochemical and biological species in complex sample matrices relied on tedious procedures including cumbersome separations and complex chemical reactions or growth incubations. The convenient and rapid mean to biochemical detection came in existence in 1970s with introduction of biosensors (Bhalla et al. 2016). Today, biosensors have become an indispensable tool alternative to the large, expensive and sophisticated analytical instruments used for detection and quantification of a multitude of chemical and biochemical analytes in health care, pharmaceutical, forensic, food quality management, and many other fields. These analytical devices are composed of a biorecognition unit that binds the analyte, and a transducer to convert the biorecognition event (analyte-target binding) into a measurable signal proportion to analyte concentration (Bhalla et al. 2016). Biosensors based on multiple detection strategies (electrochemical, optical, and piezoelectric) and various biorecognition molecules (enzymes, antibodies, whole cells, oligonucleotides and aptamers) gained dominating role in clinical medicine due to their high performance, ease of operation, portability and cost-effectiveness. However, many of the conventional biosensors in healthcare settings required biological samples collected through invasive procedure (e.g., blood), which limits their widespread use among patients, particularly in aged and infants. Efforts to overcome these limitations through changes in biosensor design and fabrication and improvement in data collection and operational ease is enabling the development of miniaturized devices with enhanced specificity, sensitivity, cost-effectiveness and

Department of Environmental Biochemistry, ICMR-National Institute for Research in Environmental Health, Bhopal Bypass Road, Bhauri, Bhopal-462030, India.
* Correspondence: r.ahirwar.nireh@gov.in

response time. Cutting edge nanotechnology be stowed novel nanomaterials such as nanoparticles, nanotubes, nanorods, and nanowires to improve sensing (Sotiropoulou et al. 2003, Malik et al. 2013). Real-time continuous health monitoring for early disease diagnosis and tracking the disease progression and drug efficiency are being made possible through the next generation wireless nanosensors (Roham et al. 2008, Akyildiz and Jornet 2010, Darwish and Hassanien 2011, Kassal et al. 2018). This chapter comprehends the basics of the conventional biosensors and next-generation nanosensors and discusses the design, components, characteristics, and applications of wireless nanosensor based systems in personalized health monitoring.

1.1 Conventional Biosensors

The term "biosensor" has evolved over the last many decades from combined efforts of researchers from different background, mainly chemistry, physics, microbiology, molecular biology and of course advanced electrical engineering and nanotechnology. The early inventions that paved the way for today's biosensors include mainly the conceptualization of pH (hydrogen ion concentration) in 1909, use of electrodes for pH measurements by W.S. Hughes in 1922, demonstration of enzyme immobilisation on aluminium hydroxide and charcoal by Griffin and Nelson, and development of first 'true' biosensor for oxygen detection by Leland C. Clark, Jr. in 1956 (Nelson and Griffin 1916, Hughes 1922, Clark 1956). Subsequently, an enzyme electrode for the amperometric determination of glucose and urea-specific enzyme electrode for urea detection was demonstrated by Guilbault and Montalvoin early 1970s (Guilbault and Montalvo 1969, Guilbault and Lubrano 1973). Eventually, the first commercial biosensor, the Yellow Springs Instrument Company analyzer (Model 23A), was developed in 1975 for the direct measurement of glucose based on Clark's technology.

"Biosensors are nowadays ubiquitous in biomedical diagnosis as well as a wide range of other areas such as point-of-care monitoring of treatment and disease progression, environmental monitoring, food control, drug discovery, forensics and biomedical research" (Kissinger 2005). Biosensors detect and quantify biological or chemical analytes by generating signals proportional to the concentration of an analyte in the reaction. A typical biosensor has following components: (i) bioreceptor-a biomolecule (e.g., enzymes, aptamers, cells, deoxyribonucleic acid (DNA) and antibodies) that specifically recognize the analyte (e.g., glucose) and generate signal in the form of charge, pH, light, heat, or mass change (ii) a transducer: electronic element that converts bioreceptor generated signal into measurable electrical or optical signals, (iii) a signal processing electronics—the electronic circuitry for signal conditioning such as amplification and conversion of signals from analogue into the digital form, and (iv) a display unit—the user interface that converts the transducer-generated analogue values to results easily understandable by users (e.g., computer LCD or printer). The various components of a typical conventional biosensor are shown in Figure 1a.

As depicted in Figure 1a, the process of analyte recognition (binding) by the immobilized bioreceptor and subsequent generation of light, heat, pH, charge or mass change-based signal is called biorecognition. Bioreceptor molecules can be

(a)

(c)

(b)

(d)

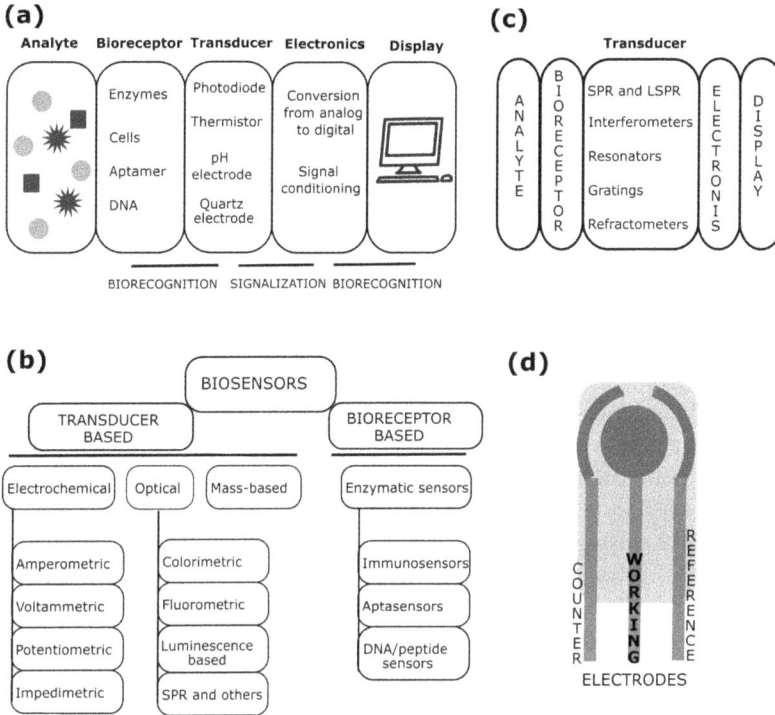

Fig. 1. Schematic representation of tools and techniques in biosensing. (a) Representation of biosensor components. Biorecognition (the process of signal generation in the form of light, heat, pH, charge or mass change), signalisation (conversion of received biorecognition signal into optical or electronic signal), and quantification (convert processed signal into user understandable values) are the three major processes of biosensing. (b) Classification of biosensors, (c) representation of components of an optical biosensor, and (d) a typical 3 electrode system used in voltammetric electrochemical biosensing. Figure (a) (reprinted from Bhalla Nikhil, *Introduction to biosensors* (Essays in Biochemistry 2016), 60(1): 1–8). Figure (c) (reprinted with permission from Damborský Pavel, *Optical biosensors* (Essays in biochemistry 2016), 60 (1): 91–100).

immobilized onto biosensor electrode surface or nanomaterials interface through adsorption, covalent bonding, cross-linking, microencapsulation, or entrapment (Sassolas et al. 2012, Bhardwaj 2014). The transducers used during construction of a biosensor depend on the type of biochemical reactions. For example, biorecognition processes yielding signal in the form of light, charge, or mass uses optical transducers (e.g., resonators, interferometers, refractometers, SPR and LSPR), electrochemical transducers (amperometric, potentiometric and conductometric) and piezoelectric transducers (e.g., quartz), respectively. The conversion process produces signals proportional to the amount of analyte present in the sample through the analyte-bioreceptor interaction at the biorecognition unit. As shown in Figure 1b, the conventional biosensors were categorized into various groups based on transducer used in the biosensor (electrochemical, optical and mass-based biosensors), and the nature of bioreceptor used in these biosensors (enzymatic biosensors, immunosensors, aptasensors, and nucleic acid/DNA/peptide sensors) (Rastislav et al. 2012).

1.2 Optical Biosensors

Optical sensing is the most widely used method for measurement of analytes in simple transparent liquid samples. The optical biosensors are very common due to their real-time detection characteristic along with advantages of high specificity, sensitivity, small size and cost-effectiveness (Damborský et al. 2016). However, they are used mostly in transparent liquids and analysis of analytes in complex and non-transparent samples preferably require electrochemical and mass-sensitive biosensors. The optical biosensors are constructed by compacting biorecognition probe which is either labelled or label-free with optical transducer like interferometers, resonators, gratings, refractometers, and surface plasmon resonance to measure absorbance, reflectance or fluorescence emissions that correlates with analyte concentrations (Figure 1c) (Brecht and Gauglitz 1995, Choi 2004, Damborský et al. 2016). Many optical-based biosensors have been developed for the detection of myriad targets of clinical applications.

Surface plasmon resonance (SPR) biosensors work on the principle that illuminating metallic surfaces by a polarized light directed at a specific angle generate surface plasmons that provides key information on the proportionate mass attached to the surface of the transducer (Singh 2016). Although the SPR based biosensors provide direct information on biomolecular interaction without the use of any labelling strategies, these biosensors suffer nonspecific binding and limited mass transfer limitations. Nonetheless, SPR based optical biosensors received a profound application in clinical diagnosis and biological sciences, drug development, and food industry and many more (Mariani and Minunni 2014, Nguyen et al. 2015).

Localized surface plasmon resonance (LSPR), a technique similar to the SPR, is based on the unique optical properties of nanostructures such as Au and Ag nanoparticles under different dispersion states (Cao et al. 2011). The major difference between SPR and LSPR is that the oscillation of plasmonsis confined locally on the nanostructure surface rather than along the metal/dielectric interface as in SPR. The AuNPs at nanomolar concentrations exhibits vibrant colours in visible region due to very high molar extinction coefficients (107–1011), for example, 10–50 nm AuNPs are bright ruby red with absorption maxima at 520 nm, and the same AuNPs appear pale blue or purple (absorption at 700 nm) upon aggregating. This change in physical state as well as colour of AuNPs can be introduced by attaching the AuNPs with recognition molecules such as aptamers to prevent their aggregation via electrostatic repulsion, which in presence of its target analyte detaches from nanoparticle surface allowing the nanoparticles to aggregate in the presence of an inducer such as salt. Colorimetric biosensors based on LSPR-based colour transition of AuNPs have been reported in clinical diagnosis of various cancer biomarkers (Ahirwar and Nahar 2016, Ranganathan et al. 2020).

Fluorescence is an optical phenomenon wherein a molecule absorbs photons of selective energy and wavelength that triggers the emission of fluorescent photons of lesser energy and longer wavelength. Fluorescence-based biosensors measure fluorescence intensity produced as a result of target-analyte interactions to determine the analyte concentrations (Strianese et al. 2012).

1.3 Electrochemical Biosensors

Electrochemical biosensors use recognition elements coated electrodes (transducers) made of metals (platinum, gold, silver, stainless steel), carbon-based materials (graphite, carbon black, carbon fibre), or conducting composites (as in interdigited electrodes) for selective recognition and detection of target analytes (Ronkainen et al. 2010, Zhu et al. 2015). They measure electrons or the resistance in electron flow as a result of biorecognition event at electrode surface. The field of electrochemical biosensors received first major attention based on the study on oxygen electrode by Clark in 1956 (Clark 1956). Since then, electrochemical biosensors have been widely used to determine the analyte concentrations, both in research and commercial applications. These biosensors allow simplistic and cost-effective detection of target analytes in small sample volumes with low detection limits, good stability, reproducibility, and wide linear response range. Based on the detection principle (transducer-based), they are categorized into amperometric, voltammetric, potentiometric, conductometric, and impedimetric biosensors (Mehrotra 2016). Also, based on the nature of bioreceptor used in these biosensors, they can be further categorized to enzymatic biosensors, immunosensors, aptasensors, and nucleic acid/DNA/peptide sensors (Asal et al. 2018).

Voltammetric methods measure current response as a result of applied potential. The potential can be varied step by step or continuously between a working and reference electrode. Three electrodes—working, auxiliary/counter, and reference (Figure 1d)—are mostly used in voltammetric aptasensors for accurate and stable application of potentials and the current measurement (Manurung et al. 2012). Cyclic voltammetry (CV) and pulse voltammetry (e.g., Differential Pulse Voltammetry) are two commonly used voltammetric techniques. Cyclic voltammetry technique involves varying the applied potential in forward and reverse directions at the working electrode and measuring the current. The resulting current is plotted against the applied potential to produce a CV graph. Contrarily, the DPV uses a series of potential pulses of fixed amplitude which is superimposed on a slowly changing base potential and measures current at two points for each pulse—just before the application of the pulse and at the end of the pulse (Bertrand 1998).

Amperometric biosensors measure the current against time over a constant electric potential. Sensors based on amperometry are fabricated by mounting electroactive biological element on the electrode's surface where the electric current, in proportion to target concentration, is produced as a result of oxidative or reductive action of biorecognition event (Dzyadevych et al. 2008, Sadeghi 2013). Glucose biosensor is the most studied amperometric biosensor system. In this system, glucose oxidase catalyzes the reaction of glucose with oxygen to produce gluconolactone and hydrogen peroxide (Lobo et al. 1997, Dzyadevych et al. 2008, Murugaiyan et al. 2014). The signal is depicted as current produced from the redox reaction of a mediator or hydrogen peroxide (fall in O_2 tension or production of H_2O_2) at the working electrode against concentration of glucose (Figure 2). Many mediators such as potassium ferricyanide, tetrathiafulvalene, tetracyanoquinodimethane, and ferrocenes have been widely used in construction of electrochemical amperometric

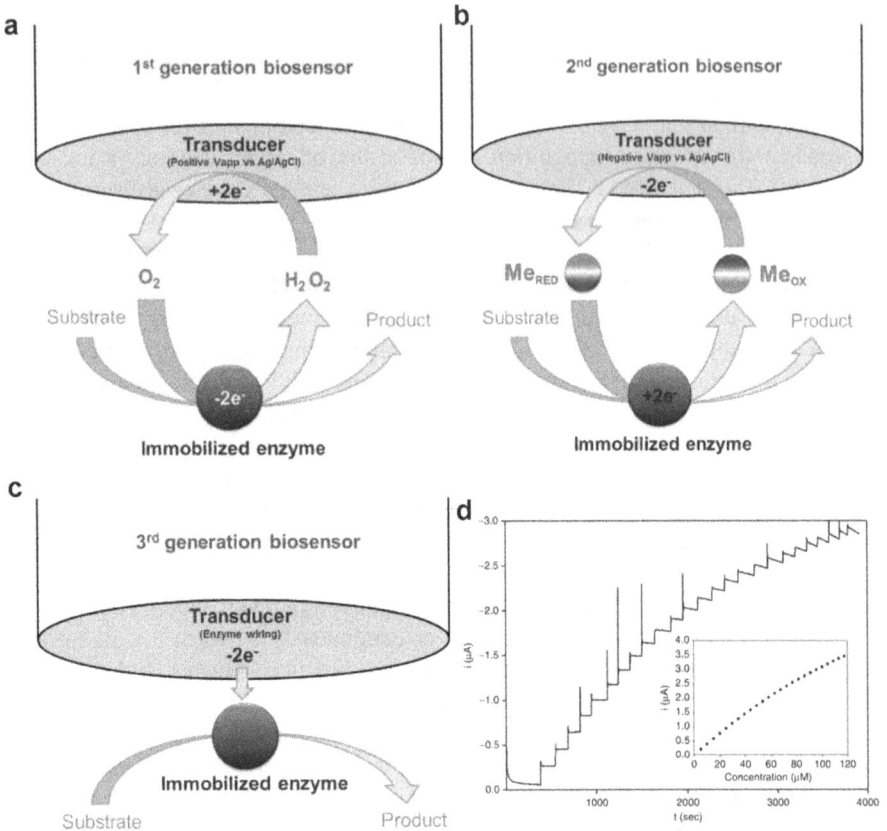

Fig. 2. Schematic representation of amperometric biosensor generations (a–c) and a typical current-time response curve (d) of a chronoamperometry experiment. Inset shows the calibration curve. Figure (a–c) (reprinted with permission from Rocchitta Gaia, *Enzyme Biosensors for Biomedical Applications: Strategies for Safeguarding Analytical Performances in Biological Fluids* (Sensors 2016), 16 (6): 780).

biosensors (Scheller et al. 1991, Chaubey and Malhotra 2002). The amperometric enzymatic biosensors are divided into three categories—first, second and third generation biosensors based on the mechanism of the electronic transport adopted during past years (Figure 2).

The first-generation biosensors measure targeted analytes based on enzymatic reactions whose product diffuses to transducer surface to produce electrical response in a mediator (electron carriers)-less manner. These biosensors used electrode-coated enzymes belonging to mainly oxidases and dehydrogenases categories which require coenzymes such as FAD, FADH, NADPH, NADH, NADP+, NAD+, and ATP during catalysis (Rocchitta et al. 2016). Second generation amperometric biosensors use mediators (oxidizing agents) for carrying electrons from the redox centre of the enzyme to the electrode for target analyte detection. They are also called mediator amperometric biosensors (Scheller et al. 1991, Chaubey and Malhotra 2002). The most common mediators reported in literature are "ferricyanide and ferrocene, methylene blue, phenazines, methyl violet, alizarin yellow, Prussian blue, thionin, azure A and C, toluidine blue and inorganic redox ions" (Chaubey and Malhotra 2002, Rocchitta et al. 2016). Unlike the first or second generation biosensors, the third generation amperometric biosensors involve direct transfer of electrons between the enzyme and electrode (Murugaiyan et al. 2014). These biosensors are still under development and not commonly used for analysis. However, future developments in nanotechnology and polymer science are believed to evolve third generation biosensors as promising devices due to their short response times and independence on oxygen/cofactor concentrations.

Other commonly studied electrochemical biosensors are conductometric, potentiometric and impedimetric biosensor (Rastislav et al. 2012). A conductometric biosensor measures conductivity change from production or consumption of ionic species involved in the metabolic process. These biosensors require no reference electrode in the system, and hence, can be miniaturized. However, as all charge carriers may lead to a change of conductivity, this directly affects the device selectivity. A potentiometric biosensor is based on the potential difference between working and reference electrodes. These biosensors measure the potential difference between two ion-selective electrodes under the conditions of no current flow, but at different analyte concentrations. The electric response depends on the activity of the species in comparison to the reference electrode, with the output signal recorded in voltage units. Layer-by-layer technique is widely used for surface modifications. Impedimetric biosensors based on electrochemical impedance spectroscopy measures the resistive and capacitive properties of materials by applying an AC potential to an electrochemical cell and then measuring the current through the cell (Lorenz and Schulze 1975). Impedimetric detection is primarily used for monitor immunological binding events (e.g., an antigen-antibody interaction) on an electrode surface, where the small changes in impedance are proportional to the concentration of the measured species.

1.4 Mass-based (piezoelectric) Biosensors

Piezoelectric biosensors work on the principle of change in mass and/or change in oscillation of piezoelectric substance upon interaction of analyte to the bioreceptor immobilized to a piezoelectric surface (Figure 3). Piezoelectric transducer (quartz crystal microbalance) converts change in pressure or mass to an electrical field. A piezoelectric biosensor is constructed by immobilizing a bioreceptor (e.g., antibody) to a piezoelectric electrode (e.g., quartz crystal microbalance) where the binding of target analyte causes a change in mass, which is detected by the electrode to produce an electrical field which changes the resonating frequency of the crystal and is noted as a beacon to show the binding of the analyte.

The other classification based on type of bioreceptor molecule used in fabricating the biosensor categorizes them into mainly enzymatic biosensors, immunosensors, aptasensors, and DNA peptide and cell-based biosensors.

Enzymatic biosensors use enzymes such as globular proteins, nucleases and ribozymes or DNAzymes as bioreceptor in the electrochemical or optical sensing applications (Wilson and Hu 2000, Rocchitta et al. 2016). Enzymatic electrochemical

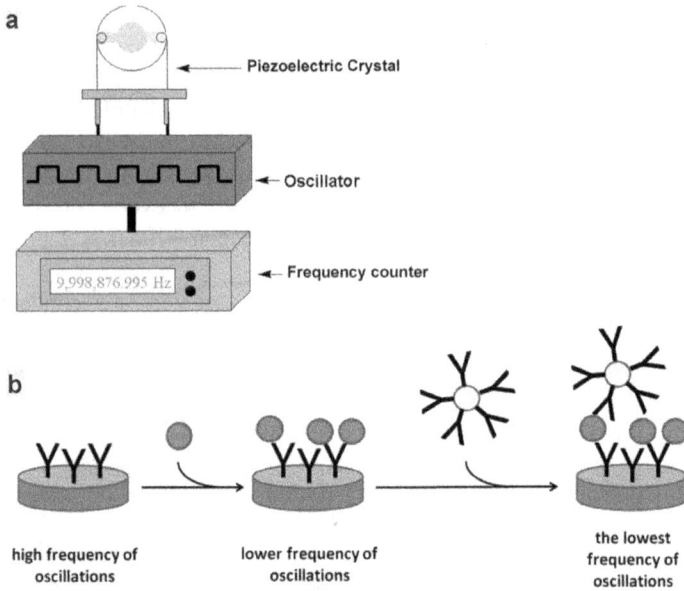

Fig. 3. Schematic representation of the construction and working principle of piezoelectric immunosensors. (a) A piezoelectric immunosensor is constructed by connecting the piezoelectric crystal (e.g., quartz crystal microbalance) to oscillator and a frequency counter. (b) Biorecognition event: binding of target antigen (dark balls) and secondary antibody coated nanoparticles (white ball) on the piezoelectric crystal surface through the immobilized bioreceptor (primary antibody) that increases oscillations in mass-dependent manner, thereby allowing sensitive determination of antigen. (Figure (b) reprinted with permission from Pohanka Miroslav, *Overview of Piezoelectric Biosensors, Immunosensors and DNA Sensors and Their Applications* (Materials 2018), 11(3): 448.)

biosensors have been constructed in three major formats: oxygen-based (first generation), mediator-based (second generation), and direct electrochemistry-based (third generation). Despite these biosensors being the most commonly used,they usually have low stability, require stringent operational conditions, and can produce error prone results upon variations in pH/temperature which limit their detection ability.

Biosensors based on the use of antibody or antibody fragment as a bioreceptor are referred as immunosensors (Prodromidis 2010, Pohanka 2018). Due to high affinity interactions among antigen-antibody pairs, these are often considered as a highly specific biosensor. Also, immunosensors gained a remarkable popularity in particularly in the clinical applications due to their high sensitivity and selectivity. Various immunosensors based on miniaturized transducers are being developed. The few limitations of immunosensors include poor solubility and limited thermal stability of antigen-antibodies (aggregate or denature by changes in temperature), and retention of binding affinities at higher temperatures.

Aptamer that represents short single stranded oligonucleotides with ability to bind a vast variety of targets is an important affinity reagent and a cost-effective substitute to antibodies in diagnostic applications (Ahirwar and Nahar 2016, Ahirwarl et al. 2019). Biosensors based on aptamer as bioreceptor are termed as aptasensors.

Applying aptamers in biosensing applications inherit the advantages of high thermal stability, cheaper production, and easy chemical modification. Benefiting the unique structure changing ability of aptamers, multiple sensing strategies (aptasensors) based on electrochemical, optical (fluorimetric and colorimetric), and piezoelectric readout systems have been developed for detection of different biomolecules representing human disease biomarkers (Ahirwar 2021, Ahirwar et al. 2021).

Peptides owing to their self-assembling nature (1D, 2D, or 3D structures) have been used in fabricating various flexible and supramolecular frameworks for a variety of applications including the biosensor. Peptide based biosensors are constructed by immobilizing the peptides to electrodes via adsorption, non-covalent interactions (H-bonding, electrostatic, aromatic, π-stacking, hydrophobic, and Van der Waals), covalent attachment, or self-assembled monolayers (Karimzadeh et al. 2018). Peptide-biosensors have added advantage of better conductivity due to helical conformation, easy synthesis, and excellent biocompatibility. Peptide-based biosensors are used in detection of analytes such as small molecules, proteins, and cells.

1.5 Characteristics of a Biosensor

The characteristics of a biosensor that govern its utility in clinical or other fields of application are specificity, sensitivity, linearity, reproducibility, and stability.

1.5.1 Specificity

It is the ability of a biosensor to detect a specific target analyte amongst a diverse pool of sample containing mixtures of unwanted contaminants. Specific recognition of an antigen by its antibody represents a classical example of high specificity or selectivity.

1.5.2 Sensitivity

It refers to the lowest detection limit of an analyte by a biosensor. This may range from nanogram to femtogram per millilitre. It is also known as limit of detection. Sensors with lower detection limits are considered more valuable for their ability to detect or quantify the least amounts of target analytes.

1.5.3 Linearity

It represents the accuracy of the obtained output within a working range where the concentration of the analyte in the sample is directly proportional to the measured signal.

1.5.4 Precision and accuracy

It is the ability of a biosensor to yield identical end results regardless of the number of times experiment is repeated. Precision and accuracy of the results of a biosensor are measured in terms of reproducibility. Reliability of biosensor output is highly dependent on the reproducibility of the biosensor devices.

1.5.5 Stability

It refers to the ability of biosensors to circumvent ambient disturbances that are likely to alter the desired output response during measurement. Harsh environmental conditions such as high temperature and humidity, and fouling of membranes by various biological or non-biological substances mostly influence the stability of a biosensor.

2. Smart Wireless Nanosensors Systems

Unlike the conventional biosensors, the nanosensors are extremely small (i.e., dimensions of 1–100 nm) integrated devices engineered from nanomaterials or biological materials. They work the same way as conventional sensors and have sizes and shapes of different dimensions depending on the application and place of installation on human body. For example, the size of the nanosensors used for biomedical applications like monitoring internal body parameters and processes via invasive measures are extremely small compared to the one used to record non-invasive parameters. The major application of nanosensors in the healthcare domain is monitoring, detection, and treatment of diseased body functions. For example, a biotransferrable graphene wireless nanosensor created for enamel allow detection of chemical compounds in concentrations as low as 1 ppb or the presence of different infectious agents like bacteria at single cell level (Mannoor et al. 2012). Furthermore, these nanosensors can be armoured with nanocommunication units to provide wireless readout. Thus, the major difference between a standard biosensor and a smart wireless nanosensor is the intelligence capabilities, wireless communication, and nanometer size of the later. Nanosensors are fabricated by integrating together a sensing unit, actuation unit, power unit, processing unit, memory storage unit, and a communication unit (Figure 4A). The size constraints of nanomaterials used in making nanosensor confer these devices physicochemical properties highly different from the same materials at the bulk scale. Several nanomaterials have been explored for construction of nanosensors with improved biorecognition and transduction abilities. Some of the widely used nanomaterials are nanoparticles, nanotubes, nanowires, nanorods, and thin films made up of nanocrystalline matter (Table 1).

2.1 Components of a Smart Wireless Nanosensor System

2.2.1 Sensing unit

The nanosensing unit integrated in the architecture of a nanosensor mostly sense three types of biochemical or biophysical process—external force, chemical substances and biomolecules. Nanosensors can be categorized into physical, chemical and biological nanosensors based on the sensing principle (Figure 4B).

Physical nanosensors measure mass, pressure, force, or displacement based on change in electronic properties of nanomaterials such as nanotubes and nanoribbons, which change upon getting bent or deformed (deformation of nanotubes/nanoribbons changes the on/off threshold voltage of the transistor) (Hierold et al. 2007). Different types of physical nanosensors or the nanoelectromechanical systems such as pressure

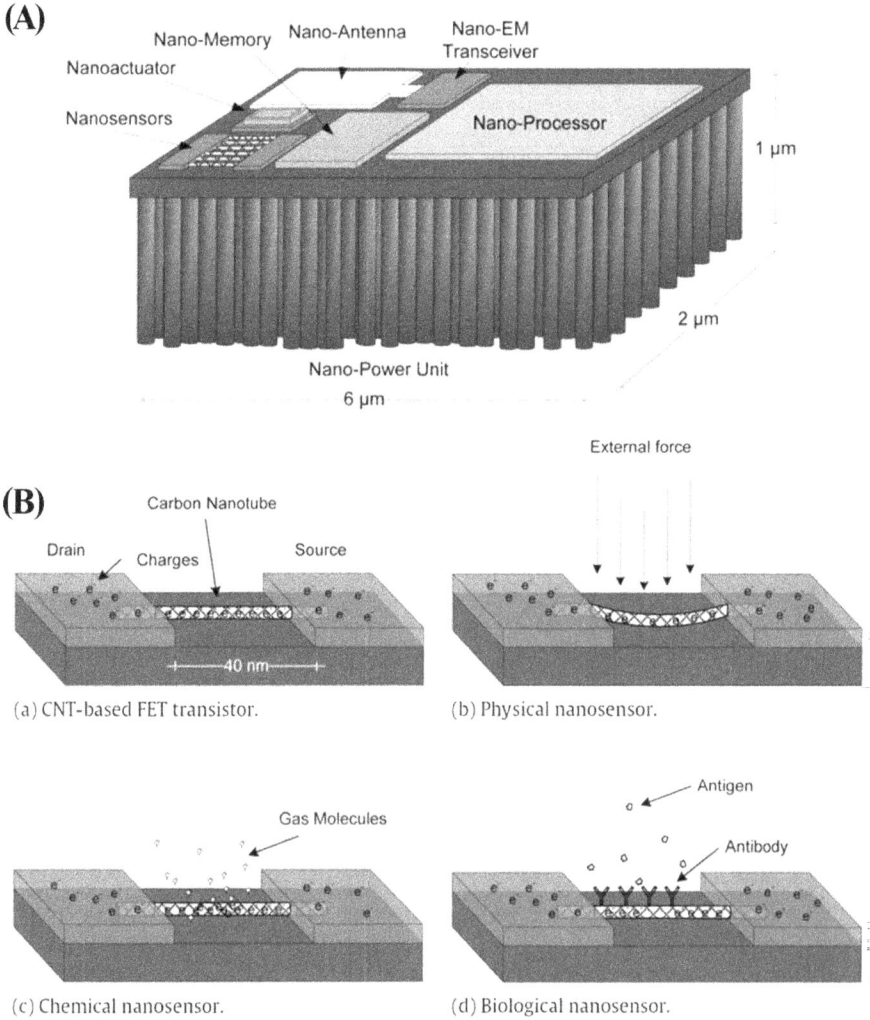

Fig. 4. Representative image of an integrated nanosensor device (A) and illustration of the working principle of carbon nanotube-based physical, chemical and biological nanosensors (Bb–d). Image B(a) shows the architecture of a CNT-based field effect transistor. (Reprinted with permission from Akyildiz Ian F., *Electromagnetic wireless nanosensor networks* (Nano Communication Networks 2010), 1(1): 3–19.)

nanosensors, force nanosensors or displacement nanosensors based on this simple principle have been proposed in the literature (Stampfer et al. 2006a,b,c).

Chemical nanosensors made of CNTs and GNRs measure presence and/or concentration of gases, and specific biochemical or molecular components of a substance based on change in their electronic properties through increase or decrease in the number of electrons able to move through the carbon lattice upon adsorption of different molecules to sensor surface (Bondavalli et al. 2009). Multiple chemical nanosensors have been reported manufactured for different gases like NO_2, NH_3, CO_2 and many more (Bondavalli et al. 2009).

Table 1. Overview of nanomaterials used for improving biosensor technology.

Nanomaterials	Properties conferred
Nanotubes	Improved enzyme loading, higher aspect ratios, ability to be functionalized, and better electrical communication
Nanoparticles	Aid in immobilization, enable better loading of bioanalyte, and also possess good catalytic properties
Quantum dots	Excellent fluorescence, quantum confinement of charge carriers, and size tunable band energy
Nanowires	Highly versatile, good electrical and sensing properties for bio- and chemical sensing; charge conduction is better
Nanorods	Good plasmonic materials which can couple sensing phenomenon well and size tunable energy regulation, can be coupled with MEMS, and induce specific field responses

Source: Data from Malik Parth, "Nanobiosensors: Concepts and Variations", *ISRN Nanomaterials* (2013): 327435.

Biological nanosensors are miniaturized electrochemical and/or optical biosensors composed of a bioreceptor and transducer to detect target analytes such as antigen and antibodies, DNA and enzymes based on biosignalling event occurring at the surface of nanotube/nanorods mount in the sensing unit of the smart nanosensor system. The biological nanosensors work similar to conventional electrochemical and optical biosensors.

2.2.2 Actuation unit

The actuation unit allows nanosensors to interact with the ambient environment. It assists in collecting physical and physiological data from the human body for monitoring, diseases detection and intervention purposes. Use of physical, chemical and biological nanoactuators has been reported in the literature. Physical nanoactuators mounted in the nanosensor device are based on nanoelectromechanical systems and work the same way as physical sensors (Li et al. 2008). Nanotweezer composed of two multi-walled carbon nanotubes is an example of physical nanoactuator.

2.2.3 Power unit

The power units supply uninterrupted power to the nanosensor system. Earlier attempts of making nanobatteries using nanomaterials like lithium required periodic recharging, thus limiting its usefulness in realistic nanosensors systems. Recently, the concept of self-powered nanodevices that harness electrical energy from conversion of mechanical energy (e.g., produced by human body movements, or muscle stretching), vibrational energy (generated by acoustic waves, mechanical and thermal noise, etc.) and hydraulic energy (produced by body fluids/blood flow) have been introduced (Wang 2008, Yang et al. 2009). Piezoelectric effect is one way to obtain such an energy conversion (e.g., zinc oxide nanowires upon mechanical deformation produce voltage).

2.2.4 Processing unit

Nanoscale processors (nanotransistors) based on nanomaterials like carbon nanotubes, graphene nanoribbon, single phosphorous atom and gold nanorods have been reported for building transistors in the nanometer scale. Despite successful testing of individual transistors, integrating these nanotransistors in processor architectures remains the major challenge.

2.2.5 Storage unit

The concept of nanomemories utilizing a single atom to store a single bit of information was introduced by Richard Feynman in 1959 where he suggested a memory unit composed of 125 atoms to prevent potential interference between adjacently stored bits and was comparable to the 32 atoms that store one bit of information in DNA (Bennewitz et al. 2002). Several type of atomic memories have been proposed in the literature, for example, atomic memories using silicon atom and magnetic and gold-based memories (Parkin et al. 2008).

2.2.6 Communication unit

Nanocommunication is the exchange of information (data measured by nanosensors) at nanoscale between other nanosensors to work in a synchronous, supervised and cooperative manner. Several nanosensors can be connected together through nanorouters that rout measured data to other nanosensors or external devices such as mobile phones, creating an interconnected cluster of nanosensors which is often termed as a nanonetwork. However, for the time being, nanocommunication is still limited to theoretical models. Generally, four different communication paradigms are envisioned for nanocommunication among the nanonetworks, namely, nanoelectromagnetic, molecular, acoustic, and nanomechanical (Freitas 2005).

Molecular communication proposes transmission and reception of information encoded in molecules. The molecular communication systems are designed and engineered from biological mechanisms and materials (Akan et al. 2017). For example, small particles such as molecules or lipid vesicles released into the fluidic or gaseous medium via the nanotransmitter propagate in the medium until they arrive at a receiver that upon detecting the small molecules decodes the information encoded in them. Messages can be encoded in different properties such as concentration, number, type, release timing, and/or a ratio of molecules (Akan et al. 2017). A list of various components and materials used for developing molecular communication systems is provided in Table 2. Sender and receiver bio-nanomachines require chemical functionality for effective communication, with the sender bio-nanomachines being able to synthesize, store, and release information molecules, while receiver bio-nanomachines need to capture and react to specific information molecules. Information molecules propagate information from a sender to receiver bio-nanomachines. Guide and transport molecules help in propagating the information by directing the molecule toward target locations. Interface molecules allow bio-nanomachines to transport variety of information molecules using the same communication mechanism. Addressing molecules allow nanosensors to transport a variety of information molecules using the same communication mechanism.

Table 2. Design and engineering of molecular communication components.

Components	Example categorized by material
Sender and receiver bio-nanomachines	Modified cells (cells genetically engineered or with modified metabolic pathways) and artificial cells (cell like structures created by imbedding proteins in vesicle)
Information molecules	Synthetic particles (e.g., nanoparticles), mediator biomolecules (e.g., cytokines, neurotransmitter, DNA/RNA)
Guide and transport molecules	Gap junction channels, molecular motors and filaments, neurons
Interface molecules	Nanoscale capsules (e.g., liposome)
Addressing molecules	DNA sequences

Source: Data from Nakano T. 2012. Molecular communication and networking: opportunities and challenges. IEEE Transactions on NanoBioscience 11(2): 135–148.

Nano-electromagnetic communication is the transmission and reception of electromagnetic radiation between nanosensors placed in human body to macro world devices and vice versa using nanoantennas and electromagnetic transceiver (Rutherglen and Burke 2009). The carbon-nanotube and graphene-based electronics provide a vast opportunity for electromagnetic communication among nanodevices in the terahertz (THz) band. Recent advances allowing refinement of the existing architectures and the utilization of new technologies enabled THz communication paradigm to the verge of reality. For effective intra-body nanocommunication among devices, THz transmitters are required to be compact and capable of providing high levels of average output power at lower THz frequencies (Rizwan et al. 2018). For example, the miniaturized FinFET and the 3D tri-gate transistor transistors mitigate the undesirable behaviour of the short channel effect and increase the transistor channel dimension. Nano-antenna made of carbon nanotube and graphene or metallic plasmonic nanoantenna have been proposed in the literature for intra-body nanonetworks (Gul et al. 2010, Nafari and Jornet 2015).

2.3 Construction of Nanosensor Systems

Manufacturing and integrating various components of nanosensor devices is carried out broadly based on three approaches—top-down, bottom-up and bio-hybrid approach (Figure 5). The top-down approach relies on downscaling the existing micro-scale level components' development to nanoscale objects by using advanced manufacturing techniques, such as electron beam lithography and micro-contact printing (Lee et al. 2004, Chen 2015). The resulting nanodevices keep the architecture of pre-existing micro-scale components. Nano-machines, such as nano-electromechanical systems components, are being developed using top-down approach (Goldstein 2005). This approach is, however, at an early stage of development, and only simple mechanical structures, such as nano-gears have been suggested fabricated using this approach (Ju Yun et al. 2007). Contrarily, the bottom-up approach relies on constructing nano-machines using individual molecules as building blocks. Nano-machines such as molecular differential gears and pumps have been theoretically designed using this approach (Peterson 2000). Still, molecular

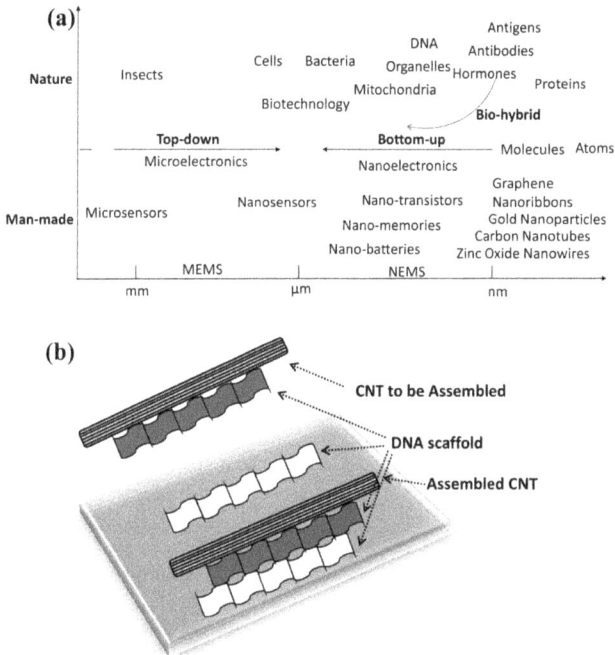

Fig. 5. Approaches for the fabrication and integration of nanosensors (a) and a representative bottom-up approach showing two carbon nanotubes precisely connected over a silicon substrate by mean of DNA scaffolding (b). (Reprinted with permission from Akyildiz Ian F., *Electromagnetic wireless nanosensor networks* (Nano Communication Networks 2010), 1(1): 3–19.)

manufacturing technologies able to assemble nano-machines molecule by molecule do not exist. The bio-hybrid approach that focuses mainly on biological nano-machines found in cells proposes their use as models to develop new nanomachines or as building blocks for more complex systems like nano-robots. Use of a biological nano-motor to power a nano-device and use of bacteria for the transport of micro-scale objects are few examples reported on this approach (Soong et al. 2000, Behkam and Sitti 2007).

3. Smart Wireless Nanosensors for Remote Health Monitoring

Wireless nanosensor systems have multiple applications in the biomedical field. Monitoring of sodium, glucose and other ions in blood, cholesterol, cancer biomarkers or the presence of different infectious agents is one key area in smart nanosensor based health care monitoring (Li et al. 2003, Dubach et al. 2007, Tothill 2009, Tallury et al. 2010). For example, tattoo-like sensors can be used for real-time monitoring of glucose levels in the blood of diabetic people, saving them from pricking their fingers several times a day (Dubach et al. 2007). Different nanosensors distributed around the body (body area nanosensor network) can collect information on physiological parameters and can transmit in wireless-manner between other nanosensors or micro-device such as a cell phone or specialized medical equipment to forward to the healthcare provider such as doctors (Figure 6).

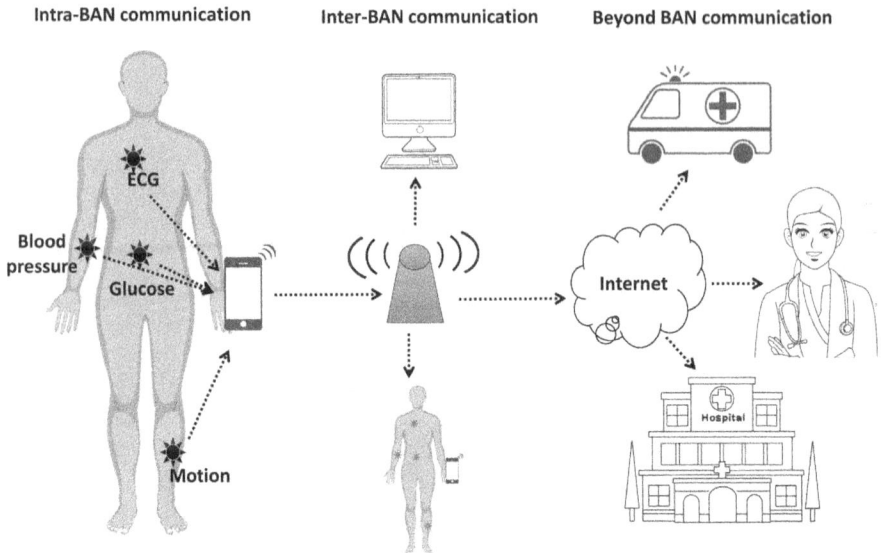

Fig. 6. Generalized architecture of a body area network.

As the name implies, these smart nanodevices can also be used to both monitor the level of a specific substance as well as to provide appropriate intervention (release a specific drug in human body) (Fernández-Pacheco et al. 2009). For example, by means of coordinated action of nanosensors and nanoactuators, these smart nanodevices can take decisions on the in conditions such as releasing or not a drug to dissolve a clot in an artery, or an engineered antibody to improve the immunologic system of humans. Dae-Hyeong Kim's group in the Republic of Korea developed a smart wireless nanosensor "diabetes patch" for continuous sweat-based diabetes monitoring and feedback therapy (Lee et al. 2016). This wearable patch (skin-mounted) of graphene-hybrid device has humidity, glucose and pH sensors (for sweat-based glucose and pH monitoring), a heater, temperature, and polymeric microneedles (for controlled transcutaneous drug delivery through bioresorbable temperature responsive microneedles), and wireless connectivity to a portable device like Smartphone for continuous monitoring and control (Figure 7a). Non-invasive epidermal electrochemical monitoring for glucose monitoring has been reported through skin interstitial fluid and sweat (Figure 7b). "Glucose in these bio-fluids can bediffused from blood vessels through the endothelium or sweat glands, reflecting blood glucose concentration by the body-compliant wearable platforms such as a patch, wrist-band or temporarytattoo" (Kim et al. 2018). However, these systems require further development, evaluation and validation for widespread implementation for improved management of diabetes and patient outcomes.

Broadly, the end applications of smart wireless nanosensor systems can be classified into three main subfields: biophysical monitoring (heart rate/pulse, human motion, and temperature), biochemical monitoring (biomolecule, blood glucose, and pH), and real-time environmental information detection (gas molecules and humidity).

(a)

(b)

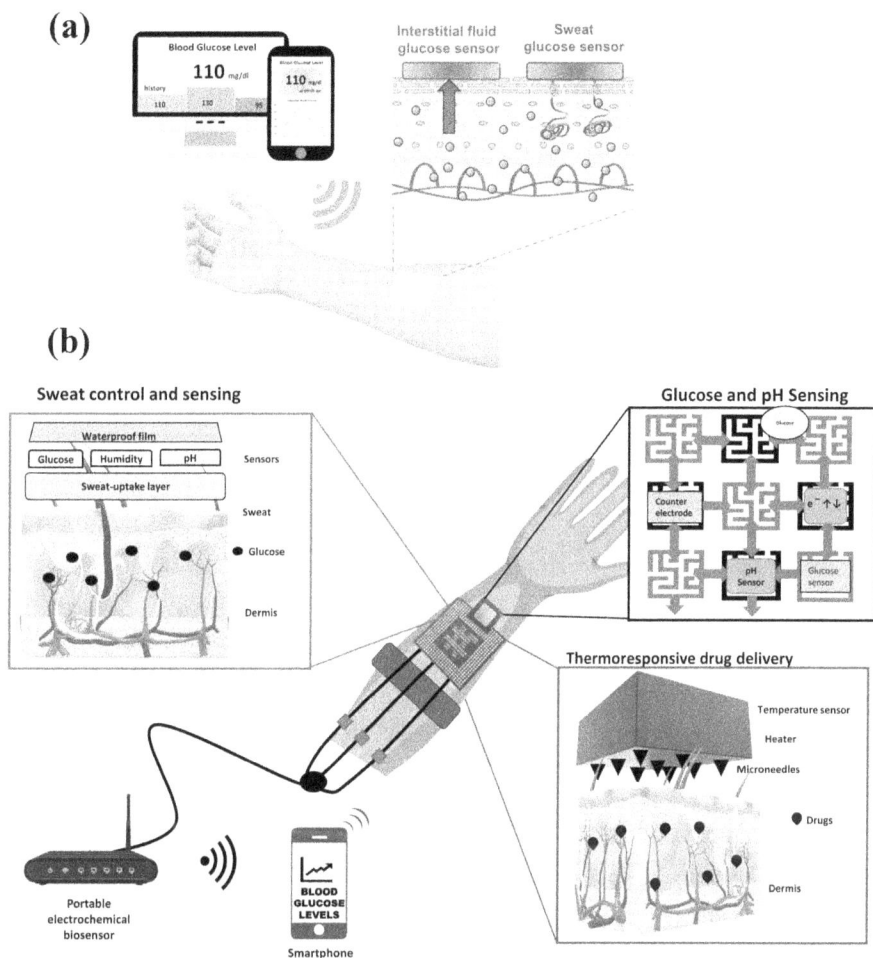

Fig. 7. Schematic representation of a wearable glucose sensor (a) and a glucose graphene skin sweat sensor and drug delivery chip (b). Image (a) (reprinted with permission from KimJayoung, *Wearable non-invasive epidermal glucose sensors: A review* (Talanta 2018), 177: 163–170.) Image (b) (reprinted with permission from LeeHyunjae, *A graphene-based electrochemical device with thermoresponsive microneedles for diabetes monitoring and therapy* (Nature Nanotechnology 2016), 11(6): 566–572.)

Remote health monitoring had been a long envisioned paradigm to support particularly the elderly people, infants, and those suffering from chronic health conditions to regularly monitor their health conditions in comfortable home environment instead of visiting expensive healthcare facilities such as hospitals or nursing homes. Nanosensor devices that support non-invasive and unobtrusive monitoring of important physiological signs and activities in a real-time have attracted significant attention in recent years. A variety of wearable sensors for physiological and activity monitoring have been proposed in the literature. Apart from that, various wearable devices and garments for physiological parameters, monitoring hydration status and metabolism, monitoring cardio-respiratory functions, and for monitoring and promoting better sleep (Table 3) are now rapidly developing, with some even reached to the market. These nanodevices include mainly the "smart watches, bands, garments, and patches with embedded sensors, small portable devices and mobile applications for feedback on cardiorespiratory function, movement patterns, sweat analysis, tissue oxygenation, sleep, emotional state, and changes in cognitive function following concussion" (Peake et al. 2018).

Detection of diseases at an early stage is a challenging task that requires continuous or periodic monitoring of selected biomarkers in body. Blood, the most predominant body fluid, is an excellent carrier of metabolites, metal ions and other biomarkers. However, collecting blood is an invasive procedure that poses risk of possible infection, therefore not an ideal matrix for continuous monitoring. Alternatively, urine, sweat, saliva, and tears have been reported to provide some of the information otherwise collected through blood based assessments. Since these non-invasive matrices contain target analytes (disease biomarkers) in relatively low concentrations, they need sophisticated techniques for sensitive and specific detection in the health care setting.

Detection of saccharides, particularly the monosaccharides (e.g., glucose), disaccharides (e.g., lactose), and polysaccharides (e.g., glycogen) in human body is an important health aspect. Glucose, which has a key role in diabetes, has been studied in biosensing applications since decades. Recently, smart nanosensor systems have been devised by various scientific groups across the world. An indium oxide (In_2O_3) based field-effect-transistor biosensor for glucose detection in sweat is depicted in Figure 8(1) (Liu et al. 2018). Similar to glucose, detection of lactic acid (lactate) in blood also provides critical information on its built up in human body in various diseases. An organic matter based electrochemical nanosensor developed for lactate detection is depicted in Figure 8(2) (Khodagholy et al. 2012). A fully bio interfaced sensing platform based on self-assembly of antimicrobial peptides onto graphene was developed by (Mannoor et al. 2012) to selectively detect bacteria at single-cell levels (Figure 8(3)).

4. Challenges in Wireless Nanosensing for Human Health Monitoring

Sensing biological information from human body using wearable and implantable sensors and communicating it (wirelessly) to remote destinations for diagnostic and therapeutic purposes need to satisfy various requirements, including mainly the unobtrusiveness, security, interoperability, reliability, and technical advancement. Unobtrusive continuous monitoring requires the wireless nanosensors to be light weight and miniature size. This in turn depends on the size and weight of nanosensor components such as nanobattery and nanoantenna. Recent technical advancements in microelectronics, system-on-chip design and low power wireless communication enable developing small size, high energy flexible and printed batteries for wearable devices, as in skin patches for transdermal drug delivery, patient temperature sensors, or RFID tracking (Nia et al. 2015). Individual's health data must be kept secure and only shared with authorized person in remote destinations. The coordination between the hardware components and the software program is fundamental to providing

Table 3. Devices for monitoring hydration status and metabolism, cardio-respiratory functions, and for monitoring and promoting better sleep.

Product category	Product name	Technical characteristics	Company
A. Devices for monitoring hydration status and metabolism			
Smart watch	Hydra Alert HRM Hydration monitor	Monitors hydration status. Sensors for detecting temperature and humidity. MAX MET (VO2 max) calibration. Heat index calculator. Standard heart rate monitor and interval timer/countdown.	Acumen™
Smart watch	Halo edge	Monitors hydration status, activity levels, and environmental conditions.	Halo wearables
Strap/band/patch	Humon hex	Strap with sensor paired to wristwatch and mobile application to measure muscle oxygenation levels.	DynometricsInc
Strap/band/patch	Nobo B60	Strap with sensor monitor hydration status	NoboInc
Strap/band/patch	ECHO™ smart patch	Wearable device to measure hydration, sodium, glucose, metabolites, various molecules, and proteins.	Kenzen
Strap/band/patch	BSX Insight	Sleeve with near infrared sensors to detect muscle oxygenation and lactate levels. Useful for determining lactate thresholds non-invasively. Connectable to ANT+ fitness tracking watches, mobile application, and computers.	BSX Athletics
Wearable device	Portamon	Measures oxy- deoxy- and total hemoglobin, blood volume and blood flow, as well as tissue saturation in muscle tissue using near infrared spectroscopy. Bluetooth (150m) or on-board data collection.	Artinis Medical systems
Wearable device	Moxy	Uses near infrared spectroscopy to measure muscle oxygenation levels in muscle tissue. Lightweight (40 g) and water resistant. On-board data collection and wireless data transmission.	Fortiori Design
Non-wearable device	Breezing	Device linked to mobile application to measure respiratory quotient in exhaled breath as a measure of the balance of carbohydrate and fat metabolism. Measures and records history of energy expenditure.	Breezing
Non-wearable device	The LEVL Device	Device linked to mobile application to measure the acetone content of exhaled breath as a measure of fat metabolism.	Medamonitor LLC

Table 3 contd. ...

...Table 3 contd.

B. Devices and garments for monitoring cardio-respiratory functions

Product category	Product name	Technical characteristics	Company
Smart watch	Helo	Monitors blood pressure, heart rate, ECG, blood temperature and O2 saturation, sleep cycle, breathing rate, calories, mood, and physical activity levels. Germanium, Hematite and Himalayan Salt plates to improve blood circulation, eliminate toxins, and purify cells.	HELO
Smart watch	E4 Wristband	Contains a photoplethysmography sensor that records blood pulse volume (from which heart rate and heart rate variability can be derived), a 3-axis accelerometer for recording activity, an electrodermal sensor to measure activity of the sympathetic nervous system (to derive features related to stress, engagement, and excitement), an infrared thermophile to record skin temperature. Connected to a mobile application and data stored in a cloud.	Empatica Inc.
Smart watch	Amiigo	Monitors heart rate, heart rate variability, blood pressure variations, pulse volume variations, respiratory rate, skin temperature, arterial blood O2 saturation, sleep time/quality, restful sleep, calories burned. Connected to mobile application.	Amiigo
Smart watch	Reign Active Recovery Band	Records type and amount of activity, calories burned, heart rate variability (through two metal sensors). Calculates a "Go-Zone" based on heart rate variability to determine personal fatigue and recovery. Training recommendations based on heart rate variability. Records habitual sleep patterns (through an accelerometer) to determine personal "Ideal Sleep" hours; makes recommendations for sleep. Connects to mobile application.	Jaybird
Smart watch	Mio SLICE™	Monitors physical activity levels and heart rate. Calculates Personal Activity Intelligence (PAI) score to match physical activity and heart rate to health assessment.	Mio™
Strap/band/ patch	Lief	Patch that monitors heart rate and breathing rate. Provides haptic signals to the user following extended periods of stress. Associated mobile application records various emotions to create a mood rating and provides cognitive behavioural therapy for emotional regulation.	Lief Therapeutics
Strap/band/ patch	Zephyr™	Sensor connected to a strap around the chest or imbedded within a singlet. Measures physiological data including heart rate, breathing rate, heart rate variability, estimated body temperature, calories burned, blood pressure, arterial blood O2 saturation.	Medtronic

Strap/band/ patch	Biostrap	Wristband that captures high-fidelity raw photoplethysmography waveforms to evaluate heart health. Connected to mobile application.	Biostrap USA, LLC
Wearable device	CorSense HRV monitor	Portable device placed on the finger and connected to a mobile application to measure heart rate variability, provide a readiness score, guide to stress and recovery.	CorSense
Non-wearable devices	MyCalmBeat	Near infrared pulse meter to assess personal best breathing rate when calm and train breathing at that rate. Consciously monitoring and adjusting breathing rate improves heart rate variability, leading to greater resilience, better pain management, and improved sense of wellbeing, enhanced ability to focus and think clearly. Connected to mobile application.	MyBrainSolution
Garments	Hexoskin	Singlet garment containing an ECG sensor, a breathing sensor and an accelerometer; measures: heart rate, heart rate variability, breathing rate, tidal volume, minute ventilation, steps, cadence, estimated calories burned. Connected to mobile application.	Carre Technologies Inc. (Hexoskin)©
Mobile application and non- wearable device	OmegaWave	Evaluates heart rate variability, neuromuscular, sensorimotor, and physical work capacity. Data derived to determine Windows of "Trainability TM for "readiness" of central nervous, cardiac, energy supply, gas exchange/pulmonary and hormonal systems and detoxification. Sensors placed on the body to record ECG and DC potential. Team and individual athlete's analysis packages.	OmegaWave
C. Wearable devices and equipment for monitoring and promoting better sleep			
UP™		Wristband connected to a mobile application. Activity tracker to measure light, deep and rapid eye movement sleep. Measures heart rate.	Jawbone
Fitbit Flex™		Wristband connected to a mobile application. Activity tracker to total sleep time, time in bed.	FitBit
FitBit Charge2™		Wristband connected to a mobile application. Activity tracker to total sleep time, time in bed.	FitBit
OURA		Ring with 3D accelerometer and gyroscope to measure light, deep, and rapid eye movement sleep. Measures heart rate.	OURA
Dreem		Headband that transmits sound simulations through bone conduction technology that synchronizes with sleep. Miniaturized EEG sensors provide feedback on sleep through mobile application.	Rythm
Plex®Sleep scanner		Chest strap that measures breathing patterns, pulse and oxygen levels during sleep. Connects to mobile application.	Sonmology

Table 3 contd. ...

...Table 3 contd.

Product category	Product name	Technical characteristics	Company
Sleep Profiler PSG2	EEG sleep monitor. Three channels of frontal EEG. Pulse rate and optional ECG. Monitors head movement and position. Provides data on total time and percentage sleep, rapid eye movement and slow wave sleep, sleep efficiency and average number of cortical, sympathetic and behavioural arousals. Recording device connects to computer to download data.	Advanced Brain Monitoring
Zmachine®	Three skin sensors placed behind each ear and the back of the neck are connected to a device for recording EEG. Records periods of light sleep, deep sleep, rapid eye movement, arousals, sleep period time, total sleep time, sleep efficiency, latency to sleep persistency, wake after sleep onset and time spent out of bed. Recording device connects to computer to download data. Two models (Insight and Synergy) available.	General Sleep Corporation
Somte PSG	Headband device with 6-channel EEG for polysomnography (PSG) assessment. Enable to simultaneously record oculomotor activity and ECG. Bluetooth wireless connection to computer software for sleep staging and events.	Compumedics®
Sleep Shepherd	Fabric headband that monitors EEG signals and sends audio sounds to reduce brain activity to a level conducive to sleep. Mobile application tracks sleep and provides alarm to lift brain out of sleep before the user wakes up.	Sleep Shepherd LLC
Re-Timer	Eyewear that projects green-blue light. Designed to be worn for 30min in morning or afternoon. Used to re-train timing of sleep onset. Online calculator available for sleep schedules and adjustment to jet lag.	Re-Tym Pty Ltd
AYO	Eyewear containing sensors to detect ambient light and projects blue light. Connected to mobile application to deliver blue light at the best time of day or night according to personal preferences and lifestyle (e.g., known periods of sleepiness or low energy); programmable to match different time zones.	Novology
illumy Sleep Smart Mask	Mask that uses gently dimming red light to promote sleep and gently brightening blue light to wake up. Sleep and wake times programmed into mobile application and synched to mask.	Headwaters Inc.
HUSH	Wireless ear plugs connected to a mobile application that plays soothing music to encourage sleep or wakefulness.	Hush technology Inc.
Kokoon	Headphones that mould to the shape of the user's head. Detects EEG signals and movement to find the lightest point of the user's natural sleep cycle during which to wake up. Active noise cancellation and white noise.	Kokoon

Source: Data from Peake Jonathan M., "A Critical Review of Consumer Wearables, Mobile Applications, and Equipment for Providing Biofeedback, Monitoring Stress, and Sleep in Physically Active Populations", Frontiers in Physiology (2018): 9: 743.

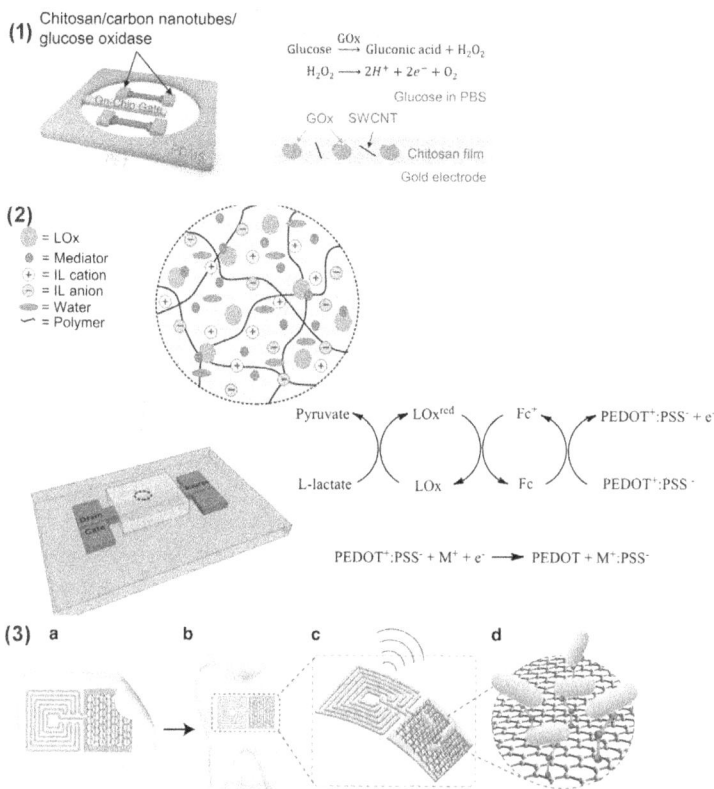

Fig. 8. Schematic representation of the In_2O_3-based FET biosensor and its working principle (1), lactate sensor and its working principle (2), and the graphene-based wireless nanosensor for bacteria detection on tooth enamel (3). Image (1) (reprinted with permission from LiuQingzhou, *Highly Sensitive and Wearable In_2O_3 Nanoribbon Transistor Biosensors with Integrated On-Chip Gate for Glucose Monitoring in Body Fluids* (ACS Nano 2018), 12(2): 1170–1178.) Image (2) (reprinted with permission from Khodagholy Dion, *Organic electrochemical transistor incorporating an ionogel as a solid state electrolyte for lactate sensing* (Journal of Materials Chemistry 2012), 22(10): 4440–4443.) Image (3) (reprinted with permission from Mannoor Manu S., *Graphene-based wireless bacteria detection on tooth enamel*(Nature Communications 2012), 3 (1): 763.)

secure and reliable communication. Further, a malicious user must not be capable of disrupting or harmfully affecting communication or quality of service provided by either nanodevices or nanonetworks (Atlam et al. 2018). Users should have options to choose who can access the data, for instance, allowing access only to their doctors. Interoperability in healthcare is the extent to which various systems and devices can interpret data and display it in a user friendly manner. It ensures that the personal health information will be available to patients whenever and wherever they need. This also helps in saving total energy expenditures by reducing the high demands on the communication channel. End-to-end reliability in wireless nanosensor networks guarantees that the nanosensor devices will not be easily physically damaged, and minimal transient molecular interference from sudden burst of molecules that can create temporal disconnections of the network at different points. Also, while the progress in manufacturing wearable nanodevices is speeding rapidly, and already few smart wireless nanosensors have reached the market, the advancement in implantable medical devices is still very slow. As the latter can provide long-term real-time monitoring of the state of tissues, organs, system, while also assisting in diagnosis and therapeutics, they are much appreciated in remote health monitoring. However, the major challenges for implantable devices are biocompatibility, biofouling, and adequate power supply.

5. Conclusion

Current healthcare system is based on a reactive approach where the diseases, infections or injuries are diagnosed once they have already occurred or shown clear symptoms noticed by patients. The time-gap between diagnosis and treatment is crucial, particularly for those remotely located with lack of healthcare facilities. An earlier detection can prompt the intervention and increase chances of treatment for better cure. The need of a proactive healthcare system (still futuristic) which will inculcate predictive, participatory, preventive and personalized (P4) healthcare seems possible with recent advancement in the smart wireless nanosensor systems. Wearable and implantable nanosensor systems empowered with wireless connectivity for efficient communication of individual's health data, and the capability to take decisions for therapeutic interventions will have great impact in the healthcare field. Optimal diabetes management through continuous monitoring of blood glucose levels, continual analysis of athletes' fitness levels, regular and real-time monitoring of pathogens in physiological fluids and tracking drug efficiency will be the few benefits of these nanosensor systems. This chapter covers some of the relevant topics on smart wireless nanosensor systems, such as the state-of-art in the area of conventional biosensors and next generation smart wireless nanosensors, their architecture, and key applications. Future research to solve the highlighted challenges and achieving new heights in biocompatible nanomaterials' manufacturing for development of implantable nanosensor systems may revolutionize the human healthcare system.

Acknowledgement

The authors would like to thank Dr. Saroj Kumar from School of Biosciences, Apeejay Stya University, Gurgaon, India for reviewing the chapter. Authors also thank Subbiah Rajasekaran and Kamini Arya for their help in data collection. Work in the laboratory of the corresponding author is supported by fund from the Indian Council of Medical Research (65/2/AKT/NIREH/2018-NCD-II) and Science and Engineering Research Board (Grant no 742 ECR/2017/003179). Authors declare no conflict of interest.

References

Ahirwar, R. and P. Nahar. (2016). Development of a label-free gold nanoparticle-based colorimetric aptasensor for detection of human estrogen receptor alpha. *Analytical and Bioanalytical Chemistry* 408(1): 327–332.

Ahirwar, R., A. Dalal, J. G. Sharma, B. K. Yadav, P. Nahar, A. Kumar and S. Kumar. (2019). An aptasensor for rapid and sensitive detection of estrogen receptor alpha in human breast cancer. *Biotechnology and Bioengineering* 116(1): 227–233.

Ahirwar, R. (2021). A review of analytical merit of nanomaterials-based immunosensors and aptasensors for HER2 assessment in breast cancer. *Microchimica Acta* 188.

Ahirwar, R., N. Khan and S. Kumar. (2021). Aptamer-based sensing of breast cancer biomarkers: A comprehensive review of analytical figures of merit. Expert Review of Molecular Diagnostics 21.

Akan, O. B., H. Ramezani, T. Khan, N. A. Abbasi and M. Kuscu. (2017). Fundamentals of molecular information and communication science. *Proceedings of the IEEE* 105(2): 306–318.

Akyildiz, I. F. and J. M. Jornet. (2010). Electromagnetic wireless nanosensor networks. *Nano Communication Networks* 1(1): 3–19.

Asal, M., Ö. Özen, M. Şahinler and İ. Polatoğlu. (2018). Recent developments in enzyme, DNA and immuno-based biosensors. *Sensors* 18(6).

Atlam, H., R. Walters and G. Wills. (2018). Internet of Nano things: Security issues and applications. In Proceedings of the 2018 International Conference on Cloud and Big Data Computing 7.

Behkam, B. and M. Sitti. (2007). Bacterial flagella-based propulsion and on/off motion control of microscale objects. *Applied Physics Letters* 90(2): 023902.

Bennewitz, R., J. N. Crain, A. Kirakosian, J. L. Lin, J. L. McChesney, D. Y. Petrovykh and F. J. Himpsel. (2002). Atomic scale memory at a silicon surface. *Nanotechnology* 13(4): 499–502.

Bertrand, M. J. (1998). Handbook of instrumental techniques for analytical chemistry Edited by Frank A. Settle. Prentice Hall: Upper Saddle River. 1997. xxi + 995 pp. ISBN 0-13-177338-0. *Journal of the American Chemical Society* 120(26): 6633–6633.

Bhalla, N., P. Jolly, N. Formisano and P. Estrela. (2016). Introduction to biosensors. Essays in Biochemistry 60(1): 1–8.

Bhardwaj, T. (2014). A review on immobilization techniques of biosensors. *International Journal of Engineering and Technical Research* 3: 294–298.

Bondavalli, P., P. Legagneux and D. Pribat. (2009). Carbon nanotubes based transistors as gas sensors: State of the art and critical review. *Sensors and Actuators B: Chemical* 140(1): 304–318.

Brecht, A. and G. Gauglitz. (1995). Optical probes and transducers. *Biosensors and Bioelectronics* 10(9): 923–936.

Cao, J., E. K. Galbraith, T. Sun and K. T. V. Grattan. (2011). Comparison of surface plasmon resonance and localized surface plasmon resonance-based optical fibre sensors. *Journal of Physics: Conference Series* 307: 012050.

Chaubey, A. and B. D. Malhotra. (2002). Mediated biosensors. *Biosensors and Bioelectronics* 17(6): 441–456.

Chen, Y. (2015). Nanofabrication by electron beam lithography and its applications: A review. *Microelectronic Engineering* 135: 57–72.

Choi, M. M. F. (2004). Progress in enzyme-based biosensors using optical transducers. *Microchimica Acta* 148(3): 107–132.

Clark, L. C. J. (1956). Monitor and control of blood and tissue oxygen tensions. *ASAIO Journal* 2(1): 41–48.

Damborský, P., J. Švitel and J. Katrlík. (2016). Optical biosensors. *Essays in Biochemistry* 60(1): 91–100.

Darwish, A. and A. E. Hassanien. (2011). Wearable and implantable wireless sensor network solutions for healthcare monitoring. *Sensors* 11(6): 5561–5595.

Dubach, J. M., D. I. Harjes and H. A. Clark. (2007). Fluorescent ion-selective nanosensors for intracellular analysis with improved lifetime and size. *Nano Letters* 7(6): 1827–1831.

Dzyadevych, S. V., V. N. Arkhypova, A. P. Soldatkin, A. V. El'skaya, C. Martelet and N. Jaffrezic-Renault. (2008). Amperometric enzyme biosensors: Past, present and future. *IRBM* 29(2): 171–180.

Fernández-Pacheco, R., J. G. Valdivia and M. R. Ibarra. (2009). Magnetic nanoparticles for local drug delivery using magnetic implants. *Methods Mol Biol* 544: 559–569.

Freitas, R. A., Jr. (2005). Nanotechnology, nanomedicine and nanosurgery. *Int J Surg* 3(4): 243–246.

Goldstein, H. (2005). The race to the bottom [consumer nanodevice]. *IEEE Spectrum* 42(3): 32–39.

Guilbault, G. G. and J. G. Montalvo, Jr. (1969). A urea-specific enzyme electrode. *J Am Chem Soc* 91(8): 2164–2165.

Guilbault, G. G. and G. J. Lubrano. (1973). An enzyme electrode for the amperometric determination of glucose. *Analytica Chimica Acta* 64(3): 439–455.

Gul, E., B. Atakan and O. B. Akan. (2010). NanoNS: A nanoscale network simulator framework for molecular communications. *Nano Communication Networks* 1(2): 138–156.

Hierold, C., A. Jungen, C. Stampfer and T. Helbling. (2007). Nano electromechanical sensors based on carbon nanotubes. *Sensors and Actuators A: Physical* 136(1): 51–61.

Hughes, W. S. (1922). The potential difference between glass and electrolytes in contact with the glass. *Journal of the American Chemical Society* 44(12): 2860–2867.

Ju Yun, Y., C. Seong Ah, S. Kim, W. Soo Yun, B. Chon Park and D. Han Ha. (2007). Manipulation of freestanding Au nanogears using an atomic force microscope. *Nanotechnology* 18(50): 505304.

Karimzadeh, A., M. Hasanzadeh, N. Shadjou and M. d. l. Guardia. (2018). Peptide based biosensors. *TrAC Trends in Analytical Chemistry* 107: 1–20.

Kassal, P., M. D. Steinberg and I. M. Steinberg. (2018). Wireless chemical sensors and biosensors: A review. *Sens. Actuators, B* 266: 228.

Khodagholy, D., V. F. Curto, K. J. Fraser, M. Gurfinkel, R. Byrne, D. Diamond, G. G. Malliaras, F. Benito-Lopez and R. M. Owens. (2012). Organic electrochemical transistor incorporating an ionogel as a solid state electrolyte for lactate sensing. *Journal of Materials Chemistry* 22(10): 4440–4443.

Kim, J., A. S. Campbell and J. Wang. (2018). Wearable non-invasive epidermal glucose sensors: A review. *Talanta* 177: 163–170.

Kissinger, P. T. (2005). Biosensors—a perspective. *Biosensors and Bioelectronics* 20(12): 2512–2516.

Lee, H. H., E. Menard, N. Tassi, J. Rogers and G. Blanchet. (2004). Large Area Microcontact Printing Presses for Plastic Electronics. MRS Proceedings 846.

Lee, H., T. K. Choi, Y. B. Lee, H. R. Cho, R. Ghaffari, L. Wang, H. J. Choi, T. D. Chung, N. Lu, T. Hyeon, S. H. Choi and D.-H. Kim. (2016). A graphene-based electrochemical device with thermoresponsive microneedles for diabetes monitoring and therapy. *Nature Nanotechnology* 11(6): 566–572.

Li, C., E. T. Thostenson and T.-W. Chou. (2008). Sensors and actuators based on carbon nanotubes and their composites: A review. *Composites Science and Technology* 68(6): 1227–1249.

Li, J., T. Peng and Y. Peng. (2003). A cholesterol biosensor based on entrapment of cholesterol oxidase in a silicic sol-gel matrix at a prussian blue modified electrode. *Electroanalysis* 15(12): 1031–1037.

Liu, Q., Y. Liu, F. Wu, X. Cao, Z. Li, M. Alharbi, A. N. Abbas, M. R. Amer and C. Zhou. (2018). Highly Sensitive and Wearable In2O3 Nanoribbon Transistor Biosensors with Integrated On-Chip Gate for Glucose Monitoring in Body Fluids. *ACS Nano* 12(2): 1170–1178.

Lobo, M. J., A. J. Miranda and P. Tuñón. (1997). Amperometric biosensors based on NAD(P)-dependent dehydrogenase enzymes. *Electroanalysis* 9(3): 191–202.

Lorenz, W. and K. D. Schulze. (1975). Application of transform-impedance spectrometry. *Journal of Electro-analytical Chemistry* 65(1): 141–153.

Malik, P., V. Katyal, V. Malik, A. Asatkar, G. Inwati and T. K. Mukherjee. (2013). Nanobiosensors: Concepts and Variations. *ISRN Nanomaterials* 2013: 327435.

Mannoor, M. S., H. Tao, J. D. Clayton, A. Sengupta, D. L. Kaplan, R. R. Naik, N. Verma, F. G. Omenetto and M. C. McAlpine. (2012). Graphene-based wireless bacteria detection on tooth enamel. *Nature Communications* 3(1): 763.

Manurung, R. V., E. Kurniawan and C. Risdian. (2012). The Electropolymerization of Conductive Polymer Ppy-PANi on Gold Electrodes for Uric Acid Biosensor.

Mariani, S. and M. Minunni. (2014). Surface plasmon resonance applications in clinical analysis. *Analytical and Bioanalytical Chemistry* 406(9-10): 2303–2323.

Mehrotra, P. (2016). Biosensors and their applications—A review. *Journal of Oral Biology and Craniofacial Research* 6(2): 153–159.

Murugaiyan, S. B., R. Ramasamy, N. Gopal and V. Kuzhandaivelu. (2014). Biosensors in clinical chemistry: An overview. *Advanced Biomedical Research* 3: 67–67.

Nafari, M. and J. M. Jornet. (2015). Metallic plasmonic nano-antenna for wireless optical communication in intra-body nanonetworks. Proceedings of the 10th EAI International Conference on Body Area Networks. Sydney, New South Wales, Australia, ICST (Institute for Computer Sciences, Social-Informatics and Telecommunications Engineering): 287–293.

Nelson, J. M. and E. G. Griffin. (1916). Adsorption of invertase. *Journal of the American Chemical Society* 38(5): 1109–1115.

Nguyen, H. H., J. Park, S. Kang and M. Kim. (2015). Surface plasmon resonance: A versatile technique for biosensor applications. *Sensors* (Basel, Switzerland) 15(5): 10481–10510.

Nia, A. M., M. Mozaffari-Kermani, S. Sur-Kolay, A. Raghunathan and N. K. Jha. (2015). Energy-Efficient Long-term Continuous Personal Health Monitoring. *IEEE Transactions on Multi-Scale Computing Systems* 1(2): 85–98.

Parkin, S. S. P., M. Hayashi and L. Thomas. (2008). Magnetic domain-wall racetrack memory. *Science* 320(5873): 190–194.

Peake, J. M., G. Kerr and J. P. Sullivan. (2018). A critical review of consumer wearables, mobile applications, and equipment for providing biofeedback, monitoring stress, and sleep in physically active populations. *Frontiers in Physiology* 9(743).

Peterson, C. (2000). Taking technology to the molecular level. *Computer* 33(1): 46–53.

Pohanka, M. (2018). Overview of piezoelectric biosensors, immunosensors and dna sensors and their applications. *Materials* (Basel, Switzerland) 11 DOI: 10.3390/ma11030448.

Prodromidis, M. I. (2010). Impedimetric immunosensors—A review. *Electrochimica Acta* 55(14): 4227–4233.

Ranganathan, V., S. Srinivasan, A. Singh and M. C. DeRosa. (2020). An aptamer-based colorimetric lateral flow assay for the detection of human epidermal growth factor receptor 2 (HER2). *Analytical Biochemistry* 588: 113471.

Rastislav, M., S. Miroslav and Š. Ernest. (2012). Biosensors—classification, characterization and new trends. *Acta Chimica Slovaca* 5(1): 109–120.

Rizwan, A., A. Zoha, R. Zhang, W. Ahmad, K. Arshad, N. A. Ali, A. Alomainy, M. A. Imran and Q. H. Abbasi. (2018). A Review on the role of nano-communication in future healthcare systems: A big data analytics perspective. *IEEE Access* 6: 41903–41920.

Rocchitta, G., A. Spanu, S. Babudieri, G. Latte, G. Madeddu, G. Galleri, S. Nuvoli, P. Bagella, M. I. Demartis, V. Fiore, R. Manetti and P. A. Serra. (2016). Enzyme biosensors for biomedical applications: Strategies for safeguarding analytical performances in biological fluids. *Sensors* (Basel, Switzerland) 16(6): 780.

Roham, M., J. M. Halpern, H. B. Martin, H. J. Chiel and P. Mohseni. (2008). Wireless amperometric neurochemical monitoring using an integrated telemetry circuit. *IEEE Trans. Biomed. Eng.* 55: 2628.

Ronkainen, N. J., H. B. Halsall and W. R. Heineman. (2010). Electrochemical biosensors. *Chem. Soc. Rev.* 39: 1747.

Rutherglen, C. and P. Burke. (2009). Nanoelectromagnetics: Circuit and electromagnetic properties of carbon nanotubes. *Small* 5(8): 884–906.

Sadeghi, S. J. (2013). Amperometric biosensors. Encyclopedia of Biophysics. G. C. K. Roberts. Berlin, Heidelberg, Springer Berlin Heidelberg: 61–67.

Sassolas, A., L. J. Blum and B. D. Leca-Bouvier. (2012). Immobilization strategies to develop enzymatic biosensors. *Biotechnology Advances* 30(3): 489–511.

Scheller, F. W., F. Schubert, B. Neumann, D. Pfeiffer, R. Hintsche, I. Dransfeld, U. Wollenberger, R. Renneberg, A. Warsinke, G. Johansson, M. Skoog, X. Yang, V. Bogdanovskaya, A. Bückmann and S. Y. Zaitsev. (1991). Second generation biosensors. *Biosensors and Bioelectronics* 6(3): 245–253.

Singh, P. (2016). SPR biosensors: Historical perspectives and current challenges. *Sensors and Actuators B: Chemical* 229: 110–130.

Soong, R. K., G. D. Bachand, H. P. Neves, A. G. Olkhovets, H. G. Craighead and C. D. Montemagno. (2000). Powering an Inorganic Nanodevice with a Biomolecular Motor. *Science* 290(5496): 1555–1558.

Sotiropoulou, S., V. Gavalas, V. Vamvakaki and N. A. Chaniotakis. (2003). Novel carbon materials in biosensor systems. *Biosensors and Bioelectronics* 18(2): 211–215.

Stampfer, C., T. Helbling, D. Obergfell, B. Schöberle, M. K. Tripp, A. Jungen, S. Roth, V. M. Bright and C. Hierold. (2006a). Fabrication of single-walled carbon-nanotube-based pressure sensors. *Nano Letters* 6(2): 233–237.

Stampfer, C., A. Jungen and C. Hierold. (2006b). Fabrication of discrete nanoscaled force sensors based on single-walled carbon nanotubes. *IEEE Sensors Journal* 6(3): 613–617.

Stampfer, C., A. Jungen, R. Linderman, D. Obergfell, S. Roth and C. Hierold. (2006c). Nano-electromechanical displacement sensing based on single-walled carbon nanotubes. *Nano Letters* 6(7): 1449–1453.

Strianese, M., M. Staiano, G. Ruggiero, T. Labella, C. Pellecchia and S. D'Auria. (2012). Fluorescence-based biosensors. *Methods Mol Biol* 875: 193–216.

Tallury, P., A. Malhotra, L. M. Byrne and S. Santra. (2010). Nanobioimaging and sensing of infectious diseases. *Advanced Drug Delivery Reviews* 62(4-5): 424–437.

Tothill, I. E. (2009). Biosensors for cancer markers diagnosis. *Semin Cell Dev. Biol.* 20(1): 55–62.

Wang, Z. L. (2008). Towards self-powered nanosystems: From Nanogenerators to Nanopiezotronics. *Advanced Functional Materials* 18(22): 3553–3567.

Wilson, G. S. and Y. Hu. (2000). Enzyme-based biosensors for *in vivo* measurements. *Chemical Reviews* 100(7): 2693–2704.

Yang, R., Y. Qin, C. Li, G. Zhu and Z. L. Wang. (2009). Converting Biomechanical Energy into Electricity by a Muscle-Movement-Driven Nanogenerator. *Nano Letters* 9(3): 1201–1205.

Zhu, C., G. Yang, H. Li, D. Du and Y. Lin. (2015). Electrochemical sensors and biosensors based on nanomaterials and nanostructures. *Anal. Chem.* 87: 230.

CHAPTER 14

Nanosensors and their Potential Role in Internet of Medical Things

Priya Rani

1. Introduction

In the 21st century, the healthcare scenario has been rapidly changing due to various technological advancements, such as wireless communications, and miniaturization and integration of devices, leading to micro-and nano-technology, which has further paved the way for the latest emerging field called the Internet of Things (IoT). IoT is known as the next revolution in technology and has been applied in many different fields, such as smart cities, smart agriculture, smart homes, etc. It is also making its way into the healthcare industry with smart sensing units, cloud servers, and end-users. The networking of multiple devices has created a huge market for IoT, whose value is estimated to be more than $14 trillion (Hamidi and Fazeli 2018). Thus, the idea of telemedicine, which was theoretical a few decades ago, has been developed to become a reality and is being extended to various other health sectors, such as hospitals, medical care units, field military camps, ambulances, and smart homes. This has led to the definition of a new by-product of IoT, called Internet of Medical Things (IoMT), which can be exclusively used to define connectivity of one or several medical devices to a healthcare system through wireless networks, and involves human-to-machine and machine-to-machine communication (Basatneh et al. 2018, Mora et al. 2017, Rodrigues et al. 2018). This has been made easier with the introduction of off-the-shelf sensors and medical devices in smart homes, which can be used for self-monitoring of various body parameters and diagnosis of health-related conditions. However, there has been a tremendous increase in the development of biosensors, especially nano-biosensors, which holds great opportunity for IoT to be utilized for medical data collection and analysis for diagnostic, prognostic, and therapeutic purposes. It has been reported in literature that the integration of IoT

Applied Artificial Intelligence Institute, Deakin University, 221 Burwood Hwy, Burwood, Victoria 3125 Australia.
Email: priya.rani@ieee.org

tools into healthcare could lower the annual costs of management of chronic diseases by close to one-third (Basatneh et al. 2018).

This chapter will discuss different biosensors, followed by nano-biosensors and their recent developments in the first section. The second section will give an introduction to IoMT, detailing its architecture. The third section will explain the applications of IoMT in different fields of healthcare—smart healthcare monitoring systems, voice-driven technologies, wearable and assistive medical devices, IoT-based rehabilitation systems, and other applications. The fourth section will give an insight into future trends in IoMT, while the fifth and sixth sections will discuss the advantages and challenges in IoMT, respectively, followed by conclusion.

2. Biosensors—from History to Present

Biosensors were first conceptualized by Leland C. Clark with the discovery of enzyme electrodes in 1962 (Nayyar et al. 2019). It brought a revolution in multiple scientific fields and scientists came together to develop highly efficient bio-sensing devices for applications such as military, agriculture, environment, biology, robotics, and healthcare. In simple words, biosensors can be defined as small analytical devices that detect any biologically active compound with a signal transducer and an electronic amplifier (Nayyar et al. 2019). To be applicable in smart healthcare, an ideal smart biosensor should have certain attributes and satisfy certain conditions, other than being biocompatible. It should have high response specificity towards the analyte, high recurrence of the reaction, and short recuperation time. It should be sensitive to low levels or volume of the analyte, be versatile in utility and be able to transmit the data wirelessly to another device for reporting and processing. These attributes can be easily achieved in nanotechnology-based sensors as it enables the manipulation of materials at the nanoscale, which not only helps to enhance selectivity and sensitivity of the materials but also lowers the cost of diagnosis (Vashistha et al. 2018).

While nanosensors are still undergoing intensive research to explore their full possibilities to be applied in healthcare, nanotechnology has been widely used in healthcare applications in the last two decades, such as potent pharmaceuticals, drug delivery, implants and devices, and diagnostic and imaging applications (Bawa et al. 2016, Miller and Kearnes 2012). Examples of nano-based drug delivery systems, which are under research and development, are nanosuspensions, polymeric nanoparticles, liposomes, dendrimers, carbon nanotubes, fullerenes, and other inorganic nanomaterials, some of which will be discussed in the next few paragraphs. Nanomaterials and particles can be used as implants for hard and soft tissue implants, dental restoratives, bone substitute materials, etc. Nanobionics is an emerging field and needs a lot more research for it to be made commercially available. Nanodiagnostics hold great promise in healthcare by monitoring and measuring specific health and physiological data, most of which will be discussed later in this chapter, as part of IoMT applications (Omanović-Mikličanin et al. 2015).

Nanosensors can be classified into chemical (atomic or molecular energies), physical (thermal, mechanical, acoustical and magnetic), and biological sensors (enzymatic reaction, DNA interaction, and antibody/antigen interaction) (Omanović-Mikličanin et al. 2015). Engineered nanosensors and nanoparticles have been

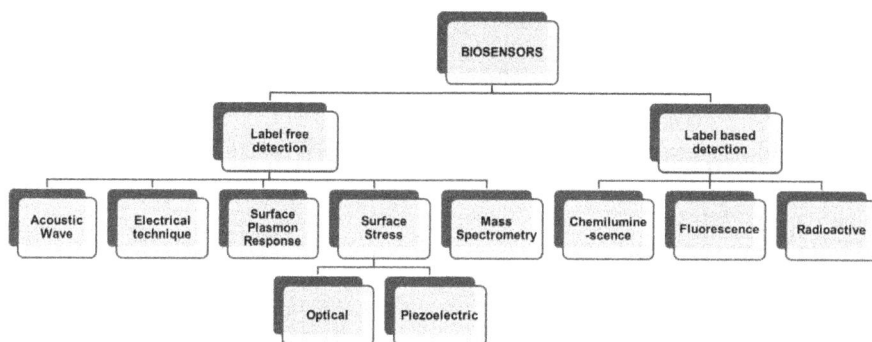

Fig. 1. Classification of biosensors based on their transduction properties (reprinted with permission from *Vashistha, Rajat, Arun Kumar Dangi, Ashwani Kumar, Deepak Chhabra, and Pratyoosh Shukla, Futuristic biosensors for cardiac health care: an artificial intelligence approach (Biotech 2018), 358–369).*

synthesized from organic, inorganic, synthetic, or biological materials. The promising key points for the use of nanoparticles are low production cost, high surface area, their specificity to target microbial cells, tissues, and biomolecules, and to detect toxins. The interaction of nanoparticles with biological systems mainly happens because the sizes of most of the functional elements in the human body are on nano-meter scale, which proves to be advantageous in their application to healthcare. Nanoparticles also have novel optical, magnetic, electronic, and structural properties compared to the same materials of larger dimensions. Due to these enhanced properties, nanoparticles can be precisely tuned to explore their applications in diagnostic and point of care monitoring devices (Vogel 2019, Lee et al. 2009).

The advantages of using nanosensors over microsensors or larger dimensional sensors in biomedical applications are greater signal-to-noise ratio, higher current density, shorter steady-state lags, faster response times, and smaller Ohmic drops (Moody et al. 2012). Inert noble metals such as gold and silver have been considered for creating biosensor electrodes which can also be used in *in vivo* systems. These metals can be constructed into nanoparticles, nanowires, or nanotubes, based on the application. Among non-metals, carbon nanoparticles have been used as nanotubes and nanofibers to function as nanoelectrodes, having obvious advantages of being cheaper and more resistant to biofouling over the noble metals (Moody et al. 2012).

The inorganic-based piezoelectric thin films are actively used in biomedical applications because of their pliable, slim, piezoelectric, lightweight, and biocompatible properties. These inherent properties make them ideal to be developed as 'self-powered' energy harvesters (also called nanogenerators) in implantable devices and as nanosensors into *in vivo* diagnostic/therapeutic systems. They are quite sensitive and can even detect mechanical nanoscale movements, for instance, the vibrations caused by acoustic deformation and resonance of biological cells (Hwang et al. 2015). The nanogenerators rely on the piezoelectric potential created by an external strain in the nanowires, which results in a transient flow of electrons in the external load due to the driving force of the piezoelectric potential. When used as nanogenerators, these nanowires can generate electric power by physical or mechanical movements, which is strong enough to turn on conventional liquid crystal displays (LCDs) or light-emitting diodes (LEDs). The frequency required for triggering these materials can be 1 Hz to thousands of Hz, which is ideal for harvesting random energy such as body motion, tiny vibrations, or gentle air movements (Pan et al. 2011, Wang 2012). Hence, when fully implemented, the implantable self-powered electronics would be fully equipped with data processing, information memory, energy harvesting/storage, and wireless communication, while providing medical diagnosis or treatment (Hwang et al. 2015).

For medical diagnostic or therapeutic applications, nanoparticles are being produced from polymers or lipids. Novel nanoparticles can be designed to carry therapeutic moieties or contrast agents in various diagnostic imaging technologies, such as computed tomography (CT) or magnetic resonance imaging (MRI) since they have high payload capacity and can increase the time available for imaging. The contrast agents usually used for MRI are gadolinium, manganese oxide, or iron particles, and for CT, the agents are iodine, barium, krypton, or xenon. To increase the imaging time, the basic idea adopted was to increase the size of the contrast agents to nanoparticle range using different nanoparticles, typically in the size of 1–100 nm, which would slow down its clearance from the blood. Diagnostic or therapeutic moieties can be included in the lipid nanoparticles in mainly four ways- by conjugating on the surface, by filling into the internal lipophilic space (as micelles), by embedding in the lipophilic membrane (liposomes), or by encapsulating in the core of liposomes (Bawa et al. 2016).

Nanomaterials are also being used in optical imaging, which is widely used in clinical applications due to the obvious advantages of being rapid, non-invasive, inexpensive, and highly sensitive. Nano-sized fluorescent imaging agents have large surface-to-volume ratio, are more photostable and brighter, making optical imaging more efficient. Every nanoparticle of the imaging agents could carry a large number of fluorescent species that would be insulated from complex biological environments. Among the many parameters considered for the synthesis of these imaging agents, particle size is an important one. The optimum size reported for these particles for achieving higher contrast is between 10 to 200 nm, depending on the imaging site (Peng and Chiu 2015). For instance, for *in vitro* targeting of tumour tissues, particle size in the range of 30–200 nm was considered appropriate for imaging, based on enhanced permeation and retention effect (Albanese et al. 2012). Hence, nanoparticles have been tested and validated to target brain tumour cells in mice and rats to perform *in vivo* brain tumour imaging (Wu et al. 2011).

Point of care (POC) diagnosis has improved with the development of graphene-metal nanoparticles and carbon nanotubes (Zhou et al. 2018). POC-based applications can be classified into lab on a chip, label-free, labelled, nanomaterial-based wireless, or wearable sensors (Quesada-González and Merkoçi 2018). The lab on a chip is an easy substitute for heavy machines and complex pathologies in which the biomarker is sensed via micro- or nano-transduction mechanisms. Based on the technical strategies involved, the nano-biosensors can perform label-based or label-free detections. The target detection for label-based biosensors is based on specific properties of label compounds which are fabricated with an immobilized target protein, hence making them highly reliable. The label-free biosensors, on the other hand, can detect molecules that are not labelled or are difficult to tag, which makes them highly applicable in medicine and healthcare. The detection mechanisms for wearable sensors can be calorimetric, optical or electrochemical. Conductive ink in intelligent tattoos and patches and on-screen printed electrodes on e-textiles have coating or threads of nanomaterials added, which makes them capable of sensing a few micro-fluids as bio-samples on the dermal layer of the skin (Mostafalu et al. 2017). Programmable bio-nanochip (p-BNC) systems are also a recent advancement in the field with the capability of learning, by producing an immunofluorescent signal corresponding to small quantities of the analyte (patient's sample). The p-BNC systems require microfluidic cartridges, portable analyzers, automated data analysis software, and inbuilt mobile health interfaces (Gaikwad and Banerjee 2018). While most of the nanoscale sensors are still far from deployment, they show huge potential to be implemented in IoT settings, as they are small, precise, more sensitive, would decrease the physical damage at the implemented site and would improve the interfacing between the body and the electronics.

3. Internet of Medical Things (IoMT)

The architecture of an IoMT-based system consists of micro or nano-sensors as the primary layer to collect information or biological signals from different parts of the body. The sensor devices can be incorporated into smartwatches, wristbands, or wearable sensors such as headgears, and can measure various types of health data or signals such as blood pressure, heart rate, body temperature, respiratory rate, ECG, EEG, EMG, etc., which will be discussed later in this chapter (Dorj et al. 2017). Each of these devices is allocated a unique identifier (UID) and are interconnected with each other by Ethernet with Transmission Control Protocol/Internet Protocol (TCP/IP). The biosensors embedded in these convert the biological signals into electrochemical (in case of analytics such as glucose, lactate, cholesterol, urea, etc.), electrical, optical or piezoelectric signals (Hamidi and Fazeli 2018).

These signals are then transmitted by a local area network (LAN) to the hosting layer such as smartphones, digital assistants, tablets, or laptops, where some local pre-processing such as filtering and amplification on these signals can be performed, if necessary. They can also have dedicated applications to detect general health abnormalities such as abnormal heart rate, blood pressure, or body temperature. The LAN interface comprises of short-to-medium-range communication protocols such as Bluetooth, Zigbee, LoWPAN, and Z-wave. The hosting layer then transmits the received information on to a cloud server through the Wide Area Network (WAN), which uses long-range communications such as 4G, 5G cellular or Wi-Fi technologies. Once the data is received on the cloud, several steps are performed on it, as described below (Kadhim et al. 2020b, Alhussein et al. 2018).

The cloud consists of the cloud manager, feature extraction and detection servers, classification server and data centre. The cloud manager performs multiple important tasks- it authenticates the details of the received data (such as whether the user is registered with a smart healthcare provider), controls the flow of data to and from the different servers, manages communication and data storage, and verifies the identity of all stakeholders in the system (patients, healthcare professionals, hospital representatives, etc.). The data is then sent to the feature extraction and

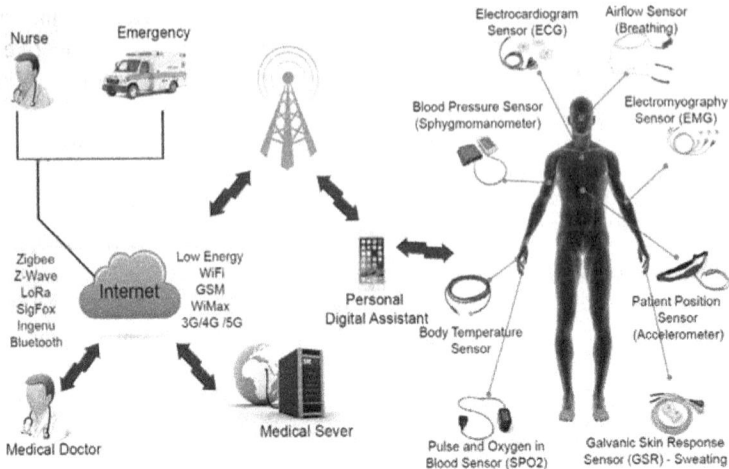

Fig. 2. A general architecture of IoMT-based systems (reprinted with permission from *Rodrigues, Joel JPC, Dante Borges De Rezende Segundo, Heres Arantes Junqueira, Murilo Henrique Sabino, Rafael Maciel Prince, Jalal Al-Muhtadi, and Victor Hugo C De Albuquerque, Enabling technologies for the internet of health things (IEEE Access 2018), 13129–41).*

detection servers to extract and detect features, and to classify them, if required. All the data, including the features and classification results, get stored in the data centre. On the other end of the cloud are the healthcare service providers who can analyze the patients' health records including raw and processed data to make necessary informed decisions (Alhussein et al. 2018, Yuehong et al. 2016). All the stakeholders of IoMT settings such as patients and healthcare professionals are authorized and managed by radio-frequency identification (RFID) tags. Thus, IoMT-based systems are enabled by various technologies, namely, communication technologies such as radio-frequency identification (RFID), cellular networks and Wi-Fi, location technology such as global positioning system (GPS), identification technology such as a unique identifier (UID), and architecture technology such as service-oriented architecture (SOA) (Fan et al. 2014, Lu and Liu 2011). Although most of the cloud-based services explained here are not fully functional yet, these hold great promise in delivering better healthcare services when fully implemented.

4. Applications of IoMT

IoMT is still in its infancy and is being used primarily for remote monitoring of patients with chronic diseases or providing rehabilitation or ambient assisted living environments by using wearable devices and mobile health (mHealth) technologies (Miller et al. 2015, Armstrong et al. 2017). All kinds of IoMT-based activities can be grouped into four basic modes: first is 'Store and forward' mode, where the data such as video, audio, images, etc., can be collected and stored at one location and later transmitted to another location depending on the need. The second one is 'Real-time', where clinicians and patients interact with each other via telecommunication services in real-time. The third one is 'remote patient monitoring', where the patients provide the monitored data from a remote location such as their homes, and the doctor diagnoses the data and sends treatment recommendations from the other end. The fourth one is 'Remote training', where care and support are provided to the patients over the internet through mobile health applications, especially the elderly or those with chronic conditions (Nazir et al. 2019).

4.1 Smart Healthcare Monitoring Systems

Smart healthcare monitoring systems can be adopted by clinicians, homecare, and hospitals to remotely monitor the vital signs of the patients. Remote monitoring can avoid patients being bed-ridden in severe cases and can allow them to move around the bedside monitors. Remote monitoring can also be of great help in cases where the patients have undergone surgery and need their conditions checked regularly but are unable to visit hospitals frequently. Several technologies have been designed for continuous monitoring of vital health parameters such as blood pressure, heart rate, respiration rate, blood oxygen, body temperature, etc., as these parameters are important to be monitored for common as well as chronic diseases (Kumar and Pandey 2018, Divakaran et al. 2017, Singh et al. 2019, Lavanya et al. 2017, Bayo-Monton et al. 2018). Also, it was found that monitoring of multiple parameters was beneficial compared to just one or two parameters, as it could give more accurate and added information to the experts to make informed decisions (Kadhim et al. 2020a, Nayyar et al. 2019). Examples of systems that have some unique features are discussed in the following paragraphs.

In 2017, XPRIZE DeepQ Tricorder biosensor was developed as a portable device, easily operable by users at home and equipped with artificial intelligence to accurately capture five vital signs in real-time and diagnose 12 common diseases (Chang et al. 2017). It comprised of a symptom checking module which would predict the potential diseases based on a series of questions asked to the users about their symptoms, and then recommended appropriate tests to them (Figure 3). It performed continuous monitoring of five important vital parameters—respiration rate, oxygen saturation, heart rate/variability, blood pressure, and body temperature. It also performed detection and diagnosis of sleep apnoea, atrial fibrillation, otitis media,

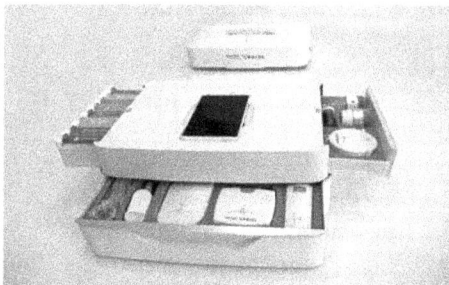

Fig. 3. DeepQ Tricoder consists of four compartments—the HTC mobile phone on the top acts as a data hub and runs the symptom checker, the lower-front drawer contains breath sense and vital sense, the right-hand-side drawer contains optical sense while the left-hand side contains urine/blood sense (reprinted with permission from *Chang, Edward Y, Meng-Hsi Wu, Kai-Fu Tang Tang, Hao-Cheng Kao, and Chun-Nan Chou, Artificial intelligence in xprize deepq tricorder (Proceedings of the 2nd international workshop on multimedia for personal health and health care 2017), 11–18).*

melanoma, anaemia, diabetes, stroke, urinary tract infection, chronic obstructive pulmonary disease, leukocytosis, pneumonia, tuberculosis, and hepatitis A.

An IoT-based monitoring system for pervasive heart diseases was developed to continuously monitor seven vital signs—ECG, heart rate, oxygen saturation, blood pressure, pulse rate, blood fat and blood glucose (blood fat and blood glucose were collected at intervals) and other environmental indicators (such as the location of the patient to assist with ambulance in case of emergencies) (Li et al. 2017). The system had four different data transmission modes to balance the need and demand of healthcare, based on patients' risk levels, which was the key to control the transmission and was a unique feature of this system. The first mode of transmission was the continuous transmission of all parameters to the remote server and displayed the data to the clinicians in real-time. It was the highest level of monitoring and was designed for monitoring patients who were at really high risk for heart diseases. Since this mode would introduce a higher demand for human resources, network quality, and a huge amount of data on the server's side, this mode would be used only for a limited number of patients. The second mode of transmission was continuous transmission during special periods, such as 3 or 4 o'clock in the afternoons or 1 or 2 hours after waking up, when the chances of heart attacks are maximum. So, this mode would transmit the data continuously during a few random periods, as well as during these special periods. The third transmission mode was event-triggered. The gateway storing all the data would compare the newly received, sampled data with the corresponding range of normal values predetermined and stored in it. If the sampled data was beyond its normal range, an event would be triggered and the transmission from the gateway to the server would be activated. A sequence of data, including the pre-and post-five-minute signals, would be sent, which would significantly reduce the amount of data sent over the server. This mode would be suitable for middle-risk patients with heart diseases. The fourth mode was transmission on patient's demand, where the monitored data would get stored on the gateway at the patient's side and would be transmitted only when the patient would feel uncomfortable and requested the clinicians for diagnosis. This was the lowest level of monitoring for low-risk patients. Thus, this system balanced the healthcare need and demand for computing and communication resources efficiently, providing pervasive healthcare service. There are many vital signs monitoring applications available in the market; however, one of the most popular platforms in the market is VitalSync™ Virtual Patient Monitoring platform, developed by Medtronic, USA, which enables remote monitoring by integrating critical patient information from the patients' devices and transmitting it to the hospitals' servers (Medtronic 2020). The software installed on hospitals' servers receives this information, formats it into Health Level 7 (HL7) protocol (latest standard for sharing and storing health data) and writes it into the hospitals' electronic medical records (EMRs).

Remote monitoring has also been applied for patients with chronic gastrointestinal diseases, such as inflammatory bowel disease, liver diseases, etc., using a combination of wearable devices, mobile applications, communication technologies, and EMRs (Yin et al. 2019, Rodrigues et al. 2018). The patient would make his fitness data available through the mobile application which would be linked with his/her EMR. This data would be sent for predictive analytics, which would be then analyzed by the clinicians and the patients would get follow-up calls, if needed (Riaz and Atreja 2016). The interventions required to control these diseases and the risk of symptom flare are greatly influenced by the patient's behaviours such as stress levels, smoking, adherence to medication, depression, etc. Hence, remote monitoring is ideal for patients with gastrointestinal diseases to improve self-care and management.

Posture recognition and body pressure mapping are important to prevent ulcers and bed sores for bed-ridden patients. Hence, posture recognition system was developed as an application to remote monitoring by measuring the body pressure distribution using a pressure sensing mattress, named Tactilus from Sensor Products Inc. (Matar et al. 2016). Raw and binary pressure distribution images were obtained as minimal information from the pressure sensor nodes in the mattress and were sent to the workstation by Bluetooth, where the data was processed using supervised machine learning algorithms. The data was classified into three classes of sleeping postures—supine and prone positions as one position (s/p position), left lateral and right lateral positions and was later sent out to the clinicians for further diagnosis.

Another application was the development of an automatic system to perform the monitoring of knee flexion at homes for patients with total knee arthroplasty (Msayib et al. 2017). The system consisted of master and slave sensor units that would attach to the thigh and shin, above and below the operated region, using Velcro straps. The system would measure and record the angles of knee flexion when the patient would perform the prescribed exercises and would transmit the data to a hospital server using Global System for Mobile Communications (GSM) infrastructure. Hence, this system had the necessary communications technology integrated into the device, which eliminated the need for a mobile phone, computer, or internet access by the patients.

An intelligent non-invasive monitoring system for infants, named xVLEPSIS, has been developed to predict potentially hazardous events related to infants (Pateraki et al. 2020). It comprises of a 'smart' bed mattress, which has pressure sensors embedded in it and records ballistocardiogram (BCG)—sudden blood ejections from the heart into the great vessels. This recording generates a plot representing the repetitive body movements during sleep. The system also has temperature and humidity sensors embedded under the mattress and is equipped with video and audio recording. The notification system to the parents is facilitated by a mobile or smartwatch application, and the recorded data can be sent to hospitals, as part of regular check-ups or in case of emergencies.

Remote patient monitoring has also been applied to check dementia to prevent patients from getting lost (Sposaro et al. 2010), to check for stroke and anticipate falls (Edgar et al. 2010), bipolar disorders to detect depression and mania (Puiatti et al. 2011), asthma patients to check for their adherence to inhalers (Chan et al. 2015), postsurgical monitoring in wearable artificial kidney cases (Rosner and Ronco 2011), monitoring and treatment of obstructive sleep apnoea (Bsoul et al. 2010), and predicting leakage from pressure readings in aneurysm repair cases (Parsa et al. 2010), among many other applications.

4.2 Voice-Driven Technologies for Healthcare Delivery

All voice-driven technologies use voice-activated commands which have become an integral part of all intelligent personal assistance such as Google Now, Cortana, Alexa, Siri, etc. These devices can learn an individual's voice and use it to interact with other applications, for instance, clinicians communicating with electronic medical records directly, or clinicians recording their words on voice-enabled medical transcriptions (Happe et al. 2003). These features can facilitate in creating self-care solutions, for instance, ordering medications, timely drug delivery reminders, getting health care information regularly, and providing communication between clinicians and patients (Basatneh et al. 2018).

Voice-enabled home assistants can be used for better management of diseases without having to visit hospitals regularly. A US-based company named Orbita has designed various voice platforms using chatbot and voicebot solutions to improve interactions between healthcare organizations and patients and reducing clinical burden concerning time and hospital visits of patients. Through conversational artificial intelligence (voice search, chatbots, and voice search engine optimization), patients can search for doctors and services such as appointment scheduling or triaging, sitting at home, through their app called OrbitaENGAGE. They have further developed apps such as OrbitaASSIST, a virtual bedside assistant, to improve COVID-19 response by automating delegation and eliminating unnecessary trips to bedside, streamlining hands-free communication between patients and care workers, thus protecting the health of the front-line staff. It provides an interactive environment for patients to confirm requests for help and combat social isolation, delivers answers to frequently asked questions, provides a zero-touch interface to communicate safely, and offers surveys and assessment tools for feedback and analysis. There is an additional app exclusively for healthcare workers with features such as call bell and electronic health records and is integrated with alerts and escalation to help them support patients effectively (Orbita 2020).

Basatneh et al. have conceptualized a voice-enabled technology for managing diabetic foot ulcers (DFUs) which is similar to the apps developed by the company Orbita (USA). The technology can be designed to instruct the patients or their caregivers on how to manage their ulcers, change their dressings, take care of their feet, and provide guidelines on maintaining dietary restrictions, reminders to take medications, and perform foot checks to prevent further ulcerations (Basatneh et al. 2018). Based on the above idea, they outlined five features to assist in managing DFUs that a voice-enabled technology could have—the first feature could be a 'doctor-patient interface', which would facilitate refiling medications, scheduling appointments, obtaining additional information about care and management of ulcers (e.g., Please ask my doctor for the next appointment). The second feature could be 'caregiver-patient interface', which could facilitate communication between caregivers and patients, such as record a message from the patient and send it to the caregiver via voicemail or text message about the status of the ulcers (e.g., Please tell my son that I have too much pain in my feet today.). The third feature could be 'sending notifications or alerting', which could notify the patients on taking medications in a timely fashion, changing dressing, or improving the lifestyle of the patients to provide better management of ulcers (e.g., Ms. Natalie, make sure you wear your prescribed shoes when you are standing or walking). The fourth feature could be 'self-care', which could help the patients to self-care by letting them ask questions related to their ulcers or health at any point in time (e.g., Which fruits are good for people with diabetes?). The fifth feature could be 'gamification', which could engage and motivate the patients to take better care of themselves (e.g., Congratulations! Your doctor confirmed that your wound is healing fast).

4.3 Wearable and Assistive Medical Devices

Several wearable technologies have proven to be promising in monitoring and diagnosis of several diseases. Several smartwatches or smartphone-based wearable devices have been designed for continuous monitoring of body parameters and can be easily integrated into IoMT settings. A smartwatch based self-management application was designed to record blood glucose, carbohydrate intake, insulin units, and specific physical activities for individuals with diabetes. It helps individuals to review their glucose levels easily and quickly along with other measured parameters. It also provides bidirectional communication between the smartwatch and smartphone to enable users to read and transfer the self-monitored values from the smartwatch to mobile phone (Årsand et al. 2015).

Smart dressings for wounds were shown to monitor their healing and detect risks associated with events such as infection, bacterial growth, the temperature of the wound-bed, pressure of the bandage, wound pH, moisture level, etc. (Farrow et al. 2012, Sharp 2013). A low-power telemetric system was also developed to measure these parameters and transmit the real-time information to a portable receiver from within the wound dressing (Mehmood et al. 2015). This technology could be further

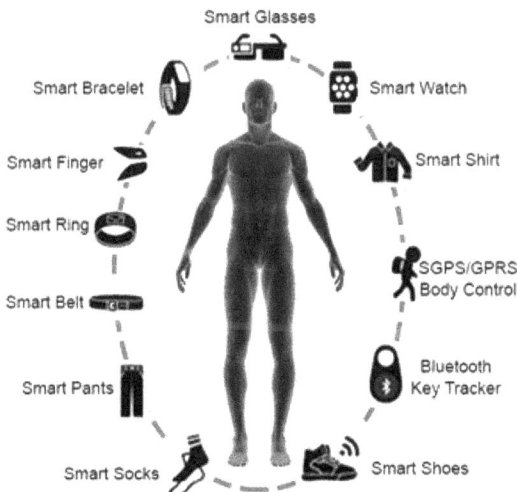

Fig. 4. Illustration of different types of wearable technology (reprinted with permission from *Rodrigues, Joel JPC, Dante Borges De Rezende Segundo, Heres Arantes Junqueira, Murilo Henrique Sabino, Rafael Maciel Prince, Jalal Al-Muhtadi, and Victor Hugo C De Albuquerque, Enabling technologies for the internet of health things (IEEE Access 2018), 13129–41).*

extended to embed temperature or pressure sensors in the insoles of the shoes of patients to monitor the temperature or pressure of feet regularly. The SurroSense Rx smart insole system consists of two pressure sensors and uses a smartwatch as the display device to alert the users when the plantar pressure exceeds the safe threshold so that the user can take appropriate measures such as sit down or adjust the position of their feet to relieve the sustained pressure (Najafi et al. 2018). At each alert, the smartwatch shows a pressure map of each foot with the locations of sustained pressure occurring over a period of time; the threshold to alert the user is determined based on pressure data integrated over this period to identify the tissues at risk. Another application was the development of a sweat-based wearable watch using nanosensors which would monitor biomarkers for chronic health conditions (Munje et al. 2017). Zinc oxide nano porous sensor arrays have been used to validate the room temperature ionic liquids through detection of interleukin-6 (IL-6) biomarker in perspired sweat for up to a week, with intact signal integrity. IL-6 is an important biomarker that can be used to monitor immune response in cancer treatment and to detect high levels of cortisol and acute stressors during psychological stress. There have also been a few successful mobile applications developed for rheumatoid arthritis (RA), such as MyRA and myVectra, to help patients track their RA symptoms and obtain disease activity status while storing all the information on the cloud (Kataria and Ravindran 2018).

In 2014, the Swiss pharma, Novartis collaborated with Google to co-develop smart contact lens and wearable device to measure blood glucose levels non-invasively for diabetic patients and to correct vision in individuals with presbyopia (Senior 2014). This 'smart' lens, currently under development, would comprise of a lens made with conventional hydrogel material, but the parts which would make it 'smart' are a wireless chip, a glucose sensor, and a battery, all in their miniaturized forms, embedded carefully between the two layers of the lens material, to avoid the iris and pupil. The tear fluid from the eye would seep into the sensor through a pinhole in the lens, which would generate the glucose readings. These readings could be then transmitted to a smartphone and the clinicians for monitoring glucose levels at-risk. However, the extent of the correlation of glucose in blood and tear fluid remains unclear. Similarly, Apple has recently developed smartwatches that help in monitoring blood glucose levels (Peckham 2019). A body sensor is supposed to be worn around the abdomen which measures the blood glucose levels every five minutes. It transmits this information to a remote handheld device placed within 20 feet of the user, which communicates with the iPhone to send the information to the Apple Watch to be displayed. The apps have been designed by DexCom- one of them can be installed on the user's device for the data to be viewed by the user, and the second one can be installed on another person's device (a clinician or a caregiver) for monitoring purposes.

One of the smart home technologies in the market to predict foot ulceration using IoMT is a 'smart mat' developed by the US-based company, Podimetrics (Podimetrics 2020). The temperature of patients' feet is monitored remotely on an everyday basis and clinicians are sent alerts when preventive measures are needed. It requires no setup or configuration by the users who just need to step on the mat with both feet for approximately 20 seconds (Figure 5). The customized thermograms

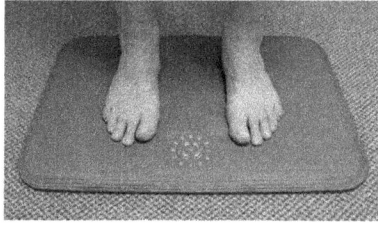

Fig. 5. The wireless thermometric mat developed for monitoring inflammatory foot diseases (reprinted with permission from *Frykberg, Robert G, Ian L Gordon, Alexander M Reyzelman, Shawn M Cazzell, Ryan H Fitzgerald, Gary M Rothenberg, Jonathan D Bloom, Brian J Petersen, David R Linders, and AksoneNouvong, Feasibility and efficacy of a smart mat technology to predict development of diabetic plantar ulcers (Diabetes care 2017), 973–80).*

are analyzed using machine learning algorithms to detect changes in temperature on the ipsilateral foot by using a threshold of greater than 2.22 degrees Celsius on the corresponding locations on the contralateral foot. This temperature rise reflects early signs of inflammation on the ipsilateral foot and hence the mat provides clinical decision support to prevent foot complications (Frykberg et al. 2017). The Podimetrics SmartMat has been able to detect 97% of ulcers on an average of five weeks before they presented themselves clinically (Podimetrics 2020).

Wearable devices have been used for a long time for continuous monitoring of heart rate (Pevnick et al. 2018, Yang et al. 2016) but they have not been very sensitive for the diagnosis of cardiovascular diseases (CVD), as two of the following three conditions should be detected for a CVD—characteristic chest pain, alterations in ECG or elevation in blood biomarker levels (Vashistha et al. 2018). Different mobile-based paradigms have been developed to monitor ECG signals (Machado et al. 2016), but it was also reported that monitoring only ECG could be misleading as half of the patients admitted to emergency departments tend to have normal ECG patterns, which makes CVD diagnosis difficult (Vashistha et al. 2018). Hence, more reliable, sensitive, and cost-effective diagnostic platforms are being developed for the diagnosis of CVD.

For long-term physiological monitoring of chronic heart conditions, the idea of e-textiles or smart textiles has been implemented by integrating nanostructured sensors with traditional fabric materials such as cotton, nylon, and polyester to monitor ECG and blood pressure (Shyamkumar et al. 2014). Textiles can be a preferred platform for integrating sensors as these are natural materials that come in close contact with the skin. Additionally, nanosensors allow for low contact impedance due to higher surface area with a small footprint, compared to plantar sensors, making nanosensors a desirable choice for cardiovascular monitoring. Hence, the authors developed a multichannel wearable wireless system in the form of an inner vest (named as 'E-bro' for men and 'E-bra' for women) which incorporated nanosensors, such as nanostructured textile electrodes with composite piezoelectric films or gold nanowire electrodes (Figure 6). The system consisted of a compression inner vest with the sensors and printed connection traces which connected the sensors to a sensor electronics module (SEM). The SEM consisted of filter circuits, an amplifier, a microcontroller, and a Zigbee wireless radio. The vest also had a near-infrared emitter-detection system for plethysmography as a removable armband on the vest. The data was received on a PC, and the installed software received and plotted the incoming data. The electrodes

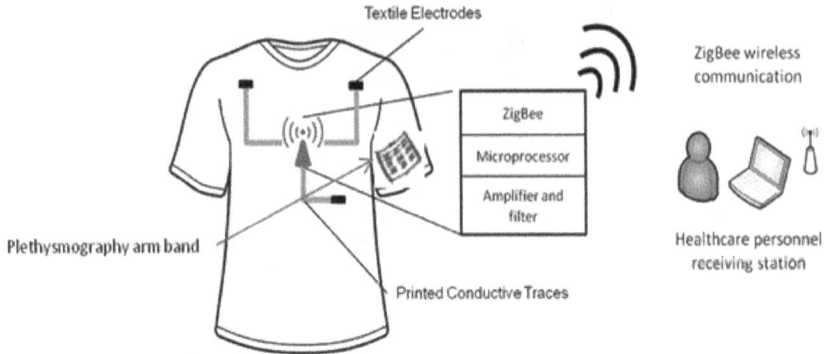

Fig. 6. The ECG acquisition system using smart textile named e-bro (reprinted with permission from *Shyamkumar, Prashanth, Pratyush Rai, Sechang Oh, MouliRamasamy, Robert E Harbaugh, and Vijay Varadan, Wearable wireless cardiovascular monitoring using textile-based nanosensor and nanomaterial systems (Electronics 2014), 504–20).*

and the armband were connected to the SEM through conductive traces, made using conductive inks which were formulated using silver nanoparticle fillers.

A self-adhesive cardiac patch, named 'ZIO Patch', was developed by iRhythm Technologies, USA to monitor the patients with transient ischemic attack or stroke for atrial fibrillation (Tung et al. 2015). It was a single-use, water-resistant, 14-day use patch placed over the patient's left pectoral region, for continuous cardiac rhythm monitoring. The patient could press a button on the patch to mark a symptomatic episode, and at the end of the monitoring period, the patient would mail the device and diary to a data processing centre where the data would be analyzed and diagnosed by clinicians to generate the corresponding clinical report.

An epileptic detection and monitoring system was proposed by Alhussein et al., where scalp EEG would be recorded by smart EEG sensors. Simultaneously, other physiological and psychological signals would also be recorded using other smart sensors, such as a patient's gestures, movements, and facial expressions to understand and determine their state of being. Once recorded, all these signals would be transmitted to the cloud where the cognitive system would continuously calculate the patient's state and make the decision to send the data to the signal processing and deep learning modules for seizure detection if it believed the data to contain seizures. The deep learning module would classify the data as seizure or non-seizure with a probability score, which could then be transmitted to clinicians to make informed decisions to assist the patient, in case of emergencies. The main challenges with EEG data are low signal-to-noise ratio because it is sensitive to external and internal noise, and artifacts introduced due to muscle movements and eye blinking. Moreover, the seizure patterns also differ from patient to patient and a generic automated model would not function accurately for seizure detection. Hence, the authors developed an efficient deep learning model that overcame the above challenges by achieving promising accuracy on patient-specific features and integrated it with the smart epileptic seizure detection and monitoring system (Alhussein et al. 2018).

Assistive care in various medical areas is also being given by intelligent robots equipped with cameras, sensors, microphones, etc. In the UK, a robot named Pepper is under development for providing help to the elderly which would decipher voice tones and expression and determine their state of well-being. A personalized health robot on the market has been developed by Pillo Health. This can answer health-related queries, store, dispense, and manage medications with a pharmacy. It synchronizes with wearables and smart home devices and has the voice and facial recognition systems installed in it (Kataria and Ravindran 2018). A robot masseuse has also been deployed in Singapore which uses a robotic arm with silicon tips that mimic human palm and thumb. The robot has sensors and diagnostics to measure the stiffness of the patient's muscles and it provides a session almost indistinguishable from a professional masseuse (Kataria and Ravindran 2018). Surgical robots and surgical prosthetics supported with artificial intelligence are becoming a common reality and will become more prevalent soon (Kadhim et al. 2020b).

Pateraki et al. have proposed the 'SMART BEAR' project, which aims to provide personalized digital health solutions to elderly people with five health-related conditions—hearing loss, cognitive impairments, cardiovascular diseases, balance disorders, and mental health issues. The platform would integrate various off-the-shelf smart sensors and assistive devices, which would collect and analyze data from the activities and modules of elderly people to design personalized interventions and promote independent and healthy living (Pateraki et al. 2020). Some of the other relevant technologies have been briefly described in Table 1 (Rodrigues et al. 2018).

Table 1. Some relevant IoMT technologies providing services as ambient assisted living/healthcare solutions using smartphones/remote healthcare monitoring/wearable systems.

Company	Product	Service	Description
Assisted Living Technologies Inc.	BeClose Remote Monitoring System	Ambient assisted living system	It provides independence and a sense of comfort to both the caregivers and the individuals who need their support.
Fade	Fade: Fall Detector app	Ambient assisted living system/Healthcare solutions using smartphones	It is an Android-based application for mobile phones and can detect falls and send alarm messages.
Mcare	Mcare app	Healthcare solutions using smartphones	It is a bracelet to alert parents when the child moves away.
Safe Heart	iOximeter app	Healthcare solutions using smartphones	It is an oximeter for smartphones.
EarlySense	EarlySense All-in-One	Remote healthcare monitoring system	It provides continuous monitoring of heart and respiratory rate, early detection of patient deterioration, fall prevention and pressure ulcer prevention.
NovaSom	AccuSom	Remote healthcare monitoring system	It is used to monitor sleeping.
Apple	Apple Watch	Wearable system	It allows the developer community to build new applications for healthcare.
Bittium	Enterprise	Wearable system	It provides customized and secure IoT solutions for healthcare, industrial and wearable sports device manufacturers.

Source: Data from *Rodrigues, Joel JPC, Dante Borges De Rezende Segundo, Heres Arantes Junqueira, Murilo Henrique Sabino, Rafael Maciel Prince, Jalal Al-Muhtadi, and Victor Hugo C De Albuquerque, "Enabling technologies for the internet of health things", IEEE Access 6(2018): 13129–41.*

4.4 IoT-based Rehabilitation Systems

Mobile health (mHealth) technology can also play a significant role in rehabilitation as it can provide tools to support and monitor home exercise programmes, measure vital signs, provide feedback to patients, provide necessary educational information to the patients and facilitate consultation with therapists and clinicians (Dicianno et al. 2015, Odendaal et al. 2020). Since mobile applications provide a two-way interface, it can also help therapists to provide the necessary support to their patients and provide intervention at the right time. The applications can collect information such as pain being experienced by the patients, updates on the exercise regimens being adhered to by the patients, etc. (Dicianno et al. 2015). A mobile health application named SkinCare was developed to support self-care and management of skin problems for individuals who are vulnerable to chronic skin complications, especially ones with spina bifida (Parmanto et al. 2015). The application facilitated self-care, scheduled task and reminders, secure patient-clinician messaging interface, monitor skin conditions and communicate, for instance, consultation with a clinician, information about skin problems, etc.

The wearable devices such as Fitbits, Surge, Blazer, etc., have allowed users to obtain objective measures about their daily activities and get detailed feedback on the performance of their physical activities (Martin et al. 2015, Gaur et al. 2019). These wrist-based wearable devices use an optical heart rate sensor and a three-dimensional accelerometer to monitor heart rate, sleep, physical activities, sedentary time, etc. These devices not only self-monitor the health parameters but also enable remote monitoring and personalized interventions by supporting the exchange of real-time information between users and healthcare providers (Dimitrov 2016, Haghi et al. 2017). Thus, a simple yet innovative mHealth platform named iCardia was developed based on open-service architecture, using Fitbit wearable devices, smartphones, and personalized text messages, to support near real-time monitoring and coaching of patients undergoing cardiac rehabilitation (Kitsiou et al. 2017). Exercise and physical activity are core components of cardiac rehabilitation and must be monitored regularly, just like other risk factors. Since Fitbit detects and

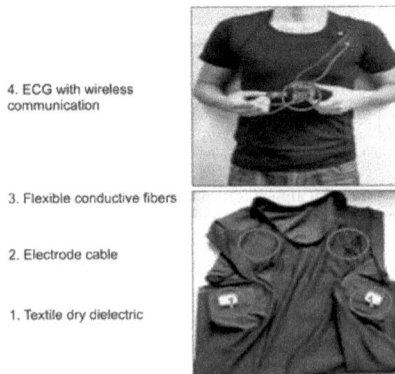

4. ECG with wireless communication

3. Flexible conductive fibers

2. Electrode cable

1. Textile dry dielectric

Fig. 7. Washable smart clothing system for ECG monitoring (reprinted with permission from *Haghi, Mostafa, Kerstin Thurow, and Regina Stoll, Wearable devices in medical internet of things: scientific research and commercially available devices (Healthcare informatics research 2017), 4–15).*

continuously records high movement activities and exercises through "SmartTrack", iCardia would collect this data from the Fitbit cloud server and would display it in informative ways using visualization tools and data analytics to support remote monitoring of these patients. Mobile phone text messaging has been used as an intervention tool to deliver educational and self-management components to the patients by rehabilitation specialists, based on behavioural theory models. Another interesting medical rehabilitation application of an existing device was 'The Myo', which was originally a motion controller for games but started being used in orthopaedics, for patients to monitor their progress on exercise and for clinicians to measure the angle of movement (Dimitrov 2016).

4.5 Other Applications

Pharma IoT is another emerging field in IoMT which aims to digitize medical outcomes during drug development and delivery, and related care using smart medical devices to create new treatment possibilities. One of the solutions on patient care using Pharma IoT has been to provide medication management to patients with multiple sclerosis and Parkinson's disease using connected wearable sensors, which were found to improve patient outcomes and quality of life (Van Uem et al. 2016). Another interesting application was to add existing medical device products such as insulin pens and inhalers to the sensors and connectivity technologies using smartphones, to monitor asthma symptoms and stress, to enhance care analytics and personalized therapy (Dzubur et al. 2015). Apart from development, the management of medicines can also be facilitated by patients using IoT systems. The essential medicines can be identified, and prediction, quantification, storage, and distribution of medicines can be managed by the IoT systems (Kadhim et al. 2020b, Murfin 2013, Mickan et al. 2013). There are also various mobile drug reference applications such as Skyscape RxDrugs/Omnio, Epocrates, Micromedex, etc., which are used to access drug-related information by clinicians and patients, for instance, drug names, dosages, indications and contraindications, cost, pharmacology, and identification guides (Aungst 2013, Mosa et al. 2012).

One of the main applications of IoMT is the adoption of electronic health records (EHRs) or electronic medical records (EMRs) which have proven to be advantageous over paper records in many ways (Dimitrov 2016, Hong et al. 2017). EHRs are digitized and keep all information in one place, are easy to share with other healthcare providers, and superseded handwritten records, which tend to be lost at times by patients or clinicians. Generally, the working of EHRs is quite basic-the individual has an identity card linked to a secure cloud network that stores the electronic record of the health, which includes lab results, vital signs of the patient, all prescriptions, and medical history. Clinicians can scan this card to get access to all the stored information on their tablets, smartphones, or laptops (Riaz and Atreja 2016). There are also mobile applications available such as PatientKeeper (Mosa et al. 2012) or TeamViewer (Yoo 2013), which allow clinicians to enter and access patients' information from the EHRs or EMRs. Specialized apps such as Mobile MIM (free app for iPads and iPhones) are also available for remote viewing of x-rays and other imaging scans when the clinicians do not have access to their imaging workstations (O'Neill et al. 2013, Ozdalga et al. 2012).

From healthcare professionals' point of view, smart devices and applications are being used mainly for administration, reference and information gathering, health record access and management, medical education, consulting, and communications (Ventola 2014). Mobile applications and cloud-based systems have made their lives easier as information and time management can be easily done by them anywhere and anytime, based on their need and convenience. Popular applications such as Evernote and Notability enable users to gather information by dictating or writing notes, saving photographs, recording audio, etc., which prove very advantageous for healthcare professionals (O'Neill et al. 2013). Health record access and management are done by various apps that support access to EHRs or EMRs, such as PatientKeeper, TeamViewer, Mobile MIM, etc. Communication and consulting with different people involved in healthcare or with patients is a very important part of their profession, which is facilitated through various apps enabled with voice and video calling, multimedia messaging, etc. There have also been "non-disclosed" professional networking applications, exclusively for healthcare professionals, such as Doximity, which allow them to network with each other and share patient-related information (Yoo 2013). There are also evidence-based mobile apps such as John Hopkins Antibiotic Guide (JHABx), UpToDate, eProcates ID, Dynamed, etc., which provide important information on diseases, diagnosis, treatment, and other related topics, and help the clinicians to learn and provide better clinical decisions (Mosa et al. 2012, Aungst 2013). Clinical monitoring systems and applications such as iWander, HanDBase, and many other such applications discussed in detail in the above sections have helped clinicians to perform effective diagnosis and treatment, even from remote areas (Ozdalga et al. 2012).

5. Future trends in IoMT

Ingestible sensor technology has been implemented in the form of smart pills, where the pills are miniature capsule-like devices that are taken through the mouth and get digested like normal medications (Kiourti and Nikita 2017, Chai et al. 2016). They consist of components found in food and get activated only upon ingestion. A micro-biosensor was developed by Proteus Digital Health, USA for evaluating medical ingestion. The sensor would be operated only when its outer layer came in contact with the gastric fluid in the stomach, which would enable the layer to act as a battery, collecting the current for powering the device. The active ingestible sensor would then communicate with the wearable receiver worn by the patient and would transmit a unique code, which would be identified by a mobile application to confirm the ingested medication (Aldeer et al. 2018). Medical adherence of the body can also be monitored along with its vital signs using ingestible sensors (Hafezi et al. 2014, Dua et al. 2017, DiCarlo et al. 2012). However, ingestible biosensors are still in their infancy and are undergoing active research, especially after the introduction of *in vivo* communications technology (Shubair and Elayan 2015, Demir et al. 2016). Other technologies in infancy, which hold great potential in revolutionizing healthcare, are nanotechnology-based fields such as nano-based implants and nanobionics.

6. Advantages of IoMT

The use of IoT with cloud computing has provided an opportunity to obtain real-time, high-quality, and low-cost technologies, and made patient-centric smart healthcare possible. Smart healthcare can help in reducing the cost of healthcare, providing easily accessible, highly efficient, and reduced hospital visits in non-emergency cases, thus enhancing the quality of life. It can be highly beneficial in rural/isolated regions, where the hospital services are too far away to be quickly accessible by patients (Yogaraj et al. 2017). Remote patient monitoring and telemedicine have been found to be effective in reducing the mortality rate and emergency hospital visits by significantly improving the ways of caring compared to the conventional techniques (Nakamura et al. 2014, McLean et al. 2011). The use of mobile devices and related applications have also been shown to have a positive effect on patient care outcomes and have enhanced the quality of life of the patients (Divall et al. 2013, Mickan et al. 2013).

7. Challenges in IoMT

IoT devices generally have low processing power capability and limited storage. When incorporated with cloud computing, large storage and better processing capabilities are possible. However, with these advantages, there also comes big challenges and limitations in the use and implementation of IoMT systems. These systems lead to big data being collected from homes, communities, and hospitals and are being updated and processed in real-time. For most of the cases, the cost of collection and storage of this huge amount of data is high. Hence, to remove redundant data, the IoT community is facing the challenge of implementing edge computing (Ray et al. 2019) and developing highly efficient brain-powered cognitive IoT systems which would possess a high level of intelligence and could make its own decisions (Alhussein et al. 2018, Chen et al. 2018). This means that the smart sensors connected to the patient's body should be able to not only sense and monitor the signals but also deduce the state of the patient by making intelligent decisions and deciding the future course of activities by involving other stakeholders of the smart healthcare system (i.e., primary and secondary caregivers, clinicians, and other healthcare workers).

Another challenge with big data is the mining of important information from the generated data. Effective data mining algorithms are required to derive a knowledge base from big data which would act as a supplement for doctor's experience. Due to the specialty and complexity of healthcare data, effective mining methods are still absent in this field (Yuehong et al. 2016). For instance, even with similar symptoms, the conditions of patients vary from one to another, which makes self-learning or data mining tasks extremely difficult for healthcare data. Advanced self-learning models such as deep learning and genetic algorithms are being developed to provide robust solutions to healthcare data.

There are also challenges to the hardware, in the case of wearable devices. An example would be the integration of multiple sensors into one, where not all sensors are of the optimal miniaturized form. This would be one of the main concerns for individuals using these devices for long periods, because these sensors may not be comfortable and unobtrusive in their current form. The use of lighter materials such as carbon fibre or fabric can prove to be promising in this regard (Yuehong et al. 2016). An added challenge is supplying power to the sensor devices, as most of these are based on rechargeable batteries. Research is being actively done to develop low energy consumption sensors or sensors that would use solar power or other sustainable energy sources for power supply. The diversity of devices being used for a specific application is another challenge, as many device manufacturers do not have an agreed set of communication standards and protocols, which would make the communication difficult between multiple integrated devices (Rghioui and Oumnad 2018).

The privacy and security of users' data should be of utmost concern for all IoMT systems. Since the data is transmitted to the internet through wireless communications, stored on cloud, and accessible by all stakeholders, there are wide possibilities of the data being collected inappropriately or being misused. Moreover, most of the devices being used are low energy and have poor computing capabilities, hence they cannot ensure the security of the data (Kadhim et al. 2020b). In 2015, Intel, Oregon Health and Science University launched a high-performance analytics platform called 'Collaborative Cancer Cloud' to collect and store private health records securely which can be used for cancer research and intended to extend this cloud network to other diseases too (Dimitrov 2016). Thus, dedicated research is needed in security and privacy management to prevent unauthorized identification, tracking, and misuse of patients' data and privacy (Nanayakkara et al. 2019).

Another important factor to note here is providing proper and mandatory training to patients, their caregivers, and all healthcare professionals on how to use and manage the IoT devices and services correctly, to get the most benefit out of these. Moreover, the added concern is that clinicians lack the means to assess whether the patients are following the prescribed treatment and medications. This lack of adherence might increase the risk of hospitalization among patients (Rghioui and Oumnad 2018). Therefore, considerable research efforts need to be exerted to establish a successful smart healthcare system.

8. Conclusion

As the global population is increasing at a tremendous rate, several diseases and disorders are also becoming prevalent, which calls for the development of better preventive, diagnostic, and therapeutic measurements. Nanotechnology has revolutionized different industries; however, it has been slow in its application to healthcare. The progress in bringing nanomedicine to clinics for various potential applications has been slower than expected. The majority of the well-proven research has been done in targeting tumour tissues with nanoparticles, and most of the other applications are still in theory or undergoing intensive research to be implemented in clinical settings.

As IoT technology continues to expand and develop, an increased potential can be seen in facilitating effective and timely management and diagnosis of chronic conditions remotely from home, reducing the time and expense of visiting the clinics and hospitals. The use of digital health assistants, for instance, can reduce the cost and burden of care and can be enabled to meet the specific needs of the patients to provide personalized care. Hence, smart healthcare is getting great attention from private companies, government organizations, medical centres, and researchers because of its benefits in economic and social settings. However, multiple challenges need to be addressed by the research community to achieve secure and effective IoMT tools. Also, many technologies reported in literature should be validated in larger clinical settings, to be effective in healthcare. Thus, nano-biosensors along with artificial intelligence and bioengineering principles can open new avenues for IoMT-based tools to assist in providing better healthcare facilities and enhancing the comfort and quality of life.

Acknowledgements

I would like to sincerely thank my mentors, Mr. Enn Vinnal and Dr. E. R. Rajkumar, for their constant support and guidance in writing this chapter.

References

Albanese, A., P. S. Tang and W. C. W. Chan. (2012). The effect of nanoparticle size, shape, and surface chemistry on biological systems. *Annual Review of Biomedical Engineering* 14: 1–16.

Aldeer, M., M. Javanmard and R. P. Martin. (2018). A review of medication adherence monitoring technologies. *Applied System Innovation* 1: 14.

Alhussein, M., G. Muhammad, M. S. Hossain and S. U. Amin. (2018). Cognitive IoT-cloud integration for smart healthcare: Case study for epileptic seizure detection and monitoring. *Mobile Networks and Applications* 23: 1624–35.

Armstrong, D. G., B. Najafi and M. Shahinpoor. (2017). Potential applications of smart multifunctional wearable materials to gerontology. *Gerontology* 63: 287–98.

Årsand, E., M. Muzny, M. Bradway, J. Muzik and G. Hartvigsen. (2015). Performance of the first combined smartwatch and smartphone diabetes diary application study. *Journal of Diabetes Science and Technology* 9: 556–63.

Aungst, T. D. (2013). Medical applications for pharmacists using mobile devices. *Annals of Pharmacotherapy* 47: 1088–95.

Basatneh, R., B. Najafi and D. G. Armstrong. (2018). Health sensors, smart home devices, and the internet of medical things: An opportunity for dramatic improvement in care for the lower extremity complications of diabetes. *Journal of Diabetes Science and Technology* 12: 577–86.

Bawa, R., G. F. Audette and I. Rubinstein. (2016). Handbook of clinical nanomedicine: Nanoparticles, imaging, therapy, and clinical applications (CRC Press).

Bayo-Monton, J., A. Martinez-Millana, W. Han, C. Fernandez-Llatas, Y. Sun and V. Traver. (2018). Wearable sensors integrated with Internet of Things for advancing eHealth care. *Sensors* 18: 1851.

Bsoul, M., H. Minn and L. Tamil. (2010). Apnea MedAssist: Real-time sleep apnea monitor using single-lead ECG. *IEEE Transactions on Information Technology in Biomedicine* 15: 416–27.

Chai, P. R., R. K. Rosen and E. W. Boyer. (2016). Ingestible biosensors for real-time medical adherence monitoring: MyTMed. In 2016 49th Hawaii International Conference on System Sciences (HICSS), 3416–23. IEEE.

Chan, A. H. Y., J. Harrison, P. N. Black, E. A. Mitchell and J. M. Foster. (2015). Using electronic monitoring devices to measure inhaler adherence: A practical guide for clinicians. *The Journal of Allergy and Clinical Immunology: In Practice* 3: 335–49. e5.

Chang, E. Y., M. Wu, K. T. Tang, H. Kao and C. Chou. (2017). Artificial intelligence in xprize deepq tricorder. In Proceedings of the 2nd International Workshop on Multimedia for Personal Health and Health Care, 11–18.

Chen, M., F. Herrera and K. Hwang. (2018). Cognitive computing: architecture, technologies and intelligent applications. *IEEE Access* 6: 19774–83.

Demir, A. F., Z. E. Ankarali, Q. H. Abbasi, Y. Liu, K. Qaraqe, E. Serpedin et al. (2016). *In vivo* communications: Steps toward the next generation of implantable devices. *IEEE Vehicular Technology Magazine* 11: 32–42.

DiCarlo, L., G. Moon, A. Intondi, R. Duck, J. Frank, H. Hafazi et al. (2012). A digital health solution for using and managing medications: Wirelessly observed therapy. *IEEE Pulse* 3: 23–26.

Dicianno, B. E., B. Parmanto, A. D. Fairman, T. M. Crytzer, D. X. Yu, G. Pramana et al. (2015). Perspectives on the evolution of mobile (mHealth) technologies and application to rehabilitation. *Physical Therapy* 95: 397–405.

Dimitrov, D. V. (2016). Medical internet of things and big data in healthcare. *Healthcare Informatics Research* 22: 156–63.

Divakaran, S., L. Manukonda, N. Sravya, M. M. Morais and P. Janani. (2017). IOT clinic-Internet based patient monitoring and diagnosis system. In 2017 IEEE international conference on power, control, signals and instrumentation engineering (ICPCSI). *IEEE*, 2858–62.

Divall, P., J. Camosso-Stefinovic and R. Baker. (2013). The use of personal digital assistants in clinical decision making by health care professionals: A systematic review. *Health Informatics Journal* 19: 16–28.

Dorj, U., M. Lee, J. Choi, Y. Lee and G. Jeong. (2017). The intelligent healthcare data management system using nanosensors. *Journal of Sensors* 2017.

Dua, A., W. A. Weeks, A. Berstein, R. G. Azevedo, R. Li and A. Ward. (2017). An *in-vivo* communication system for monitoring medication adherence. In 2017 IEEE Wireless Communications and Networking Conference (WCNC). *IEEE*, 1–6.

Dzubur, E., M. Li, K. Kawabata, Y. Sun, R. McConnell, S. Intille et al. (2015). Design of a smartphone application to monitor stress, asthma symptoms, and asthma inhaler use. *Annals of Allergy, Asthma & Immunology* 114: 341–42. e2.

Edgar, S. R., T. Swyka, G. Fulk and E. S. Sazonov. (2010). Wearable shoe-based device for rehabilitation of stroke patients. In 2010 Annual International Conference of the IEEE Engineering in Medicine and Biology. *IEEE* 3772–75.

Fan, Y. J., Y. H. Yin, L. D. Xu, Y. Zeng and F. Wu. (2014). IoT-based smart rehabilitation system. *IEEE Transactions on Industrial Informatics* 10: 1568–77.

Farrow, M. J., I. S. Hunter and P. Connolly. (2012). Developing a real time sensing system to monitor bacteria in wound dressings. *Biosensors* 2: 171–88.

Frykberg, R. G., I. L. Gordon, A. M. Reyzelman, S. M. Cazzell, R. H. Fitzgerald, G. M. Rothenberg et al. (2017). Feasibility and efficacy of a smart mat technology to predict development of diabetic plantar ulcers. *Diabetes care* 40: 973–80.

Gaikwad, P. S. and R. Banerjee. (2018). Advances in point-of-care diagnostic devices in cancers. *Analyst* 143: 1326–48.

Gaur, B., V. K. Shukla and A. Verma. (2019). Strengthening people analytics through wearable IOT device for real-time data collection. In 2019 international conference on automation, computational and technology management (ICACTM). *IEEE*, 555–60.

Hafezi, H., T. L. Robertson, G. D. Moon, K. Au-Yeung, M. J. Zdeblick and G. M. Savage. (2014). An ingestible sensor for measuring medication adherence. *IEEE Transactions on Biomedical Engineering* 62: 99–109.

Haghi, M., K. Thurow and R. Stoll. (2017). Wearable devices in medical internet of things: Scientific research and commercially available devices. *Healthcare Informatics Research* 23: 4–15.

Hamidi, H. and K. Fazeli. (2018). Using Internet of Things and biosensors technology for health applications. *IET Wireless Sensor Systems* 8: 260–67.

Happe, A., B. Pouliquen, A. Burgun, M. Cuggia and P. L. Beux. (2003). Automatic concept extraction from spoken medical reports. *International Journal of Medical Informatics* 70: 255–63.

Hong, J., P. Morris and J. Seo. (2017). Interconnected personal health record ecosystem using IoT cloud platform and HL7 FHIR. In 2017 IEEE International Conference on Healthcare Informatics (ICHI). *IEEE*, 362–67.

Hwang, G., M. Byun, C. K. Jeong and K. J. Lee. (2015). Flexible piezoelectric thin-film energy harvesters and nanosensors for biomedical applications. *Advanced Healthcare Materials* 4: 646–58.

Kadhim, K. T., A. M. Alsahlany, S. M. Wadi and H. T. Kadhum. (2020a). Monitoring vital signs of human hear based on IOT. *Al-Furat Journal of Innovations in Electronics and Computer Engineering* 1: 9–13.

Kadhim, K. T., A. M. Alsahlany, S. M. Wadi and H. T. Kadhum. (2020b). An overview of patient's health status monitoring system based on Internet of Things (IoT). *Wireless Personal Communications* 1–28.

Kataria, S. and V. Ravindran. (2018). Digital health: A new dimension in rheumatology patient care. *Rheumatology International* 38: 1949–57.

Kiourti, A. and K. S. Nikita. (2017). A review of in-body biotelemetry devices: Implantables, ingestibles, and injectables. *IEEE Transactions on Biomedical Engineering* 64: 1422–30.

Kitsiou, S., M. Thomas, G. E. Marai, N. Maglaveras, G. Kondos, R. Arena et al. (2017). Development of an innovative mHealth platform for remote physical activity monitoring and health coaching of cardiac rehabilitation patients. In 2017 IEEE EMBS International Conference on Biomedical & Health Informatics (BHI). *IEEE*, 133–36.

Kumar, S. and P. Pandey. (2018). A smart healthcare monitoring system using smartphone interface. In 2018 4th International Conference on Devices, Circuits and Systems (ICDCS). *IEEE*, 228–31.

Lavanya, S., G. Lavanya and J. Divyabharathi. (2017). Remote prescription and I-Home healthcare based on IoT. In 2017 International Conference on Innovations in Green Energy and Healthcare Technologies (IGEHT). *IEEE*, 1–3.

Lee, K. B., A. Solanki, J. D. Kim and J. Jung. (2009). Nanomedicine: Dynamic integration of nanotechnology with biomedical science. *In*: Nanomedicine: A Systems Engineering Approach (Pan Stanford Publishing Pte. Ltd.).

Li, C., X. Hu and L. Zhang. (2017). The IoT-based heart disease monitoring system for pervasive healthcare service. *Procedia Computer Science* 112: 2328–34.

Lu, D. and T. Liu. (2011). The application of IOT in medical system. In 2011 IEEE International Symposium on IT in Medicine and Education. *IEEE*, 272–75.

Machado, F. M., I. M. Koehler, M. S. Ferreira and M. A. Sovierzoski. (2016). An mHealth remote monitor system approach applied to MCC using ECG signal in an android application. *In*: New Advances in Information Systems and Technologies (Springer).

Martin, S. S., D. I. Feldman, R. S. Blumenthal, S. R. Jones, W. S. Post, R. A. McKibben et al. (2015). 'mActive: A randomized clinical trial of an automated mHealth intervention for physical activity promotion. *Journal of the American Heart Association* 4: e002239.

Matar, G., J. Lina, J. Carrier, A. Riley and G. Kaddoum. (2016). Internet of Things in sleep monitoring: An application for posture recognition using supervised learning. In 2016 IEEE 18th International Conference on e-Health Networking, Applications and Services (Healthcom). *IEEE*, 1–6.

McLean, S., U. Nurmatov, J. L. Y. Liu, C. Pagliari, J. Car and A. Sheikh. (2011). Telehealthcare for chronic obstructive pulmonary disease. Cochrane Database of Systematic Reviews.

Medtronic. (2020). Vital Sync Virtual Patient Monitoring Platform. https://www.medtronic.com/covidien/en-us/products/health-informatics-and-monitoring/vital-sync-virtual-patient-monitoring-platform.html.

Mehmood, N., A. Hariz, S. Templeton and N. H. Voelcker. (2015). A flexible and low power telemetric sensing and monitoring system for chronic wound diagnostics. *Biomedical Engineering Online* 14: 17.

Mickan, S., J. K. Tilson, H. Atherton, N. W. Roberts and C. Heneghan. (2013). Evidence of effectiveness of health care professionals using handheld computers: a scoping review of systematic reviews. *Journal of Medical Internet Research* 15: e212.

Miller, G. and M. Kearnes. (2012). Nanotechnology, ubiquitous computing and the internet of things. *Council of Europe Report.*

Miller, J. D., B. Najafi and D. G. Armstrong. (2015). Current standards and advances in diabetic ulcer prevention and elderly fall prevention using wearable technology. *Current Geriatrics Reports* 4: 249–56.

Moody, B., M. K. Zachek and G. S. McCarty. (2012). Devices and Sensors for Bioelectric Monitoring and Stimulation. *Biomedical Nanosensors*: 273.

Mora, H., D. Gil, R.M. Terol, J. Azorín and J. Szymanski. (2017). An IoT-based computational framework for healthcare monitoring in mobile environments. *Sensors* 17: 2302.

Mosa, A. S. M., I. Yoo and L. Sheets. (2012). A systematic review of healthcare applications for smartphones. *BMC Medical Informatics and Decision Making* 12: 67.

Mostafalu, P., A. S. Nezhad, M. Nikkhah and M. Akbari. (2017). Flexible electronic devices for biomedical applications. *In*: Advanced Mechatronics and MEMS Devices II (Springer).

Msayib, Y., P. Gaydecki, M. Callaghan, N. Dale and S. Ismail. (2017). An intelligent remote monitoring system for total knee arthroplasty patients. *Journal of Medical Systems* 41: 90.

Munje, R. D., S. Muthukumar, B. Jagannath and S. Prasad. (2017). A new paradigm in sweat based wearable diagnostics biosensors using Room Temperature Ionic Liquids (RTILs). *Scientific Reports* 7: 1–12.

Murfin, M. (2013). Know your apps: An evidence-based approach to evaluation of mobile clinical applications. *The Journal of Physician Assistant Education: The Official Journal of the Physician Assistant Education Association* 24: 38–40.

Najafi, B., C. B. Chalifoux, J. B. Everett, J. Razjouyan, E. A. Brooks and D. G. Armstrong. (2018). Cost effectiveness of smart insoles in preventing ulcer recurrence for people in diabetic foot remission. *WCM*, 1: 1–7.

Nakamura, N., T. Koga and H. Iseki. (2014). A meta-analysis of remote patient monitoring for chronic heart failure patients. *Journal of Telemedicine and Telecare* 20: 11–17.

Nanayakkara, N., M. Halgamuge and A. Syed. (2019). Security and Privacy of Internet of Medical Things (IoMT) Based Healthcare Applications: A Review.

Nayyar, A., V. Puri and N. G. Nguyen. (2019). Biosenhealth 1.0: A novel internet of medical things (iomt)-based patient health monitoring system. In International Conference on Innovative Computing and Communications, 155–64. Springer.

Nazir, S., Y. Ali, N. Ullah and I. García-Magariño. (2019). Internet of Things for Healthcare using effects of mobile computing: A systematic literature review. Wireless Communications and Mobile Computing, 2019.

O'Neill, K. M., H. Holmer, S. L. Greenberg and J. G. Meara. (2013). Applying surgical apps: Smartphone and tablet apps prove useful in clinical practice. *Bulletin of the American College of Surgeons* 98: 10–18.

Odendaal, W. A., J. A. Watkins, N. Leon, J. Goudge, F. Griffiths, M. Tomlinson et al. (2020). 'Health workers' perceptions and experiences of using mHealth technologies to deliver primary healthcare services: A qualitative evidence synthesis. Cochrane Database of Systematic Reviews.

Omanović-Mikličanin, E., M. Maksimović and V. Vujović. (2015). The future of healthcare: Nanomedicine and internet of nano things. Folia Medica Facultatis Medicinae Universitatis Saraeviensis, 50.

Orbita. (2020). Conversational AI that Redefines the Patient Journey. https://orbita.ai/.

Ozdalga, E., A. Ozdalga and N. Ahuja. (2012). The smartphone in medicine: A review of current and potential use among physicians and students. *Journal of Medical Internet Research* 14: e128.

Pan, C., Z. Li, W. Guo, J. Zhu and Z. L. Wang. (2011). Fiber-based hybrid nanogenerators for/as self-powered systems in biological liquid. *Angewandte Chemie International Edition* 50: 11192–96.

Parmanto, B., G. Pramana, X. Y. Daihua, A. D. Fairman and B. E. Dicianno. (2015). Development of mHealth system for supporting self-management and remote consultation of skincare. *BMC Medical Informatics and Decision Making*, 15: 1–8.

Parsa, C. J., M. A. Daneshmand, B. Lima, K. Balsara, R. L. McCann and G. C. Hughes. (2010). Utility of remote wireless pressure sensing for endovascular leak detection after endovascular thoracic aneurysm repair. *The Annals of Thoracic Surgery* 89: 446–52.

Pateraki, M., K. Fysarakis, V. Sakkalis, G. Spanoudakis, I. Varlamis, M. Maniadakis et al. (2020). Biosensors and Internet of Things in smart healthcare applications: Challenges and opportunities. *In*: Wearable and Implantable Medical Devices (Elsevier).

Peckham, J. (2019). Your Apple Watch may soon get a better accessory to monitor your glucose levels. In: TechRadar.

Peng, H. and D. T. Chiu. (2015). Soft fluorescent nanomaterials for biological and biomedical imaging. *Chemical Society Reviews* 44: 4699–722.

Pevnick, J. M., K. Birkeland, R. Zimmer, Y. Elad and I. Kedan. (2018). Wearable technology for cardiology: An update and framework for the future. *Trends in Cardiovascular Medicine* 28: 144–50.

Podimetrics. (2020). One Step Closer to Healthier Feet. https://podimetrics.com/.

Puiatti, A., S. Mudda, S. Giordano and O. Mayora. (2011). Smartphone-centred wearable sensors network for monitoring patients with bipolar disorder. In 2011 Annual International Conference of the IEEE Engineering in Medicine and Biology Society. *IEEE*, 3644–47.

Quesada-González, D. and A. Merkoçi. (2018). Nanomaterial-based devices for point-of-care diagnostic applications. *Chemical Society Reviews* 47: 4697–709.

Ray, P. P., D. Dash and D. De. (2019). Edge computing for Internet of Things: A survey, e-healthcare case study and future direction. *Journal of Network and Computer Applications* 140: 1–22.

Rghioui, A. and A. Oumnad. (2018). Challenges and opportunities of internet of things in healthcare. *International Journal of Electrical & Computer Engineering* (2088–8708), 8.

Riaz, M. S. and A. Atreja. (2016). Personalized technologies in chronic gastrointestinal disorders: Self-monitoring and remote sensor technologies. *Clinical Gastroenterology and Hepatology* 14: 1697–705.

Rodrigues, J. J. P. C., D. B. D.R. Segundo, H. A. Junqueira, M. H. Sabino, R. M. Prince, J. Al-Muhtadi et al. (2018). Enabling technologies for the internet of health things. *IEEE Access* 6: 13129–41.

Rosner, M. H. and C. Ronco. (2011). Remote monitoring for the wearable artificial kidney. *In*: Hemodialysis (Karger Publishers).

Senior, M. (2014). Novartis signs up for Google smart lens. *In*: Nature Publishing Group.

Sharp, D. (2013). Printed composite electrodes for *in-situ* wound pH monitoring. *Biosensors and Bioelectronics* 50: 399–405.

Shubair, R. M. and H. Elayan. (2015). *In vivo* wireless body communications: State-of-the-art and future directions. In 2015 Loughborough Antennas & Propagation Conference (LAPC). *IEEE*, 1–5.

Shyamkumar, P., P. Rai, S. Oh, M. Ramasamy, R. E. Harbaugh and V. Varadan. (2014). Wearable wireless cardiovascular monitoring using textile-based nanosensor and nanomaterial systems. *Electronics* 3: 504–20.

Singh, B., S. Urooj, S. Mishra and S. Haldar. (2019). Blood pressure monitoring system using wireless technologies. *Procedia Computer Science* 152: 267–73.

Sposaro, F., J. Danielson and G. Tyson. (2010). iWander: An Android application for dementia patients. In 2010 Annual International Conference of the IEEE Engineering in Medicine and Biology. *IEEE*, 3875–78.

Tung, C. E., D. Su, M. P. Turakhia and M. G. Lansberg. (2015). Diagnostic yield of extended cardiac patch monitoring in patients with stroke or TIA. *Frontiers in Neurology* 5: 266.

Van U., J. MT, K. S. Maier, S. Hucker, O. Scheck, M. A. Hobert et al. (2016). Twelve-week sensor assessment in Parkinson's disease: Impact on quality of life. *Movement Disorders* 31: 1337–38.

Vashistha, R., A. K. Dangi, A. Kumar, D. Chhabra and P. Shukla. (2018). Futuristic biosensors for cardiac health care: An artificial intelligence approach. 3 Biotech, 8: 358.

Ventola, C. L. (2014). Mobile devices and apps for health care professionals: Uses and benefits. *Pharmacy and Therapeutics* 39: 356.

Vogel, V. (2019). Nanosensors and particles: A technology frontier with pitfalls. *Journal of Nanobiotechnology* 17: 111.

Wang, Z. L. (2012). Self-powered nanosensors and nanosystems. *Advanced Materials* 24: 280–85.

Wu, C., S. J. Hansen, Q. Hou, J. Yu, M. Zeigler, Y. Jin et al. (2011). 'Design of highly emissive polymer dot bioconjugates for *in vivo* tumor targeting. *Angewandte Chemie* 123: 3492–96.

Yang, Z., Q. Zhou, L. Lei, K. Zheng and W. Xiang. (2016). An IoT-cloud based wearable ECG monitoring system for smart healthcare. *Journal of Medical Systems* 40: 286.

Yin, A. L., D. Hachuel, J. P. Pollak, E. J. Scherl and D. Estrin. (2019). Digital health apps in the clinical care of inflammatory bowel disease: Scoping review. *Journal of Medical Internet Research* 21: e14630.

Yogaraj, A., M. R. Ezilarasan, R. Anuroop, C. S. Sivanthiram and S. K. Thakur. (2017). IOT based smart healthcare monitoring system for rural/isolated areas. *International Journal of Pure and Applied Mathematics* 114: 679–88.

Yoo, J. (2013). The meaning of information technology (IT) mobile devices to me, the infectious disease physician. *Infection & Chemotherapy* 45: 244–51.

Yuehong, Y. I. N., Y. Zeng, X. Chen and Y. Fan. (2016). The internet of things in healthcare: An overview. *Journal of Industrial Information Integration* 1: 3–13.

Zhou, W., K. Li, Y. Wei, P. Hao, M. Chi, Y. Liu et al. (2018). Ultrasensitive label-free optical microfiber coupler biosensor for detection of cardiac troponin I based on interference turning point effect. *Biosensors and Bioelectronics* 106: 99–104.

CHAPTER 15

Microfluidic-Chip Technology for Disease Diagnostic Applications via Dielectrophoresis

Soumya K. Srivastava[1,*] and *Anthony T. Giduthuri*[2,3]

1. Introduction

The human body embodies various systems that naturally work together for maintaining the overall health and wellbeing. At the body's optimal internal and external environmental conditions, these systems thrive so well that the need for troubleshooting diseases' etiology does not necessarily arise. However, a deviation from good health occurs when either or both internal and external body conditions become adverse. Many factors can cause a departure from good health, such as autoimmune disorders, disruption of the balance in the human body system, cancers, bacteria, viruses, fungal and other infections (Rahman et al. 2017).

Each time the World Health Organization (WHO) releases its annual reports, point-of-care diagnostic devices' importance keeps resurfacing. Among the 194 member-states of the World Health Organization (WHO), many avoidable deaths keep reoccurring in a progressively increasing manner. Of the 56.4 million deaths recorded in 2015, more than 54% were health-related, with developing countries taking the larger share. This trend calls for urgent and more focused attention on developing health technologies, especially disease diagnostic methodologies, that would stop this transmission.

In the early 20th century, diagnostic activities were carried out solely in pre-defined spaces, and patients had to travel miles for a medical test. Most essential diagnostic tools were bulky and occupied volumes as large as 180 cubic feet. Besides,

[1] Department of Chemical and Biomedical Engineering, West Virginia University.
[2] Bioproducts, Sciences and Engineering Laboratory, Washington State University, Tri-Cities, Richland, WA 99354.
[3] The Gene and Linda Voiland School of Chemical Engineering and Bioengineering, Washington State University, Pullman, WA 99163.
* Corresponding author: soumya.srivastava@mail.wvu.edu

they were mostly immobile, and diagnostic tests were carried out in confined spaces. However, with technological advances, some of the bottlenecks bedeviling diagnostics started fading away with the introduction of mobile diagnostic devices. Some of these devices are being used not only for accurate diagnoses but also to screen large numbers of apparently healthy individuals.

Current observations in the miniaturization of medical devices do not come without the integrated field of micro- nano-fluidics. Microfluidics is a multidisciplinary field intersecting engineering, physics, chemistry, and nanotechnology, with practical applications to systems design. Very low volumes of samples are processed to achieve multiplexing, automation, and high-throughput screening, collectively known as lab-on-a-chip (LOC). Since the dawn of microfluidics, various attempts have been made to transform the standard immobile laboratory setting into different carry-on easy-to-use devices known broadly as point-of-care (POC) devices. The first analytical laboratory miniaturization was a gas chromatograph system fabricated on a silicon chip by Terry et al. in 1979 at Stanford University. This discovery brought microfluidics into the frontline in the early 80s. Advances in molecular analysis, biodefense, molecular biology, and microelectronics have substantially improved microfluidics' relevance to disease diagnostics.

Various techniques are available to detect diseases, including dye marker labeling, enzyme-based labeling assays, and nucleic acid-based assays (Rahman et al. 2017). All of these methods form the basis for nearly all of the pathology laboratory tests in hospitals. Specific assays use fluorescence substances to label the antibodies and antigens to quantify them. Some use the antibody to label the antigen. These methods feature high sensitivity and specificity values; however, conducting these tests often requires trained clinical personnel and expensive equipment.

Furthermore, while patients are on the verge of the right treatment, these tests are time-consuming, which could be a significant drawback of these techniques. Economically, these tests are considered expensive to be run in small laboratories, typically located in remote villages; thus, there is the need to send the samples to a central laboratory. This procedure takes time and causes a delay in the treatment given to the patients. Besides, any contaminants or analytical errors during the process will lead to false-negative results, placing technicians in charge of handling techniques at a disadvantage.

One of the powerful techniques that overcome the drawbacks, as mentioned earlier, is to use microfluidics to manipulate the particles flowing within the microfluidic channels by utilizing an electric field (uniform or non-uniform) to direct particles according to their intrinsic electric characteristics. Usually, the electric field is set up with platinum (Pt) or gold (Au) electrodes. Microfluidic devices that utilize a uniform electric field are electrophoretic (EP) in nature, while those that employ a non-uniform electric field are dielectrophoretic (DEP). This chapter primarily focuses on the dielectrophoretic aspect, i.e., non-uniform electric field gradient of microfluidics as applicable to disease diagnostics.

1.1 Electrokinetics

Electrokinetics, a combination of electrostatics and hydrodynamics, is the science that governs the flow of fluid in a microfluidic channel under the influence of the electric field. R. J. Hunter, in 'Foundation of Colloid Science', defines electrokinetics as "those processes in which the boundary layer between one charge phase and another is forced to undergo some sort of shearing process. The charge attached to one phase will move in one direction, and that associated with the adjoining phase will move in the opposite direction". Before this technique is discussed in relation to its operability, it is vital to understand the various phenomenological concepts that come into play prior to the effective orientation of bioparticles with respect to their trajectories.

Therefore, this chapter starts with the concept of electrostatics: the principle behind charge-environment interrelation for statics. The static charge analysis is then extended to a moving-charge scenario through the principle of electrodynamics. The electroosmosis technique (EO) as a driving force for fluid is then discussed in two folds, i.e., with respect to the bulk medium and the channel wall regions. Electrophoresis (EP), the study of motion of suspended charged particles under a uniform electric field, is also discussed. The chapter further gives adequate attention to the principles behind dielectrophoresis (manipulating particles through non-uniform electric field) and how these electrokinetic forces (EO and EP) are regulated for maximum dielectrophoretic impact on the particles.

1.1.1 The Origin of Charges and their effects

Every matter is made up of particles, and each particle contains protons, neutrons, and electrons. Whenever the number of protons (positive charges) in a particle equals that of the electrons (negative charges), the particle is electrically neutral. When such a neutral atom gains an electron, it becomes negatively charged but positively charged when it loses an electron. Both protons and electrons of a neutral atom are termed as 'bound charges' since they are held in position and could only vibrate about their mean positions. In electrically charged particles, however, there exist certain charges which provide valence to the particles, also known as 'free charges'. Hence, electrically charged particles could be represented by both free charges and bound charges.

In microfluidics, all the effects associated with bound charges are accounted for by electrical permittivity. This, therefore, allows enough focus on the free charges, which contribute to the effects the particle would have on its environment or vice versa. When a charged particle is introduced into a liquid medium, it creates both electric field displacement (D) and polarization of the medium. Usually, the electric field component of displacement is small, making the polarization part dominant. Therefore, considerable attention is given to the polarization effects caused by introducing a charged particle into a medium.

If a charged particle is placed in free space, it induces an electric field with no polarization due to the absence of medium. But if the charged particle is in a liquid, say water, it creates electric field displacement and causes water to polarize. This is because, in water, the hydrogen ion (H^+) or hydroxonium ion (H_3O^+) has a partial positive charge while the oxygen ion (O^{2-}) has a partial negative charge. This molecule possesses an inherent dipole, which is organized with some amount of hydrogen bonding. Statistically speaking, its orientation in space is random. Therefore, if a negatively charged particle, for instance, is introduced into water, all the water atoms will rotate to make the H^+ or H_3O^+ in water closer to the introduced charge. The two components of displacement can be represented in three different forms, as shown below:

$$\vec{D} = \vec{E} + \vec{P} \tag{1}$$

where 'E' is the electric field and 'P' the polarization.

$$\vec{D} == \varepsilon_o\, \vec{E}\, (1+X_e) \tag{2}$$

where, X_e is the electric susceptibility of the medium and ε_o is the permittivity of the vacuum.

$$\vec{D} == \varepsilon\vec{E} \tag{3}$$

where, ε is the electrical permittivity of the medium.

While Eq. 1 describes the physics of the problem, Eq. 2 allows the description of the part played by free space and the medium, i.e., $D = \varepsilon_o\, \vec{E}$ (for free space) $+ \varepsilon_o\, \vec{E}\, X_e$ (for the medium). In free space with an absence of medium, the electric susceptibility becomes zero. Equation 3 shows that the displacement (what describes the effects of the charge on its environment) is linearly dependent on the electric field $\vec{E}$ (which describes how the environment acts back on the charge). If the electric permittivity is assumed constant, it implies that the medium responds instantaneously to the charge, and the response is linear. Equation 3 is the most commonly used relation for displacement and electric field. For the expression of displacement in Eq. 3, two fundamental Eqs. (4 and 5) are usually applied:

$$\text{Gauss' Law for electricity, } \nabla.\,\vec{D} = \rho_E \tag{4}$$

where ρ_E is the net charge density.

$$\text{Volumetric force on a charge fluid } f = \rho_E\,\vec{E} \tag{5}$$

Gauss' law is one of the four Maxwell equations used in electromagnetism. Usually, in microfluidics, the magnetic effect is often neglected since its effect is infinitesimal. Moreover, the charges in the electrolyte solution are carried by the moving ions, which are slow in movement than electrons moving in metals. So, if any attempt is made to create a magnetic field by ions moving in water, it will be found that the field is highly negligible. If the particle is not charged, there would be a need to make it charged for creating the displacement effects discussed above. This induced-charge generation process can be accomplished using an extrinsic electric field, as discussed in the next section. Particles that can be polarized by an applied electric field are referred to as dielectric particles, and they (including charged particles) are usually employed in dielectrophoretic separations.

1.2 Electro-osmosis

Having known that a charged particle (or dielectric particles under the influence of electric field) placed in a medium can cause the medium to polarize, it is now necessary to understand how the bulk of the medium travels along the microchannel under the influence of the electric field. This section describes what happens at the boundary between the medium and the channel walls and how this wall-medium interrelation plays a vital role in the bulk motion of the medium and particle transport. The utilization of the boundary conditions in obtaining the velocity profile of the flowing fluid is also discussed. The transport of the bulk fluid medium along the microchannel under the electric field's influence is called electro-osmosis.

1.2.1 Fluid Flow Generation in Microfluidic Devices

Two methods accomplish fluid flow in a microfluidic channel. One method involves the use of an external pumping device called a micropump. This arises due to pressure-driven fluid flow, and the velocity profile associated with such pressure-driven flow is parabolic (Figure 1A). Another method is the use of electro-osmosis, usually referred to as electroosmotic pumping. This electroosmotic-driven flow stems from what is happening at the interface of the medium and the channel wall. It is essential to know what is happening at the microchannel boundary because the information obtained from these boundary conditions will help solve the Navier-Stokes equation, the equation governing the fluid flow.

If the voltage at the wall and the voltage of the bulk medium of electrolyte flowing in a microfluidic channel are measured, a potential difference is observed. Since the chemical potential of the ions at the wall and the bulk are not the same, it means the potential must be different if the system must remain at equilibrium. This potential difference has been found to decay exponentially from the wall to the bulk solution. This decaying potential is responsible for the shape of the velocity profile observed at some distance from the channel walls, as seen in Figure 1B.

Microfluidic channels are usually made of glass or poly(dimethylsiloxane) (PDMS) materials. Each of these materials consists of component structures that offer them some form of surface charge when in contact with a liquid medium (or electrolyte). For instance, glass has a surface terminal dominated by the silanol (SiOH) group, as shown in Figure 2.

Suppose a pH-7 electrolyte solution is placed in contact with glass. In that case, the SiOH group behaves like weak acid and thus ionizes and releases H^+ or H_3O^+ into the electrolyte solution leaving SiO^- group at the glass surface (Figure 3). This SiO^- group at the glass surface makes the glass wall negatively charged. Hence, a negative charge density exists at the wall, and this is associated with a relatively lower electric potential at the wall.

The dipole molecules' re-orientation causes this lower potential at the wall in the polar electrolyte (polar electrolytes are usually used in microfluidic channels). A good observation is that if a voltmeter is used to measure the voltage at the wall and in the bulk medium, the voltage difference decays in a very short length scale (~10 nm). At ~10 nm from the wall, the intrinsic electric potential in the whole microchannel becomes uniform. This distance at which the electrical potential becomes uniform is

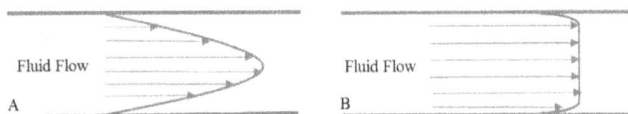

Fig. 1. The velocity profile of fluid flow in microchannel. (A) Flow profile for pressure-driven flow. (B) The flow profile for electroosmotic flow.

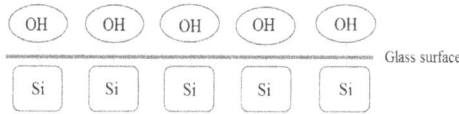

Fig. 2. The structural representation of the glass surface.

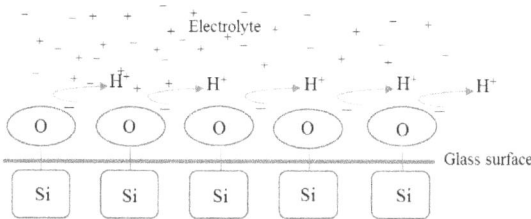

Fig. 3. The behavior of silanol group on contact with a neutral electrolyte.

called Debye length. For any wide microchannel (> 25 μm), the details of the wall region can be neglected, and the intrinsic potential difference be ignored.

Since the glass surface is negatively charged due to the ionization, positive charges from the polar electrolytes become tightly adsorbed on the glass surface and form a stable layer with the SiO⁻ group. This layer is termed as the 'stern layer'. The remaining negative charges in the polar electrolyte become attracted to the positive charges on the stern layer by electrostatically induced Coulomb force, thus creating a second layer of ions, diffuse layer, on the glass surface (Figure 4). The bond on the stern layer (i.e., between SiO⁻ and the positive charges of the polar medium) is stronger than the Coulomb force created between the positive surface of the stern layer and the diffuse layer. The negative ions on the positive surface of the stern layer are swept away by other negative charges in the medium. This creates the fluid motion near the walls and transfers via viscous forces into the convective motion of the bulk fluid. This is the process by which the bulk movement of the fluid is achieved in a microchannel.

The combination of both the stern and diffuse layer generates the electric double layer at the glass surface. This is the underpinning principle that governs the electrokinetic transportation of materials within any microchannel. Thus, a potential exists between the stern and diffuse layers, called Zeta potential, and it is a function

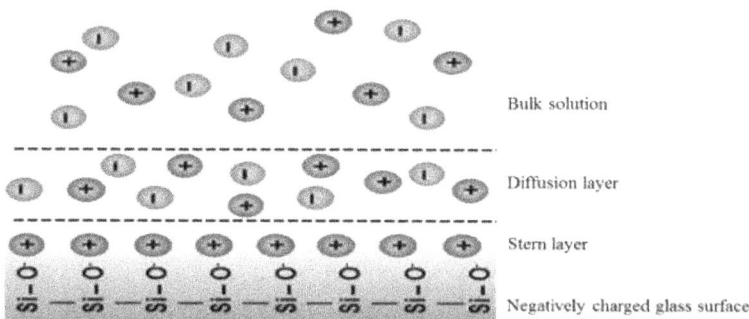

Fig. 4. Formation of stern and diffusion layer through silanol ionization and Columbic interaction.

of the materials used in fabricating the microchannel (Tandon et al. 2008). Suppose the suspended particles in the medium are negatively charged. In that case, some will be attracted by the weak coulombic force and will be seen moving along the channel wall's surface while the bulk of the electrolyte is moving. This phenomenon is referred to as electroosmotic pumping.

1.2.2 Integration of Electric Field into Electroosmotic Flow Profile

In any microfluidic device, fluid flow and electric field analyses are usually integrated. Fluid flow analysis is generally through the Navier-Stokes equation in conjunction with the continuity equation. Mathematically, electroosmotic fluid flow in the microfluidic channel can be analyzed as described below.

Given an elemental volume of fluid (very close to the channel wall) with $\phi = 0$ as the electric potential of the bulk of the solution, $\phi = \phi_o$ as the specified potential at the wall (Figure 5), the distance 'h' relatively small such that $\phi = 0$ at level B, one can consider a flow parallel to the wall with the no-slip condition at the wall ($u = 0$) and a parallel extrinsically applied electric field (this electric field (E_{ext}) moving in the x-direction is distinguished from the intrinsic electric field moving in the y-direction as result of the voltage difference between points A and B). The Navier-Stokes equation for an electroneutral fluid is thus given by:

$$\rho \frac{\delta u}{\delta t} + \rho u . \nabla u + \nabla . p = \mu \nabla^2 u \qquad (6)$$

where, ρ and μ are the density and viscosity of the fluid, respectively, 'p' is the associated pressure, and 'u' is the velocity of the flowing fluid. Assuming there exists no pressure gradient and that the flow is unidirectionally at a steady-state, the Navier-Stokes equation can be simplified as:

$$\mu \nabla^2 u = 0 \qquad (7)$$

Since the electric field is being applied, electroneutrality of the fluid cannot be assumed, hence

$$\mu \nabla^2 u + \rho_E E_{ext} = 0 \qquad (8)$$

Equation 8 signifies an electric field pushing on any local control volume of fluid, and it is being counterbalanced by the net flux of viscous momentum.

Recall the critical relation between charge density and electric field (Eq. 5) that is missing in Figure 5, i.e., absence of ρ_E in the boundary condition. Hence, there is a need to relate Eq. 8 to the parameters in the boundary condition of Figure 5. This is done by invoking the Gauss Law, which states that the charge density gives the

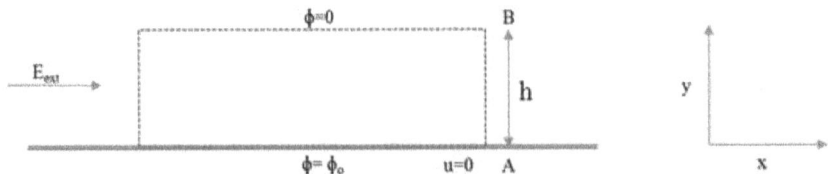

Fig. 5. Demonstration of potential change between the fluid-glass interface and the bulk solution.

divergence of the electric displacement as in Eq. 4. Since D = εE (Eq. 3), and electric field is given as the negative of the gradient of the electric potential, i.e.,

$$E = -\nabla\phi \qquad (9a)$$

then, substituting Eq. 3 and Eq. 9a into Eq. 4 gives:

$$-\varepsilon\nabla^2\phi = \rho_E \qquad (9b)$$

provided that ε is spatially uniform. Thus, Eq. 8 can be modified using Eq. 9b as:

$$\mu\nabla^2 u = \varepsilon E_{ext}\nabla^2\phi \qquad (10)$$

The electroosmotic velocity profile of the moving fluid is then obtained by solving Eq. 10 in any desired coordinate system.

1.3 Electrophoresis

This section focuses on the fundamental mechanisms that govern the transport of charged particles within the microchannel under the electric field's influence, called electrophoresis (EP).

Electrophoresis is the movement of charged particles (suspended in a medium) along a microchannel. When a charged (e.g., negative) particle is introduced into a polar electrolyte solution in a microchannel, two things happen. First, the surface charge on the channel wall becomes ionized, and the electrolyte solution is affected with reference to its orientation, thus generating an electroosmotic force that pushes forward the bulk of the electrolyte solution in the direction of an applied external electric field. Second, the negatively charged particle causes the polarization of the medium such that positive charges from the medium re-orientate themselves around the particle generating the stokes frictional force (the force acting on the interface between the fluid and the particle).

When an electric field is applied, the negatively charged particle moves towards the anode of the electric field source through electrophoresis. However, the positive charges surrounding the negatively charged particle tend to move towards the cathode (through electroosmosis). Thus, electroosmosis forces the particle to move forward along the external electric field's direction, creating a drag force on the charged particle (in a direction opposite to the electrophoretic line of action). Therefore, it is evident that under the electric field's influence, a charged particle will move towards the exit by the combined effects of electroosmosis and electrophoresis (Figure 6). These combined effects are referred to as electrokinetics.

1.4 Dielectrophoresis

So far, it has been shown that electrokinetics forces (EO and EP) dominate the flow in a microfluidic channel. In this section, the particle motion trajectory will be examined when the applied electric field is non-uniform. The utilization of a non-uniform electric field in a microchannel for manipulating the particle trajectories within the channel is referred to as dielectrophoresis (DEP). This technique relies on the electrical property changes between the particles and the medium.

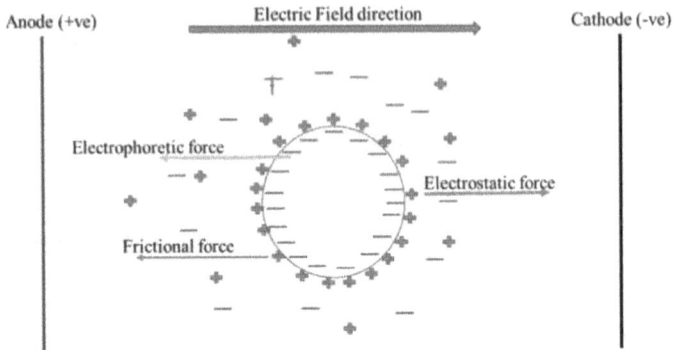

Fig. 6. The schematic diagram showing the forces acting on a negatively charged particle flowing under a uniform electric field within a microchannel.

1.4.1 The Concepts of Dielectrophoresis (DEP)

When a polarizable particle is subjected to a non-uniform electric field, the force exerted on the particle causes it to move towards high or low field density regions; this behavior is termed 'dielectrophoresis'. DEP is observed only when a non-uniform electric field is exerted on any particle because coulomb forces generated on both sides of the particle are different, thus facilitating the motion of the particle towards regions of electric field maxima or minima. Both alternating current (AC) and direct current (DC) electric fields can be applied to nonlinear channel geometries to generate non-uniform columbic forces across a particle, thus yielding particle motion by DEP.

Dielectrophoretic phenomena have traditionally been associated with applying AC voltage and frequency in conjunction with spatially non-uniform electrode geometries to create novel electric field gradients to manipulate particles of interest. However, in a new research area, DC voltage has also been applied to achieve dielectrophoretic separation, but the need to create a spatially non-uniform field is still evident. Dielectrophoresis depends on a wide range of medium and particle properties. Due to the observed differences in physical and chemical properties, one of the two phenomenological DEP effects first explained by H.A. Pohl are usually observed: positive DEP and negative DEP (Pohl 1978). These concepts are discussed in the next two subsections in concert with a discussion on classical dielectrophoresis (AC-DEP) or electrode DEP and the newly explored direct current dielectrophoresis (DC-DEP) or electrodeless DEP or insulator DEP (iDEP).

1.4.2 Alternating Current Dielectrophoresis (AC-DEP)

This is the classical DEP technique employing AC voltages and frequency to manipulate particles of interest. Embedded microelectrodes positioned in a spatially non-uniform manner are used to achieve particle separation (Figure 7), trapping, and focusing by applying AC electric fields from kHz—GHz range (Li 2006). When a microparticle is suspended in a highly conductive medium, the particles experience less polarization than the medium. In such cases, the particles/cells exhibit negative dielectrophoresis (nDEP), wherein the particles move away from the high field density regions. In the case of positive dielectrophoresis (pDEP), the particles are

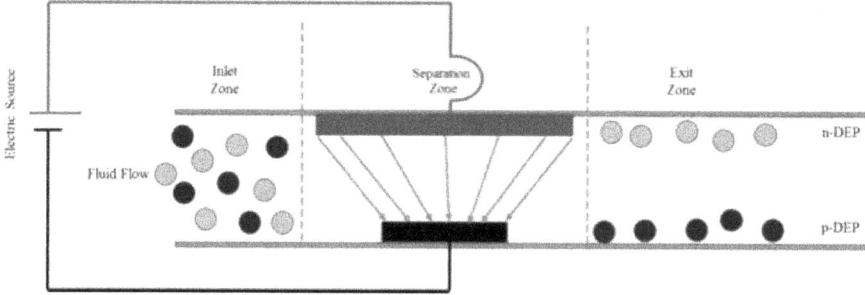

Fig. 7. A typical representation of particle separation through AC-DEP. Electrodes are embedded within the microchannel.

attracted to the high field density regions by a translational force due to the particles' polarizability being more significant than that of the medium.

The transition between negative DEP and positive DEP as one moves up in AC frequency is known as the cross-over frequency. At this frequency, the force experienced by the particle is zero. Crossover frequency is dependent on particle dielectric properties and medium properties. It is different for each particle or cell; this information can optimize trapping or separation schemes. Some complex cells exhibit multiple crossover frequencies. Most biomolecules exhibit nDEP, but mixed responses might also be possible due to the complex make-up of their structure, composition, and charge distribution.

The net dielectric force, $\vec{F}_{DEP}$, resulting from transient polarization of cells and the electric field is given by

$$\vec{F}_{DEP} = 2\pi r^3 \, \epsilon_m \alpha \nabla \vec{E^2} \tag{11}$$

where 'r' is the particle radius, 'ε_m' is the medium permittivity, '$\vec{E}$' is the applied electric field, and 'α' is the real part of the Clausius-Mossotti (CM) factor, which is the effective polarizability of the particle relative to the suspending medium and is frequency dependent. α is given by:

$$\alpha = Re[K(\omega)] \tag{12}$$

$$\text{where } K(\omega) = \frac{\varepsilon_p^* - \varepsilon_m^*}{\varepsilon_p^* + 2\varepsilon_m^*} \tag{13}$$

$$\text{where } \varepsilon^* = \varepsilon - \left(\frac{i\sigma}{\omega}\right) = \varepsilon_\infty + \frac{\varepsilon_s - \varepsilon_\infty}{1 + i\omega\tau} \tag{14}$$

where ε^* denotes complex permittivity, and the subscript 'p' refers to a lossless dielectric sphere particle suspended in a medium 'm'. The complex permittivity ε^* given by Eq. 14 is a function of permittivity, ε, medium electrical conductivity, σ, and the angular frequency, ω. The passive permittivity is also a function of ε_s, the permittivity at the low-frequency limit, ε_∞, the permittivity at the high-frequency limit, and τ, the charge relaxation time constant (the time taken by the dielectric to return to its equilibrium state after being disturbed by an external electric field).

The transient polarization of particles results in their movement in the electric field that scales between two extremes depending on the exciting AC frequency. In the seminal text *"Dielectrophoresis: The behavior of neutral matter in nonuniform electric fields,"* Herbert Pohl defined these two phenomenological extremes as positive dielectrophoresis and negative dielectrophoresis (Pohl 1978). These two cases arise because of the polarizability of a uniform composition particle being greater or lesser than the medium's polarizability in which it is suspended. If the real part of the effective polarizability, i.e., of the particle, is greater than that of the medium, then the electric field lines pass through the particle, causing a polarization that is slightly skewed due to the spatially varying electric field lines. A resultant force directs the particle to high field density regions, and this observed movement is known as 'positive dielectrophoresis' (pDEP). If the effective polarizability, of the particle is less than that of the medium in which it is suspended, spatially non-uniform electric field lines divert around the outside of the particle, causing ion depletion at the particle poles and subsequent polarization. The resulting force directs the particle to the low field density region, termed 'negative dielectrophoresis' (nDEP).

1.4.3 Direct Current Dielectrophoresis (DC-DEP)

Direct current dielectrophoresis is a novel technique developed in the last decade (Srivastava et al. 2011, Adekanmbi and Srivastava 2016). It employs insulating objects or hurdles fabricated by various microfabrication methods within the channel to create spatial field non-uniformities in the separation zone, as shown in Figure 8. It is also known as insulator-based dielectrophoresis (iDEP) or electrodeless dielectrophoresis (eDEP) because electrodes are placed far outside the channel in inlet and outlet ports. The electrodes are immersed in the supporting media but are not in direct contact with the observed particles. This is a massive advantage over the traditional AC dielectrophoretic technique. In DC-iDEP, there is no frequency dependency involved; such aspatial electric field non-uniformity within the microchannel is solely responsible for the DEP force experienced by the polarizable particles.

The force exerted on the particle impels the particle to move away from the insulating obstacle region, thus undergoing nDEP phenomena. In the case of pDEP, the particles get trapped at sharp points or constrictions in the insulating region, which is the region of high field maxima in DC-iDEP devices. A particle in the insulating obstacle region experiences dielectrophoretic and electrokinetic forces; the relative magnitude of each determines whether the particle is trapped or flows through the constriction in a specific fluid flow streamline.

The observed cell motion in iDEP devices depends on two forces: electrokinetics (EK) and dielectrophoresis (DEP).

$$\vec{J} \propto \vec{u}_{EK} + \vec{u}_{DEP} \tag{15}$$

where $\vec{J}$ is the particle flux, $\vec{u}_{EK}$ the electrokinetic velocity (expressed as the sum of electrophoretic $\vec{u}_{EP}$, and electroosmotic $\vec{u}_{EO}$ velocities), and $\vec{u}_{DEP}$ the dielectrophoretic velocity of the particle.

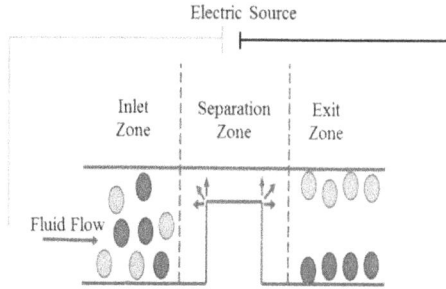

Fig. 8. The basic idea behind DC-DEP particle separation: An insulating hurdle in the separation zone is being used to generate electric field non-uniformity at the separation zone.

Electrokinetic velocity can be expressed as the sum of electro-osmotic and electrophoretic mobilities:

$$\vec{u}_{EK} = \mu_{EK}\,\vec{E} = (\mu_{EP} + \mu_{EO})\,\vec{E} \tag{16}$$

where u_{EK} is the electrokinetic mobility, u_{EP} electrophoretic mobility, u_{EO} electroosmotic mobility, and $\vec{E}$ applied electric field to create non-uniformities in the channel. Neglecting the frequency component for strict DC-iDEP, the CM factor in Eq. 13 is modified to:

$$\alpha = \frac{\sigma_p - \sigma_m}{\sigma_p + 2\sigma_m} \tag{17}$$

where σ_p is the conductivity of the particle, and σ_m the conductivity of the medium. This simplification is substituted into Eq. 11 yielding

$$\vec{F}_{DEP} = \frac{1}{2}\,V\,\frac{\sigma_p - \sigma_m}{\sigma_p + 2\sigma_m}\,\varepsilon_m \nabla \vec{E}^2 \tag{18}$$

where V the volume of the particle, ε_m permittivity of the medium, and $\vec{E}$ the magnitude of the applied DC electric field.

From Eq. 17, if the conductivity of the particle is greater than the medium, the CM factor gives positive values, and the dielectrophoretic force pushes the particle towards high field density regions, thus trapping them, i.e., the particle gets attracted towards an insulating obstacle region, whereas if the conductivity of the particle is less than that of the medium, the particles are repelled from the high field density regions, thus yielding in negative values of CM factor and therefore the movement of particles in the fluid streamlines is observed, i.e., particles are repelled from the insulating obstacle regions. The conductivity of the particle (σ_p) is given as a function of surface conductivity and bulk conductivity as (Ozuna-Chacón et al. 2008):

$$\sigma_p = \sigma_b + \frac{2K_s}{r} \tag{19}$$

where σ_b is the bulk conductivity, K_s is the surface conductance, and 'r' the particle's radius.

DC-iDEP techniques are advantageous compared to AC DEP separation. It is a very nascent technique and has the potential to improve cell manipulation efficiency.

The advantages of using such insulating obstacles include less affected by fouling than embedded electrodes, fabrication is more straightforward because no embedded metal components are required, devices are robust and chemically inert, electrolysis gas evolution at the remotely located metal electrodes causes minimal or no bubbles inside channels, and microdevices are continuous flow systems and can be operated with complex electrical signals.

Due to the electrodes placed in the large reservoirs at the channel inlet and outlet ports, they often cause re-dilution of the concentrated samples with some Joule heating and electrolysis at the electrodes. A simple, robust device was designed to mitigate these effects where the electrodes are not in direct contact with the sample. This technique is referred to as contactless dielectrophoresis (cDEP), wherein the electric field is generated by placing the electrodes in two conductive microchambers separated by thin insulating barriers from the main channel (Shafiee et al. 2010). cDEP is particularly well-suited for manipulating sensitive biological particles. Another manipulation technique, which evolved from DC-iDEP, is the use of both AC and DC fields in the insulating obstacle geometry to achieve trapping. Using both AC and DC signals, high voltage DC amplitudes could be reduced and replaced by low AC frequencies on the order of < 100 kHz, thus reducing the potential problem of Joule heating and sample degradation.

1.4.4 DC-biased AC Dielectrophoresis (AC-iDEP)

Compared to classical DEP (AC DEP), where only AC fields are employed to focus, sort, trap cells, iDEP could be operated utilizing either only DC (DC-iDEP) or with both AC and DC field components (DC-biased AC fields; AC-iDEP) (Srivastava et al. 2011).

From Eq. 11, the DEP force $\vec{F}_{DEP}$ on a spherical particle under DC-biased AC field (AC-iDEP) conditions is given by:

$$\vec{F}_{DEP} = 2(\alpha^2 + 1)\pi\varepsilon_m r^3 \,\mathfrak{Re}[\alpha]\nabla|\vec{E}_{DC}|^2 \tag{20}$$

where 'r' is the radius of the particle, the real component $\mathfrak{Re}[\alpha]$ is the particle's induced dipole in a DEP field and the effective polarizability of the particle. AC-iDEP is operated under the combination of AC and DC voltage defined by β (the ratio of AC to DC electric field amplitudes (Hawkins et al. 2007, Kirby 2010):

$$\beta = \frac{|\vec{E}_{AC}|}{|\vec{E}_{DC}|} \tag{21}$$

2. Particle Modeling

When bioparticles are under the influence of an external electric field, their movement can be interpreted by applying diverse analytical models (Irimajiri et al. 1979, Asami et al. 1989, Turcu and Lucaciu 1989, Miller and Jones 1993, Becker et al. 1994, Raicu et al. 1996, Vrinceanu and Gheorghiu 1996, Pethig and Markx 1997, Gheorghiu and Asami 1998, Feldman et al. 2003, Hayashi et al. 2003, Sancho et al. 2003). Depending on the morphology of the bioparticles, i.e., sphere, ellipse, cylinder, or rod with single or multiple shells, the models could be fine-tuned to extract the electrical properties (Irimajiri et al. 1979, Asami et al. 1989). To overcome computational limitations, several researchers have resorted to simplifying these models to obtain reasonably accurate characterization properties. For example, *Escherichia coli* is a rod-shaped gram-negative bacterium usually modeled as a spherical or ellipsoidal bioparticle (Irimajiri et al. 1979, Asami et al. 1989). Besides, numerical computation tools such as COMSOL Multiphysics software are not currently equipped with functionalities for handling biological models of diverse shapes. Besides, deriving an analytical model representation for different shapes of bioparticles is currently a herculean task. Until the foreseeable future, spherical shell models will dominate the domain of bioparticle characterization. Details of each shell model have been reported elsewhere (Jubery et al. 2014, Adekanmbi and Srivastava 2016, 2019a,b, Giduthuri et al. 2021).

With an adequate understanding of different data collection techniques and shell modeling, it is less cumbersome to obtain the much-needed electrical properties of bioparticles. In this section, a step-by-step procedure to obtain the electrical properties are presented. Once the required data is obtained from DEP experiments, there are several ways by which they could be analyzed. While many commercial software packages could be utilized to acquire these properties, the example here will involve using a readily available versatile software package like Microsoft Excel. Suppose a set of crossover frequency *vs.* suspending medium conductivity data is collected from the DEP experiments and the specific membrane capacitance (C_{mem}) and conductance (G_{mem}) are required to be estimated from an ellipsoidal single-shell model using Excel, the procedure published in (Adekanmbi and Srivastava 2019a) could, amongst other methods, be used.

3. Device Modeling

Applying DEP in studying biological cells and subsequently applying it in disease diagnostics involves a series of steps, of which the primary step is to characterize the cells of interest for their electrical signature. This biophysical marker affects the cell's behavior in the presence of a non-homogenous electric field. This eliminates the need to use cell labeling techniques, which are expensive, laborious, and time-consuming. After the cells have been characterized, the determined electrical properties should be used to model and develop a suitable device design that can discriminate the target cells from a mixture of cells. This usually involves solving a variety of higher-order equations involving boundary conditions for fluid transport, i.e., fluid drag and mass transport involving diffusion and the electrical component, which affects the cell membrane and fluid interface. The math behind modeling a real-time device is typically not possible without making certain assumptions. The equations with the necessary assumptions can be solved using several commercially available packages to simulate the fluid flow numerically, particle trajectory, and electrical strength profiles, primarily based on electrode shape, spacing, size, and orientation. This numerical modeling approach saves time, effort, and research expenditure over experimental trial and error testing of the device prior to final experimental validation. Several reported DEP-based numerical modeling involved using commercial packages such as MATLAB, COMSOL (Adekanmbi et al. 2020b,

Giduthuri et al. 2021), CFD-ACE⁺ (Jen et al. 2011), OpenFOAM (Kazemi and Darabi 2018).

Modeling and designing devices is a crucial step in disease diagnostics application of DEP (Cetin and Li 2011) to optimize the device design and effectively manipulate/separate cells using minimal electric field strength. A successful numerical simulated device will then have to be experimentally validated for real-time usage.

4. Electrophysiological Characterization

Electrical properties of bioparticles (cells, in particular) define their response to the external electric fields (Miklavcic 2006). Characterizing bioparticles and identifying these dielectric properties is very important in predicting bioparticle pathways, discerning infections, recognizing phenotypes and genotype, pneumographic analysis, and biosensing. On the engineering side of developing and designing point-of-care diagnostic platforms, recognizing these dielectric properties will aid in the simulation of bioparticle-electromagnetic field interaction (Miklavcic 2006). These electrophysiological properties are also essential in the development of diagnostic and therapeutic procedures (Miklavcic 2006). For instance, when a bioparticle is suspended in an external electric field (AC or DC), the electric current densities (Pethig 2010), electric current pathways (Chizmadzhev et al. 1998, Vahey et al. 2013, Trainito 2015), and energy absorption depend primarily on the intrinsic electrical characteristics of the bioparticle. Such intrinsic electrical properties include permittivity, conductivity, impedance, conductance, capacitance, resistance, etc.

To measure these dielectric properties, three primary methods are most commonly used: (1) impedance cytometry, (2) electrorotation (ROT), and (3) dielectrophoresis for both single and multiple cell analyses (Jaffe and Voldman 2018). In impedance cytometry, an AC electric potential is applied between a pair of electrodes, and the resulting current flowing through the system is measured (Holmes and Webb 2012). The impedance of the system is then found as a function of the applied voltage and the electric current passing through the system. Data from such impedance experiments are then fitted with appropriate shell models, i.e., single and multiple shells, to extract these bioparticles' dielectric properties (Holmes and Morgan 2010, Sun and Morgan 2010, Zhang et al. 2019). However, impedance cytometry is usually limited to two frequencies per single bioparticle unless combined with other methods, such as fluorescence (Jaffe and Voldman 2018) and optics (Haandbæk et al. 2016). Apart from this holdup, cytometry is notorious for rolling out an overwhelming amount of information that can be unnecessary. Also, the quantification of these properties itself is too slow compared with other methods, and it is often limited to particles < 70 μm in size (Piyasena et al. 2012).

Electrorotation (ROT), on the other hand, commonly uses a four-phase quadrature applied to quadrupolar electrodes to generate electrical torque from the rotating electric field that induces the rotation of a bioparticle (suspended between the electrodes) (Huang et al. 1992, Fuhr and Hagedorn 1996, Gasperis et al. 1998, Esch et al. 1999, Dalton et al. 2001, Voyer et al. 2010, Ren et al. 2011, Bonincontro and Risuleo 2015, Lin et al. 2017, Huang et al. 2019). The magnitude of the rotating particle velocity is a function of the electrical characteristics, i.e., dielectric properties of the bioparticle. The bioparticle's electrical properties can be obtained (Goater and Pethig 1998, Cristofanilli et al. 2002, Dalton et al. 2004, Benhal et al. 2015, Michálek et al. 2019). A major disadvantage of the electrorotation technique is its lower throughput because of the time required to obtain bioparticle velocity data (Jaffe and Voldman 2018).

The third characterization method, dielectrophoresis (DEP), involves a spatially non-uniform electric field as a driving force to induce translational motion of the bioparticle (Xiao-Bo et al. 1997, Jones 2003, Chen and Yuan 2019). DEP is the focus of this work. In DEP, neutral or charged bioparticles in applied alternating current (AC) electric fields become polarized and form dipoles, which move according to their electrical properties and other surrounding factors like the field gradient, particle size, and properties of the suspending medium. Information from such translational orientation of bioparticles can reveal their electrical properties (Gascoyne et al. 1994, 1997). DEP has the advantage of obtaining a quick response of bioparticles to electric field effects. This is a huge benefit as the bioparticle viability tends to be preserved when they are not exposed to the electric field for longer than 5 min (Lu et al. 2012). In terms of throughput, DEP tends to be on the lower side than impedance cytometry, although higher than electrorotation (Jaffe and Voldman 2018). Cost advantage, ease of operation, flexibility in fabrication, and a balance between testing time and throughput are the rationales behind using DEP as a sorting or trapping technique.

4.1 Electrorotation

Electrorotation (ROT), a classical cell characterization technique, utilizes the imaginary part of the Clausius-Mossotti (CM) factor, where the dipole and the field are out of phase (Arnold and Zimmermann 1988). CM factor relates the permittivity and conductivity of the medium to that of the bioparticle.

In classical electrorotation, three or more electrodes are each charged with AC voltage and frequency at different phases to generate a rotating electric field, thus setting up an electrical torque (Figure 9) (Liu et al. 2017, Adekanmbi and Srivastava 2019a).

However, it is now possible to achieve electrorotation using only two electrodes (Huang et al. 2019). When a bioparticle (e.g., spherical) is suspended within this rotating field, it becomes polarized, inducing a dipole. This induced dipole within the bioparticle rotates with (or against) the electric field at a certain velocity. The bioparticle's rotation with (or against) the electric field depends on the sign (positive or negative) of the imaginary part of the Clausius-Mossotti factor in Eq. 12–14 (Ramos 2011). Further, the rotation of the particle depends on whether the charge relaxation time constant, τ (Eq. 14), for the bioparticle is greater than that of the suspending medium or vice versa (Ramos 2011). Usually, the particle will rotate against (or with) the direction of the field if the charge relaxation time constant for the particle is more (or less) than that of the medium, and the angular velocity of rotation, ω, will be constant if the particle is suspended in a liquid of known viscosity (Ramos 2011). However, the multiphase nature of the driving force on the four electrodes causes the

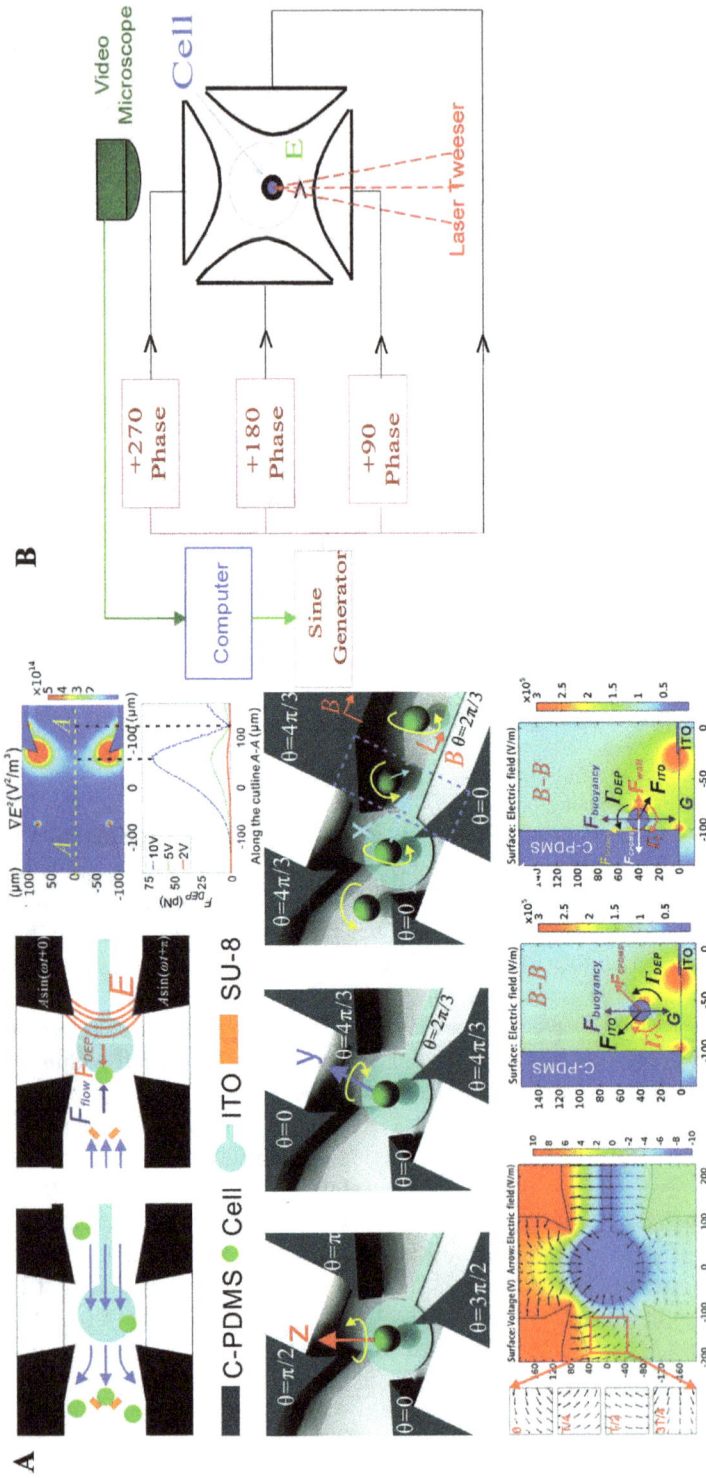

Fig. 9. The working principle of electrorotation. A bioparticle can rotate with the rotating field, and its angular velocity can be related to the frequency of the electric field signals. (A) In the first row, one single cell flowing with the solution streamline was trapped hydrodynamically, released by backflow as predicted by simulation. In the second row, the cell displaying different rotation patterns in various locations along the channel. The third row shows the simulation results of the rotating electric field and its distribution. (B) The demonstration of the physics of electrorotation in a 4-electrode arrangement with 90° phase difference. (Reprinted with permission from Adekanmbi, E. O. and S. K. Srivastava. 2019. Dielectric characterization of bioparticles via electrokinetics: The past, present, and the future. Applied Physics Reviews 6(4): 041313 (Adekanmbi and Srivastava 2019a).

particle to lag the field by a factor that depends on the frequency of the rotating field. Since the torque determines the particle velocity in the rotating electric field, the particle's electric properties can be extracted by measuring the torque's dependence on the field frequency.

Suppose the dipole caused by the rotating electric field is perfectly in phase with the field. In that case, no torque is generated since the Clausius-Mossotti factor is valid only for the out of phase (imaginary part) mode (Kirby 2010). When a torque is generated, it is usually dependent on the bioparticle size, the particle's dielectric properties, and the suspending medium (Huang et al. 2018). It is important to note that at higher frequencies, the obtained electrical properties from electrorotation and impedance cytometry are alike, as evident from many reports (Arnold and Zimmerman 1982, Pastushenko et al. 1985, Arnold et al. 1987, Donath et al. 1990, Wang et al. 1993, Zhou et al. 1995, Burt et al. 1996, Gimsa et al. 1996, Sukhorukov and Zimmermann 1996). One of the main advantages of electrorotation is its versatility in obtaining the dielectric (interior and membrane) properties of a single bioparticle or a cell. However, the possibility of cell lysis at high frequency and the fabrication of the device for submicron particles can be a challenge (Arnold and Zimmerman 1982, Donath et al. 1990, Wang et al. 1994, Egger and Donath 1995, Huang et al. 1995, Gimsa et al. 1996, Sukhorukov and Zimmermann 1996).

4.2 DEP Crossover Frequency

The frequency at which the field switches from negative DEP to positive DEP is known as the crossover frequency (Adekanmbi and Srivastava 2019a,b). The crossover frequency is unique to the particle and its medium, making it possible to manipulate or sort particles into subgroups based on their physiological makeup. The CM factor can be used to find the crossover frequency when Eq. 12 approaches to zero, i.e., $\Re e[K(\omega)] = 0$. The particle's total conductivity depends on the surface conductance and the bulk conductivity, as shown in Eq. 19. The surface conductance is the conductance from the flow of charge in the electrical double layer (EDL). The dipole model assumes that the EDL is much thinner than the particle's diameter and that the EDL is formed instantaneously.

The dipole model is useful to make trend predictions and for qualitative analysis on predicting the particle movement, i.e., either nDEP or pDEP. However, it does have its limitations and does not account for particle size or when the electric field is not uniform. When particles are larger or non-uniform, there is a deviation from predicted values using the dipole method. While the dipole model is useful for smaller particles, it does not accurately predict the crossover frequency on larger particles. In such cases, the Maxwell stress tenor technique can be used for calculating the DEP force on particles of various sizes more accurately (Weng et al. 2016). The force on a particle can be found by integrating the stress tensor, T, over the particle volume described by:

$$T = \varepsilon[EE - \frac{1}{2}(E * E)I] \tag{22}$$

E represents the electric field of the particle, and I is the unit tensor. The DEP force on the particle can be expressed as (Weng et al. 2016):

$$F = \int f dV = \int \nabla T dV = \int T n dS \tag{23}$$

where V and S is the volume and surface enclosing the volume, respectively, n is the unit normal vector of S. These integrations can be used to find the DEP force on particles of various sizes and shapes more approximately than the dipole model. The specific volume electromagnetic force for a cylindrical axisymmetric coordinate reduces to the partial equation, as shown below:

$$f = \varepsilon_p \left[\left(\frac{\partial E_z}{\partial z} + \frac{\partial E_r}{\partial r} \right) E_z + \left(\frac{\partial E_z}{\partial r} - \frac{\partial E_r}{\partial z} \right) E_r \right] \tag{24}$$

This method can be numerically modified using the finite element method for any arbitrary particle shape and size, thus providing new insight into the interpretation of dielectrophoretic applications using lab-on-a-chip systems.

DEP is introduced to cells employing an electrode or multiple electrodes geometry. To accomplish this, an electrode array is fabricated by using a thin platinum wire to act as the electrode(s), surrounded by a boundary of poly (dimethylsiloxane) (PDMS), creating a microwell. Figure 10 depicts the electrode geometry for a point and planar microwell and a quadrupole electrode.

Biological cells suspended in medium are then pipetted into the microwell, also seen in the above figures. The entire apparatus is then placed under a microscope for video visualization and attached to probes that deliver the electrical signals, which induces DEP. Figure 11 depicts a fully set up microwell device for a point and planar electrode and quadruple electrode apparatus, respectively. Video taken through the microscope allows for the visualization of the cells moving either in nDEP (away from the electrode) or pDEP (toward the electrode) regime. The measured frequency (in Hz) when the cells exhibit a switch in DEP behavior is termed as the crossover frequency.

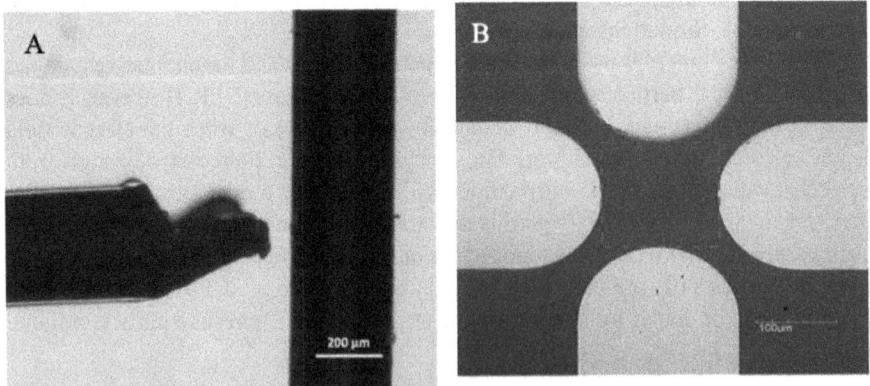

Fig. 10. (A) Point and planar electrode microwell for DEP crossover frequency quantification, (B) Quadrupole electrode for DEP electrorotation experiments. (Reprinted with permission from Zhou, T., Y. Ming, S. F. Perry and S. Tatic-Lucic. 2016. Estimation of the physical properties of neurons and glial cells using dielectrophoresis crossover frequency. Journal of Biological Physics 42(4): 571–586.)

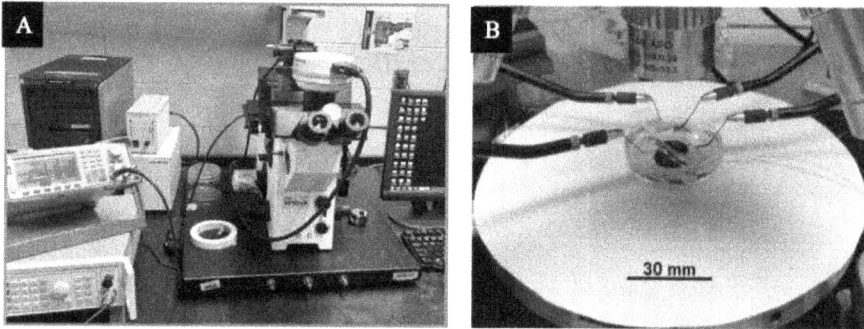

Fig. 11. (A) Point and planar electrode microwell apparatus set-up, (B) Quadrupole electrode set-up for electrorotation experiments (Zhou et al. 2016). (Reprinted with permission from Zhou, T., Y. Ming, S. F. Perry and S. Tatic-Lucic. 2016. Estimation of the physical properties of neurons and glial cells using dielectrophoresis crossover frequency. Journal of Biological Physics 42(4): 571–586.)

5. Biological Applications of Dielectrophoresis

DEP has a wide range of applications from biological cell characterization to cell manipulation and sorting to help medical diagnostics holistically. Over the past two decades, significant advancements in electrode fabrications like 3D electrodes (Xing and Yobas 2016), passivated electrodes (Soltanian-Zadeh et al. 2017), carbon electrodes (Xie et al. 2014), etc., have increased throughput, sensitivity, and efficiency of DEP devices (Hawkins et al. 2020). Apart from biomedical applications, DEP has also been applied for industrial applications such as to separate water droplets from water in oil emulsions using membrane filtration (Molla et al. 2005), and for separation of minerals (Ballantyne and Holtham 2010). This chapter focuses on cancer cells, stem cells, viruses, bacteria, yeast, and DNA to limit the discussion to disease diagnostics.

5.1 Oncology

Oncology, the study of cancer, is an emerging field focusing on cancer diagnoses. Early cancer diagnosis helps eliminate the tumorous cells in the affected organ without metastasis through various therapies available such as immunotherapy, chemotherapy, and many more. DEP-based studies on cancer cells show promising results,aiding in diagnosing cancer at an early stage. Several cancer researchers have utilized DEP to study breast cancer cells (Adekanmbi et al. 2020b), glioblastoma (brain cancer) (Alinezhadbalalami et al. 2019), prostate cancer cells (Salmanzadeh et al. 2012, Yang et al. 2013), circulating tumor cells (Gascoyne and Shim 2014) and many more.

Breast cancer cells were dielectrically characterized, and a sorting device platform was developed and optimized for the separation of adenocarcinoma cells from peripheral blood cells, i.e., leukocytes recently (Adekanmbi et al. 2020b) as shown in Figure 12. Another similar study focused on separating healthy monocytes from cancer cells (Zahedi Siani et al. 2020).

One of the early DEP-based studies on cancer cells attempts to separate human breast cancer cells from blood at a sorting rate of 10^3 cells/sec

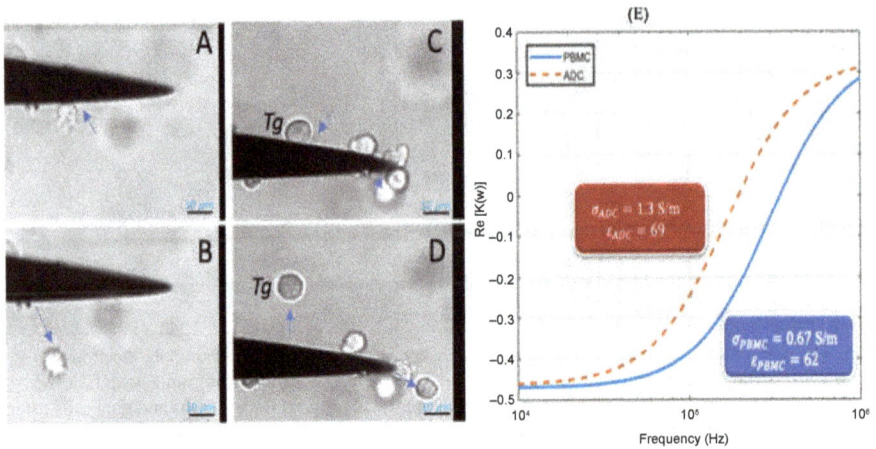

Fig. 12. Breast cancer (Adenocarcinoma: ADCs) experiencing DEP at varying AC fields. (A) cells undergo pDEP, (B) cells experience nDEP, (C, D) are the images while tracking the labeled cells, and (E) plots showing varying CM factor with frequency (Adekanmbi et al. 2020b). (Reprinted with permission from Adekanmbi, E. O., A. T. Giduthuri and S. K. Srivastava. 2020. Dielectric characterization and separation optimization of infiltrating ductal adenocarcinoma via insulator-dielectrophoresis. Micromachines (Basel) 11(4): 340.)

(Gascoyne et al. 1997b). Human prostate cancer cells and colorectal cancer cells, which are similar in their size, were successfully separated using AC-DEP (Yang et al. 2013). Distinguishing cancer cells from a completely different cell line is relatively more straightforward than differentiating cancer cells of a particular cell line from another type of cancer cell. This precision of distinguishing cells of the same kind based on their physiological make-up is the most sought property of DEP. These oncological studies' advances using DEP show a promising future for developing DEP-based cancer detection enabling early detection.

5.2 Stem Cells

Stem cells are undifferentiated cells capable of developing into different tissues (Łos et al. 2019). They can self-renew and proliferate, which are highly sought properties for tissue engineering applications. Properties of stem cells vary widely based on their source type. Current challenges for using stem-cell-based therapies in a clinical-based setting are the lack of efficient sorting or enrichment mechanisms. Commercialized stem cell separation techniques like Fluorescent Activated Cell Sorting (FACS) are tedious and are based on labeling cells using fluorescent and magnetic methods, altering the cellular membrane (Cumova et al. 2010, Mayer et al. 2010).

Stem cells have been extensively manipulated recently due to their ability to differentiate and proliferate to become other functional cells and their applications in wound healing, regenerative medicine, and organ-on-chips (Giduthuri et al. 2020). DEP has also been evaluated in a clinical setting using adipose-derived stem cells (Wu and Morrow 2012). Scientists have consistently proved that a link exists between electrophysiological properties and conductivity/permittivity of these stem cells (Hildebrandt et al. 2010, Hirota and Hakoda 2011, Nourse et al. 2014, Manczak et al. 2019). Furthermore, several hybrid DEP stem cells sorting platforms have been developed for label-free, rapid enrichment for clinical applications (Vykoukal et al. 2008, Simon et al. 2014, Song et al. 2015, Jiang et al. 2019). DEP offers a label-free, rapid, and high-throughput way of sorting stem cells from a heterogenous population.

5.3 *Viruses*

Viruses are pathogenic agents whose origin remains unclear (Wessner 2010) and rely on host cells to multiply, infect and destroy them. Some examples of fatal viruses are influenza, human immunodeficiency virus (HIV), polio, Ebola, and many more. Viruses can cause an outbreak of infectious diseases leading to an epidemic or a pandemic as the 1918 Spanish Flu and the recent 2019 Coronavirus, which is fatal, wiping away thousands of humans. Viruses can infect cells of animals, plants, and humans, causing severe health conditions that require rapid testing to determine the type of infection and to proceed with the treatment method. For example, the Hepatitis B virus (HBV) is responsible for the death of at least 600,000 annually due to health complications of HBV virus (Guirgis et al. 2010). With health, time plays a critical factor in saving lives. Current detection methods are based on detecting antigens, antibodies, RNA-isolation, and polymerase chain reaction (PCR) (Storch 2000, Greiser-Wilke et al. 2007). Developing rapid testing methods capable of accurately detecting minimal viral loads will advance the field of detecting diseases causing viruses (serology). DEP is a potential tool in understanding, characterizing, and manipulating viruses, as seen in several reported DEP based viral studies. DEP studies on viruses were first reported in 1996 (Müller et al. 1996), which demonstrated trapping of *Influenza* and *Sendai* viruses using high-frequency electric currents. Several other studies have been reported in the later years (Schnelle et al. 1996, Morgan and Green 1997, Hughes et al. 1998, Morgan et al. 1999). One of the early studies of viral particles using DEP focuses on the separation of a mixture of two different viruses, tobacco mosaic virus (TMV), a plant virus, and Herpes simplex virus, a mammalian virus, by subjecting the viral particles to electrical signal of 20 V_{pp} over the frequency ranging from 1 kHz – 20 MHz. TMV virus was also characterized using DEP modeling earlier (Morgan and Green 1997). Table 1 below

Table 1. DEP application to trap, sort, and enrich viruses.

Virus	Shape	Signal	DEP Application	References
Tobacco mosaic virus	Rod shape	100 kHz, 4 V_{pp}	Trapping and Separation	(Morgan and Green 1997)
Herpes simplex virus	Spherical	1 kHz-20 MHz, 20 V_{pp}	Separation	(Morgan et al. 1999)
Sindbis virus	Icosahedron	70 V_{DC} (iDEP)	Concentrating	(Ding et al. 2016)
Influenza (strain A)	-	-	Trapping and Enrichment	(Müller et al. 1996)
Sendai (strain Z)	-	-	Trapping	(Müller et al. 1996)
Vesicular stomatitis	-	8V_{pp}	Capture	(Docoslis et al. 2004)
Vaccinia	-	8 V_{pp}, 1 kHz	Capture and Detection	(Madiyar et al. 2017)

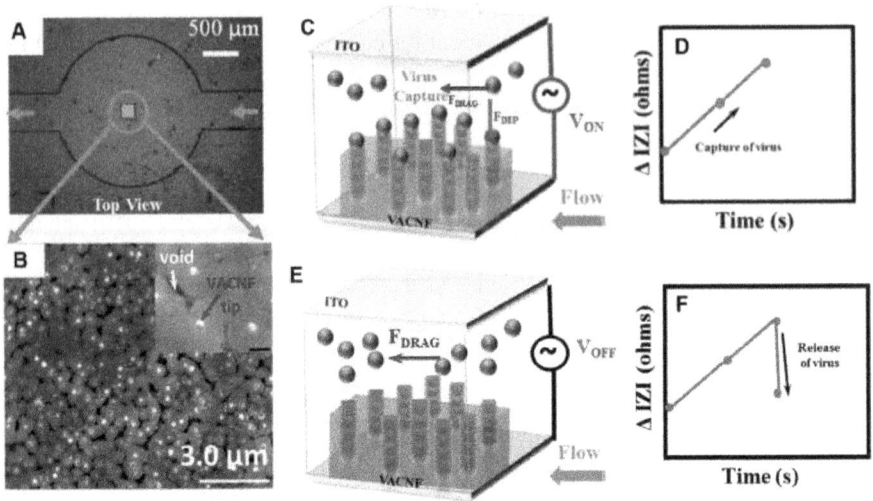

Fig. 13. Schematic diagram of DEP device used to capture and detect Vaccinia viral particles using vertically aligned carbon nanofibers (VACNFs). (A) Microscopic image of the device design. (B) Scanning electron microscopic image where bright dots represent the unexposed VACNF tips exposed, and the large irregular dark spots are the voids in the SiO_2 matrix. (C) and (E) are the schematic diagrams of working principle of while AC voltage is on and off, respectively. (D) and (F) shows the trend of change in impedance when the AC voltage is on and off (Madiyar et al. 2017). (Reprinted with permission from Madiyar, F. R., S. L. Haller, O. Farooq, S. Rothenburg, C. Culbertson and J. Li. 2017. AC dielectrophoretic manipulation and electroporation of vaccinia virus using carbon nanoelectrode arrays. Electrophoresis 38(11): 1515–1525.)

includes a few of the virus-based studies that employ DEP as a mechanism to trap, sort, and concentrate. Vaccinia virus was successfully captured at low AC voltage (8 V_{pp}) and 1 kHz using AC-DEP on an array of nanoelectrodes as shown in Figure 13 (Madiyar et al. 2017).

The above studies show that DEP trapping voltage can be harnessed to target viral species to concentrate the specific species of interest. Further, DEP has also been used to manipulate exosomes, which are similar to viruses except that they do not replicate and are released by prokaryotes and eukaryotes (Pethig 2016). They play a critical role in cellular communication to facilitate immune response, coagulation, and cell apoptosis. Recently, exosomes have been separated at higher DC voltages, i.e., 800 – 2000 V by few research groups (Ayala-Mar et al. 2019a,b) and at lower DC voltages as well using a micropipette conical design (Shi et al. 2019).

5.4 Bacteria

Inception of DEP began with studying yeast cells, a type of bacteria cells by Herbert A. Pohl (Pohl and Hawk 1966), where live and dead cells were separated. A similar study focused on using insulator-based DC DEP to concentrate and separate live and dead *E. coli* bacteria (Lapizco-Encinas et al. 2004). Several bacteria have been studied using DEP, such as *Cupriavidus necator* (Adekanmbi et al. 2019b, 2020a,

Giduthuri et al. 2021), *Babesiabovis* (Adekanmbi et al. 2016), *E. coli* (Bai et al. 2007), *Salmonella Enteritidis* (Chiok et al. 2018), *C. elegans* (Zhu et al. 2018), and many more. *E. coli* is one of the most extensively studied bacteria using DEP for characterization and as a model. A detailed review of microfluidics and DEP for bacteria-based applications has been discussed (Hanson et al. 2016). DEP can also identify isogenic strain differences in gram-negative bacteria (using *E. coli* as a model) (Castellarnau et al. 2006). This can be further extended to several other gram-negative bacterial species, which can be an important aspect in evaluating bacterial populations. Additionally, dielectrophoresis has been used to differentiate bacteria at the strain level, i.e., gram-negative and gram-positive, based on their varying dielectric properties at a low electric field of 10 V/mm as shown in Figure 14 (Braff et al. 2013). Commercializing DEP systems for bacterial analysis remains challenging, with many issues to be addressed (Páez-Avilés et al. 2016). DEP has a great potential to be used as a rapid diagnostic tool that can detect mutations as demonstrated by (Braff et al. 2013) and can replace the expensive conventional techniques such as sequencing and rRNA fingerprinting.

5.5 Yeasts

Separation of live and dead yeast cells was one of DEP's first applications in 1968, employing microbes in a pin-plate electrode system mounted in a shallow cylindrical well (Crane and Pohl 1968). Yeasts are complex eukaryotes that vary in cell volume and geometry throughout the cell division cycle due to cell growth and budding. DEP has been successfully applied to analyze and synchronize yeast cell division (Valero et al. 2011). More recently, *Saccharomyces cerevisiae*, a type of yeast cells, were subjected to pDEP and were captured at 90% efficiency using a 'barrier contactless DEP' where a thin dielectric coating of silicon carbide was used to passivate the

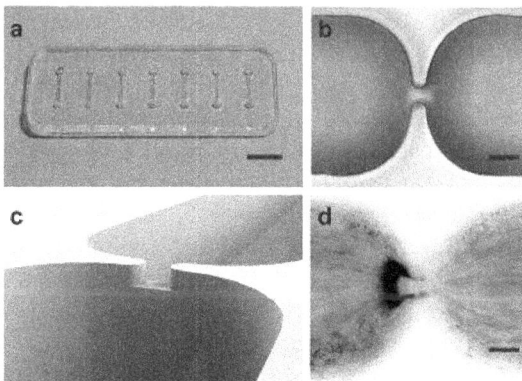

Fig. 14. Images of the microfluidic chip used to discriminate bacteria based on their surface properties. (A) Fabricated device (B) Micrograph of the contraction region (C) CAD image highlighting the constricted region (D) Inverted fluorescent image of S. mitis trapping (Braff et al. 2013). (Reprinted with permission from Braff, W. A., D. Willner, P. Hugenholtz, K. Rabaey and C. R. Buie. 2013. Dielectrophoresis-based discrimination of bacteria at the strain level based on their surface properties. PLoS One 8(10): e76751.)

electrodes (Podoynitsyn et al. 2019). Yeast cells are considered a perfect candidate to apply the multiple shell model since they possess various sub-cellular components like vacuoles, nuclei, mitochondria, etc. Dielectric properties for double yeast cells were obtained using two V-shaped electrodes, and the technique was compared to the reported values in the literature for single yeast cells to validate the instrument (García-Diego et al. 2019). Further, the same group evaluated yeast cells' dielectric properties exposed to thermal shock and compared it with live yeast cells using a combination of DEP and Stokes' drag force (Fernando-Juan et al. 2019). Using a combined DEP technique, the sorting platforms' efficiency has been dramatically improved and are more suitable to biologically and geometrically distinguish yeast cells than the state-of-the-art tools like electrorotation.

5.6 DNA

DNA is a complex cellular entity that plays a vital role in genetics. Earlier studies of DEP demonstrated that non-uniform electric fields polarize DNA and cause them to cluster, increasing the concentration of DNA by atleast 60-fold (Asbury et al. 2002). One of the reported studies uses negative DEP spectroscopy as a transduction sensor to accurately detect single nucleotide polymorphism in short DNA strands, capable of being developed into a gene sequencing tool (Gudagunti et al. 2019). Size-based separation of DNA using insulator DEP reported efficiency of up to 92% (Jones et al. 2017). Size based trapping of DNA strands ranging from 27 bp–8 kbp using DEP were studied, including the effects of size and gap on trapping efficiency over a frequency range of 0.2–10 MHz with immobilization of thiol modified DNA (Tuukkanen et al. 2007). Light-induced electrodes are also seen to have the capability to manipulate DNA, and this technique is also referred to as light-induced DEP (Hoeb et al. 2007). Conventional DEP uses metal electrodes to create a non-homogenous electric field to polarize DNA and manipulate them. Hybrid techniques such as light-induced DEP, as discussed above, and electrodeless DEP, further negate the effects of electrodes inside the channel, i.e., electrode fouling and gas evolution, avoiding electrolysis, and do not require complex fabrication techniques (Chou et al. 2002). A recent review focuses on the various DNA-based applications like trapping, enrichment, and molecular binding detection utilizing both classical and insulator-based DEP techniques (Viefhues and Eichhorn 2017).

DEP has always been popular for being a label-free technique. However, most DNA-based studies require the DNA to be fluorescently tagged, which is a huge drawback since the purity of DNA enriched or trapped is compromised and is considered no longer viable for medical applications. Alternative mechanisms that rely on impedance or capacitance-based detection could be used that depend on charged particles present in the analyzed area of the microchip (Basuray et al. 2009, Li et al. 2013).

These drawbacks definitely outweigh the advantages of using DEP since DEP is based on dielectric property change without using any matrices like affinity-based matrices making it label-free. The field of DEP has been evolving since its inception in the 1950s and will continue to progress. DEP's DNA-based applications will advance DNA science and potentially revolutionize the biomedical and molecular biology fields through DEP-based DNA sequencing (Mahshid et al. 2018).

6. Summary

Dielectrophoresis, a label-free technique, relies on the particles' electrophysiological or dielectric properties, suspending medium, and the electrodes' geometry or insulating structures embedded in the micron-scaled device that introduces motion of the particles under electric field. Smaller particles are harder to manipulate and to maintain viability. Advances in photolithography and rapid prototyping have aided in utilizing biocompatible electrodes like gold and handling a wide range of particles from nano to micron sizes. With the increasing utilization of artificial intelligence (AI) and deep learning algorithms towards precision medicine and health monitoring, researchers have a blooming opportunity to adapt some of these AI tools towards data collection and analysis to improve the sensitivity of dielectrophoretic particle characterization and sorting.

Acknowledgements

The authors would like to thank all the graduate and undergraduate students who have contributed towards experiments and modeling of the diagnostic platforms in Microfluidics & Electrokinetics bioSeparations & Analysis laboratory. A part of this chapter was adapted from the M.S. Thesis of Dr. Ezekiel Adekanmbi, titled: Applications of Electrokinetics for Disease Diagnostics.

References

Adekanmbi, E. and S. Srivastava. (2019a). Applications of Electrokinetics and Dielectrophoresis on Designing Chip-Based Disease Diagnostic Platforms. Bio-Inspired Technology. R. Srivastava. London, UK, IntechOpen.

Adekanmbi, E. O. and S. K. Srivastava. (2016). Dielectrophoretic applications for disease diagnostics using lab-on-a-chip platforms. *Lab Chip* 16: 2148–2167.

Adekanmbi, E. O., M. W. Ueti, B. Rinaldi, C. E. Suarez and S. K. Srivastava. (2016). Insulator-based dielectrophoretic diagnostic tool for babesiosis. *Biomicrofluidics* 10: 033108–033108.

Adekanmbi, E. O. and S. K. Srivastava. (2019). Dielectric characterization of bioparticles via electrokinetics: The past, present, and the future. *Applied Physics Reviews* 6: 041313.

Adekanmbi, E. O., A. T. Giduthuri, S. Waymire and S. K. Srivastava. (2019b). Utilization of dielectrophoresis for the quantification of rare earth elements adsorbed on cupriavidus necator. *ACS Sustainable Chemistry & Engineering* 8: 1353–1361.

Adekanmbi, E. O., A. T. Giduthuri, B. A. C. Carv, J. Counts, J. G. Moberly and S. K. Srivastava. (2020a). Application of dielectrophoresis towards characterization of rare earth elements biosorption by Cupriavidus necator. *Anal Chim Acta* 1129: 150–157.

Adekanmbi, E. O., A. T. Giduthuri and S. K. Srivastava. (2020b). Dielectric characterization and separation optimization of infiltrating ductal adenocarcinoma via insulator-dielectrophoresis. *Micromachines* (Basel) 11: 340.

Alinezhadbalalami, N., T. A. Douglas, N. Balani, S. S. Verbridge and R. V. Davalos. (2019). The feasibility of using dielectrophoresis for isolation of glioblastoma subpopulations with increased stemness. *Electrophoresis* 40: 2592–2600.

Arnold, W. M. and U. Zimmerman. (1982). Rotating-field-induced rotation and measurement of the membrane capacitance of single mesophyll cells of Avena sativa Zeitschrift fuer Naturforschung 37c: 908–915.

Arnold, W. M., H. P. Schwan and U. Zimmermann. (1987). Surface conductance and other properties of latex particles measured by electrorotation. *Journal of Physical Chemistry* 91: 5093–5098.

Arnold, W. M. and U. Zimmermann. (1988). Electro-rotation: Development of a technique for dielectric measurements on individual cells and particles. *Journal of Electrostatics* 21: 151–191.

Asami, K., Y. Takahashi and S. Takashima. (1989). Dielectric properties of mouse lymphocytes and erythrocytes. *Biochim Biophys Acta* 1010: 49–55.

Asbury, C. L., A. H. Diercks and G. Van den Engh. (2002). Trapping of DNA by dielectrophoresis. *Electrophoresis* 23: 2658–2666.

Ayala-Mar, S., R. C. Gallo-Villanueva and J. González-Valdez. (2019a). Dielectrophoretic manipulation of exosomes in a multi-section microfluidic device. *Materials Today: Proceedings* 13: 332–340.

Ayala-Mar, S., V. H. Perez-Gonzalez, M. A. Mata-Gómez, R. C. Gallo-Villanueva and J. González-Valdez. (2019b). Electrokinetically Driven Exosome Separation and Concentration Using Dielectrophoretic-Enhanced PDMS-Based Microfluidics. *Analytical Chemistry* 91: 14975–14982.

Bai, W., K. Zhao and K. Asami. (2007). Effects of copper on dielectric properties of *E. coli* cells. *Colloids and Surfaces B: Biointerfaces* 58: 105–115.

Ballantyne, G. R. and P. N. Holtham. (2010). Application of dielectrophoresis for the separation of minerals. *Minerals Engineering* 23: 350–358.

Basuray, S., S. Senapati, A. Aijian, A. R. Mahon and H.-C. Chang. (2009). Shear and AC field enhanced carbon nanotube impedance assay for rapid, sensitive, and mismatch-discriminating DNA hybridization. *ACS Nano* 3: 1823–1830.

Becker, F. F., X.-B. Wang, Y. Huang, R. Pethig, J. Vykoukal and P. R. C. Gascoyne. (1994). The removal of human leukaemia cells from blood using interdigitated microelectrodes. *Journal of Physics D: Applied Physics* 27: 2659–2662.

Benhal, P., G. Chase, P. Gaynor, B. Oback and W. Wang. (2015). Multiple-cylindrical electrode system for rotational electric field generation in particle rotation applications. *International Journal of Advanced Robotic Systems* 12: 84.

Bonincontro, A. and G. Risuleo. (2015). Electrorotation: A spectroscopic imaging approach to study the alterations of the cytoplasmic membrane. *Advances in Molecular Imaging* 5: 1–15.

Braff, W. A., D. Willner, P. Hugenholtz, K. Rabaey and C. R. Buie. (2013). Dielectrophoresis-based discrimination of bacteria at the strain level based on their surface properties. *PLOS ONE* 8: e76751.

Burt, J. P., K. L. Chan, D. Dawson, A. Parton and R. Pethig. (1996). Assays for microbial contamination and DNA analysis based on electrorotation. *Ann Biol Clin* (Paris) 54: 253–257.

Castellarnau, M., A. Errachid, C. Madrid, A. Juarez and J. Samitier. (2006). Dielectrophoresis as a tool to characterize and differentiate isogenic mutants of *Escherichia coli*. *Biophys J* 91: 3937.

Cetin, B. and D. Li. (2011). Dielectrophoresis in microfluidics technology. *Electrophoresis* 32: 2410–2427.

Chen, Q. and Y. J. Yuan. (2019). A review of polystyrene bead manipulation by dielectrophoresis. *RSC Advances* 9: 4963–4981.

Chiok, K. L., N. C. Paul, E. O. Adekanmbi, S. K. Srivastava and D. H. Shah. (2018). Dimethyl adenosine transferase (KsgA) contributes to cell-envelope fitness in Salmonella Enteritidis. *Microbiol Res* 216: 108–119.

Chizmadzhev, Y. A., A. V. Indenbom, P. I. Kuzmin, S. V. Galichenko, J. C. Weaver and R. O. Potts. (1998). Electrical properties of skin at moderate voltages: Contribution of appendageal macropores. *Biophysical Journal* 74: 843–856.

Chou, C.-F., J. O. Tegenfeldt, O. Bakajin, S. S. Chan, E. C. Cox, N. Darnton, T. Duke and R. H. Austin. (2002). Electrodeless dielectrophoresis of single- and double-stranded DNA. *Biophysical Journal* 83: 2170–2179.

Crane, J. S. and H. A. Pohl. (1968). A study of living and dead yeast cells using dielectrophoresis. *Journal of the Electrochemical Society* 115: 584.

Cristofanilli, M., G. De Gasperis, L. Zhang, M.-C. Hung, P. R. C. Gascoyne and G. N. Hortobagyi. (2002). Automated electrorotation to reveal dielectric variations related to HER-2/neu overexpression in MCF-7 sublines. *Clinical Cancer Research* 8: 615.

Cumova, J., L. Kovarova, A. Potacova, I. Buresova, F. Kryukov, M. Penka, J. Michalek and R. Hajek. (2010). Optimization of immunomagnetic selection of myeloma cells from bone marrow using magnetic activated cell sorting. *Int J Hematol* 92: 314–319.

Dalton, C., A. D. Goater, J. P. H. Burt and H. V. Smith. (2004). Analysis of parasites by electrorotation. *Journal of Applied Microbiology* 96: 24–32.

Dalton, C., A. D. Goater, J. Drysdale and R. Pethig. (2001). Parasite viability by electrorotation. *Colloids and Surfaces A: Physicochemical and Engineering Aspects* 195: 263–268.

Ding, J., R. M. Lawrence, P. V. Jones, B. G. Hogue and M. A. Hayes. (2016). Concentration of Sindbis virus with optimized gradient insulator-based dielectrophoresis. *The Analyst* 141: 1997–2008.

Docoslis, A., L. Tercero, B. Israel, N. Abbott and P. Alexandridis. (2004). Dielectrophoretic capture of viral particles from media of physiological ionic strength. AIChE Annual Meeting, Conference Proceedings.

Donath, E., M. Egger and V. P. Pastushenko. (1990). Dielectric behavior of the anion-exchange protein of human red blood cells: Theoretical analysis and comparison to electrorotation data. *Journal of Electroanalytical Chemistry and Interfacial Electrochemistry* 298: 337–360.

Egger, M. and E. Donath. (1995). Electrorotation measurements of diamide-induced platelet activation changes. *Biophysical Journal* 68: 364–372.

Esch, M., V. L. Sukhorukov, M. Kürschner and U. Zimmermann. (1999). Dielectric properties of alginate beads and bound water relaxation studied by electrorotation. *Biopolymers* 50: 227–237.

Feldman, Y., I. Ermolina and Y. Hayashi. (2003). Time domain dielectric spectroscopy study of biological systems. *IEEE Transactions on Dielectrics and Electrical Insulation* 10: 728–753.

Fernando-Juan, G.-D., M. Rubio-Chavarría, P. Beltrán and F. J. Espinós. (2019). Thermal shock response of yeast cells characterised by dielectrophoresis force measurement. *Sensors* 19: 5304.

Fuhr, G. and R. Hagedorn. (1996). Cell electrorotation. Electrical Manipulation of Cells. P. T. Lynch and M. R. Davey. Boston, MA, Springer.

García-Diego, F.-J., M. Rubio-Chavarría, P. Beltrán and F. J. Espinós. (2019). Characterization of simple and double yeast cells using dielectrophoretic force measurement. *Sensors* (Basel, Switzerland) 19: 3813.

Gascoyne, P., R. Pethig, J. Satayavivad, F. F. Becker and M. Ruchirawat. (1997a). Dielectrophoretic detection of changes in erythrocyte membranes following malarial infection. *Biochimica et Biophysica Acta (BBA) - Biomembranes* 1323: 240–252.

Gascoyne, P. R. and S. Shim. (2014). Isolation of circulating tumor cells by dielectrophoresis. *Cancers* 6: 545.

Gascoyne, P. R. C., J. Noshari, F. F. Becker and R. Pethig. (1994). Use of dielectrophoretic collection spectra for characterizing differences between normal and cancerous cells. *IEEE Transactions on Industry Applications* 30: 829–834.

Gascoyne, P. R., X. B. Wang, Y. Huang and F. F. Becker. (1997b). Dielectrophoretic separation of cancer cells from blood. *IEEE Trans Ind Appl* 33: 670.

Gasperis, G. D., X. Wang, J. Yang, F. F. Becker and P. R. C. Gascoyne. (1998). Automated electrorotation: Dielectric characterization of living cells by real-time motion estimation. *Measurement Science and Technology* 9: 518–529.

Gheorghiu, E. and K. Asami. (1998). Monitoring cell cycle by impedance spectroscopy: Experimental and theoretical aspects. *Bioelectrochemistry and Bioenergetics* 45: 139–143.

Giduthuri, A. T., S. K. Theodossiou, N. R. Schiele and S. K. Srivastava. (2020). Dielectrophoresis as a tool for electrophysiological characterization of stem cells. *Biophysics Reviews* 1: 011304.

Giduthuri, A. T., E. O. Adekanmbi, S. K. Srivastava and J. G. Moberly. (2021). Dielectrophoretic ultra-high-frequency characterization and *in silico* sorting on uptake of rare earth elements by Cupriavidus necator. *Electrophoresis* 42: 656–666.

Gimsa, J., T. Muller, T. Schnelle and G. Fuhr. (1996). Dielectric spectroscopy of single human erythrocytes at physiological ionic strength: Dispersion of the cytoplasm. *Biophys J* 71: 495–506.

Goater, A. D. and R. Pethig. (1998). Electrorotation and dielectrophoresis. *Parasitology* 117 Suppl: S177–189.

Greiser-Wilke, I., S. Blome and V. Moennig. (2007). Diagnostic methods for detection of Classical swine fever virus—Status quo and new developments. Vaccine 25: 5524–5530.

Gudagunti, F. D., L. Velmanickam, D. Nawarathna and I. T. Lima, Jr. (2019). Nucleotide Identification in DNA Using Dielectrophoresis Spectroscopy. *Micromachines* 11: 39.

Guirgis, B. S. S., R. O. Abbas and H. M. E. Azzazy. (2010). Hepatitis B virus genotyping: Current methods and clinical implications. *International Journal of Infectious Diseases* 14: e941–e953.

Haandbæk, N., S. C. Bürgel, F. Rudolf, F. Heer and A. Hierlemann. (2016). Characterization of single yeast cell phenotypes using microfluidic impedance cytometry and optical imaging. *ACS Sensors* 1: 1020–1027.

Hanson, C., M. Sieverts, K. Tew, A. Dykes, M. Salisbury and E. Vargis. (2016). The use of microfluidics and dielectrophoresis for separation, concentration, and identification of bacteria, SPIE.

Hawkins, B. G., A. E. Smith, Y. A. Syed and B. J. Kirby. (2007). Continuous-flow particle separation by 3D insulative dielectrophoresis using coherently shaped, DC-biased, AC electric fields. *Anal Chem* 79: 7291–7300.

Hawkins, B. G., N. Lai and D. S. Clague. (2020). High-sensitivity in dielectrophoresis separations. *Micromachines* (Basel) 11: 391.

Hayashi, Y., L. Livshits, A. Caduff and Y. Feldman. (2003). Dielectric spectroscopy study of specific glucose influence on human erythrocyte membranes. *Journal of Physics D: Applied Physics* 36: 369–374.

Hildebrandt, C., H. Büth, S. Cho, Impidjati and H. Thielecke. (2010). Detection of the osteogenic differentiation of mesenchymal stem cells in 2D and 3D cultures by electrochemical impedance spectroscopy. *Journal of Biotechnology* 148: 83–90.

Hirota, Y. and M. Hakoda. (2011). Relationship between dielectric characteristic by DEP levitation and differentiation activity for stem cells. *Key Engineering Materials* 459: 84–91.

Hoeb, M., J. O. Rädler, S. Klein, M. Stutzmann and M. S. Brandt. (2007). Light-induced dielectrophoretic manipulation of DNA. *Biophysical Journal* 93: 1032–1038.

Holmes, D. and H. Morgan. (2010). Single cell impedance cytometry for identification and counting of CD4 T-cells in human blood using impedance labels. *Analytical Chemistry* 82: 1455–1461.

Holmes, D. and B. L. J. Webb. (2012). Electrical Impedance Cytometry. *Encyclopedia of Nanotechnology*. B. Bhushan. Dordrecht, Springer Netherlands 662–671.

Huang, L., W. He and W. Wang. (2019). A cell electro-rotation micro-device using polarized cells as electrodes. *Electrophoresis* 40: 784–791.

Huang, L., P. Zhao and W. Wang. (2018). 3D cell electrorotation and imaging for measuring multiple cellular biophysical properties. *Lab on a Chip* 18: 2359–2368.

Huang, Y., R. Holzel, R. Pethig and X.-B. Wang. (1992). Differences in the AC electrodynamics of viable and non-viable yeast cells determined through combined dielectrophoresis and electrorotation studies. *Physics in Medicine and Biology* 37: 1499–1517.

Huang, Y., X.-B. Wang, R. Holzel, F. F. Becker and P. R. C. Gascoyne. (1995). Electrorotational studies of the cytoplasmic dielectric properties of Friend murine erythroleukaemia cells. *Physics in Medicine and Biology* 40: 1789–1806.

Hughes, M. P., H. Morgan, F. J. Rixon, J. P. Burt and R. Pethig. (1998). Manipulation of herpes simplex virus type 1 by dielectrophoresis. *Biochim Biophys Acta* 1425: 119–126.

Irimajiri, A., T. Hanai and A. Inouye. (1979). A dielectric theory of "multi-stratified shell" model with its application to a lymphoma cell. *Journal of Theoretical Biology* 78: 251–269.

Jaffe, A. and J. Voldman. (2018). Multi-frequency dielectrophoretic characterization of single cells. *Microsystems & Nanoengineering* 4: 23.

Jen, C.-P., C.-H. Weng and C.-T. Huang. (2011). Three-dimensional focusing of particles using negative dielectrophoretic force in a microfluidic chip with insulating microstructures and dual planar microelectrodes. *Electrophoresis* 32: 2428–2435.

Jiang, A. Y. L., A. R. Yale, M. Aghaamoo, D. H. Lee, A. P. Lee, T. N. G. Adams and L. A. Flanagan. (2019). High-throughput continuous dielectrophoretic separation of neural stem cells. *Biomicrofluidics* 13: 064111.

Jones, P. V., G. L. Salmon and A. Ros. (2017). Continuous separation of DNA molecules by size using insulator-based dielectrophoresis. *Analytical Chemistry* 89: 1531–1539.

Jones, T. B. (2003). Basic theory of dielectrophoresis and electrorotation. *IEEE Engineering in Medicine and Biology Magazine* 22: 33–42.

Jubery, T., S. K. Srivastava and P. Dutta. (2014). Dielectrophoresis separation of bioparticles in microdevices: A review. *Electrophoresis* 35: 691–713.

Kazemi, B. and J. Darabi. (2018). Numerical simulation of dielectrophoretic particle separation using slanted electrodes. *Physics of Fluids* 30: 102003.

Kirby, B. J. (2010). Micro- and Nanoscale Fluid Mechanics: Transport in Microfluidic Devices. Cambridge, Cambridge University Press.

Lapizco-Encinas, B. H., B. A. Simmons, E. B. Cummings and Y. Fintschenko. (2004). Dielectrophoretic concentration and separation of live and dead bacteria in an array of insulators. *Analytical Chemistry* 76: 1571–1579.

Li, P. C. H. (2006). Microfluidic Lab-on-a-Chip for Chemical and Biological Analysis and Discovery. Boca Raton, FL, CRC press.

Li, S., Q. Yuan, B. I. Morshed, C. Ke, J. Wu and H. Jiang. (2013). Dielectrophoretic responses of DNA and fluorophore in physiological solution by impedimetric characterization. *Biosens Bioelectron* 41: 649–655.

Lin, Y.-S., S. Tsang, R. Ghasemi, S. Bensalem, O. Français, F. Lopes, H.-Y. Wang, C.-L. Sun and B. L. Pioufle. (2017). Dielectric characterisation of single microalgae cell using electrorotation measurements. Proceedings 1.

Liu, W., Y. Ren, Y. Tao, Y. Li and X. Chen. (2017). Controllable rotating behavior of individual dielectric microrod in a rotating electric field. *ELECTROPHORESIS* 38: 1427–1433.

Łos, M. J., A. Skubis and S. Ghavami. (2019). Chapter 2—Stem Cells. Stem Cells and Biomaterials for Regenerative Medicine. M. J. Łos, A. Hudecki and E. Wiecheć, Academic Press 5–s16.

Lu, J., C. A. Barrios, A. R. Dickson, J. L. Nourse, A. P. Lee and L. A. Flanagan. (2012). Advancing practical usage of microtechnology: A study of the functional consequences of dielectrophoresis on neural stem cells. *Integrative Biology* 4: 1223–1236.

Madiyar, F. R., S. L. Haller, O. Farooq, S. Rothenburg, C. Culbertson and J. Li. (2017). AC dielectrophoretic manipulation and electroporation of vaccinia virus using carbon nanoelectrode arrays. *Electrophoresis* 38: 1515–1525.

Mahshid, S., J. Lu, A. A. Abidi, R. Sladek, W. W. Reisner and M. J. Ahamed. (2018). Transverse dielectrophoretic-based DNA nanoscale confinement. *Scientific Reports* 8: 5981.

Manczak, R., S. Saada, T. Provent, C. Dalmay, B. Bessette, G. Begaud, S. Battu, P. Blondy, M. Jauberteau, C. B. Kaynak, M. Kaynak, C. Palego, F. Lalloue and A. Pothier. (2019). UHF-dielectrophoresis crossover frequency as a new marker for discrimination of glioblastoma undifferentiated cells. *IEEE Journal of Electromagnetics, RF and Microwaves in Medicine and Biology*, 1–1.

Mayer, G., M. S. Ahmed, A. Dolf, E. Endl, P. A. Knolle and M. Famulok. (2010). Fluorescence-activated cell sorting for aptamer SELEX with cell mixtures. *Nat Protoc* 5: 1993–2004.

Michálek, T., A. Bolopion, Z. Hurák and M. Gauthier. (2019). Control-oriented model of dielectrophoresis and electrorotation for arbitrarily shaped objects. *Physical Review E* 99: 053307.

Miklavcic, D., N. Paveselj and F. X. Hart. (2006). Electrical properties of Tissues. *Encyclopedia of Biomedical Engineering*. M. A. (Ed). https://doi.org/10.1002/9780471740360.ebs0403.

Miller, R. D. and T. B. Jones. (1993). Electro-orientation of ellipsoidal erythrocytes. Theory and experiment. *Biophys J* 64: 1588–1595.

Molla, S., J. Masliyah and S. Bhattacharjee. (2005). Simulation of dielectrophoretic membrane filtration process for removal of water droplets from water-in-oil emulsions. *Journal of Colloid and Interface Science* 287: 338–350.

Morgan, H. and N. G. Green. (1997). Dielectrophoretic manipulation of rod-shaped viral particles. *Journal of Electrostatics* 42: 279–293.

Morgan, H., M. P. Hughes and N. G. Green. (1999). Separation of submicron bioparticles by dielectrophoresis. *Biophysical Journal* 77: 516–525.

Müller, T., S. Fiedler, T. Schnelle, K. Ludwig, H. Jung and G. Fuhr. (1996). High frequency electric fields for trapping of viruses. *Biotechnology Techniques* 10: 221–226.

Nourse, J. L., J. L. Prieto, A. R. Dickson, J. Lu, M. M. Pathak, F. Tombola, M. Demetriou, A. P. Lee and L. A. Flanagan. (2014). Membrane biophysics define neuron and astrocyte progenitors in the neural lineage. *Stem Cells* 32: 706–716.

Ozuna-Chacón, S., B. H. Lapizco-Encinas, M. Rito-Palomares, S. O. Martínez-Chapa and C. Reyes-Betanzo. (2008). Performance characterization of an insulator-based dielectrophoretic microdevice. *Electrophoresis* 29: 3115–3122.

Páez-Avilés, C., E. Juanola-Feliu, J. Punter-Villagrasa, B. Del Moral Zamora, A. Homs-Corbera, J. Colomer-Farrarons, P. L. Miribel-Català and J. Samitier. (2016). Combined dielectrophoresis and impedance systems for bacteria analysis in microfluidic on-chip platforms. *Sensors* (Basel, Switzerland) 16: 1514.

Pastushenko, V. P., P. I. Kuzmin and Y. A. Chizmadshev. (1985). Dielectrophoresis and electrorotation: A unified theory of spherically symmetrical cells. *Stud Biophys* 110: 51–57.

Pethig, R. and G. H. Markx. (1997). Applications of dielectrophoresis in biotechnology. *Trends Biotechnol* 15: 426–432.

Pethig, R. (2010). Review article-dielectrophoresis: Status of the theory, technology, and applications. *Biomicrofluidics* 4: 022811.

Pethig, R. (2016). Review—Where is Dielectrophoresis (DEP) going? *Journal of The Electrochemical Society* 164: B3049–B3055.

Piyasena, M. E., P. P. Austin Suthanthiraraj, R. W. Applegate, A. M. Goumas, T. A. Woods, G. P. López and S. W. Graves. (2012). Multinode acoustic focusing for parallel flow cytometry. *Analytical Chemistry* 84: 1831–1839.

Podoynitsyn, S. N., O. N. Sorokina, M. A. Klimov, I. I. Levin and S. B. Simakin. (2019). Barrier contactless dielectrophoresis: A new approach to particle separation. *Separation Science Plus* 2: 59–68.

Pohl, H. A. and I. Hawk. (1966). Separation of living and dead cells by dielectrophoresis. *Science* 152: 647–649.

Pohl, H. A. (1978). Dielectrophoresis the behavior of neutral matter in nonuniform electric fields. Cambridge, Cambridge University Press.

Rahman, N. A., F. Ibrahim and B. Yafouz. (2017). Dielectrophoresis for biomedical sciences applications: A Review. *Sensors* 17: 449.

Raicu, V., G. Raicu and G. Turcu. (1996). Dielectric properties of yeast cells as simulated by the two-shell model. *Biochimica et Biophysica Acta (BBA) - Bioenergetics* 1274: 143–148.

Ramos, A. (2011). Electrokinetics and Electrohydrodynamics in Microsystems, Springer.

Ren, Y. K., D. Morganti, H. Y. Jiang, A. Ramos and H. Morgan. (2011). Electrorotation of metallic microspheres. *Langmuir* 27: 2128–2131.

Salmanzadeh, A., L. Romero, H. Shafiee, R. C. Gallo-Villanueva and M. A. Stremler. (2012). Isolation of prostate tumor initiating cells (TICs) through their dielectrophoretic signature. *Lab Chip* 12: 182.

Sancho, M., G. Martínez and C. Martín. (2003). Accurate dielectric modeling of shelled particles and cells. *Journal of Electrostatics* 57: 143–156.

Schnelle, T., T. Müller, S. Fiedler, S. G. Shirley, K. Ludwig, A. Herrmann, G. Fuhr, B. Wagner and U. Zimmermann. (1996). Trapping of viruses in high-frequency electric field cages. *Naturwissenschaften* 83: 172–176.

Shafiee, H., M. B. Sano, E. A. Henslee, J. L. Caldwell and R. V. Davalos. (2010). Selective isolation of live/dead cells using contactless dielectrophoresis (cDEP). *Lab Chip* 10: 438–445.

Shi, L., D. Kuhnell, V. J. Borra, S. M. Langevin, T. Nakamura and L. Esfandiari. (2019). Rapid and label-free isolation of small extracellular vesicles from biofluids utilizing a novel insulator based dielectrophoretic device. *Lab on a Chip* 19: 3726–3734.

Simon, M. G., Y. Li, J. Arulmoli, L. P. McDonnell, A. Akil, J. L. Nourse, A. P. Lee and L. A. Flanagan. (2014). Increasing label-free stem cell sorting capacity to reach transplantation-scale throughput. *Biomicrofluidics* 8: 064106–064106.

Soltanian-Zadeh, S., K. Kikkeri, A. N. Shajahan-Haq, J. Strobl, R. Clarke and M. Agah. (2017). Breast cancer cell obatoclax response characterization using passivated-electrode insulator-based dielectrophoresis. *Electrophoresis* 38: 1988–1995.

Song, H., J. M. Rosano, Y. Wang, C. J. Garson, B. Prabhakarpandian, K. Pant, G. J. Klarmann, A. Perantoni, L. M. Alvarez and E. Lai. (2015). Continuous-flow sorting of stem cells and differentiation products based on dielectrophoresis. *Lab on a Chip* 15: 1320–1328.

Srivastava, S. K., A. Gencoglu and A. R. Minerick. (2011). DC insulator dielectrophoretic applications in microdevice technology: A review. *Anal Bioanal Chem* 399: 301–321.

Storch, G. A. (2000). Diagnostic virology. *Clinical Infectious Diseases* 31: 739–751.

Sukhorukov, V. L. and U. Zimmermann. (1996). Electrorotation of erythrocytes treated with dipicrylamine: Mobile charges within the membrane show their "Signature" in rotational spectra. *The Journal of Membrane Biology* 153: 161–169.

Sun, T. and H. Morgan. (2010). Single-cell microfluidic impedance cytometry: A review. *Microfluidics and Nanofluidics* 8: 423–443.

Tandon, V., S. K. Bhagavatula, W. C. Nelson and B. J. Kirby. (2008). Zeta potential and electroosmotic mobility in microfluidic devices fabricated from hydrophobic polymers: 1. The origins of charge. *Electrophoresis* 29: 1092–1101.

Trainito, C. (2015). Study of cell membrane permeabilization induced by pulsed electric field—electrical modeling and characterization on biochip, Université Paris-Saclay.

Turcu, I. and C. M. Lucaciu. (1989). Dielectrophoresis: A spherical shell model. *Journal of Physics A: Mathematical and General* 22: 985–993.

Tuukkanen, S., A. Kuzyk, J. J. Toppari, H. Häkkinen, V. P. Hytönen, E. Niskanen, M. Rinkiö and P. Törmä. (2007). Trapping of 27 bp-8 kbp DNA and immobilization of thiol-modified DNA using dielectrophoresis. *Nanotechnology* 18: 10 p.

Vahey, M. D., L. Quiros Pesudo, J. P. Svensson, L. D. Samson and J. Voldman. (2013). Microfluidic genome-wide profiling of intrinsic electrical properties in Saccharomyces cerevisiae. *Lab on a Chip* 13: 2754–2763.

Valero, A., T. Braschler, A. Rauch, N. Demierre, Y. Barral and P. Renaud. (2011). Tracking and synchronization of the yeast cell cycle using dielectrophoretic opacity. *Lab on a Chip* 11: 1754–1760.

Viefhues, M. and R. Eichhorn. (2017). DNA dielectrophoresis: Theory and applications a review. *Electrophoresis* 38: 1483–1506.

Voyer, D., M. Frénéa-Robin, F. Buret and L. Nicolas. (2010). Improvements in the extraction of cell electric properties from their electrorotation spectrum. *Bioelectrochemistry* 79: 25–30.

Vrinceanu, D. and E. Gheorghiu. (1996). Shape effects on the dielectric behaviour of arbitrarily shaped particles with particular reference to biological cells. *Bioelectrochemistry and Bioenergetics* 40: 167–170.

Vykoukal, J., D. M. Vykoukal, S. Freyberg, E. U. Alt and P. R. C. Gascoyne. (2008). Enrichment of putative stem cells from adipose tissue using dielectrophoretic field-flow fractionation. *Lab on a Chip* 8: 1386–1393.

Wang, X. B., Y. Huang, R. Holzel, J. P. H. Burt and R. Pethig. (1993). Theoretical and experimental investigations of the interdependence of the dielectric, dielectrophoretic and electrorotational behaviour of colloidal particles. *Journal of Physics D: Applied Physics* 26: 312–322.

Wang, X.-B., Y. Huang, P. R. C. Gascoyne, F. F. Becker, R. Hölzel and R. Pethig. (1994). Changes in Friend murine erythroleukaemia cell membranes during induced differentiation determined by electrorotation. *Biochimica et Biophysica Acta (BBA) - Biomembranes* 1193: 330–344.

Weng, P.-Y., I. A. Chen, C.-K. Yeh, P.-Y. Chen and J.-Y. Juang. (2016). Size-dependent dielectrophoretic crossover frequency of spherical particles. *Biomicrofluidics* 10: 011909–011909.

Wessner, D. R. (2010). The Origins of Viruses. *Nature Education* 3: 37.

Wu, A. Y. and D. M. Morrow. (2012). Clinical use of Dieletrophoresis separation for live Adipose derived stem cells. *Journal of Translational Medicine* 10: 99.

Xiao-Bo, W., H. Ying, P. R. C. Gascoyne and F. F. Becker. (1997). Dielectrophoretic manipulation of particles. *IEEE Transactions on Industry Applications* 33: 660–669.

Xie, H., R. Tewari, H. Fukushima, J. Narendra, C. Heldt, J. King and A. R. Minerick. (2014). Development of a 3D graphene electrode dielectrophoretic device. *J Vis Exp* e51696.

Xing, X. and L. Yobas. (2016). A single-cell impedance flow cytometry microsystem based on 3D silicon microelectrodes. 2016 IEEE 29th International Conference on Micro Electro Mechanical Systems (MEMS).

Yang, F., X. Yang, H. Jiang, W. M. Butler and G. Wang. (2013). Dielectrophoretic separation of prostate cancer cells. *Technol Cancer Res Treat* 12: 61–70.

Zahedi Siani, O., M. Zabetian Targhi, M. Sojoodi and M. Movahedin. (2020). Dielectrophoretic separation of monocytes from cancer cells in a microfluidic chip using electrode pitch optimization. *Bioprocess and Biosystems Engineering* 43: 1573–1586.

Zhang, Y., Y. Zhao, D. Chen, K. Wang, Y. Wei, Y. Xu, C. Huang, J. Wang and J. Chen. (2019). Crossing constriction channel-based microfluidic cytometry capable of electrically phenotyping large populations of single cells. *Analyst* 144: 1008–1015.

Zhou, T., Y. Ming, S. F. Perry and S. Tatic-Lucic. (2016). Estimation of the physical properties of neurons and glial cells using dielectrophoresis crossover frequency. *Journal of Biological Physics* 42: 571–586.

Zhou, X.-F., G. H. Markx, R. Pethig and I. M. Eastwood. (1995). Differentiation of viable and non-viable bacterial biofilms using electrorotation. *Biochimica et Biophysica Acta (BBA)—General Subjects* 1245: 85–93.

Zhu, Z., W. Chen, B. Tian, Y. Luo, J. Lan, D. Wu, D. Chen, Z. Wang and D. Pan. (2018). Using microfluidic impedance cytometry to measure C. elegans worms and identify their developmental stages. *Sensors and Actuators B: Chemical* 275: 470–482.

CHAPTER 16

Nanosensor Arrays
Innovative Approaches for Medical Diagnosis

Naumih M. Noah[1,*] and *Peter M. Ndangili*[2]

1. Introduction

In medical diagnosis, one problem that is faced is that symptoms of some conditions can only arise after a certain amount of time and by the time the symptoms come to the surface, the underlying condition will have advanced to a stage at which its treatment is much more complicated than it would have been had the problem been discovered earlier (McIntosh 2016). Therefore, diseases' early diagnosis and monitoring of physical conditions are very important for better-quality health management which is crucial in providing better health care to reduce mortality rates and medical care costs (Bhardwaj and Kaushik 2017, Kim et al. 2017, Kaushik and Mujawar 2018, Kaushik et al. 2018, Noah and Ndangili 2019). This can be accomplished by making opportune decisions based on fast diagnostics, smart data analysis, and informatics analysis (Kaushik and Mujawar 2018). Early detection is very crucial for many diseases to provide better treatment since it can increase the probability of curing diseases and significantly improve the mortality rate (Bellah et al. 2012). This demands smart therapeutics, analytical tools, and diagnostics systems to enhance health wellness (Kaushik et al. 2014, Kaushik and Mujawar 2018). Management of a disease progression and monitoring evaluation effectively is important for understanding and controlling the disease and depends on the optimization of therapeutics (Kaushik and Mujawar 2018). Therefore, the development of smart diagnostic systems for personalized health care such as point-of-care devices is imperative. Point-of-care testing ensures fast detection of analytes near to the patient, thereby facilitating a better disease diagnosis, monitoring, and management. It also enables quick medical

[1] School of Pharmacy and Health Sciences, United States International University-Africa (USIU-A) P.O Box 14634-00800, Nairobi, Kenya.
[2] School of Chemistry and Material Science (SCMS), Technical University of Kenya P.O. Box 52428 – 00200 Nairobi- Kenya.
   Email: ndangilipeter0@gmail.com
* Corresponding author: mnoah@usiu.ac.ke, noahnaumih@gmail.com

decisions since the diseases can be diagnosed at an early stage leading to improved health outcomes for the patients and enable them to start early treatment (Vashist 2017).

Many diseases can be accompanied by characteristic odors, whose recognition can provide diagnostic clues, can also guide the laboratory evaluation, and affect the choice of immediate therapy (Center 2015). For example, the study of the chemical composition in human breath using gas chromatography/mass spectrometry (GC/ MS) has been found to show a correlation between volatile compounds and the manifestation of certain illnesses (Center 2015). The existence of those specific compounds can provide a warning to physiological malfunction and support the disease diagnosis (Center 2015). This condition requires an analytical tool with very high sensitivity for measurement (Center 2015). With the advent of personalized medicine, novel diagnostic tools that capture molecular activity at a disease site *in vivo* are needed to better understand an individual patient's disease state, inform clinical decision-making, and monitor therapeutic response (Schürle-Finke 2020).

Nanotechnology has been able to improve the sensitivity and linear ranges of various detection methods improving assessment of the disease onset and its progression and help to plan for treatment of many diseases. Therefore, the development of nanosensors for point-of-care diagnostics is an important area of research. Nanosensors are defined as sensing devices with at least one of their sensing dimensions being up to100 nanometer (nm) (Munawar et al. 2019, Abdel-Karim et al. 2020). Their specificity is imparted by targeting ligands which are directly conjugated to the nanomaterials and depending on the functionality of the ligand, it attracts a particular marker of interest (analyte), while the nanomaterials contribute the sensitivity, converting the signals from one form to the other or act as a detector for generated signals (Munawar et al. 2019). Nanosensors are instrumental for (a) detecting physical and chemical changes, (b) monitoring biomolecules and biochemical changes in cells, and (c) measuring toxic and polluting materials presented in the industry and environment (Abdel-Karim et al. 2020). Nanosensors with high sensitivity normally utilize electrical, optical, and acoustic properties to improve the detection limits of analytes (Munawar et al. 2019). The unique and exceptional properties of nanomaterials (large surface area to volume ratio, composition, charge, reactive sites, physical structure, and potential) are exploited for sensing purposes. The high-sensitivity in analyte recognition is achieved by preprocessing of samples, signal amplification, and applying different transduction approaches (Munawar et al. 2019). Nanosensors have shown promise as theranostic tools and as nanoprobes for diagnostic imaging, disease monitoring, and the detection of pathogens and environmental contaminants and can provide real-time information of a patient's condition which can be of significant benefit (Munawar et al. 2019, Abdel-Karim et al. 2020). For example, the application of nanoelectronic biosensors in medicine is a huge area of research due to the ability of nanoparticles to proficiently target and permeate specific tissues within the body (Miranda 2019).

Nanosensors can be classified according to their energy source, structure, and applications (Abdel-Karim et al. 2020). The nanostructured materials used in the manufacturing of nanosensors are such as nanoscale wires, which are capable of high detection sensitivity, carbon nanotubes with very high surface area and high electron conductivity, thin films, metal and metal oxides' nanoparticles that have better absorption and scattering abilities, polymer, and biomaterials (Miranda 2019, Abdel-Karim et al. 2020). In the next sections of this chapter, we describe the development of the different types of nanosensors, their characterization and validation, their fabrication into nanosensor arrays, and finally the application of the nanosensor arrays in medical diagnostics of cancers, diabetes, malaria, and Human Immunodeficiency Virus (HIV).

2. Development of Nanosensors

With the rising need in the field of point-of-care and on-field diagnostics, the need for portable, sensitive, and specific sensors has led to the development of new sensors that fundamentally function by the integration of bioreceptor molecules against various specific biomarkers (targeted antigenic biomolecules against which a bioreceptor can be generated) with transducer elements coupled with nanostructures (Tuteja et al. 2019). Advanced analytical techniques then utilize these biomarkers for the diagnosis of different clinical/environmental analytes (Tuteja et al. 2019). Due to their distinctive electronic, optical, and catalytic properties of nanomaterials such as gold nanoparticles (AuNPs), silver nanoparticles (AgNPs), carbon nanotubes (CNTs), graphene oxide (GO), gold nanorods, quantum dots (QDs), and silica nanoparticles have been used as transducers, sensor platforms or labels in the development of nanosensors for environmental pollutants or clinically important biomarkers (Tuteja et al. 2019). A typical nanosensor normally consists of the analyte, sensors surface, transducer, and a detector with feedback from the detector going to the sensor (AzoNano 2019) as shown in Figure 1. The sensitivity, specificity, and ease of execution are the main goals in designing a nanosensor. The sensitivity of the sensor platform significantly depends on the characteristics of nanomaterials used as well as the functionalization of sensor surfaces with biorecognition elements (Tuteja et al. 2019). Nanosensors typically work by monitoring electrical changes in the sensor materials (AzoNano 2019, Rong et al. 2019). For example, in carbon nanotube-based sensors when a molecule of nitrogen dioxide (NO_2) is present, it will strip an electron from the nanotube, which in turn causes the nanotube to be less conductive. If ammonia (NH_3) is present, it reacts with water vapor and donates an electron to

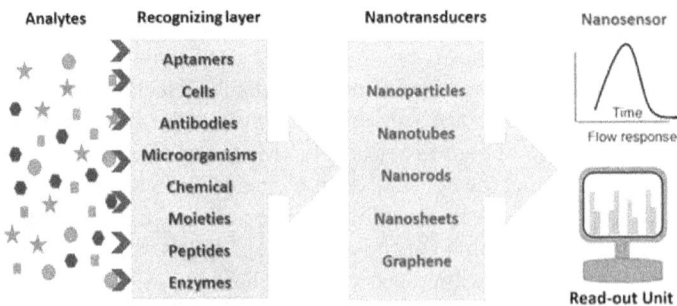

Fig. 1. Schematic diagram showing essential components of a nanosensor and their widely used examples.

the carbon nanotube, making it more conductive. By treating the nanotubes with various coating materials, they can be made sensitive to certain molecules and immune to others (Rong et al. 2019). Nanosensors have been developed to the point of measurement at the single-molecule level (Rong et al. 2019).

2.1 Classification of Nanosensors

There are different ways in which nanosensors are classified. They can be classified into chemical and mechanical nanosensors according to their energy source, structure, and applications (Critchley 2018, Abdel-Karim et al. 2020). A chemical nanosensor detects chemicals by measuring the change in the electrical conductivity of the nanomaterial once an analyte has been detected (Critchley 2018). Since many nanomaterials have a high electrical conductivity, which reduces upon binding or adsorption of a molecule, it is that detectable change that is measured (Critchley 2018). One dimensional (1D) materials, such as nanowires and nanotubes, are excellent examples of chemical nanosensors, as their electrically confined structure can act as both the transducer and the electronic wires once an analyte has been detected (Critchley 2018). Mechanical nanosensors also work by detecting a change in the electrical conductivity of a material. However, the nanomaterials in this case change their electrical conductivity when the material is physically manipulated, and this physical change invokes a detectable response (Critchley 2018). This response can also be measured using an attached capacitor, where the physical change creates a measurable change in the capacitance (Critchley 2018).

Nanosensors can also be classified based on the nature of the receptor molecules, structure, and application (Khanna 2011, Saini et al. 2017, Munawar et al. 2019). Based on the receptor molecules, there can be affinity-based nanosensors and catalytic based nanosensors. The affinity-based nanosensors normally use receptors such as antibodies, hormones, and nucleic acids which bind molecules of interest irreversibly and non-catalytically, while catalytic based nanosensors use receptors such as enzymes and microbiological cells which bind molecules of interest and catalytically convert them into recognizable products (Khanna 2011, Saini et al. 2017). Based on the structure, there can be optical nanosensors that measure amplitude, energy, polarization delay time, as well as decay phase, and electrochemical nanosensors which measure electrochemical and mass transduction mechanisms (Khanna 2011, Saini et al. 2017, Munawar et al. 2019). Lastly, based on applications, we have chemical nanosensors which consist of capacitive cantilevers and electronics which can detect single chemical and biological molecules, bionanosensors which consist of biomolecules for the detection of cancers, deoxyribonucleic acid (DNA) among others, electrometers which consist of torsional mechanical resonators and deployable nanosensors used in military and other security applications. Figure 2 below summarizes these nanosensors (Saini et al. 2017).

3. Fabrication of Nanosensors Arrays

Nanomaterials possess optoelectronic, electrocatalytic, optical, magnetic, and magnetic properties which depend on their size, shape, and inter-particle spacing (Bayati et al. 2010). Each of these properties can be used to trigger the analytical signals in sensor

Nanosensors

Based on Receptor Molecules
- Affinity based nanosensors
- Catalytic based nanosensors

Based on Structure or detection mechanism
- Optical nanosensors
- Electrochemical nanosensors
- Mechanical nanosensors

Based on Applications
- Chemical nanosensors
- Bionanosensors
- Deplyable nanosensors
- Electrometers

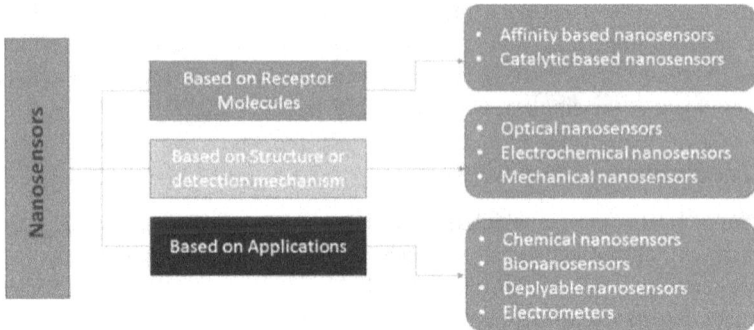

Fig. 2. Classification of nanosensors based on receptor molecules, structure, and applications.

technology. Further, nanomaterials can be functionalized to add more and desirable properties depending on the intended application. The incorporation of multiplex sensing functionalities into a nanosensing platform allows for the development of easily integrated light devices (Zilberstein et al. 2017) whose structure, function, and programmability are highly improved (Ayerden and Wolffenbuttel 2017). Therefore, nanomaterials have been used to fabricate smart sensors and multi-sensing platforms whose multiple signal emission allows for simultaneous detection of a matrix mixture of complex analytes, each with a definite and separate signal (Zhang and Gao 2019). The interface between the analyte, nanomaterial, and the transducer plays a key role in sensing and significantly influences the sensor sensitivity. Therefore, the fabrication of the nanosensor should be such that a high surface area is retained on the sensor surface to assure effective binding of the analyte. Since some sensors use biomolecules (biosensors), the sensor assembly should also ensure that the active sites of the biomolecules are retained. This consideration affects the choice of the method used to assemble the nanomaterial on the sensing platform. An example of fabricated nanosensor arrays is shown in Figure 3 below.

Two main methods are used to form nanoparticle assemblies on sensing platforms. One of them is the synthesis of nanomaterials separately, then introduced into the substrate physically (drop-coating) or chemically by use of chemical bonds between the substrate and functional groups on the nanomaterial. Drop-coating suffers from disadvantages such as falling off the substrate, resulting in lack or retention of the sensing surface throughout the measurement. Nanomaterials immobilized by this method are also said to aggregate at high temperatures (Zhang and Gao 2019). The other method is the *in situ* growth of nanoparticles on the surface of the substrate. This method has higher retention, results in geometrically well-ordered nanostructures and nanoarrays with 0 dimensions (quantum dots), 1 dimensional and 2 dimensional and whose thermal stability is high (Zhu et al. 2015). Several nanomaterials are suitable for the fabrication of nanoarrays. These include metal nanoparticles (Lin et al. 2016, Wang et al. 2018), carbon nanotubes (Yang et al. 2010), silicon (Kwon et al. 2018), hybrid nanomaterials (composed of two or more different types of nanoparticles or polymeric nanomaterials) (Yang et al. 2010, Zang et al. 2014) as well as III-V or II-VI based hetero-nanoarrays (Zhang and Gao 2019). These nanomaterials are

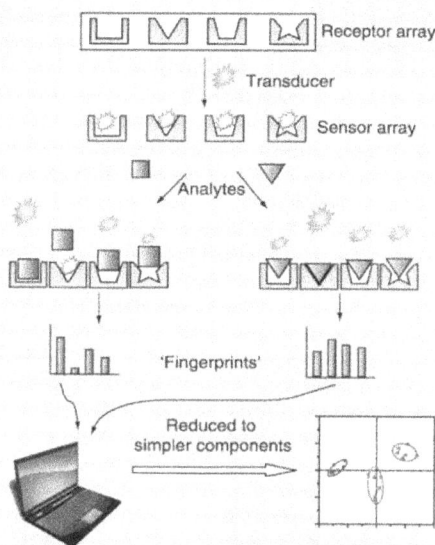

Fig. 3. A sensor array composed of different recognition elements that possess varying binding affinities to different analytes provides the basis of this method. As a result, distinct patterns of analytes are generated based on different binding affinities of analytes toward the sensor array. These patterns (fingerprints of each analyte) are processed by multivariate analysis methods. After reducing the data matrix dimensions, different clusters of the analytes can be visualized, demonstrating a successful classification (Reproduced with permission from (Le et al. 2014)).

assembled into nanoarrays using several methods, which include electrodeposition, hydrothermal deposition, template-assisted method, precipitation, green synthesis, and chemical vapor deposition (Simeonidis et al. 2019). Each of these methods is briefly discussed below.

3.1 Electrodeposition

Nanocomposites synthesized by this method are electrodeposited on the surface of a suitable substrate using electrochemical techniques such as electrolysis. The nanomaterial grows on the surface of a conductive substrate from a solution containing the respective salts. This method allows uniform coating of the nanomaterial on the surface of the substrate irrespective of its geometrical conformation. However, the method is limited by the requirement that the substrate must be conductive. The method has been used for the synthesis of nanoarrays of palladium nanoparticles (Pd) on carbon-coated titanium dioxide (TiO_2) (Cheng et al. 2012), Au nanorods and nanopillars on nanoporous anodic aluminum oxide membranes (Shaban 2016), porous cobalt hydroxide ($Co(OH)_2$) and manganese ($Mn(OH)_2$) on Co_3O_4 core nanowires to form core/shell arrays (Xia et al. 2012), vertically aligned gold nanotube arrays and gold nanowire arrays on aluminum, oxide templates (Tian et al. 2016), and Co-doped cerium oxide (CeO_2) nanorod arrays on fluorine-doped tin oxide substrate (Younis et al. 2013) among others.

3.2 Hydrothermal Deposition

This method is highly studied for the synthesis of iron oxide nanoparticles (Wetterskog et al. 2014). It involves the temperature-driven breakdown of organometallic precursors mixed with stabilizers and in the presence or absence of oxygen to give metal oxides. It is a two-step synthesis, characterized by seed deposition followed by hydrothermal growth. The deposited seeds could sometimes initiate nucleation and growth of nanoarrays on substrates. The organometallic precursors are prepared by mixing suitable salts of the metal with a soluble organic salt. Common organometallic precursors used for the synthesis of iron oxide nanoparticles are iron oleate (Kovalenko et al. 2007, Wetterskog et al. 2014, Hufschmid et al. 2015, Zhou et al. 2015), iron acetate (Sun and Zeng 2002), and iron acetylacetonate (Kovalenko et al. 2007). In the absence of oxygen, the nanoparticles formed are characterized by crystalline defects, poor magnetic properties, and a mixture of oxide phases (Unni et al. 2017). This method is suitable for the synthesis of 2D planar and 3D honeycomb structured nanoarrays on substrates of various shapes (Zhang and Gao 2019).

3.3 Template-assisted method

This method involves direct growth of particularly oriented nanoarrays on solid surfaces mostly using anodized aluminum oxide membrane (Shaban 2016, Wen et al. 2017, 2018) or polycarbonate tracketch membrane (Zhang and Gao 2019). The method allows for regulation of the size, orientation, and structure periodicity. Binary pore anodized aluminum oxide templates enable the formation of binary heterogeneous nano architectures through precise control of size, morphology as well as a heterogeneous pattern (Wen et al. 2018). The method has been used for the synthesis of Au nanorods and nanopillars on nanoporous anodic aluminum oxide membranes (Shaban 2016).

3.4 Precipitation

Precipitation synthesis processes involve ion exchange reactions in which one of the products formed is insoluble. These reactions are not complex and mostly proceed spontaneously since one species in the reaction is strong enough to displace the other from its compound. For instance, the reaction between sodium hydroxide (NaOH) and zinc sulfate ($ZnSO_4$) forms insoluble zinc hydroxide ($Zn(OH)_2$), which can further be heated to give zinc oxide (ZnO) nanoparticles. This method has been used to prepare both doped and undoped ZnO nanoparticles such ZnO (Sani et al. 2017), $ZnWO_4$ (Hosseinpour-Mashkani et al. 2016), Ce-doped ZnO (Lang et al. 2016), and many other different types of metal oxide nanoparticles. Magnetic metal particles are known to aggregate to form bulk materials, which are thermodynamically stable (Ma et al. 2013). This aggregation decreases their surface area and hence the adsorption sites. Synthesis procedures for these particles therefore should include steps that ensure post-synthesis non-aggregation. This can be done using surface modification techniques such as coating the nanoparticle with a more stable shell; for instance, a metal oxide layer. This is because metal oxide particles once hydrated form hydroxyl groups. These hydroxyl groups can undergo protonation and deprotonation reactions depending on the pH to form negative charges or positive charges on the surface of the nanoparticle. A positive charge density is developed at pH values less than the point of zero charges (PZC) pH whereas a negative charge density develops at pH values higher than PZC pH (Ma et al. 2013). A dense negative charge on a nanomaterial surface creates electrostatic repulsive forces on the individual particles, hence keeping them apart. The same principle applies for a dense positive charge on nanoparticle surface.

Co-precipitation is also possible for the synthesis of metal nanocomposites. It has been applied for the synthesis of metal nanocomposites such as manganese ferrite ($MnFe_2O_4$) @Mn–Co oxide core-shell nanoparticles (co-precipitation) (Ma et al. 2013), magnetite–silica (Kokate et al. 2013). Co-precipitation is said to be a simple and highly scalable synthesis method, which gives nanosized super magnetic metal nanoparticles such as iron oxide. These nanoparticles, however, have poor crystallinity due to agglomeration (aggregation). This can be addressed through surface modification as discussed above. Co-precipitation makes it possible to achieve the synthesis of the core material as well as the surface modification in one step (Kokate et al. 2013).

3.5 Green Synthesis

Green synthesis uses reducing agents obtained from plant extracts to reduce aqueous species of the metal in solution. An example is the reduction of Ag^+ from an aqueous solution of $AgNO_3$ using plant extracts such as *Clitoriaternatea*, plants (green tea (*Camellia sinensis*), alfalfa (*Medicago sativa*), lemongrass (*Cymbopogonflexuosus*), geranium (*Pelargonium graveolens*) (Rus et al. 2017) Citrus paradisi (Grapefruit red) (Ayinde et al. 2020) and *Solanum nigrum* as reducing agents (Krithiga et al. 2015). Other biological agents such as bacteria (*Pseudomonas stutzeri AG259, Lactobacillus* strains, etc.), fungi (*Fusarium oxysporum, Aspergillus flavus*), algae (*Lyngbya majuscule, Spirulina subsalsa, Rhizocloniumheiroglyphicum, Chlorella vulgaris*) may also be used as reducing agents for silver (Rus et al. 2017). These agents present cost-effective and environmentally friendly pathways for nanoparticle synthesis (Mayedwa et al. 2017). It has been used for the synthesis of nanomaterials such as copper oxide (CuO) (Ganesan et al. 2020), ZnO (Mayedwa et al. 2017), nickel-cobalt (Ni-Co)hydroxide nanoarrays with facet engineering on carbon chainlike nanofibers (Lu et al. 2020), gold nanoparticles using different leaf extracts of *Ocimumgratissimum* (Arti et al. 2019), gold nanoparticles with *Zingiberofficinale* extract (Kumar et al. 2011a) and iron oxide nanorods using *Withaniacoagulans* extract (Qasim et al. 2020) among others.

3.6 Vapor Condensation

Chemical vapor condensation, also known as chemical vapor deposition, is a thermodynamic process whose chemical reactions are under specific conditions of temperature, pressure, reaction rates, and momentum, mass, and energy transport (Pedersen and Elliott 2014, Abegunde et al. 2019). It involves the deposition of solid material from vapor by a chemical reaction occurring within the vicinity of the heated substrate. The solid material that results is in the form of films, crystal, or powder.

This method allows one to tune a wide range of physical, chemical, and tribological properties of the material synthesized by varying conditions such as substrate temperature, substrate material, total pressure gas flow as well the composition of the reaction gas mixture (Carlsson and Martin 2010, Pedersen and Elliott 2014). All chemical vapor deposition reactions involve several steps. The main steps are precursor, generation of active gaseous reactant species; transport, delivering the precursor into the reaction chamber; adsorption of the precursor onto the hot surface; decomposition of the precursor to give the atom needed for the film and organic waste; migration of atoms to a strong binding site; nucleation that leads to the growth of the thin film; desorption of unwanted side products; removal of unwanted products (Binions and Parkin 2011, Ogawa et al. 2018). However, it is possible to achieve this in two steps especially using rotary chemical vapor deposition (RCVP) (Zhang and Goto 2015).

4. Characterization of Nanosensor Arrays

The characterization of a sensor platform is a very crucial step in sensor design since it confirms the actual sizes and orientation of the synthesized nanoarrays. It also confirms the retention of the properties to be used for signal detection. The methods used for characterization therefore partially depend on the signal transduction method to be used. The surface area available for binding with the target analyte and characterization serves to confirm proper immobilization of the biomolecule on the sensor surface and retention of its active sites is informed through characterization. Since disease diagnostic relies on biomarkers from biological fluids, its detection platforms mostly contain one or more biomolecules. In this case, it is important to confirm the retention of the biomolecule on the sensor surface while ensuring the orientation of its active sites towards the target analytes. The following section discusses various methods used to characterize nanosensor array platforms used for disease diagnostics.

4.1 Electrochemical methods

Electrochemical methods of disease diagnostic rely on electrical signals resulting from the interaction of the disease biomarker with the sensor platform. The electrochemical characterization methods are used to establish successful electrode modification, electron transport as well as retention of electrocatalytic properties (Obisesan et al. 2019). This is achieved through electrochemical studies at each stage of nanosensor fabrication using techniques such as cyclic voltammetry, square wave voltammetry, differential pulse voltammetry as well as electrochemical impedance spectroscopy among others.

4.2 Ultraviolet-visible (UV-visible) spectroscopy and photoluminescence (PL)

Ultraviolet-visible spectroscopy and photoluminescence spectroscopic techniques are used to study the electronic transitions between the conduction and valence bands or between the lowest unoccupied molecular orbital to the highest occupied molecular orbital. The energy difference between the two states, usually referred to as bandgap, is a measure of the energy transition. This is an important property to study in nanoarrays that use optical properties for signal transduction. Information obtained from UV-visible studies is also used to calculate optical bandgaps as well as the sizes of the nanoparticles.

4.3 X-ray diffraction (XRD) spectroscopy

X-ray diffraction spectroscopy is usually used to give information on the crystalline structure of the material synthesized and to determine the particle size. It, therefore, presents a suitable technique for studying the size modulation of nanoparticles as a function of changes in dopants, temperature and synthesize time among other parameters (Chouchene et al. 2017). Besides the crystalline structure, information obtained from XRD is also used to calculate the particle size using Scherrer's formula (Varghese et al. 2014) given below.

$$D = \frac{K\lambda}{B\cos\theta},$$

where D is the mean size of crystallites (nm), K is a factor, λ is x-ray wavelength, B is full width at half maxima, and θ is the Bragg angle.

4.4 Scanning Electron Microscopy (SEM) and Transmission Electron Microscopy (TEM)

This is used to give information about the morphology of the nanoparticles and their arrangement in the sensor array. The SEM images give information about the surface area of the nanoarray platform, whether the particles are agglomerated or independent, and their approximate size.

4.5 Fourier Transform Infrared Spectroscopy (FTIR)

This is used to study the formation of new bonds following surface functionalization of nanoparticles with desirable molecules to create a surface suitable for bioconjugation or stabilize the nanoparticles. This is inferred from the disappearance of absorption bands characteristic to certain functional groups and appearance of others or shifting of absorption bands.

5. Validation of Nanosensor Arrays

Validation of nanosensor arrays is very important to determine their performance. The performance characteristics can be determined by several parameters such as selectivity, sensitivity, the limit of detection, reproducibility, linear range, response time, accuracy, precision, and resolution (Shafiee et al. 2019). Selectivity is the ability of the sensor to detect a particular bioanalyte without reacting to other analytes it is exposed to. Sensitivity, on the other hand, is the ratio of output charge concerning the relative input charge. If the outputs versus input values are plotted, then the slope of the calibration curve shows the sensitivity of the nanosensor. One of the most important parameters used to characterize a biosensor is its limit of detection, which denotes the minimum amount of bioanalyte that the biosensor can detect. The proximity of the output reproduced by the similar input, assuming all other factors are constant, is called reproducibility. The difference between the minimum and maximum thresholds that is detectable by a sensor is called the measurement range. The necessary time for a sensor to reach a particular steady-state output value for an alterable input is called the response time. The difference between the actual value and detection amount denotes the device's 'accuracy', which is reported as a percentage of the full-scale reading and determined as a ratio. In a situation where all other conditions are equal, the reproducibility of a biosensor is referred to as its precision of measurement. Resolution refers to the smallest change in the input value that can be detected by a sensor. To eliminate the impact of environmental variation, discrepancies in sensor concentration, as well as the drift in the ratiometric signal, must be calculated by normalizing the analyte-sensitive signal to an analyte-insensitive signal, which is referred to as the ratiometric signal. Table 1 summarizes the important factors that characterize a nanosensor's performance, including popular methods and their related parameters (Shafiee et al. 2019).

6. Application of Nanosensor Arrays in Medical diagnosis

An important goal of medical diagnosis is to be able to diagnose medical problems as fast as possible, enabling clinicians to treat patients before any irreversible or long-lasting damage can occur (McIntosh 2016). As described earlier, this can be achieved by point-of-care testing which ensures fast detection of analytes near to the patient facilitating a better disease diagnosis, monitoring, and management (McIntosh 2016) and with nanomaterials, the sensitivity of the nanosensors has been tremendously improved. Nanosensors have been used to monitor the build-up of bacteria on implants and warn clinicians when treatment is required before the problem escalates (McIntosh 2016). In the next section, we discuss in detail how various nanosensor arrays have been used in the diagnosis of various medical conditions specifically cancer, diabetes, malaria, and human immunodeficiency virus (HIV).

6.1 Applications of Nanosensor Arrays in Cancer Diagnosis

Cancer is one of the leading causes of death accounting for one in every seven deaths in the world (Huber et al. 2015, Hayes et al. 2018). Over 200 different types of cancer have been reported with the most common types being breast cancer, ovarian cancer, prostate cancer, esophageal cancer, colorectal cancer, lung cancer, bladder cancer, kidney cancer, lymphoma, skin cancer, liver cancer, pancreatic cancer, and thyroid cancer (Paul et al. 2019). The most commonly reported life-threatening type of cancer in women is breast cancer with about 180, 000 new cases diagnosed every year. Ovarian cancer which also affects women follows breast cancer with more than 238,000 diagnoses worldwide, out of which 151,000 deaths occur (Jayson et al. 2014). Early screening and diagnosis are important practices that have been recognized for improving the likelihood of cancer survival and recovery, thereby leading to a significant decrease in cancer mortality rates (Wu and Qu 2015). Cancer diagnosis usually involves detecting symptoms and characteristics that signify

Table 1. Important factors used to characterize a nanosensor.

Fig of Merit	Definition	The most popular characterization method	Parameters involved
Accuracy	The closeness of the detected amount to the actual value	Reported as a percentage of the full-scale reading and determined as a ratio	Actual and detection values
Limit of detection (LOD)	Denotes the minimum amount of bioanalyte that the nanosensor can detect	Start from a minimal amount and gradually increase to obtain LOD	Minimum detection value
Linear detection range	The difference between the minimum and maximum thresholds that are detectable by the nanosensor	The lowest and highest detectable bioanalyte must be measured to calculate the range	Minimum and maximum of the detection
Sensitivity	The ratio of output charge concerning the relative input charge	Comparing output and input to calculate the ratio	Output and input
Selectivity	Selectivity is the ability of the sensor to detect a particular bioanalyte without reacting to other analytes it is exposed to	The slope of the calibration curve produced by plotting the output vs input values shows the sensitivity of the biosensor	Mostly the bioreceptor and bioanalyte
Reproducibility	The proximity of the output reproduced by the similar input, other parameters are kept constant	Multiple time the same amount of bioanalyte is introduced to the sensor and deviation of output is measured	Output
Resolution	The slightest change in the input value that can be detected by a sensor	A known amount of bioanalyte in minimal degrees is introduced to the sensor	Input slightest change
Response time	The necessary time for a sensor to reach a particular steady-state output value for an alterable input	With input, the time is measured until the moment a reliable output is obtained	Time for the reliable output value

Adapted with permission from (Shafiee et al. 2019).

the presence of anomalies, which include biomarkers (Mishra and Verma 2010). Cancer biomarkers include nucleic acids, proteins, sugars, whole cells, cytogenetic parameters as well as small metabolites founds in body fluids (Wu and Qu 2015). Of these biomarkers, proteins have received particular attention owing to their primary location in blood and urine where they can be detected with current technologies. Blood contains a wide variety of protein biomarkers with potential applications in early cancer diagnostics and detection (Kosaka et al. 2014). However, conventional blood tests for early detection of cancer biomarkers yield low sensitivities owing to the biomarkers' low concentration in the cardiovascular system (Mosayebi et al. 2018). To effectively detect biomarkers in blood, sensors whose sensitivity allows them to detect biomarkers at a million times lower than the concentration of other blood proteins are required (Kosaka et al. 2014). Currently, there exist limited devices for early screening, diagnosis, and monitoring cancer progress (Sandbhor Gaikwad and Banerjee 2018). The few devices that already exist are costly, time consuming, their use require centralized or hospital based laboratories and high expertise in operating them (Hayes et al. 2018). New devices are needed for ultrasensitive and precise point of care diagnostics for early screening and detection of cancer biomarkers even at the bedside (Sandbhor Gaikwad and Banerjee 2018). An ideal point of care diagnostic device is one that is portable and assures reliability. The currently developed diagnostic tools for cancer operate using sophisticated instrumentation, require highly skilled operators, and are costly. This calls for research on the development of new devices that would offer continuous, cost-effective real-time *in vivo* monitoring of cancer, which would provide early diagnosis, drug efficacy, and effective drug delivery (Siontorou et al. 2017). The use of nanotechnology for drug diagnosis, drug delivery, and cancer therapy has enabled the use of nanomaterials for extraction and detection of specific tumor biomarkers (Okaie et al. 2016), circulating tumor cells, or extracellular vesicles shed by the tumor (Salvati et al. 2015). Recent research has seen emergence of nano and microfabrication based technologies that are integrated with different sensing platforms (Sandbhor Gaikwad and Banerjee 2018) and molecular communication (Mosayebi et al. 2018). Various nanosensor arrays have been reported to improve the detection and diagnosis of various cancer. Since there are very many nanomaterials used to make nanosensor arrays for cancer diagnosis, in the next section we discuss in more details nanosensor arrays from carbon nanotubes, metal nanoparticles and graphene based nanosensor arrays in cancer diagnosis.

6.1.1 Carbon Nanotube-based Nanosensor Arrays in Cancer Diagnosis

Carbon nanotubes (CNTs) consist of cylindrical fabricated rolled-up graphene sheets classified as single-walled carbon nanotubes (SWCNTs) consisting of a single graphite sheet flawlessly wrapped into a cylindrical tube or multi-walled carbon nanotubes (MWCNTs) which comprise an array of such nanotube (Andrews et al. 2002, Eatemadi et al. 2014, Camilli and Passacantando 2018). The CNTs, though made of carbon with a similar dimension aspect ratio, can be either metallic or semiconducting depending on how the graphene layers are rolled up (Camilli and Passacantando 2018). They are characterized by a high aspect ratio making them suitable for functionalization through chemical or physical methods (Camilli and Passacantando 2018). They can be produced via the chemical vapor deposition method which has the advantage to be scalable, allowing large-area deposition and providing CNTs that are already attached to a substrate and hence easy to be collected (Colomer et al. 2000). They normally have amazing electrical, mechanical, and thermal properties as well as partial antibacterial activity due to their high aspect ratio and high surface area (Ursino et al. 2018). This high sensitivity and the electronic properties of nanotubes to adsorb molecules on their surface and the unparalleled unit surface area make the CNT a promising starting material for the development of super miniaturized chemical and biological sensors (Akhmadishina et al. 2013, Zaporotskova et al. 2016) as illustrated in Figure 4 which shows carbon nanotubes sensor arrays for cancer detection. The operation of the CNTs-based sensors is established on the changes in the V-I curve of the nanotube as a result of adsorption of specific molecules on their surface (Zaporotskova et al. 2016) which is one of their most promising applications in electronics. The sensors should have a high sensitivity as well as fast response and recovery (Zaporotskova et al. 2016).

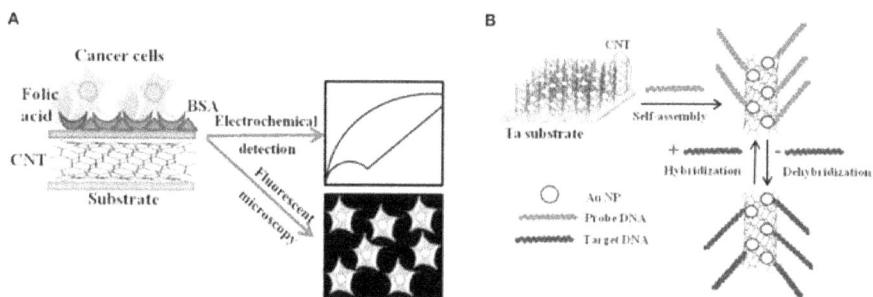

Fig. 4. (A) Schematic illustration of carbon nanotubes immobilized on a sensor surface for enhanced electrochemical detection of cancer cells. (B) Schematic representation of gold nanoparticles/aligned CNTs immobilized for an electrochemical DNA biosensor for cancer detection (Adapted from (Tîlmaciu and Morris 2015), an open-access article).

We give a few examples where carbon nanotube sensor arrays have been used in the detection or diagnosis of cancer.

A sensitive implantable carbon nanotube sensor has been reported by Williams et al. (Williams et al. 2018), where they engineered the first *in vivo* prototype optical sensor composed of an antibody-functionalized carbon nanotube complex, which responds quantitatively to human epididymis protein 4 (HE4), a biomarker for ovarian cancer, via modulation of the nanotube optical bandgap (Williams et al. 2018). This was a noninvasive cancer biomarker detection in orthotopic models of disease where the complexes were found to measure HE4 with nanomolar sensitivity to differentiate disease from benign patient biofluids(Williams et al. 2018).

Carbon nanotube sensor arrays have also been reported in the detection of lung cancer (Huang et al. 2018). In their study, Huang and coworkers developed a breath test for the detection of lung cancer using carbon nanotube chemical sensor arrays and a machine learning technique (Huang et al. 2018). They used the electronic nose (E-nose) chemical sensor Cyranose 320 (Sensigent, Baldwin Park, CA, USA), composed of 32 nanocomposites conducting polymer (CP) sensors consisting of highly sensitive carbon nanotubes to analyze the breath samples (Lu et al. 2006, Huang et al. 2018). Using a total of 117 cases and 199 controls,the diagnostic accuracy was assessed using the pathological reports as the reference standard (Huang et al. 2018). Their study showed that with the use of a highly sensitive (at the ppb level) chemical sensor and an advanced data analysis technique, the E-nose can be used to diagnose early-stage lung cancer with high accuracy (Huang et al. 2018). Another study by (Okuno et al. 2007) reports the fabrication of a label-free electrochemical immunosensor using microelectrode arrays modified with single-walled carbon nanotubes (SWNTs) for the detection of the total prostate-specific antigen (T-PSA), a cancer marker using differential pulse voltammetry (DPV) where current signals, derived from the oxidation of tyrosine (Tyr), and tryptophan (Trp) residues, increased with the interaction between T-PSA on T-PSA-mAb covalently immobilized on SWNTs (Okuno et al. 2007).

6.1.2 Metal Nanoparticle-based Nanosensor Arrays in Cancer Diagnosis

Metal nanoparticles, including gold (Au), silver (Ag), platinum (Pt), palladium (Pd), copper (Cu), cobalt (Co), normally have unique physical and chemical properties making them exceptionally suitable for designing new and improved sensing devices such as electrochemical sensors and biosensors (Luo et al. 2006, Le et al. 2014, Maduraiveeran and Jin 2017). In many cases, these metal nanoparticles can be used in the immobilization of biomolecules, catalysis of electrochemical reactions, enrichment of electron transfer between electrode surfaces and proteins, labeling of biomolecules as well as acting as a reactant (Luo et al. 2006). The metal nanoparticles can also be used as analytical transducers in various sensing principles as well as signal amplification elements (Pallares et al. 2019). This combination of nanoparticle sensing principles with recognition elements has occasioned their use in bioassays with rapid responses and visual outcome, suitable for use in resource-constrained environments (Noah 2018, Pallares et al. 2019). Nanoparticle-based sensor arrays can provide versatile and attractive alternative strategies for specific biomarker-focused cancer diagnosis—a necessary development in the current healthcare system to better provide personalized treatments (Le et al. 2014). For example, in a study reported by Bajaj and coworkers, gold nanoparticles were conjugated to the fluorescent polymer to develop an array for the detection of different cancer cell types. In their work, they used cationic gold nanoparticles as recognition elements to non-covalently bind negatively charged poly(p-phenylene-ethynylene) polymer (PPE-CO2) as the transducer (Bajaj et al. 2009). They found that the electrostatic interaction between PPE-CO2 and the gold nanoparticles quenched the fluorescence of the polymers, which was recovered to varying degrees upon incubation with cells (Bajaj et al. 2009). They further observed distinct as well as differentiated patterns with human cancerous Michigan Cancer Foundation-7 (MCF-7), metastatic (MDA-MB-231), a highly destructive, aggressive and poorly differentiated triple-negative breast cancer (TNBC) cell line along with normal (MCF10A) breast cell lines (Bajaj et al. 2009). In a further study, the gold nanoparticles were used to develop gold nanoparticle-green fluorescent protein (NP-GFP) based arrays for rapid identification of mammalian cells based on cell surface properties (Bajaj et al. 2010). From their work, Bajaj et al. obtained highly reproducible characteristic patterns from different cell types which enabled the identification of cell types and cancer states (Bajaj et al. 2010). Using the arrays,they were able to differentiate between isogenic normal, cancer, and metastatic cell types using only ~ 5000 cells.

The interparticle Plasmon coupling of the nanoparticles leads to color changes which have been widely used in biosensors based on the aggregation of the nanoparticles. For example, small gold nanoparticles are red and well dispersed but turn blue or purple on aggregation (Lin et al. 2013), while silver nanoparticles are yellowish brownish when dispersed but turn black when they aggregate (Doria et al. 2012). This fundamental property of these nanoparticles has been used in the development of colorimetric sensor arrays with the potential of the rapidity of analysis, are cost-effective, and easy to use since they can provide naked-eye observations (Sun et al. 2020). Using this phenomenon, gold nanoparticles have been used to develop a colorimetric sensor array with reported protein aptamers as nonspecific receptors (Lu et al. 2013).

Silver nanoparticles have also been used to fabricate reproducible, rapid, and low-cost nanosensor arrays for the detection of Squamous cell carcinoma antigen (SCCa), a tumor biomarker, which plays an important role in adjuvant diagnosis, treatment evaluation, and prognosis prediction for cervical cancer patients (Zhao et al. 2014).

6.1.3 Grapheneoxide-based Nanosensor Arrays in Cancer Diagnosis

Graphene oxide is a hydrophilic carbon nanomaterial obtained in the oxidized form of graphene (Jhaveri and Murthy 2016a). It has been found to improve thermal and mechanical properties of polymeric membranes (Ionita et al. 2014) since it presents hydrophilic functional groups such as amine($-NH_2$), hydroxide (-OH), and sulphonic acid ($-SO_3H$) which provide a variety of surface-modification reactions (Enotiadis et al. 2012, Liu et al. 2017) that can be useful to functionalize graphene oxide- and graphene-based materials (Jhaveri and Murthy 2016b). Graphene and graphene oxide based nanomaterials are known to have extraordinary optical, electronic, and thermal properties; chemical and mechanical stability; large surface area; and good biocompatibility, and are therefore applicable alternatives as versatile platforms to detect biomarkers at the early stage of cancer (Cruz et al. 2016, Gu et al. 2019).The graphene provides a natural biocompatible immobilization carrier for more specific recognition units, consequently enriching the sensitivity of the diagnosis system (Gu et al. 2019). Their excellent electric and optical properties provide an amplified signal for sensing and imaging, further supporting a lower detection limit to biomarkers detection (Gu et al. 2019). The graphene offers flexible functionalization characteristics that can be integrated with a variety of desirable nanomaterials providing a multi-function system with a synergistic effect for a more precise and sophisticated diagnosis (Gu et al. 2019). A study by Verma and co-workers reports a fabricated electrochemical immunosensor based on a gold nanoparticles-reduced graphene oxide (AuNPs-rGO) composite material as a transducer matrix for label-free and noninvasive detection of salivary oral cancer biomarker interleukin-8 (IL8) as illustrated in Figure 5 (Verma et al. 2017). Their results indicated a synergy between rGO and AuNPs which allowed the immunosensor to have a fast response (9 mins) of IL8 and high sensitivity with a linear dynamic range of 500 femtogram/mL (fg mL^{-1}) to 4 nanograms/mL (ng mL^{-1}) and a detection limit of 72.73 ± 0.18 picogram/mL (pg mL^{-1})as well as excellent specificity towards the detection of IL8 in human saliva samples which was attributed to the improved electron transfer behavior of the composite (Verma et al. 2017).

6.2 Applications of Nanosensor Arrays in Diabetes Diagnosis

Diabetes is one of the most rampant and persistent diseases currently affecting very many people around the world and can lead to serious complications such as limb amputations, blindness, cardiovascular and kidney diseases (Cash and Clark 2010, Makaram et al. 2014, Kumar et al. 2018). It is known to be caused by an insulin disorder and based on the underlying mechanism, it is often classified as type 1 (insulin-dependent or juvenile) diabetes and type 2 (non-insulin or adult-onset) diabetes (Makaram et al. 2014). Type 1 diabetes results when there is an autoimmune attack that destroys the insulin-producing beta cells of the pancreas while type 2 results from metabolic disorder described by elevated blood glucose comprising insulin resistance and relative insulin deficiency (Makaram et al. 2014). Since diabetes has no cure, tight monitoring and controlling the blood glucose

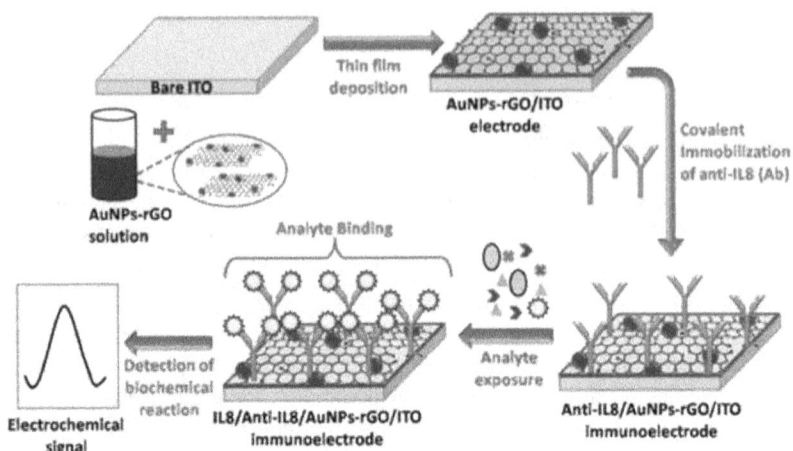

Fig. 5. A schematic diagram representing the fabrication of Au NPs-rGO-based immunoelectrode for immunosensing application (Reproduced with permission from (Verma et al. 2017), American Chemical Society, Copyright © 2017).

levels by the patients can reduce its complications (Cash and Clark 2010, Zhang et al. 2011, Metkar and Girigoswami 2019). Strategies for early detection which can prevent the progression of diabetes would make a very big difference for the patients as well as an economic benefit for a resource constraint country (Kumar et al. 2018). Monitoring the blood glucose has been considered as the gold standard method for diagnosis and self-monitoring of diabetes (Makaram et al. 2014) and studies have shown maintaining the blood glucose level in the normal range, of 4.9–6.9 millimeter (mM) in healthy individuals, can lead to improvement of complications such as nephropathy, retinopathy, coronary artery disease and stroke (Metkar and Girigoswami 2019).

However, the underlying process of blood glucose control is invasive and painful for patients since the patients have to obtain a small blood sample, normally via a finger prick, place it onto a sensor test strip, and then read by a handheld electronic reader, which reports the blood glucose concentration (Cash and Clark 2010, Zhang et al. 2011, Makaram et al. 2014). Additionally, to successfully manage this disease, the finger pricking process must be completed several times a day, and the patient must be awake for data analysis to be done (Pickup et al. 2008, Wang 2008, Cash and Clark 2010, Garg and Akturk 2017). This greatly contributes to the immense need for non-invasive monitoring options including human serums such as saliva, sweat, breath, urine, and tears since they contain trace amounts of glucose and can be easily accessed and allow minimal to non-invasive monitoring (Makaram et al. 2014). Several non-invasive techniques for glucose detection have lately gained recognition as feasible substitutes due to the introduction of nanotechnology-based sensors (Makaram et al. 2014). These nanosensors are ideal for blood glucose testing in serums other than blood due to their superior sensitivity and selectivity ranges, in addition to their size and compatibility with electronic circuitry (Makaram et al. 2014). These nanotechnology approaches are rapidly evolving leading to the development of nanosensor arrays, many of which are being used in the diagnosis of diabetes. In the next section, we describe in detail the current trends of nanosensor arrays used in the diagnosis of diabetes.

6.2.1 Nanosensor arrays for blood-based diagnosis of diabetes

Blood is used as the gold standard biomarker for the detection of glucose levels in patients mostly using electrochemical transducers since they are known to have high sensitivity, reproducibility, and manufacturability in huge volume and at low cost (Newman and Turner 2005). The glucose sensors normally use the glucose oxidase and glucose dehydrogenase, enzymes that catalyze the redox reaction of hydrogen peroxide (H_2O_2) during the biochemical reaction (Makaram et al. 2014). Nanoarrays which have been developed for the detection of blood in the diagnosis of diabetes use nanomaterials such as metal nanoparticles, nanorods, nanowires, and carbon nanotubes. These nanomaterials provide high surface area, enhanced electron transfer from enzyme to the electrode, and the ability to include additional catalytic steps which are associated with the nanomaterials that have been shown to improve the sensor's sensitivity of up to nanomolar (nM) and picoMolar (pM) (Makaram et al. 2014). For example, Yang and co-workers developed an array of platinum nanowires that were grown in polycarbonate membrane and used the nanoarrays to detect glucose in blood samples with a wide linear range of between 0.018 to 540.477 milligram/deciliter (mg/dL) (Yang et al. 2006). Likewise, sensitive nanoelectrode arrays based on single-walled CNTs (SWCNTs), grown from silicon substrate have been reported and used to detect glucose levels in the blood at concentrations as low as 1.4412 mg/dL (Lin et al. 2004).

6.2.2 Nanosensor Arrays for Urine-based Diagnosis of Diabetes

Urine as an analyte has been used for noninvasive detection of glucose since when the renal threshold for glucose is exceeded (for most individuals ~ 10 millimole/liter (mmol/L), it starts to be secreted with urine (Leroux 2001). Normally, glucose and other components are present in urine in very small quantities and glucose overflows into the urine when the blood glucose level is high, causing urine to have a sweet or fruity odor (Su et al. 2012). The existence of high levels of glucose in urine (more than 50–100 mg/dL) is considered a dangerous condition since it indicates the worsening of diabetes (Su et al. 2012). Several nanosensors for the detection of glucose in urine have been developed as alternatives to commercial low-cost colorimetric dipsticks including a reusable and stable nanosensor composed of a composite of gold-graphene oxide electrodes (Kim et al. 2014). The use of $ZnFe_2O_4$ magnetic nanoparticles has also been utilized in the development of a simple, inexpensive, highly sensitive, and selective colorimetric nanobiosensor for the detection of urine glucose using glucose oxidase (GOX) with a linear detection range of between 1.25×10^{-6} to 1.875×10^{-5} mol L^{-1} with a detection limit of 3.0×10^{-7} mol L^{-1} (Su et al. 2012). These characteristics of the nanobiosensors were attributed to the high catalytic efficiency, good stability, monodispersion, and rapid separation of the zinc-iron oxide $ZnFe_2O_4$ magnetic nanoparticles (MNPs) (Su et al. 2012).

6.2.3 Nanosensor Arrays for Saliva-based Diagnosis of Diabetes

The use of saliva as an analyte has been used as an alternative noninvasive approach to detect glucose(Witkowska Nery et al. 2016). The concentration of glucose in saliva has been found to show a high correlation with blood glucose in diabetic patients with statistically significant but a lower correlation in nondiabetic patients (Abikshyeet et al. 2012, Witkowska Nery et al. 2016). Studies have shown that different time is needed before both salivary and blood glucose reach their peak values estimated as between 15–40 min for salivary glucose and 30–60 min for blood glucose (Zhang et al. 2015). Various nanosensor arrays for the detection of glucose in saliva have been reported. For example, a low cost, accurate, disposable nanobiosensor composed of an array of single-walled carbon nanotube (SWCNTs), gold nanoparticles, glucose oxidase for real-time monitoring of glucose in saliva was reported by (Zhang et al. 2015). In their work, they used more than eight clinical trials on two healthy individuals for 2 to 3 hr periods for each trial including both regular food and standard glucose beverage intake with more than 35 saliva samples obtained. Their results indicated that the nanobiosensor could measure as low as 0.1 mg/dL and as high as 20 mg/dL glucose in saliva, which was found to be sufficient to be medically applicable for diabetic diagnosis and health surveillance (Zhang et al. 2015).

6.2.4 Nanosensor Arrays for Tears-based Diagnosis of Diabetes

The concept of testing glucose in human lachrymal liquid commonly known as tears as a substitute for blood has been proposed in many studies. It has been argued that if a good correlation between glucose in tears and blood samples can be established, then measuring glucose in tears could provide a smart alternative technique for monitoring for blood glucose levels within the normal, as well as hyperglycemic and hypoglycemic ranges (Yan et al. 2011). As such, quantification of glucose in tears has been tried by a few researchers as an invasive way of monitoring glucose though it has proved to be very difficult due to the little volumes (0.5 – 2.2 microliter/minute (μL/min) of the secreted fluid (Zhang et al. 2011). A few flexible nanosensors which can work with sample volumes as low as 100 nL have been used for the analysis of not only glucose but also dopamine and ascorbate as reported (Andoralov et al. 2013). Using a cell with a working volume of between 60–100 nL and a sampling deviation of 6.7 %, Andoralov and coworkers were able to monitor glucose in the tears with mechanical flexibility which minimized the risk of damage to the eye (Andoralov et al. 2013).

6.2.5 Nanosensor Arrays for Sweat-based Diagnosis of Diabetes

Glucose concentration has been found in interstitial fluid surrounding the cells depending on the metabolic rates of the adjacent cells, blood flow, nerve stimulation, and insulin concentration (Witkowska Nery et al. 2016). Sweat has been reported as the easiest for sampling since it is readily available in human beings (Saraoğlu and Koçan 2010). While the normal blood glucose levels range between 90 mg/dL (5 milliMolar (mM)) and about 140 mg/dL (7.75 mM) or higher, glucose in sweat has been reported in very low concentrations of five(5) to twenty (20) mg/dL (0.277–1.11 mM) (Moyer et al. 2012). Though studies have shown an average lag-time of between 8–10 min with maximum values reaching 45 min and some inconsistencies observed in cases of different sampling methods, a few invasive and noninvasive nanosensors for the detection of glucose in the interstitial fluids including sweat have been reported (Witkowska Nery et al. 2016). For example, an interesting system using a hybrid of graphene with gold as an electrode material integrated into a flexible and transparent silicon patch for sweat-based diabetes monitoring and feedback therapy for both glucose sensing and therapeutic modules has been reported (Lee et al. 2016).

6.2.6 Nanosensor Arrays for Breath-based Diagnosis of Diabetes

Exhaled breath normally comprises various molecular species whose composition can give important information about biochemical processes in the body (Miekisch et al. 2004, Risby and Solga 2006). Owing to both active co-transport and glucose leakage from plasma and the lung interstitium, glucose has been reported in the respiratory fluid with the breath-blood glucose ratio found to be between 0.08 and 0.09 depending on whether the patient has suffered from any lung disease or not (Roberts et al. 2012). Acetone (2-propanone), a ketone, is a major component in exhaled human breath since it is normally absorbed into the bloodstream and expelled in the breath along with the normal constituents of exhaled human breath (Makaram et al. 2014). It is believed that the fruity odor in the breath is due to the acetone and increases significantly during periods of glucose deficiency since the insulin does not work effectively making the body to use fat instead of glucose for energy, leading to huge amounts of acetone in the blood and breath and hence it is known as a useful biomarker for type 1 diabetes (Makaram et al. 2014).

Many initial studies on the detection of breath acetone concentrations were done using expensive and complicated gas chromatography-mass spectrometry (GC-MS), cavity ring down spectroscopy, and selected ion flow tube mass spectroscopy (SIFT-MS) (Wang and Surampudi 2008, Huang et al. 2010). However, with the advancement of nanotechnology, low-cost nanosensors for breath acetone have been developed. For example, a chemiresistor based on ferroelectric tungsten nanoparticles and a nanosensor based on tungsten oxide (WO_3) nanorods have been reported to develop nanosensors for selective detection of acetone in breath simulated media indicating that real-time monitoring of the acetone concentration is feasible and hence can be used for non-invasive diabetes diagnostic tool (Wang et al. 2009, Meng et al. 2011, Zhao et al. 2015).

A combination of metal oxides together with nanostructures has also been used to develop nanosensor arrays for sensitive detection of breath acetone. Examples of these include the combination of thin films of silica dioxide (SiO_2) doped with tungsten oxide (WO_3) nanoparticles (Righettoni et al. 2010), carbon nanotubes coated with WO_3 (Xiong and Xia 2012), photo-induced single-walled carbon nanotubes, and titanium oxide ($SWNT–TiO_2$ core/shell) hybrid nanostructures (Ding et al. 2013).

6.3 Applications of Nanosensor Arrays in Malaria Diagnosis

Malaria is an infectious hematologic disease caused by *Plasmodium falciparum*, a protozoan parasite transmitted through injection of sporozoites into human blood by female *Anopheles* mosquitoes (Krampa et al. 2020). Other species reported to cause malaria infections in humans are *Plasmodium malariae, Plasmodium ovale* (*P. ovalecurtisi* and *P. ovalewallikeri*), and *Plasmodium vivax* (Jain et al. 2014, Krampa et al. 2020). Malaria infection impacts have attracted global health concerns due to millions of worldwide infections, most of which result in deaths. The burden of malaria is significant in Africa which records nearly 1.2 million deaths from malaria annually (Dutta 2020). Over 90 % of those who die of malaria in Africa are children aged 5 years and below (De Moraes et al. 2018). A key challenge in preventing the spread of malaria is that some patients are asymptomatic. These patients exhibit mild or no clinical symptoms of malarial infection, even when infected, but can transmit the *Plasmodium falciparum* parasite to other individuals within the population (De Moraes et al. 2018). Besides spreading to the general population, malaria infection in asymptomatic patients mostly goes undetected, untreated, and progresses from early-stage malaria to cerebral malaria (Dutta 2020). To mitigate these high mortality rates and high budgetary demands needed for malaria treatment, research has focused on the development of cost-effective, user-friendly, and reliable methods that can detect early asymptomatic malaria infections. Research has shown that identifying asymptomatic patients and early detection of malaria presents an effective strategy for breaking the transmission cycle, reducing mortality rates and cutting costs involved in treatment of advanced malaria infection (Rabinovich et al. 2017). This is in line with the aim of the World Health Organization (WHO) of eliminating malaria by 2030 (Feachem et al. 2019).

Early diagnosis of malaria requires a thorough understanding of the parasites' life cycle and the time taken between the injection of sporozoites and invasion of hepatocytes. Once inside the body of the host, the sporozoites enter the liver's hepatocytes, where they multiply and differentiate into merozoites (the red blood cell—invasive form of the parasite) in the pre-erythrocytic stage. This stage of infection is asymptomatic and essential in malaria diagnosis. This is because immune responses are triggered by a high population of the parasite, a high concentration of nucleic biomarkers of the presence of soluble antigen proteins (Krampa et al. 2020). Further, the time between the injection of sporozoites and invasion of hepatocytes ranges from 30–45 min (Krampa et al. 2020). This makes it difficult to detect low numbers of sporozoites within a short time. *Plasmodium ovale* and *Plasmodium vivax* infections often form dormant and persistent hypnozoites in the liver (Mueller et al. 2009). These infections go undetected and cause recurrent infection, posing a hindrance to malaria eradication.

The diagnosis of malaria involves the identification and quantification of malaria diagnostic biomarkers, in which the target metabolites are found in body fluids such as blood, urine, saliva as well as volatile body metabolites (De Moraes et al. 2018, Hede et al. 2018, Obisesan et al. 2019). Common biomarkers used in the detection of malaria include Plasmodium falciparum histidine-rich protein 2 (PfHRP-2) (Sharma et al. 2008, 2010, de Souza Castilho et al. 2011, Hemben et al. 2017), biocrystalhemozoin (Rebelo et al. 2013), β-hematin, plasmodium lactate dehydrogenase (pLDH) (Barber et al. 2013, Dzakah et al. 2014, Hemben et al. 2017), glutamate dehydrogenase (GDH) (Zocher et al. 2012), aldolase (Barber et al. 2013, Dzakah et al. 2014, Hemben et al. 2017), to poisomerase I (Tesauro et al. 2012, Givskov et al. 2016, Hede et al. 2018), and merozoite surface proteins (MSP)

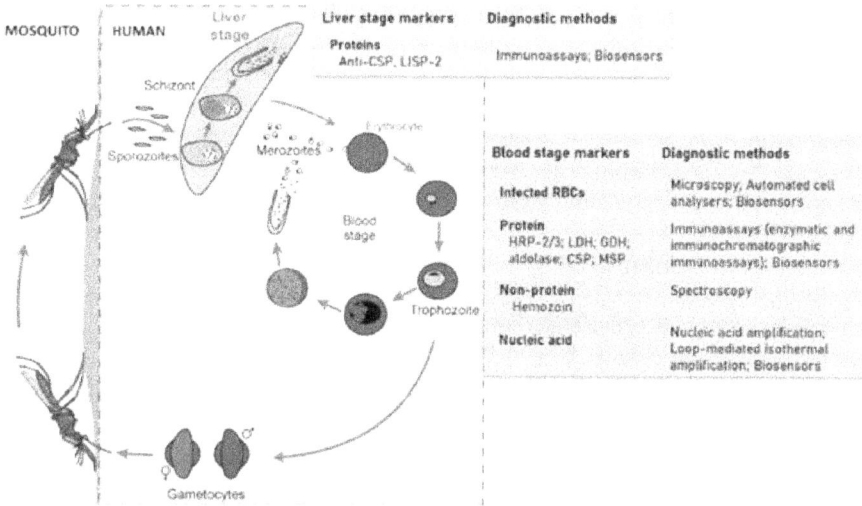

Fig. 6. Developmental cycle of human Plasmodium species in a mammalian host and the strategies used in detecting parasite-specific markers (Adopted from (Krampa et al. 2020), an open-access article distributed under the terms of the Creative Commons Attribution License).

(Hisaeda et al. 2002). The choice of the biomarker as well as the method of diagnosis depends on the developmental stage of the parasite as illustrated in Figure 6 below. In the next sections, we describe the nanosensor arrays available for the detection of these malaria biomarkers.

6.3.1 Nanosensors for the Detection of Plasmodium Falciparum Histidine-rich Protein 2 (PfHRP-2)

Histidine-rich protein 2 (HRP-2) is specific to *Plasmodium falciparum* (PfHRP-2) and is used as the main target for rapid diagnostic test (RDT) for *Plasmodium falciparum* infection (Lifson et al. 2016, Hemben et al. 2017, Krampa et al. 2020). It is the most preferred biomarker for immunosensor fabrication because of its significant expression throughout the life cycle of the parasite. Clinical samples for diagnosis of malaria using HRP-2 are blood, cerebrospinal fluid, urine, and saliva of infected patients. The use of urine and saliva presents an opportunity for painless non-invasive diagnosis, which has the advantage of attracting voluntary participation in the testing screening of malaria by public health officials. However, only trace amounts of HRP-2 antigen are found in urine and saliva. Blood, therefore, continues to be the main sample for this type of diagnosis since small blood samples contain sufficient antigen required for effective detection. Highly sensitive methods are inevitable if non-invasive diagnostic samples (saliva and urine) have to be used. Interdigitated electrodes (IDE) have been proved to provide ultrasensitive and label-free platforms for the detection of PfHRP-2 in saliva (Soraya et al. 2019).

The PfHRP-2 may also be used in combination with LDH, plasmodium-specific LDH (pLDH), and plasmodium—specific aldolase common to all species for diagnosis of *Plasmodium falciparum* alone or mixed infection (Wilson 2012). However, *Plasmodium falciparum* exhibits poor pan-aldolase enzyme expression, which confers low sensitivity in RDTs that incorporate pan-aldolase (Barber et al. 2013). The use of nanoparticles presents a promising strategy for achieving high sensitivity. Some of the nanoparticles used for malarial detection using HRP-2 include gold nanoparticles (Sharma et al. 2008), gold nanoparticle—aluminum oxide (Sharma et al. 2010), activated magnetic nanoparticles, and tosyl-activated magnetic beads (de Souza Castilho et al. 2011). The use of gold nanoparticles is preferred since they are bio-conjugable, retain bioactivity, provide a neutral environment for the biomolecules, and are capable of enhancing the electrochemical signal in an immunosensor (Hemben et al. 2017).

6.3.2 Nanosensors for the Detection of Biocrystalhemozoin and Beta-hematin

Biocrystalhemozoin, also known as malaria pigment or malaria biomarker, is a paramagnetic nanoparticle by-product of the malaria parasite (Obisesan et al. 2019), and its presence in the blood is indicative of malarial infection. Biocrystalhemozoin is highly recommended as a biomarker in the development of malarial diagnostic devices because it is localized in the parasites' digestive vacuoles, more stable, cheaper, and easily available compared to PfHRP 2 (Noah and Ndangili 2019)

Beta-hematin (β-hematin) is a ferriprotoporphyrin IX (Fe(III)PPIX) microcrystalline cyclic dimer. It is used as a biomarker for malaria diagnosis in a dense magnetic field surface (Yuen and Liu 2012) as well as an assay for malaria detection (Rebelo et al. 2013). It is also used as a biomarker in the synthesis of antimalarial drugs due to its ability to inhibit the synthesis of hemozoin's in the blood. Studies have also shown that β-hematin and hemozoin are chemically and structurally similar (Noah and Ndangili 2019). For this reason, most authors use β-hematin to mimic hemozoin in the development of malarial sensor devices.

Various metal oxide nanoparticles have been synthesized and tested for the fabrication of nanosensors for the diagnosis of malaria using β-hematin as a biomarker. The popular ones are those of CuO, Al_2O_3, and Fe_2O_3 nanoparticles (Obisesan et al. 2019). Synthesis methods, mainly microwave and chemical, were used for each of the aforementioned metal nanoparticles and applied for electrochemical nanosensor fabrication.

6.3.3 Nanosensors for the Detection of Plasmodium Lactate Dehydrogenase (pLDH)

Plasmodium lactate dehydrogenase (pLDH) is produced in the red blood cells of infected persons by metabolically active parasites. Except in *Plasmodium knowlesi*, pLDH in all other species of plasmodium contain conserved catalytic residues. The presence of pLDH in red blood cells indicates a fresh infection and therefore important for early diagnosis. It clears within 24 hours after the parasite is eliminated from the system. This makes it an important biomarker for post medication diagnosis to identify any unresolved infections.

A few reports have demonstrated the utility of pLDH as a biomarker in the fabrication of nanosensors for malaria diagnosis. An aptasensor fabricated from a 90 oligomers single-strand deoxyribonucleic acids (ssDNA) aptamer, immobilized on graphene oxide modified glassy carbon electrode was used for successful detection of pLDH (Jain et al. 2016). Using electrochemical impedance spectroscopy to monitor changes in the charge transfer resistance resulting from capture of pLDH by the immobilized aptamers, concentrations as low as 0.5 fm could be detected.

6.3.4 Nanosensors for the Detection of Glutamate Dehydrogenase (GDH)

Glutamate dehydrogenase is an enzyme in parasites that take part in the catalytic metabolism of glutamate and absorption of ammonium (Zocher et al. 2012, Jain et al. 2014). This enzyme is significantly present in soluble form throughout the sexual and asexual stage of the parasite development. The use of GDH biomarker for malaria detection has been reported in aptasensors (Singh et al. 2018, 2019) and two of them have been demonstrated. In one report, a thiolated aptamer, specifically designed to bind to *Plasmodium falciparum* glutamate dehydrogenase (pfGDH) antigen, is immobilized on a gold electrode. Voltage-induced binding between the aptamer and the pfGDH antigen is then performed in a binding buffer after which electrochemical impedance spectroscopy measurements are done to quantify the pfGDH. This aptasensor can detect pfGDH with a detection limit of 0.77 pM in serum spiked samples. However, changing the transduction method of the same sensor to a field-effect transistor, the detection limit increases to 48.6 pM.

6.3.5 Nanosensors for the Detection of Topoisomerase I

Topoisomerase I is an essential enzyme that facilitates the maintenance of the genomic topology, by relaxing helical tension through a swivel-based mechanism including transient single-stranded nicking of the DNA double helix (Givskov et al. 2016). The enzyme is ubiquitously expressed by the malaria-causing *Plasmodium* parasite (Hede et al. 2018). The use of Topoisomerase I as a biomarker for the diagnosis of malaria has not been extensively explored. Hede et al. (Hede et al. 2018) described one of the few such applications. However, the detection platform was not nano-based, although the analysis sample (saliva) presents an attractive non-invasive diagnosis path. There is a report by Jepsen et al. (Jepsen et al. 2014) of a quantum dot-based DNA sensor for Topoisomerase I, but whose target is cancer as opposed to malaria. The successful demonstration of this nanosensor assembly indicates that Topoisomerase I based nanosensors for malaria is possible.

6.3.6 Nanosensors for the Detection of Merozoite Surface Protein (MSP)

Proteins associated with the surface of merozoites may be used as biomarkers for malaria infection. Merozoite surface proteins in *Plasmodium falciparum* and *Plasmodium vivax* are structurally related (Hisaeda et al. 2002). The protein is expressed on the merozoite surface when the parasite is temporarily exposed to the host immune system and may be used to detect infections caused by both *Plasmodium falciparum* and *Plasmodium vivax* parasites. This has been demonstrated by the successful fabrication of reduced graphene oxide (rGO) based gold nanoparticles, followed by the conjugation of merozoite surface protein-1 (MSP-I) antibody specific for *Plasmodium vivax*, which is then deposited on a carbon strip (Singh et al. 2020). The nanosensor was used for electrochemical detection of malaria in *Plasmodium vivax* infected red blood cells by monitoring the current-density as well as charge transfer resistance changes resulting from interactions of the MSP-I antibodies and the infected red blood cells. This nanosensor was further tested for its suitability in detecting malaria in human's whole red blood samples from infected patients and results were obtained within 5 minutes.

6.3.7 Nanosensors for the Detection of Aldolase

Aldolase enzyme catalyzes the breakdown of fructose-1,6-bisphosphate into glyceraldehyde-3-phosphate and dihydroxyacetone phosphate in the plasmodium species. Its use as a biomarker for malaria diagnosis has been widely reported in immunochromatographic tests (ICTs) (Dzakah et al. 2014). However, the genetic coding of aldolase, especially in *Plasmodium falciparum* and *Plasmodium vivax,* is highly conserved, conferring poor sensitivity and specificity in aldolase-based ICTs. Aldolase is therefore branded a poor biomarker for malaria diagnosis. There are therefore hardly any reports on nanosensors using aldolase for malaria diagnosis.

6.4 Applications of Nanosensor Arrays in HIV Diagnosis

The human immunodeficiency virus (HIV) is a retrovirus that belongs to the lentivirus genus (Saylan and Denizli 2020), characterized by the long interval between infection and first symptom appearance. HIV is classified into two, HIV-1 and HIV-2, with HIV-1 being the most prevalent (Saylan and Denizli 2020). Once inside the bloodstream, HIV infects the CD4$^+$ T cells (Inci et al. 2013) and replicates, causing a chronic and deadly condition known as acquired immunodeficiency syndrome (AIDS). There is no known cure for HIV/AIDS and no vaccine has been developed to date, to prevent its infection. Management of AIDS is done through the lifetime use of a combination of antiretroviral drugs (ARVs) (Mamo et al. 2010). The lifetime uses of the ARVs cause devastating side effects in patients, while others develop resistance to the drugs. The long incubation periods, lack of cure/vaccine as well as the effects of lifetime use of ARVs present challenges in effective containment of HIV/AIDS. Research has shown that early diagnosis and management of HIV/AIDS is capable of maintaining low viral load in infected persons, thereby decreasing the chances of infecting other persons. Infected patients who know about their infection at early stages and follow the recommended dietary and lifestyle protocols have been shown to lead healthy lives for long. Many approaches have been explored to provide rapid point of care diagnostic tools for HIV/AIDS. Most of the current research focuses on nanotechnologies due to the numerous opportunities that these materials present. Most of these nanosensors target HIV-1 due to its high infection prevalence. For detection of HIV-1, the most relied biomarkers are anti-HIV-1 antibody, HIV-1 ribonucleic acid (HIV-1 RNA), and HIV-1 p24 capsid antigen (Tang and Hewlett 2010). Detection of p24 capsid antigen allows for diagnosis of AIDS at any stage since its levels are significantly high during the early infection stages, acute phase, and terminal stage of the disease (Tang and Hewlett 2010). It is therefore an important target for early disease diagnostic. It can also be used to monitor the disease progression once the patient is put on ARVs, to indicate the CD4$^+$ T cells count in the patients. Several nanosensors for the detection of HIV/AIDS have therefore emerged and are discussed in the following section.

Gold nanoparticles have received the most attention in nanosensor fabrication for HIV/AIDS. This is because gold nanoparticles conjugate easily with biomolecules and exhibit fluorescence, surface-enhanced Raman scattering, surface Plasmon resonance, and magnetic properties (Malefane 2020). These properties easily give detectable signals using enzyme-linked immunosorbent assay (ELISA), Plasmon labeling and imaging as well as electrochemical methods (Kumar et al. 2011b). Gold nanoparticles are suitable materials for use in signal amplification via bio barcode amplification (BCA) assay (Nam et al. 2003). This allows signal amplification at each antigen-antibody binding event in an immunoassay, giving rise to high sensitivities. An example of this approach for HIV-1 diagnosis used HIV-1 p24 capsid antigen as a biomarker (Tang and Hewlett 2010). Apart from using nanoparticles as probes, signal amplification particles of immobilization platforms, nanosensor arrays can also be formed by modulating the size of the biomolecules to the nanoscale. Arrays of antibodies with well-defined feature size and spacing are necessary for developing highly sensitive and selective immunoassays to detect macromolecules in complex solutions. For instance, the HIV-1 p24 antigen can be modulated into nanoscale patterns on a gold surface using dip-pen nanolithography (Lee et al. 2004). This is then hybridized with HIV-1 p24 antigens from the plasma of infected patients. By further hybridizing the bound protein on a gold antibody functionalized nanoparticle probe, the signal gets significantly enhanced. This nanosensor can afford detections as low as 0.025 pg ml^{-1} and presents a greatly improved sensitivity compared to the one reported for BCA.

The above two examples describe the use of gold nanoparticles for immunosensor fabrication. These particles can also be used for label-free genosensors. For instance, single-strand DNA specific to HIV-1 can be immobilized onto gold nanorods and allowed to hybridize with corresponding complementary or mismatched DNA sequences (Darbha et al. 2008). The hybridization event can then be measured using hyper Raleigh scatter intensity, arising from the nonlinear optical response, induced by the gold nanorods. The importance of this method is that it is label-free, easily detects single base pair mismatch, does not involve DNA modification, and affords high sensitivities.

Gold nanoparticles used in conjunction with carbon nanotubes such as single-wall carbon nanotubes provide superior sensing platforms with enhanced sensitivity. For instance, by modifying a gold electrode with thiolated single-wall carbon nanotubes and gold nanoparticles, a biocompatible surface is obtained. On this surface, an electroactive probe such as ferrocene pepstatin can be immobilized via self-assembly (Mahmoud et al. 2008). The electrode thus prepared can be used to detect HIV-1 proteins by monitoring the interaction between ferrocene pepstatin and HIV-1 using electrochemical impedance spectroscopy (Mahmoud and Luong 2008) or cyclic voltammetry (Mahmoud et al. 2008). Electrochemical detection of HIV-1 via direct electron transfer is also possible using gold nanoparticles (Lee et al. 2013). The modification of gold nanoparticles using suitable metal oxide nanoparticles such as indium tin oxide has been shown to improve the surface area (Lee et al. 2013) upon which antibody fragments can be self-assembled.

Carbon nanotubes possess photoluminescence properties, especially fluorescent in the near infra-red region, and are resistant to photobleaching (Harvey et al. 2019).

Their unique excitations and emissions are triggered by changes in their local dielectric environments. They can therefore be used to monitor biomolecule binding events such as hybridization reactions. Studies have shown that the formation of double-strand DNA upon hybridization changes the surface coverage of carbon nanotubes (Harvey et al. 2017). This increases the sites available for binding of surfactants such as dodecylbenzenesulfonate (SDBS) to the carbon nanotubes' surface, causing significant changes in their optical bandgap. This enhancement effect is pronounced in biological fluids such as urine and serum, which offers a possible route for the detection of short DNA sequences and microRNA. Carbon nanotubes screen printed electrodes are also suitable platforms upon which streptavidin-modified oligonucleotides specific to HIV can be immobilized for its diagnosis (Adam et al. 2010). The streptavidin is modified with paramagnetic microparticles to confer high affinity to biotin, which allows the use of biotinylated oligonucleotides.

7. Conclusions and Future Perspective

In medical diagnosis, clinicians need to be able to diagnose medical problems fast. This enables them to treat the patients before any irreversible or long-lasting damage can occur and can be achieved by point-of-care testing which ensures fast detection of analytes near to the patient, facilitating a better disease diagnosis, monitoring, and management. Nanotechnology has demonstrated a huge potential to improve the sensitivity and linear ranges of various detection methods, improving assessment of the disease onset and its progression and help to plan for treatment of many diseases. Taking advantage of the unique properties of nanomaterials in sensor devices has led to the development of nanosensor arrays with increased sensitivity, fast response, and reduced cost. There now exist proven nanosensor arrays for medical diagnosis of various diseases such as cancer, malaria, diabetes, and HIV as discussed in this chapter, which can provide real-time information of a patient's condition.

However, despite the great success in utilizing nanomaterials with enhanced physicochemical properties, certain issues are hindering the rapid development of marketable nanosensor arrays for medical diagnosis. For example, it is important to note that when using nanoparticles in nanosensor arrays, there is batch-to-batch variability which makes it difficult to produce high quality and reproducible nanosensor arrays (Salvati et al. 2015). Besides, some large metal nanoparticles have poor stability in biological fluids such as blood, saliva, and urine and therefore can affect the stability of the nanosensor arrays in many medical diagnoses. This calls for careful engineering when fabricating the nanosensor arrays to maintain the same performance in the hostile physiological conditions of many medical diagnoses. Another setback in the application of these nanosensor arrays in medical diagnosis is brought about by the type of sample that is used. For example, during the detection of diabetes, salivary glucose values with statistically significant correlation to blood glucose can only be obtained only in the case of fasting measurements and at least 2 h after a meal (Zhang et al. 2015, Rozsypal et al. 2018) and therefore additional attention would have to be given to validation of the nanoarrays used in their measurements. Also, the use of urine in diabetes diagnosis, though a noninvasive method, is extremely limited since they cannot be utilized for analysis of samples of normal or hypoglycemic glucose levels and the concentration of the glucose in urine does not reflect the current concentration in blood (Witkowska Nery et al. 2016). Therefore, the nanosensor arrays for diagnosis of diabetes using urine must be extremely sensitive which might be a challenge to achieve.

Another huge challenge facing the adoption of nanosensor arrays in medical diagnosis for disease monitoring is the normal initial high price for the smart innovative devices and the complex production processes. Although these nanosensor arrays are perceived to be of low cost, in the long run, these initial high costs prevent these nanotechnologies from being routinely used in a clinical setup. This, therefore, calls for the improvement of their performance at a level that would offset the manufacturing costs offering a solution for the nanosensor arrays to reach the market (Salvati et al. 2015). This can be achieved by using a combination approach whereby multiple nanomaterials are used at the same time amplifying the signals to achieve a multitargeted detection, producing nanosensor arrays suitable for various situations and thus more attractive for the market (Salvati et al. 2015).

It is also important to note that only diabetes diagnosis has made significant progress in introducing noninvasive nanosensors since tears, urine, and saliva, sweat and breath can be used as diagnostic samples. Cancer and malaria diagnoses have tried saliva, but with limited sensitivity. The diagnostic sample for most diseases discussed here remains largely obtained through invasive methods, which hinders voluntary participation in mass screening and testing of diseases by public health officials. Research is therefore needed to focus on non-invasive nanosensors for these diseases.

In conclusion, regardless of the mentioned challenges, nanotechnology is a growing area of research being devoted to the realization of innovative nanosensors arrays for the early diagnostic imaging, disease monitoring, and the detection of pathogens and hence provide real-time information of a patient's condition which can be of significant benefit.

Acknowledgements

The authors would like to thank the School of Pharmacy and Health Sciences—United States International University-Africa and the Technical University of Kenya for providing a conducive environment and facilities from which this work was written.

References

Abdel-Karim, R., Y. Reda and A. Abdel-Fattah. (2020). Review—nanostructured materials-based nanosensors. *Journal of the Electrochemical Society* 167: 037554.
Abegunde, O. O., E. T. Akinlabi, O. P. Oladijo, S. Akinlabi and A. U. Ude. (2019). Overview of thin film deposition techniques. *Materials Science* 6: 174–199.
Abikshyeet, P., V. Ramesh and N. Oza. (2012). Glucose estimation in the salivary secretion of diabetes mellitus patients. *Diabetes Metab Syndr Obes* 5: 149–154.
Adam, V., D. Huska, J. Hubalek and R. Kizek. (2010). Easy to use and rapid isolation and detection of a viral nucleic acid by using paramagnetic microparticles and carbon nanotubes-based screen-printed electrodes. *Microfluidics and Nanofluidics* 8: 329–339.

Akhmadishina, K. F., I. I. Bobrinetskii, I. A. Komarov, A. M. Malovichko, V. K. Nevolin, V. A. Petukhov et al. (2013). Flexible biological sensors based on carbon nanotube films. *Nanotechnologies in Russia* 8: 721–726.

Andoralov, V., S. Shleev, T. Arnebrant and T. Ruzgas. (2013). Flexible micro(bio)sensors for quantitative analysis of bioanalytes in a nanovolume of human lachrymal liquid. *Anal Bioanal Chem* 405: 3871–3879.

Andrews, R., D. Jacques, D. Qian and T. Rantell. (2002). Multiwall carbon nanotubes: Synthesis and application. *Accounts of Chemical Research* 35: 1008–1017.

Arti, G., P. Sonia, V. Bharat, S. Shailesh and Y. Jitendra Singh. (2019). Green synthesis of gold nanoparticles using different leaf extracts of ocimum gratissimum linn for anti-tubercular activity. *Current Nanomedicine* 9: 146–157.

Ayerden, N. P. and R. F. Wolffenbuttel. (2017). The miniaturization of an optical absorption spectrometer for smart sensing of natural gas. *IEEE Transactions on Industrial Electronics* 64: 9666–9674.

Ayinde, W. B., W. M. Gitari, M. Munkombwe, A. Samie and J. A. Smith. (2020). Green synthesis of AgMgOnHaP nanoparticles supported on chitosan matrix: Defluoridation and antibacterial effects in groundwater. *Journal of Environmental Chemical Engineering* 8: 104026.

AzoNano. (2019). Nanosensors: Definition, Applications and How They Work. from https://www.azonano.com/article.aspx?ArticleID=1840#:~:text=%E2%80%9CNanosensors%20are%20chemical%20or%20mechanical,%2C%20and%20other%20chemicals.%E2%80%9D%20.

Bajaj, A., O. R. Miranda, I.-B. Kim, R. L. Phillips, D. J. Jerry, U. H. F. Bunz et al. (2009). Detection and differentiation of normal, cancerous, and metastatic cells using nanoparticle-polymer sensor arrays. *Proceedings of the National Academy of Sciences of the United States of America* 106: 10912–10916.

Bajaj, A., S. Rana, O. R. Miranda, J. C. Yawe, D. J. Jerry, U. H. F. Bunz et al. (2010). Cell surface-based differentiation of cell types and cancer states using a gold nanoparticle-GFP based sensing array. *Chemical Science* 1: 134–138.

Barber, B. E., T. William, M. J. Grigg, K. Piera, T. W. Yeo and N. M. Anstey. (2013). Evaluation of the sensitivity of a pLDH-Based and an Aldolase-Based rapid diagnostic test for diagnosis of uncomplicated and severe malaria caused by PCR-Confirmed Plasmodium knowlesi, Plasmodium falciparum, and Plasmodium vivax. *Journal of Clinical Microbiology* 51: 1118.

Bayati, M., P. Patoka, M. Giersig and E. R. Savinova. (2010). An approach to fabrication of metal nanoring arrays. *Langmuir* 26: 3549–3554.

Bellah, M. M., S. M. Christensen and S. M. Iqbal. (2012). Nanostructures for medical diagnostics. *Journal of Nanomaterials* 2012: 486301.

Bhardwaj, V. and A. Kaushik. (2017). Biomedical applications of nanotechnology and nanomaterials. *Micromachines* 8: 298.

Binions, R. and I. P. Parkin. (2011). Novel Chemical Vapour Deposition Routes to Nanocomposite Thin Films.

Camilli, L. and M. Passacantando. (2018). Advances on sensors based on carbon nanotubes. *Chemosensors* 6: 62.

Carlsson, J.-O. and P. M. Martin. (2010). Chapter 7—Chemical Vapor Deposition. Handbook of Deposition Technologies for Films and Coatings (Third Edition). P. M. Martin. Boston, William Andrew Publishing 314–363.

Cash, K. J. and H. A. Clark. (2010). Nanosensors and nanomaterials for monitoring glucose in diabetes. *Trends in Molecular Medicine* 16: 584–593.

Center, A. R. (2015). Nanosensors for medical diagnosis. from https://www.techbriefs.com/component/content/article/tb/pub/techbriefs/sensors-data-acquisition/22540.

Cheng, K., D. Cao, F. Yang, L. Zhang, Y. Xu and G. Wang. (2012). Electrodeposition of Pd nanoparticles on C@TiO2 nanoarrays: 3D electrode for the direct oxidation of NaBH4. *Journal of Materials Chemistry* 22: 850–855.

Chouchene, B., T. B. Chaabane, K. Mozet, E. Girot, S. Corbel, L. Balan et al. (2017). Porous Al-doped ZnO rods with selective adsorption properties. *Applied Surface Science* 409: 102–110.

Colomer, J. F., C. Stephan, S. Lefrant, G. Van Tendeloo, I. Willems, Z. Kónya et al. (2000). Large-scale synthesis of single-wall carbon nanotubes by catalytic chemical vapor deposition (CCVD) method. *Chemical Physics Letters* 317: 83–89.

Critchley, L. (2018). Nanosensors: An introduction. from https://www.azonano.com/article. aspx?ArticleID=4933#:~:text=There%20are%20two%20types%20of,an%20analyte%20has%20 been%20detected.

Cruz, S. M. A., A. F. Girão, G. Gonçalves and P. A. A. P. Marques. (2016). Graphene: the missing piece for cancer diagnosis? *Sensors* (Basel, Switzerland) 16: 137.

Darbha, G. K., U. S. Rai, A. K. Singh and P. C. Ray. (2008). Gold-nanorod-based sensing of sequence specific HIV-1 virus DNA by using hyper-rayleigh scattering spectroscopy. *Chemistry—A European Journal* 14: 3896–3903.

De Moraes, C. M., C. Wanjiku, N. M. Stanczyk, H. Pulido, J. W. Sims, H. S. Betz et al. (2018). Volatile biomarkers of symptomatic and asymptomatic malaria infection in humans. *Proceedings of the National Academy of Sciences* 115: 5780–5785.

de Souza Castilho, M., T. Laube, H. Yamanaka, S. Alegret and M. I. Pividori. (2011). Magneto Immunoassays for Plasmodium falciparum Histidine-rich protein 2 related to malaria based on magnetic nanoparticles. *Analytical Chemistry* 83: 5570–5577.

Ding, M., D. C. Sorescu and A. Star. (2013). Photo induced charge transfer and acetone sensitivity of single-walled carbon nanotube–titanium dioxide hybrids. *Journal of the American Chemical Society* 135: 9015–9022.

Doria, G., J. Conde, B. Veigas, L. Giestas, C. Almeida, M. Assunção et al. (2012). Noble metal nanoparticles for biosensing applications. *Sensors* (Basel, Switzerland) 12: 1657–1687.

Dutta, G. (2020). Electrochemical biosensors for rapid detection of malaria. *Materials Science for Energy Technologies* 3: 150–158.

Dzakah, E. E., K. Kang, C. Ni, S. Tang, J. Wang and J. Wang. (2014). Comparative performance of aldolase and lactate dehydrogenase rapid diagnostic tests in Plasmodium vivax detection. *Malaria Journal* 13: 272.

Eatemadi, A., H. Daraee, H. Karimkhanloo, M. Kouhi, N. Zarghami, A. Akbarzadeh et al. (2014). Carbon nanotubes: Properties, synthesis, purification, and medical applications. *Nanoscale Research Letters* 9: 393.

Enotiadis, A., K. Angjeli, N. Baldino, I. Nicotera and D. Gournis. (2012). Graphene-Based Nafion Nanocomposite Membranes: Enhanced Proton Transport and Water Retention by Novel Organo-functionalized. *Graphene Oxide Nanosheets* 8: 3338–3349.

Feachem, R. G. A., I. Chen, O. Akbari, A. Bertozzi-Villa, S. Bhatt, F. Binka et al. (2019). Malaria eradication within a generation: Ambitious, achievable, and necessary. *The Lancet* 394: 1056–1112.

Ganesan, K., V. K. Jothi, A. Natarajan, A. Rajaram, S. Ravichandran and S. Ramalingam. (2020). Green synthesis of Copper oxide nanoparticles decorated with graphene oxide for anticancer activity and catalytic applications. *Arabian Journal of Chemistry* 13: 6802–6814.

Garg, S. K. and H. K. Akturk. (2017). The future of continuous glucose monitoring. *Diabetes Technology & Therapeutics* 19: S1–S2.

Givskov, A., E. L. Kristoffersen, K. Vandsø, Y.-P. Ho, M. Stougaard and B. R. Knudsen. (2016). Optimized Detection of Plasmodium falciparum Topoisomerase I Enzyme Activity in a Complex Biological Sample by the Use of Molecular Beacons. *Sensors* 16.

Gu, H., H. Tang, P. Xiong and Z. Zhou. (2019). Biomarkers-based Biosensing and Bioimaging with Graphene for Cancer Diagnosis. *Nanomaterials* (Basel, Switzerland) 9: 130.

Harvey, J. D., P. V. Jena, H. A. Baker, G. H. Zerze, R. M. Williams, T. V. Galassi et al. (2017). A carbon nanotube reporter of microRNA hybridization events *in vivo*. *Nature Biomedical Engineering* 1: 0041.

Harvey, J. D., H. A. Baker, M. V. Ortiz, A. Kentsis and D. A. Heller. (2019). HIV Detection via a Carbon Nanotube RNA Sensor. *ACS Sensors* 4: 1236–1244.

Hayes, B., C. Murphy, A. Crawley and R. O'Kennedy. (2018). Developments in point-of-care diagnostic technology for cancer detection. *Diagnostics* (Basel, Switzerland) 8: 39.

Hede, M. S., S. Fjelstrup, F. Lötsch, R. M. Zoleko, A. Klicpera, M. Groger et al. (2018). Detection of the Malaria causing Plasmodium Parasite in Saliva from Infected Patients using Topoisomerase I Activity as a Biomarker. *Scientific Reports* 8: 4122.

Hemben, A., J. Ashley and I. E. Tothill. (2017). Development of an immunosensor for PfHRP 2 as a biomarker for malaria detection. *Biosensors* 7: 28.

Hisaeda, H., A. Saul, J. J. Reece, M. C. Kennedy, C. A. Long, L. H. Miller et al. (2002). Merozoite Surface Protein 3 and Protection against Malaria in Aotus nancymai Monkeys. *The Journal of Infectious Diseases* 185: 657–664.

Hosseinpour-Mashkani, S. M., M. Maddahfar and A. Sobhani-Nasab. (2016). Precipitation synthesis, characterization, morphological control, and photocatalyst application of ZnWO4 nanoparticles. *Journal of Electronic Materials* 45: 3612–3620.

Huang, C.-H., C. Zeng, Y.-C. Wang, H.-Y. Peng, C.-S. Lin, C.-J. Chang et al. (2018). A study of diagnostic accuracy using a chemical sensor array and a machine learning technique to detect lung cancer. *Sensors* (Basel, Switzerland) 18: 2845.

Huang, Y., F. Yu, Y.-S. Park, J. Wang, M.-C. Shin, H. S. Chung et al. (2010). Co-administration of protein drugs with gold nanoparticles to enable percutaneous delivery. *Biomaterials* 31: 9086–9091.

Huber, F., H. Lang, J. Zhang, D. Rimoldi and C. Gerber. (2015). *Nanosensors for Cancer Detection.* Swiss Med. Wkly. 145:w14092.

Hufschmid, R., H. Arami, R. M. Ferguson, M. Gonzales, E. Teeman, L. N. Brush et al. (2015). Synthesis of phase-pure and monodisperse iron oxide nanoparticles by thermal decomposition. *Nanoscale* 7: 11142–11154.

Inci, F., O. Tokel, S. Wang, U. A. Gurkan, S. Tasoglu, D. R. Kuritzkes et al. (2013). Nanoplasmonic quantitative detection of intact viruses from unprocessed whole blood. *ACS Nano* 7: 4733–4745.

Ionita, M., A. M. Pandele, L. Crica and L. Pilan. (2014). Improving the thermal and mechanical properties of polysulfone by incorporation of graphene oxide. *Composites Part B: Engineering* 59: 133–139.

Jain, P., B. Chakma, S. Patra and P. Goswami. (2014). Potential biomarkers and their applications for rapid and reliable detection of malaria. *BioMed Research International* 2014: 852645.

Jain, P., S. Das, B. Chakma and P. Goswami. (2016). Aptamer-graphene oxide for highly sensitive dual electrochemical detection of Plasmodium lactate dehydrogenase. *Analytical Biochemistry* 514: 32–37.

Jayson, G. C., E. C. Kohn, H. C. Kitchener and J. A. Ledermann. (2014). "Ovarian cancer" *Lancet* 384: 1376–1388.

Jepsen, M. L., A. Ottaviani, B. R. Knudsen and Y.-P. Ho. (2014). Quantum dot based DNA nanosensors for amplification-free detection of human topoisomerase I. *RSC Advances* 4: 2491–2494.

Jhaveri, J. H. and Z. V. P. Murthy. (2016a). A comprehensive review on anti-fouling nanocomposite membranes for pressure driven membrane separation processes. *Desalination* 379: 137–154.

Jhaveri, J. H. and Z. V. P. Murthy. (2016b). Nanocomposite membranes. *Desalination and Water Treatment* 57: 26803–26819.

Kaushik, A. and M. A. Mujawar. (2018). Point of care sensing devices: Better care for everyone. *Sensors* 18: 4303.

Kaushik, A., A. Vasudev, S. K. Arya, S. K. Pasha and S. Bhansali. (2014). Recent advances in cortisol sensing technologies for point-of-care application. *Biosensors and Bioelectronics* 53: 499–512.

Kaushik, A., A. Yndart, S. Kumar, R. D. Jayant, A. Vashist, A. N. Brown et al. (2018). A sensitive electrochemical immunosensor for label-free detection of Zika-virus protein. *Scientific Reports* 8: 9700.

Khanna, V. K. (2011). Nanosensors: Physical, chemical, and biological, CRC Press.

Kim, H. Y., K. J. Jang, M. Veerapandian, H. C. Kim, Y. T. Seo, K. N. Lee et al. (2014). Reusable urine glucose sensor based on functionalized graphene oxide conjugated Au electrode with protective layers. *Biotechnology Reports* 3: 49–53.

Kim, S.-J., S.-J. Choi, J.-S. Jang, H.-J. Cho and I.-D. Kim. (2017). Innovative nanosensor for disease diagnosis. *Accounts of Chemical Research* 50: 1587–1596.

Kokate, M., K. Garadkar and A. Gole. (2013). One pot synthesis of magnetite–silica nanocomposites: Applications as tags, entrapment matrix and in water purification. *Journal of Materials Chemistry A* 1: 2022–2029.

Kosaka, P. M., V. Pini, J. J. Ruz, R. A. da Silva, M. U. González, D. Ramos et al. (2014). Detection of cancer biomarkers in serum using a hybrid mechanical and optoplasmonic nanosensor. *Nature Nanotechnology* 9: 1047.

Kovalenko, M. V., M. I. Bodnarchuk, R. T. Lechner, G. Hesser, F. Schäffler and W. Heiss. (2007). Fatty Acid Salts as Stabilizers in Size- and Shape-Controlled Nanocrystal Synthesis: The Case of Inverse Spinel Iron Oxide. *Journal of the American Chemical Society* 129: 6352–6353.

Krampa, F. D., Y. Aniweh, P. Kanyong and G. A. Awandare. (2020). Recent advances in the development of biosensors for malaria diagnosis. *Sensors* (Basel, Switzerland) 20 DOI: 10.3390/s20030799.

Krithiga, N., A. Rajalakshmi and A. Jayachitra. (2015). Green synthesis of silver nanoparticles using leaf extracts of <i>Clitoria ternatea</i> and <i>Solanum nigrum</i> and study of its antibacterial effect against common nosocomial pathogens. *Journal of Nanoscience* 2015: 928204.

Kumar, K. P., W. Paul and C. P. Sharma. (2011a). Green synthesis of gold nanoparticles with Zingiber officinale extract: Characterization and blood compatibility. *Process Biochemistry* 46: 2007–2013.

Kumar, A., B. Mazinder Boruah and X.-J. Liang. (2011b). Gold nanoparticles: promising nanomaterials for the diagnosis of cancer and HIV/AIDS. *Journal of Nanomaterials* 2011: 202187.

Kumar, V., S. Hebbar, A. Bhat, S. Panwar, M. Vaishnav, K. Muniraj et al. (2018). Application of a nanotechnology-based, point-of-care diagnostic device in diabetic kidney disease. *Kidney International Reports* 3: 1110–1118.

Kwon, Y. J., A. Mirzaei, H. G. Na, S. Y. Kang, M. S. Choi, J. H. Bang et al. (2018). Porous Si nanowires for highly selective room-temperature NO2 gas sensing. *Nanotechnology* 29: 294001.

Lang, J., J. Wang, Q. Zhang, X. Li, Q. Han, M. Wei et al. (2016). Chemical precipitation synthesis and significant enhancement in photocatalytic activity of Ce-doped ZnO nanoparticles. *Ceramics International* 42: 14175–14181.

Le, N. D. B., M. Yazdani and V. M. Rotello. (2014). Array-based sensing using nanoparticles: An alternative approach for cancer diagnostics. *Nanomedicine* (London, England) 9: 1487–1498.

Lee, H., T. K. Choi, Y. B. Lee, H. R. Cho, R. Ghaffari, L. Wang et al. (2016). A graphene-based electrochemical device with thermoresponsive microneedles for diabetes monitoring and therapy. *Nature Nanotechnology* 11: 566–572.

Lee, J.-H., B.-K. Oh and J.-W. Choi. (2013). Electrochemical sensor based on direct electron transfer of HIV-1 Virus at Au nanoparticle modified ITO electrode. *Biosensors and Bioelectronics* 49: 531–535.

Lee, K.-B., E.-Y. Kim, C. A. Mirkin and S. M. Wolinsky. (2004). The use of nanoarrays for highly sensitive and selective detection of human immunodeficiency virus type 1 in plasma. *Nano Letters* 4: 1869–1872.

Leroux, M. (2001). Laboratory Testing in Diabetes Mellitus. Diabetes and Cardiovascular Disease: Etiology, Treatment, and Outcomes. A. Angel, N. Dhalla, G. Pierce and P. Singal. Boston, MA, Springer US 359-366.

Lifson, M. A., M. O. Ozen, F. Inci, S. Wang, H. Inan, M. Baday et al. (2016). Advances in biosensing strategies for HIV-1 detection, diagnosis, and therapeutic monitoring. *Advanced Drug Delivery Reviews* 103: 90–104.

Lin, H.-J., J. P. Baltrus, H. Gao, Y. Ding, C.-Y. Nam, P. Ohodnicki et al. (2016). Perovskite nanoparticle-sensitized Ga2O3 nanorod arrays for CO detection at high temperature. *ACS Applied Materials & Interfaces* 8: 8880–8887.

Lin, M., H. Pei, F. Yang, C. Fan and X. Zuo. (2013). Applications of gold nanoparticles in the detection and identification of infectious diseases and biothreats. *Advanced Materials* 25: 3490–3496.

Lin, Y., F. Lu, Y. Tu and Z. Ren. (2004). Glucose biosensors based on carbon nanotube nanoelectrode ensembles. *Nano Letters* 4: 191–195.

Liu, G., K. Han, H. Ye, C. Zhu, Y. Gao, Y. Liu et al. (2017). Graphene oxide/triethanolamine modified titanate nanowires as photocatalytic membrane for water treatment. *Chemical Engineering Journal* 320: 74–80.

Lu, Y., C. Partridge, M. Meyyappan and J. Li. (2006). A carbon nanotube sensor array for sensitive gas discrimination using principal component analysis. *Journal of Electroanalytical Chemistry* 593: 105–110.

Lu, Y., Y. Liu, S. Zhang, S. Wang, S. Zhang and X. Zhang. (2013). Aptamer-based plasmonic sensor array for discrimination of proteins and cells with the naked eye. *Anal Chem* 85: 6571–6574.

Lu, Y., H. Zhang, Y. Du, C. Han, Z. Nie, Z. Sun et al. (2020). Structure design of ni–co hydroxide nanoarrays with facet engineering on carbon chainlike nanofibers for high-efficiency oxygen evolution. *ACS Applied Energy Materials* 3: 6240–6248.

Luo, X., A. Morrin, A. J. Killard and M. R. Smyth. (2006). Application of nanoparticles in electrochemical sensors and biosensors. *Electroanalysis* 18: 319–326.

Ma, Z., D. Zhao, Y. Chang, S. Xing, Y. Wu and Y. Gao. (2013). Synthesis of MnFe2O4@Mn–Co oxide core–shell nanoparticles and their excellent performance for heavy metal removal. *Dalton Transactions* 42: 14261–14267.

Maduraiveeran, G. and W. Jin. (2017). Nanomaterials based electrochemical sensor and biosensor platforms for environmental applications. *Trends in Environmental Analytical Chemistry* 13: 10–23.

Mahmoud, K. A., S. Hrapovic and J. H. T. Luong. (2008). Picomolar detection of protease using peptide/ single walled carbon nanotube/gold nanoparticle-modified electrode. *ACS Nano* 2: 1051–1057.

Mahmoud, K. A. and J. H. T. Luong. (2008). Impedance method for detecting HIV-1 protease and screening for its inhibitors using ferrocene–peptide conjugate/au nanoparticle/single-walled carbon nanotube modified electrode. *Analytical Chemistry* 80: 7056–7062.

Makaram, P., D. Owens and J. Aceros. (2014). Trends in Nanomaterial-based non-invasive diabetes sensing technologies. *Diagnostics* 4: 27–46.

Malefane, M. E. (2020). Applications of nanotechnology towards detection and treatment of HIV/AIDS: A review article. *Research & Development in Material Science* 12: 1315–1321.

Mamo, T., E. A. Moseman, N. Kolishetti, C. Salvador-Morales, J. Shi, D. R. Kuritzkes et al. (2010). Emerging nanotechnology approaches for HIV/AIDS treatment and prevention. *Nanomedicine* (London, England) 5: 269–285.

Mayedwa, N., A. T. Khalil, N. Mongwaketsi, N. Matinise, Z. K. Shinwari and M. Maaza. (2017). The study of structural, physical and electrochemical activity of Zno nanoparticles synthesized by green natural extracts of Sageretia Thea. Nano Research & Applications 3.

McIntosh, J. (2016). Nanosensors: The future of diagnostic medicine? from https://www.medicalnewstoday.com/articles/299663.

Meng, G.-F., Q. Xiang, Q.-Y. Pan and J.-Q. Xu. (2011). The selective acetone detection based on Fe3O4 doped WO3 nanorods. *Sensor Letters* 9: 128–131.

Metkar, S. K. and K. Girigoswami. (2019). Diagnostic biosensors in medicine—A review. *Biocatalysis and Agricultural Biotechnology* 17: 271–283.

Miekisch, W., J. K. Schubert and G. F. E. Noeldge-Schomburg. (2004). Diagnostic potential of breath analysis—focus on volatile organic compounds. *Clinica Chimica Acta* 347: 25–39.

Miranda, G. M. (2019). Nanoelectronic Biosensor Applications in Human Diseases. from https://www.news-medical.net/health/Nanoelectronic-Biosensors-Applications-in-Human-Diseases.aspx.

Mishra, A. and M. Verma. (2010). Cancer biomarkers: Are we ready for the prime time?" *Cancers* 2: 190–208.

Mosayebi, R., A. Ahmadzadeh, W. Wicke, V. Jamali, R. Schober and M. Nasiri-Kenari. (2018). Early Cancer Detection in Blood Vessels Using Mobile Nanosensors.

Moyer, J., D. Wilson, I. Finkelshtein, B. Wong and R. Potts. (2012). Correlation between sweat glucose and blood glucose in subjects with diabetes. *Diabetes Technology & Therapeutics* 14: 398–402.

Mueller, I., M. R. Galinski, J. K. Baird, J. M. Carlton, D. K. Kochar, P. L. Alonso et al. (2009). Key gaps in the knowledge of Plasmodium vivax, a neglected human malaria parasite. *The Lancet Infectious Diseases* 9: 555–566.

Munawar, A., Y. Ong, R. Schirhagl, M. A. Tahir, W. S. Khan and S. Z. Bajwa. (2019). Nanosensors for diagnosis with optical, electric and mechanical transducers. *RSC Advances* 9: 6793–6803.

Nam, J.-M., C. S. Thaxton and C. A. Mirkin. (2003). Nanoparticle-based bio-bar codes for the ultrasensitive detection of proteins. *Science* 301: 1884–1886.

Newman, J. D. and A. P. F. Turner. (2005). Home blood glucose biosensors: A commercial perspective. *Biosensors and Bioelectronics* 20: 2435–2453.

Noah, N. (2018). Green Synthesis: Characterization and Applications of Silver and Gold nanoparticles. Green Synthesis, Characterization and Applications of Nanoparticles. S. Holt, Elsevier Publishers. 1: 111–135.

Noah, N. M. and P. M. Ndangili. (2019). Current trends of nanobiosensors for point-of-care diagnostics. *Journal of Analytical Methods in Chemistry* 2019: 2179718.

Obisesan, O. R., A. S. Adekunle, J. A. O. Oyekunle, T. Sabu, T. T. I. Nkambule and B. B. Mamba. (2019). Development of electrochemical nanosensor for the detection of malaria parasite in clinical samples. *Frontiers in Chemistry*, 7.

Ogawa, F., C. Masuda and H. Fujii. (2018). *In situ* chemical vapor deposition of metals on vapor-grown carbon fibers and fabrication of aluminum-matrix composites reinforced by coated fibers. *Journal of Materials Science* 53: 5036–5050.

Okaie, Y., T. Nakano, T. Hara and S. Nishio. (2016). Target Detection and Tracking by Bionanosensor Networks.

Okuno, J., K. Maehashi, K. Kerman, Y. Takamura, K. Matsumoto and E. Tamiya. (2007). Label-free immunosensor for prostate-specific antigen based on single-walled carbon nanotube array-modified microelectrodes. *Biosens. Bioelectron* 22: 2377–2381.

Pallares, R. M., N. T. K. Thanh and X. Su. (2019). Sensing of circulating cancer biomarkers with metal nanoparticles. *Nanoscale* 11: 22152–22171.

Paul, P., A. K. Malakar and S. Chakraborty. (2019). The significance of gene mutations across eight major cancer types. *Mutation Research/Reviews in Mutation Research* 781: 88–99.

Pedersen, H. and S. D. Elliott. (2014). Studying chemical vapor deposition processes with theoretical chemistry. *Theoretical Chemistry Accounts* 133: 1476.

Pickup, J. C., Z. L. Zhi, F. Khan, T. Saxl and D. J. Birch. (2008). Nanomedicine and its potential in diabetes research and practice. *Diabetes Metab Res Rev* 24: 604–610.

Qasim, S., A. Zafar, M. S. Saif, Z. Ali, M. Nazar, M. Waqas et al. (2020). Green synthesis of iron oxide nanorods using Withania coagulans extract improved photocatalytic degradation and antimicrobial activity. *Journal of Photochemistry and Photobiology B: Biology* 204: 111784.

Rabinovich, R. N., C. Drakeley, A. A. Djimde, B. F. Hall, S. I. Hay, J. Hemingway et al. (2017). malERA: An updated research agenda for malaria elimination and eradication. *PLOS Medicine* 14: e1002456.

Rebelo, M., C. Sousa, H. M. Shapiro, M. M. Mota, M. P. Grobusch and T. Hänscheid. (2013). A novel flow cytometric hemozoin detection assay for real-time sensitivity testing of Plasmodium falciparum. *PLOS ONE* 8: e61606.

Righettoni, M., A. Tricoli and S. E. Pratsinis. (2010). Thermally stable, silica-doped ε-WO3 for sensing of acetone in the human breath. *Chemistry of Materials* 22: 3152–3157.

Risby, T. H. and S. F. Solga. (2006). Current status of clinical breath analysis. *Applied Physics B* 85: 421–426.

Roberts, K., A. Jaffe, C. Verge and P. S. Thomas. (2012). Noninvasive monitoring of glucose levels: Is exhaled breath the answer? *J Diabetes Sci Technol* 6: 659–664.

Rong, G., E. E. Tuttle, A. N. Reilly and H. A. Clark. (2019). Recent developments in nanosensors for imaging applications in biological systems. *Annual Review of Analytical Chemistry* 12: 109–128.

Rozsypal, J., D. Riman, V. Halouzka, T. Opletal, D. Jirovsky, M. Prodromidis et al. (2018). Use of interelectrode material transfer of nickel and copper-nickel alloy to carbon fibers to assemble miniature glucose sensors. *Journal of Electroanalytical Chemistry* 816: 45–53.

Rus, A., V.-D. Leordean and P. Berce. (2017). Silver Nanoparticles (AgNP) impregnated filters in drinking water disinfection. *MATEC Web of Conferences* 137: 1–6.

Saini, R. K., L. P. Bagri and A. K. Bajpai. (2017). 14—Smart nanosensors for pesticide detection. New Pesticides and Soil Sensors. A. M. Grumezescu, Academic Press 519–559.

Salvati, E., F. Stellacci and S. Krol. (2015). Nanosensors for early cancer detection and for therapeutic drug monitoring. *Nanomedicine* 10: 3495–3512.

Sandbhor Gaikwad, P. and R. Banerjee. (2018). Advances in point-of-care diagnostic devices in cancers. *Analyst* 143: 1326–1348.

Sani, H. A., M. B. Ahmad, M. Z. Hussein, N. A. Ibrahim, A. Musa and T. A. Saleh. (2017). Nanocomposite of ZnO with montmorillonite for removal of lead and copper ions from aqueous solutions. *Process Safety and Environmental Protection* 109: 97–105.

Saraoğlu, H. M. and M. Koçan. (2010). A study on non-invasive detection of blood glucose concentration from human palm perspiration by using artificial neural networks. *Expert systems* 27: 156–165.

Saylan, Y. and A. Denizli. (2020). Virus detection using nanosensors. *Nanosensors for Smart Cities*: 501–511.

Schürle-Finke, S. (2020). Nanosensors for disease diagnostics. 2020, from https://rbsl.ethz.ch/research/nanosensors-for-disease-diagnostics.html.

Shaban, M. (2016). Morphological and optical characterization of high density Au/PAA nanoarrays. *Journal of Spectroscopy* 2016: 5083482.

Shafiee, A., E. Ghadiri, J. Kassis and A. Atala. (2019). Nanosensors for therapeutic drug monitoring: Implications for transplantation. *Nanomedicine* 14: 2735–2747.

Sharma, M. K., V. K. Rao, G. S. Agarwal, G. P. Rai, N. Gopalan, S. Prakash et al. (2008). Highly sensitive amperometric immunosensor for detection of Plasmodium falciparum histidine-rich protein 2 in serum of humans with malaria: comparison with a commercial kit. Journal of Clinical Microbiology 46: 3759–3765.

Sharma, M. K., G. S. Agarwal, V. K. Rao, S. Upadhyay, S. Merwyn, N. Gopalan et al. (2010). Amperometric immunosensor based on gold nanoparticles/alumina sol–gel modified screen-printed electrodes for antibodies to Plasmodium falciparum histidine rich protein-2. *Analyst* 135: 608–614.

Simeonidis, K., C. Martinez-Boubeta, P. Zamora-Pérez, P. Rivera-Gil, E. Kaprara, E. Kokkinos et al. (2019). Implementing nanoparticles for competitive drinking water purification. *Environmental Chemistry Letters* 17: 705–719.

Singh, N. K., S. K. Arya, P. Estrela and P. Goswami. (2018). Capacitive malaria aptasensor using Plasmodium falciparum glutamate dehydrogenase as target antigen in undiluted human serum. *Biosensors and Bioelectronics* 117: 246–252.

Singh, N. K., P. D. Thungon, P. Estrela and P. Goswami. (2019). Development of an aptamer-based field effect transistor biosensor for quantitative detection of Plasmodium falciparum glutamate dehydrogenase in serum samples. *Biosensors and Bioelectronics* 123: 30–35.

Singh, P., M. Chatterjee, K. Chatterjee, R. K. Arun and N. Chanda. (2020). Design of a point-of-care device for electrochemical detection of P.vivax infected-malaria using antibody functionalized rGO-gold nanocomposite. *Sensors and Actuators B: Chemical*: 128860.

Siontorou, C. G., G.-P. D. Nikoleli, D. P. Nikolelis, S. Karapetis, N. Tzamtzis and S. Bratakou. (2017). Point-of-Care and Implantable Biosensors in Cancer Research and Diagnosis. Next Generation Point-of-care Biomedical Sensors Technologies for Cancer Diagnosis. P. Chandra, Y. N. Tan and S. P. Singh. Singapore, Springer Singapore 115–132.

Soraya, G. V., C. D. Abeyrathne, C. Buffet, D. H. Huynh, S. M. Uddin, J. Chan et al. (2019). Ultrasensitive and label-free biosensor for the detection of Plasmodium falciparum histidine-rich protein II in saliva. *Scientific Reports* 9: 17495.

Su, L., J. Feng, X. Zhou, C. Ren, H. Li and X. Chen. (2012). Colorimetric detection of urine glucose based ZnFe2O4 magnetic nanoparticles. *Analytical Chemistry* 84: 5753–5758.

Sun, J., Y. Lu, L. He, J. Pang, F. Yang and Y. Liu. (2020). Colorimetric sensor array based on gold nanoparticles: Design principles and recent advances. *TrAC Trends in Analytical Chemistry* 122: 115754.

Sun, S. and H. Zeng. (2002). Size-controlled synthesis of magnetite nanoparticles. *Journal of the American Chemical Society* 124: 8204–8205.

Tang, S. and I. Hewlett. (2010). Nanoparticle-based immunoassays for sensitive and early detection of HIV-1 capsid (p24) antigen. *The Journal of Infectious Diseases 201 Suppl.* 1: S59–S64.

Tesauro, C., S. Juul, B. Arnò, C. J. F. Nielsen, P. Fiorani, R. F. Frøhlich et al. (2012). Specific detection of topoisomerase i from the malaria causing P. falciparum parasite using isothermal Rolling Circle Amplification. 2012 Annual International Conference of the IEEE Engineering in Medicine and Biology Society.

Tian, T., J. Dong and J. Xu. (2016). Direct electrodeposition of highly ordered gold nanotube arrays for use in non-enzymatic amperometric sensing of glucose. *Microchimica Acta* 183: 1925–1932.

Tilmaciu, C.-M. and M. C. Morris. (2015). Carbon nanotube biosensors. *Frontiers in Chemistry* 3: 59–59.

Tuteja, S. K., R. Mutreja, S. Neethirajan and S. Ingebrandt. (2019). Chapter 5—bioconjugation of different nanosurfaces with biorecognition molecules for the development of selective nanosensor platforms. Advances in Nanosensors for Biological and Environmental Analysis. A. Deep and S. Kumar, Elsevier 79–94.

Unni, M., A. M. Uhl, S. Savliwala, B. H. Savitzky, R. Dhavalikar, N. Garraud et al. (2017). Thermal decomposition synthesis of iron oxide nanoparticles with diminished magnetic dead layer by controlled addition of oxygen. *ACS Nano* 11: 2284–2303.

Ursino, C., R. Castro-Muñoz, E. Drioli, L. Gzara, M. H. Albeirutty and A. Figoli. (2018). Progress of Nanocomposite membranes for water treatment. *Membranes* 8: 18.

Varghese, N., M. Hariharan, A. B. Cherian, J. Paul and Asmy Antony K. A. (2014). PVA—assisted synthesis and characterization of nano α-Alumina. *International Journal of Scientific and Research Publications* 4: 1–5.

Vashist, S. K. (2017). Point-of-care diagnostics: Recent advances and trends. *Biosensors* 7: 62.

Verma, S., A. Singh, A. Shukla, J. Kaswan, K. Arora, J. Ramirez-Vick et al. (2017). Anti-IL8/AuNPs-rGO/ ITO as an immunosensing platform for noninvasive electrochemical detection of oral cancer. *ACS Appl. Mater Interfaces* 9: 27462–27474.

Wang, C. and A. B. Surampudi. (2008). An acetone breath analyzer using cavity ringdown spectroscopy: an initial test with human subjects under various situations. *Measurement Science and Technology* 19: 105604.

Wang, J. (2008). Electrochemical glucose biosensors. *Chemical Reviews* 108: 814–825.

Wang, L., X. Yun, M. Stanacevic, P. I. Gouma, M. Pardo and G. Sberveglieri. (2009). An acetone nanosensor for non-invasive diabetes detection. *AIP Conference Proceedings* 1137: 206–208.

Wang, R., Z. Wang, X. Xiang, R. Zhang, X. Shi and X. Sun. (2018). MnO2 nanoarrays: An efficient catalyst electrode for nitrite electroreduction toward sensing and NH3 synthesis applications. *Chemical Communications* 54: 10340–10342.

Wen, L., R. Xu, C. Cui, W. Tang, Y. Mi, X. Lu et al. (2018). Template-guided programmable janus heteronanostructure arrays for efficient plasmonic photocatalysis. *Nano Letters* 18: 4914–4921.

Wen, L., R. Xu, Y. Mi and Y. Lei. (2017). Multiple nanostructures based on anodized aluminium oxide templates. *Nature Nanotechnology* 12: 244–250.

Wetterskog, E., M. Agthe, A. Mayence, J. Grins, D. Wang, S. Rana et al. (2014). Precise control over shape and size of iron oxide nanocrystals suitable for assembly into ordered particle arrays. *Science and Technology of Advanced Materials* 15: 055010.

Williams, R. M., C. Lee, T. V. Galassi, J. D. Harvey, R. Leicher, M. Sirenko et al. (2018). Noninvasive ovarian cancer biomarker detection via an optical nanosensor implant. *Science Advances* 4: eaaq1090-eaaq1090.

Wilson, M. L. (2012). Malaria rapid diagnostic tests. *Clinical Infectious Diseases* 54: 1637–1641.

Witkowska Nery, E., M. Kundys, P. S. Jeleń and M. Jönsson-Niedziółka. (2016). Electrochemical glucose sensing: Is there still room for improvement? *Analytical Chemistry* 88: 11271–11282.

Wu, L. and X. Qu. (2015). Cancer biomarker detection: Recent achievements and challenges. *Chemical Society Reviews* 44: 2963–2997.

Xia, X., J. Tu, Y. Zhang, J. Chen, X. Wang, C. Gu et al. (2012). Porous hydroxide nanosheets on preformed nanowires by electrodeposition: Branched nanoarrays for electrochemical energy storage. *Chemistry of Materials* 24: 3793–3799.

Xiong, X. and M. Xia. (2012). Carbon nanotube-based ultra-sensitive breath acetone sensor for non-invasive diabetes diagnosis. University of Bridgeport: Bridgeport, CT, USA, 2012.

Yan, Q., B. Peng, G. Su, B. E. Cohan, T. C. Major and M. E. Meyerhoff. (2011). Measurement of tear glucose levels with amperometric glucose biosensor/capillary tube configuration. *Analytical Chemistry* 83: 8341–8346.

Yang, J., L.-C. Jiang, W.-D. Zhang and S. Gunasekaran. (2010). A highly sensitive non-enzymatic glucose sensor based on a simple two-step electrodeposition of cupric oxide (CuO) nanoparticles onto multi-walled carbon nanotube arrays. *Talanta* 82: 25–33.

Yang, M., F. Qu, Y. Lu, Y. He, G. Shen and R. Yu. (2006). Platinum nanowire nanoelectrode array for the fabrication of biosensors. *Biomaterials* 27: 5944–5950.

Younis, A., D. Chu and S. Li. (2013). Stochastic memristive nature in Co-doped CeO2 nanorod arrays. *Applied Physics Letters* 103: 253504.

Yuen, C. and Q. Liu. (2012). Magnetic field enriched surface enhanced resonance Raman spectroscopy for early malaria diagnosis. *Journal of Biomedical Optics* 17: 017005.

Zang, W., Y. Nie, D. Zhu, P. Deng, L. Xing and X. Xue. (2014). Core–Shell In2O3/ZnO nanoarray nanogenerator as a self-powered active gas sensor with high H2S sensitivity and selectivity at room temperature. *The Journal of Physical Chemistry* C 118: 9209–9216.

Zaporotskova, I. V., N. P. Boroznina, Y. N. Parkhomenko and L. V. Kozhitov. (2016). Carbon nanotubes: Sensor properties. *A review. Modern Electronic Materials* 2: 95–105.

Zhang, B. and P.-X. Gao. (2019). Metal oxide nanoarrays for chemical sensing: A review of fabrication methods, sensing modes, and their inter-correlations. Frontiers in Materials 6.

Zhang, J. and T. Goto. (2015). Fabrication of Al_2O_3-Cu Nanocomposites Using Rotary Chemical Vapor Deposition and Spark Plasma Sintering. *Journal of Nanomaterials* 2015: 790361.

Zhang, J., W. Hodge, C. Hutnick and X. Wang. (2011). Noninvasive diagnostic devices for diabetes through measuring tear glucose. *Journal of Diabetes Science and Technology* 5: 166–172.

Zhang, W., Y. Du and M. L. Wang. (2015). Noninvasive glucose monitoring using saliva nano-biosensor. *Sensing and Bio-Sensing Research* 4: 23–29.

Zhao, Q., R. Duan, J. Yuan, Y. Quan, H. Yang and M. Xi. (2014). A reusable localized surface plasmon resonance biosensor for quantitative detection of serum squamous cell carcinoma antigen in cervical cancer patients based on silver nanoparticles array. *International Journal of Nanomedicine* 9: 1097–1104.

Zhao, Q., W. W. Yu, Y. Sun, R. Cong, Q. Xiang, N. Qin et al. (2015). WO3 nanoparticles based gas sensor for acetone detection with high sensitivity and fast response. *Sensor Letters* 13: 895–899.

Zhou, Z., X. Zhu, D. Wu, Q. Chen, D. Huang, C. Sun et al. (2015). Anisotropic shaped iron oxide nanostructures: Controlled synthesis and proton relaxation shortening effects. *Chemistry of Materials* 27: 3505–3515.

Zhu, Y.-P., Y.-P. Liu, T.-Z. Ren and Z.-Y. Yuan. (2015). Self-supported cobalt phosphide mesoporous nanorod arrays: A flexible and bifunctional electrode for highly active electrocatalytic water reduction and oxidation. *Advanced Functional Materials* 25: 7337–7347.

Zilberstein, G., R. Zilberstein, S. Zilberstein, U. Maor, E. Baskin, S. Zhang et al. (2017). A miniaturized sensor for detection of formaldehyde fumes. *ELECTROPHORESIS* 38: 2168–2174.

Zocher, K., K. Fritz-Wolf, S. Kehr, M. Fischer, S. Rahlfs and K. Becker. (2012). Biochemical and structural characterization of Plasmodium falciparum glutamate dehydrogenase 2. *Molecular and Biochemical Parasitology* 183: 52–62.

CHAPTER 17

Smart Nanosensors for Healthcare Monitoring and Disease Detection using AIoT Framework

Kunwar Shahbaaz Singh Sahi[1,*] and *Suresh Kaushik*[2]

1. Introduction

Artificial Intelligence (AI) is the intelligence of machines and can be defined as the study of an intelligent agent or device that can perceive and understand its surroundings and accordingly take appropriate action to maximize its chances of achieving its objectives (Wooldridge and Jennings 1995). AI involves the situation wherein machines can simulate human minds in learning and analysis, and can work in problem solving. Rapid advancement in AI is reforming almost every industry in the world and IoT has become dominant in recent years. The coupling of AI and Internet of Things (IoT) is making a new branch of emerging technology known as Artificial Intelligence of Things (AIoT). AIoT is the new combination of AI technology with IoT infrastructure to achieve more efficient IoT operations, improve human-machine interactions and enhance data management and analytics.

Artificial Intelligence has been at the forefront of medical prediction, diagnosis, and prognose (Liu et al. 2020, Abramoff et al. 2018, Das et al. 2018, Bera et al. 2019, Battineni et al. 2020) using advanced Machine Learning (ML) tools such as Deep Learning (DL) and Computer Vision (CV) since the past decade with exponential and rapid growth in the medical domain. (He et al. 2019, Wang et al. 2019). With increased demand in keeping digital records and processing, imaging data analysis, genome sequence data, and nano-bio sensors are used for physiological monitoring of parameters (Topol 2019). AI has made it possible for personalized and unique

[1] Department of Biomedical Engineering, School of Bioengineering and Biosciences, Lovely Professional University, Phagwara, India, 144001.
[2] Department of Chemistry, Indian Agricultural Research Institute, New Delhi-110012.
* Corresponding author: kunwarsahi@gmail.com

treatment given the custom datasets acquired from each patient in comparison to a general dataset for the training set. For example, Nano-robots has emerged as a promising drug delivery system and for monitoring *in vivo* healthcare parameters in depth (Hamet et al. 2017). New technologies have paved a way for futuristic healthcare monitoring and treatment in terms of advanced drug delivery, identifying viruses at the subatomic level (Mousavi et al. 2018). Cancer diagnosis using Raman spectroscopy using various techniques such as optical, magnetic, electrical and mechanical nanosensors (Lu et al. 2010). AI has been inculcated in the field of nano-biomedicine since the mid-2000s as has been seen as an emerging conjugation of nanoparticles embedded with AIoT chips. This chapter discusses the applicability of AIoT based-nanosensors in various sectors of medicine including its types for best output based on the problem statements and use cases for a majority of clinical disorders.

2. Artificial Intelligence of Things

Artificial Intelligence is now covering a wide range of healthcare applications. It has been used for signal and image processing, and for prediction of function changes (Tran et al. 2019). AI involves a system that consists of both software and hardware. From a software perspective, AI is concerned with algorithms. An artificial neural network (ANN) is a conceptual framework for executing AI algorithms (Hopfield 1982). It mimics of the human brain with interconnected network of neurons. The neural network (NN) can generate outputs as its responses to environmental stimuli, just as the brain reacts to different environmental changes. From the hardware perspective, AI is mainly concerned with the implementation of neural network algorithms on a physical computation platform (Rong et al. 2020). Due to the rapid development of AI software and hardware technologies, AI has been applied in various fields such as the Internet of Things (IoT) (Chiang and Zhang 2016), machine vision (Guo et al. 2016), autonomous driving (Yang et al. 2018), natural language processing (Alshahrani and Kapetanios 2016, Kim 2010) and robotics (Schaal 1999).

IoT devices, generally, act as passive sensors which are deployed in numbers that we can get data to do analysis or processing. But, with the introduction of AI chips the IoT devices can be more active rather than passive. Hence, AIoT enabled devices will be able to be proactive rather than reactive using AI. This means that AI enabled IoT systems will be more robust, more secure and even more scalable. AI can be used to transform IoT data into useful information for improved decision-making processes. AIoT refers to AI and the IoT working together. By pairing these and having them together we can get unbelievable insight into our machines. It will not just collect data; it will analyze the patterns and give us the information we need to make the best decisions. AIoT is the new combination of AI technology with IoT infrastructure to achieve more efficient IoT operations, improve human-machine interactions and enhance data management and analytics.

AIoT is transformational and mutually beneficial for both types of technology as AI adds value to IoT through machine learning capabilities and IoT adds value to AI through connectivity, signaling and data exchange. With AIoT, AI is embedded into infrastructure components such as programs, chipsets and Edges computing, all interconnected with IoT networks. Application Programming Interfaces (APIs) are then used to extend interoperability between components at the nanosensor devices level, software level and platform level. All these units will then focus mainly on optimizing systems and network operations as well as extracting value from data.

As the concept of AIoT is still relatively new, many possibilities exist to improve industry, particular healthcare sector, and will continue to arise with its growth. AIoT technology could provide viable solutions to solve existing operational problems and combining ML and IoT networks and systems applied to automate tasks in connected workplace to provide real-time data which is the key value of all AIoT use cases and solutions. Hence, AIoT solutions could be integrated with human resources-related platform to create an AI-Decision. Recently researchers in the biomedical areas have been actively trying to apply AI to assist improve analysis and treatment outcomes for enhancing the efficacy of the overall healthcare sector (Yu et al. 2018a, Mamoshin 2016, Peng 2010).

3. Nanosensor Fundamentals for Applied Medicine

Humans, whether in the industrial, modern, or rural sector require fast pace medical treatment which is accurate, precise, and available at affordable costs. A general norm, minimally invasive treatment is much preferred over invasive techniques for therapeutic and diagnostic purposes. There is an increased demand for better and advanced tools and technologies such as nanosensors for real-time healthcare monitoring, analysis, and smart diagnosis. The ongoing approach and blueprint define novel nano-bio sensors embedded with state-of-the-art intelligent systems for faster processing and data analysis.

The term 'Nanosensor' denotes a nano-meter dimensional electric (voltage, charge, current) or optical (refractive, radiation, fluorescence) transducer that interacts with a bio-analyte and produces a significantly measurable signal as illustrated in Figure 1 (Verma and Bhardwaj 2015, Goode et al. 2015, Saylan and Denizli 2020, Singh and Yadava 2020). Present-day nanosensors offer a low detection rate, faster response time, and high sensitivity easily integrated with microelectronics such as microprocessors, micro IoT chips, and electromechanical systems to inculcate smart lab-on-chip systems (Waggoner and Craighead 2007) and implantable *in vivo* biosensors (Ruckh and Clark 2014). This technology offers mass reproducibility, hence low-cost production and is wirelessly adjustable, giving it more novelty with the integration of AIoT systems for wireless monitoring and cloud-based storage and

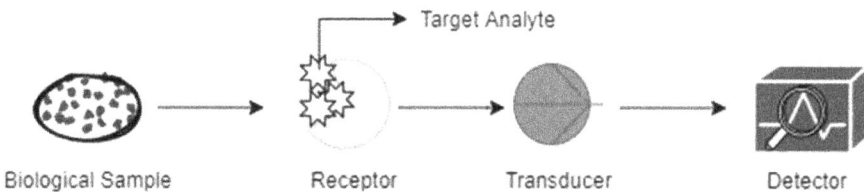

Modules of Nanosensor

Fig. 1. Illustration of Action Mechanism of Nano-biosensors.

analysis through ML algorithms from any remote location, therefore, amounting to telemedicine implementation as well.

Nanosensors for biological application have evolved and can be segregated into various types described below. These sensors typically use either of two conceptual approaches; electrochemical or mechanical. Apart from these two, magnetic and calorimetric techniques are also emerging to pave a way for advanced detection, monitoring, and therapeutic purposes.

3.1 Electrochemical Nanosensors

Electrochemical-based Nanosensors pave a way for faster processing and signal acquisition with high sensitivity for label-free screening and sensing. In this, amperometric and voltammetric methods are used to detect and measure the incoming signal from a catalytic enzyme or nanoparticle composite film is used which enables a redox reaction. More information can be gathered in a study by Ronkainen and colleagues on electrochemical nanosensors for medicinal use (Ronkainen et al. 2010, Santos et al. 2002). This specific feature of nanoparticles' adaptability and efficient working has been seen in cancer research (Perfe´zou et al. 2012) and DNA analysis (De la Escosura-Mun˜iz et al. 2009). Ion Selective field-effect transistors (ISFETs) are one of the most important architectures of electrochemical or potentiometric biosensors. This provides immense potential for label-free and lab-on-chip analytics for precise healthcare diagnosis. In a review by Kaisiti, the study described various ISFET sensors and their functional application in biomedical research labs and hospitals or clinics (Kaisiti et al. 2017, Grieshaber et al. 2008, Dzyadevych at el. 2006, Tores et al. 2006).

Electrochemical nanosensors include screen-printed electrodes and semiconductors. They can also monitor minute changes in dielectric properties, dimension, charge distribution, and shape. Further, they can be separated into three groups: amperometric, potentiometric, and impedimetric transducers which can be used to target various pathogens, microbes, and enzymes/proteins (The´venot et al. 2001, Russo et al. 2018, Yu et al. 2018b).

3.2 Nano-Mechanical Sensors

These types of nanosensors are based upon electromechanical or electroacoustic designs (Huber et al. 2015, Yadava et al. 2012, Devkota et al. 2017). These sensors operate by converting the biological signal into mechanical parameters such as surface stress, strain or elastic stiffness of analyte characteristics to mechanical measurands such as frequency, displacement, mass, velocity.

3.3 Other Nanosensors

Optical nanosensors compute the change in the reflective index of the transducer by a complex produced by target and recognition elements. They are further divided into two categories: direct and indirect nano-optical sensors. Piezoelectric sensors measure viscoelasticity and change in mass through recording frequency and altering

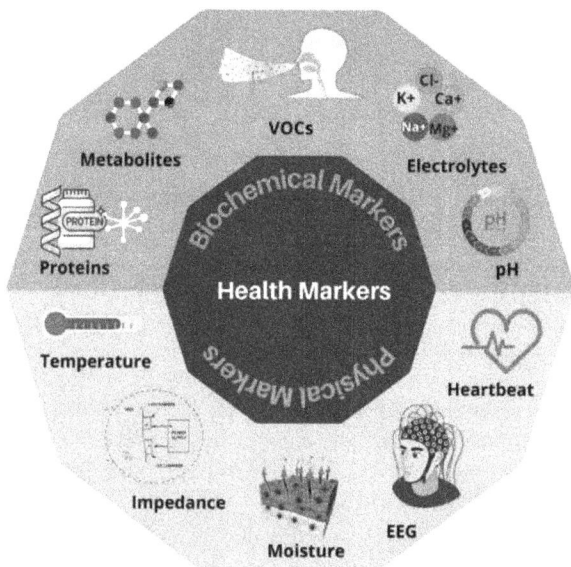

Fig. 2. Illustrates various segments in which nanosensors have application in diagnosing and prognosing through various features and techniques.

and quartz crystal resonant (Saylan et al. 2019, Zhu et al. 2018). Being highly sensitive, it requires isolation that reduces any external environmental hindrance. These nanosensors have been used extensively to identify and monitor targets. (Bakhshpour et al. 2019, Atay et al. 2016). Thermal nanosensors rely on physical heat released or absorbed by a biochemical reaction. Transducers here used are thermistors which are highly sensitive to change in heat. New advancements are bringing in more feasible Nanosensors using enzyme catalysis, flow injection, calorimetry, and immobilization on specific matrices. Magnetic nanosensors incorporate magnetic beads with outer coating labeled with a ligand. New avenues of research include hyperthermia treatment, drug delivery systems, magnetic actuation, and bioassays formed on fluorescent detectors for use in genetics and biotechnology.

3.4 Sensing Methods for Nano-Cantilevers

Nano-cantilevers offer unapparelled high sensitivity and attainable detection levels amounting to label-free detection of analytes without the requirement of complex sample collection and amplification. These can be operated in both static and dynamic modes.

3.4.1 Static Mode

In this method, sensing the cantilever surface is operationalized for analyte binding. Once the analyte is captured, a differential surface stress on thickness is generated which results in flexural bending. It's most abundantly described by the Stoney formula for the curvature of the beam (Stoney et al. 1909). The formula is described below in Eq. 1 (Singh and Yadava et al. 2018) which represents cantilever tip deflection Z_L:

$$Z_L = \frac{3(1-v)L^2}{Eh^2}\Delta\sigma \qquad (1)$$

where $\Delta\sigma$ is differential surface stress, E, h, L and represent young's modulus, thickness, length, and Poisson's Ratio respectively (Sang et al. 2014, Moulin et al. 2000). A noticeable advantage of this method is that it can be utilized in a liquid phase without the effects of a hydrodynamic environment.

3.4.2 Dynamic Mode

This methodology resembles the one mentioned above except for the output of the sensor which is analyzed by the change in cantilever resonance frequency. The properties of resonance frequency depend on intrinsic, drag, and driving conditions. However, the surface functionalization and operationalization differ in fluid, causing a shift in resonance frequency due to surface stress, mass density, and flexural rigidity, and damping losses (Eom et al. 2011).

3.5 Applications of Nanosensors in Pathogens Detection

One of the wide range applications of nanosensors is to detect pathogens and enzymes in the human body by binding to them and reflecting the measurable output signal through a transducer. For example, there are several studies to detect viral infections using nanosensor as shown in Table 1.

Table 1. Different studies of viral infections detected by various nanosensors.

S. No.	Virus	Reflection	References
1	HIV	Applied electroluminescence sensor for marking HIV-1 gene	Babamiri et al. 2018
2	Hepatitis B	Proposed a study which concluded with hybridization of DNA by electrochemical impedance spectroscopy	Hassen et al. 2008
3	Ebola	Developed a smartphone based digital Nanosensor probe for detection and continuous monitoring	Natesan et al. 2019
4	Zika	Developed graphene based nanosensor for detection of Zika Virus in human blood serum through antigen binding	Afsahi et al. 2018
5	Influenza	Theorized a technique for immobilization of DNA employing Carbon Nanotubes for detection.	Tam et al. 2009

4. Role of AI in Early Detection and Disease Diagnosis

AI, in the past decade has seen a tremendous growth, especially for prediction and diagnosis, most popularly in biomedical imaging, drug discovery, remote patient monitoring through IoT. Other domains such as genetics and biotech have also gained substantial gain from AI assistance. Upcoming technologies have also paved a way for inclusion of AI in hardware components for continuous patient monitoring, prediction through Regression and Deep Learning and diagnosis through various other algorithms discussed further in the chapter.

4.1 AI in Radiology

Due to increased demand of radiologists, AI is needed extensively for faster processing, automation, digital data handling, storage, Big Data Analytics and, precise and accurate diagnosis with more efficacy (McDonald et al. 2015, Boland et al. 2009). The first innovation in 1960s (Ledley et al. 1959, Lodwick et al. 1963) using computational and statistical models laid a foundation for the emergence of applied AI in the radiodiagnosis field (Aminder 2005, Haug 1993). There are mainly two approaches for Biomedical Imaging classification and prediction according to Hosney et al. 2018. Machine Learning (ML) approach which consists of predefined algorithms and are extensively used to classify and diagnose patients as an aid to the clinicians. However, on the contrary, these algorithms are more relied on expert predefined opinion and hence aren't so discriminant by virtue. These frameworks also are incapable of adapting to variation among different imaging modalities such as Functional Magnetic Resonance Imaging (fMRI) conjugated with Positron Emission Tomography (PET) or any other combination of radio-diagnosis acquisition system. The Deep Learning (DL) approach uses advanced technology such as TensorFlow and Computer Vision's leading domain, Convolutional Neural Networks (CNN). These algorithms and modules work similarly and require less efforts since they are fast processing and require no pre-requisite requirement of pre-defined labels by the clinicians, unlike the one mentioned above. Figure 3 represents various segments of diagnostic applications in the field of radiology by ML and DL algorithms.

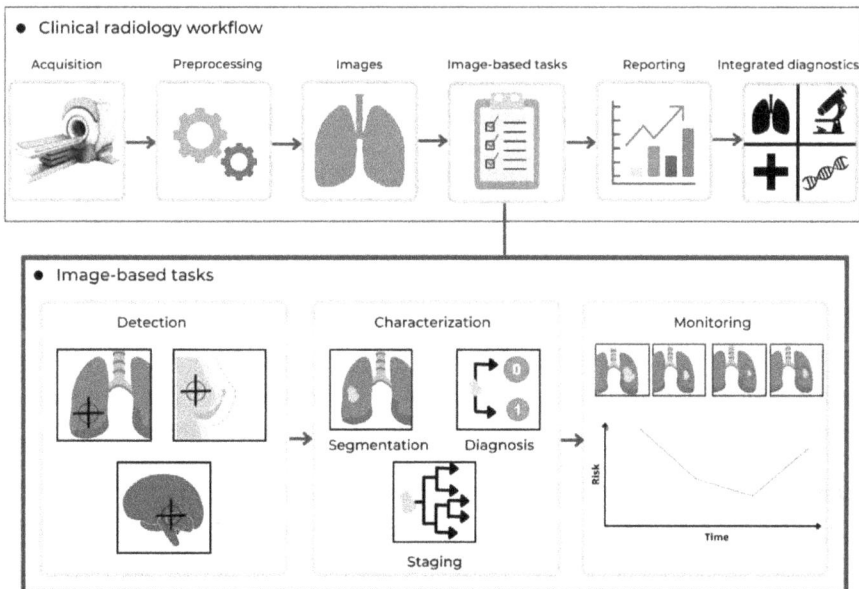

Fig. 3. Describes the workflow of AI-based diagnostic procedure with segments such as Detection, Characterization and Monitoring in Image processing tasks.

4.2 AI in Oncology

Oncology and AI closely work on three dimensions: abnormality detection, subsequent change monitoring and lastly, characterization as shown in Figure 3. Abnormality changes and lesion detection is efficiently done by trained DL algorithms such as CNNs or Recurrent Neural Networks (RNNs). However, for efficient detection, a validated dataset is hence required to meet the exact medical criteria. Due to a significant number of types of cancers, it is difficult to achieve high accuracy and a viable dataset for training and testing purposes. Nonetheless, with more advancements in the field and inclusion of AI since the past decade, more and more datasets are now available for most popular types of cancers with high precision and accuracy to aid radio-oncologists. Monitoring of change from benign to malignant tumors and subsequent prediction of metastasis involves high-end eCAD (electronic Computer Aided Detection) embedded with AI models working on functional imaging techniques such as fMRI, PET have proven to be a leading field in continuous cancer monitoring and future prediction generation from the input data of biomedical images.

Classification on the other hand works with parameters such as size, diameter, sphericity, internal texture and margins to classify the type of tumors. Classification is easily done through classification models incorporated with CNNs such as Logistic Regression, Naïve Bayes, Support Vector Machine, Random Forest etc.

4.3 AI in other Fields

It is now known that the application of AI is just not limited to Radio-based diagnosis. Due to its vast plethora of various algorithms, ML/DL and CV can now be applied to a range of medical fields apart from Radiology. In Dermatology, CV can be used to diagnose skin infections, lesions, size, shape, texture (Esteva et al. 2017). New AI models have also emerged to be used in pathology by using Predictive and Diagnostic models using ML algorithms such as regression, classification, etc. Moreover, for diagnosing a large number of patients, clustering methods have also been developed to cluster the datapoints, i.e., patients into small clusters that resemble similar properties. Clustering methods include k-means clustering, Hierarchal Clustering (Albarqouni et al. 2016, Djuric et al. 2017, Janowczyk and Madabhushi 2016, Bejnordi et al. 2017).

5. Health Monitoring Framework using Nano-Iot and Applied AI

Since now, we have discussed the usage of nanotechnology and AI in medicine for a wide area of applicability. This section describes the application of AIoT and nanosensors for health monitoring purposes in hospitals, interventional telemedicine and clinics offering Out Patient Services (OPD) for wireless monitoring of essential parametric evaluation and analysis by AI.

5.1 Different Monitoring Techniques using Nanosensors

Majorly speaking, there are four main techniques used for monitoring of continuous and functional health parameters (Javaid et al. 2021). Optical Localized Surface

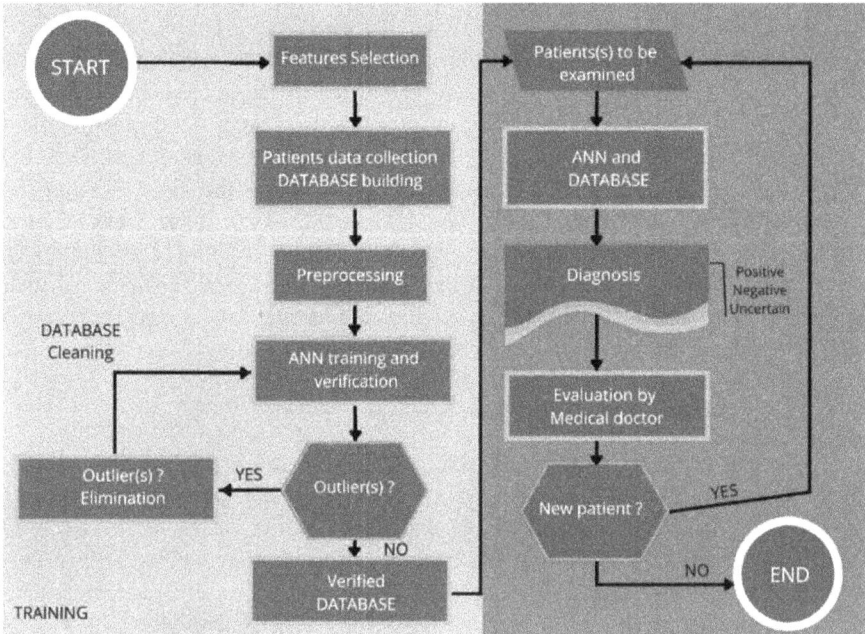

Fig. 4. Explains the Deep Learning Framework for medical diagnosis. The left side illustrates training and the right segment illustrates diagnosis through a network of ANNs with Big data Analytics.

Plasmon Resonance (LSPR) technique has been used to detect biomarkers and subsequent biomolecules for advanced diagnosis and prognoses with remarkably high efficiency (Haes et al. 2005, Chen et al. 2009, Maruvada et al. 2005, Kitano 2002). LSPR is widely used as a conjugate with antibody to detect antigens in the blood which can be also used for tumor segmentation, single protein molecule analysis, tumor necrosis factor and simple to complex binding reactions in real time. This applied framework of LSPR was given by Huang and colleagues (Huang et al. 2008) in cancer research and genetics. Chen and colleagues, put forward a novel ultrahigh sensitive Nanosensor of LSPR type with sensitivity of attomoles of analyte per cm² of area of nanoparticle (Chen et al. 2009). Apart from this, Zhou developed a system of silver nanoparticles for detection of serum levels of p53, which is a nuclear tumor suppressor protein (Zhou et al. 2011); hence indicating the wide spectrum of applicability of LSPR based-nanosensors for biomolecular analysis. As discussed above, Microcantilevers are used for mechanical methods of acquisition and have low sampling volumes and can be affected by the viscous fluids. Colorimetric technique is used in accordance with other methods such as Optical Plasmon mentioned above to target analytes of molecular size including peptides, glycan, amines, aptamer, nucleic acid (Tang and Li 2017). Surface Enhanced Raman Spectroscopy (SERS) is used for sensing of neurotransmitters inside the brain and around neural cells. This technique is more experimental in nature to study physiology of neurons (Lussier et al. 2017).

5.2 IoT based-Nanosensor Network

In an innovation by Dorj and colleagues, they developed an intelligent healthcare monitoring system by using nanosensors of various types to measure heart rate, electroencephalograph signal and electromyograph signal and blood pressure nano monitoring system for combined analysis on the medical healthcare server WBAN (Wireless Body Area Network) with an end IoT device being the smart phone of the patient, thus aiding to telemedicine as well (Dorj et al. 2017). In a review by Yang and researchers, studied biomarkers, troponin, Cerebrospinal Fluid (CSF), antigens, etc., with the help of nanosensors and concluded with a critical remark being that the sensitivity of antigens needs improvement and multiple detection signals from a single integrated chip with automated fluid monitoring is required (Yang et al. 2020).

5.3 Clinical AI in Live Health Monitoring

Clinical AI helps in wireless monitoring and cloud-based analysis of healthcare. With recent advancements, outcomes such as better prognoses, improved patient outcomes, personalized and customized experience and reduced costs. AI can be applied in various diagnostic and predictory evaluation with integrated on Nano-IoT sensors for EEG, ECG and EMG recordings along with Radiological equipment mentioned in this chapter. Classifiers such as Naïve Bayes have been widely used to diagnose a range of disorders by monitoring a range of live parameters including SpO_2, Blood Glucose level, Beats per minute (BPM), biochemical aspects, physical parameters, electrophysiological measurements, etc. Moreover, CNNs can also be used for data analyzed by the output of electrophysiological recording equipment mentioned above with IoT frameworks for faster processing and accuracy.

6. Conclusion

This chapter aimed at introducing AIoT framework with nanosensor integration to better detect, predict, diagnose and prognose a patient with various disorders using high end information systems discussed. Apart from this, it was also found that AI has better promising outcomes due to its features of faster processing, data handling and storage and extent of usability among a spectrum of medical fields such as radiology, dermatology, neurology. It is also hence proven that one could acquire a tremendous dataset which can be personalized to a patient. Elaborating on this point, we can say that a reinforcement learning system or algorithm could be made possible so to establish a customized and highly personalized dataset getting trained by every single time a recording takes place. After a benchmark is gained, the same dataset could be used on the specific patient for evaluation and monitoring. A smart analytical feature or monitoring software accompanied by AIoT. Nanosensor hardwares can be produced for greater efficacy and ultrahigh sensitivity which has a far better magnitude when compared with non-AI based systems. This might involve a ML classifier, DL and CV to work in synchronization and produce the output through IoT mechanism to the clinician as well as patient through smart telemedicine. As the AIoT technology is in nascent stage and relatively new concept, many possibilities exist to improve healthcare sector, and will continue to arise with its growth.

References

Abràmoff, M. D., P. T. Lavin, M. Birch, N. Shah and J.C. Folk. (2018). Pivotal trial of an autonomous AI-based diagnostic system for detection of diabetic retinopathy in primary care offices. *NPJ Digital Medicine* 1(1): 1–8.

Afsahi, M., B. Lerner, J. M. Goldstein, J. Lee, X. Tang, D. A. Bagarozzi, Jr. et al. (2018). Novel graphene-based biosensor for early detection of Zika virus infection. *Biosens Bioelectron* 100: 85–88.

Albarqouni, S., C. Baur, F. Achilles, V. Belagiannis, S. Demirci and N. Navab. (2016). AggNet: Deep learning from crowds for mitosis detection in breast cancer histology images. *IEEE Trans Med Imag* 35: 1313–1321.

Alshahrani, S. and E. Kapetanios. (2016). Are deep learning approaches suitable for natural language processing? pp. 343–349. *In*: Métais, E., F. Meziane, M. Saraee, V. Sugumaran and S. Vadera (eds.). Natural Language Processing and Information Systems. Cham: Springer.

Ambinder, E. P. (2005). A history of the shift toward full computerization of medicine. *J Oncol Pract* 1: 54–56.

Atay, S., K. Piskin, F. Yılmaz, C. Cakır, H. Yavuz and A. Denizli. (2016). Quartz crystal microbalance based-biosensors for detecting highly metastatic breast cancer cells via their transferrin receptors. *Anal Methods* 8: 153–161.

Babamiri, B., A. Salimi and R. Hallaj. (2018). A molecularly imprinted electrochemiluminescence sensor for ultrasensitive HIV-1 gene detection using EuS nanocrystals as luminophore. *Biosens Bioelectron* 117: 332–339.

Bakhshpour, M., A. K. Piskin, H. Yavuz and A. Denizli. (2019). Quartz crystal microbalance biosensor for label-free MDA MB 231 cancer cell detection via notch-4 receptor. *Talanta* 204: 840–845.

Battineni, G., G. G. Sagaro, N. Chinatalapudi and F. Amenta. (2020). Applications of machine learning predictive models in the chronic disease diagnosis. *Journal of Personalized Medicine* 10(2): 21.

Bejnordi, B. E., J. Lin, B. Glass, M. Mullooly, G. L. Gierach, M. E. Sherman et al. (2017). Deep learning-based assessment of tumor-associated stroma for diagnosing breast cancer in histopathology images. pp. 929–932. In 2017 IEEE 14th International Symposium on Biomedical Imaging, ISBI 2017 Melbourne, Australia.

Bera, K., K. A. Schalper, D. L. Rimm, V. Velcheti and A. Madabhushi. (2019). Artificial intelligence in digital pathology-new tools for diagnosis and precision oncology. *Nature Reviews Clinical Oncology* 16(11): 703–715.

Bolan, G. W. L., A. S. Guimaraes and P. R. Mueller. (2009). The radiologist's conundrum: Benefits and costs of increasing CT capacity and utilization. *Eur Radiol* 19: 9–12.

Chen, S., M. Svedendahl, M. Käll, L. Gunnarsson and A. Dmitriev. (2009). Ultrahigh sensitivity made simple: Nanoplasmonic label-free biosensing with an extremely low limit-of-detection for bacterial and cancer diagnostics. *Nanotechnology* 20(43): 434015.

Chiang, M. and T. Zhang. (2016). Fog and IoT: An overview of research opportunities. *IEEE Internet Things J* 3(6): 854–864.

Das, S., S. Biswas, A. Paul and A. Dey. (2018). AI doctor: An intelligent approach for medical diagnosis. In Industry Interactive Innovations in Science, Engineering and Technology (pp. 173–183). Springer, Singapore.

De la Escosura-Muniz, A., A. Ambrosi, M. Maltez, B. P. Lopez, S. Marın and A.I. Merkoc. (2009). Nanomaterial based electrochemical transducing platforms for biomedical applications. pp. 41–44. *In*: Dossel, O. and W. C. Schlegel (eds.). IFMBE Proceedings, Micro- and Nanosystems in Medicine, Active Implants, Biosensors, vol. 25/VIII, Springer.

Devkota, J., R. P. Ohodnicki and W. D. Greve. (2017). SAW sensors for chemical vapors and gases. *Sensors* 17(4): 801.

Djuric, U., G. Zadeh, K. Aldape and P. Diamandis. (2017). Precision histology: How deep learning is poised to revitalize histomorphology for personalized cancer care. *Precision Oncol* 1: 22.

Dorj, U. O., M. Lee, J. Choi, Y. K. Lee and G. Jeong. (2017). The intelligent healthcare data management system using nanosensors. *Journal of Sensors* vol. 2017, Article ID 7483075, 9 pages.

Dzyadevych, S. V., A. P. Soldatkin, A. V. El'skaya, C. Martelet and N. J. Renault. (2006). Enzyme biosensors based on ion-selective field-effect transistors. *Anal Chim Acta* 568: 248–258.

Eom, K., H. S. Park, D. S. Yoon and T. Kwon. (2011). Nanomechanical resonators and their applications in biological/chemical detection: Nanomechanics principles. *Phys Rep* 503: 115–163.

Esteva, A., B. Kuprel, R. A. Novoa, J. Ko, S. M. Swetter, H. M. Blau et al. (2017). Dermatologist-level classification of skin cancer with deep neural networks. *Nature* 542: 115–118.

Goode, J. A., J. V. H. Rushworth and P. A. Millner. (2015). Biosensor regeneration: A review of common techniques and outcomes. *Langmuir* 31: 6267–6276.

Grieshaber, D., R. MacKenzie, J. Voŕoˇs and E. Reimhult. (2008). Electrochemical biosensors-sensor principles and architectures. *Sensors* 8: 1400–1458.

Guo, Y., Y. Liu, A. Oerlemans, S. Lao, S. Wu and M. S. Lew. (2016). Deep learning for visual understanding: A review. *Neurocomputing* 187: 27–48.

Haes, J., L. Chang, W. L. Klein and R. P. Van Duyne. (2005). Detection of a biomarker for Alzheimer's disease from synthetic and clinical samples using a nanoscale optical biosensor. *Journal of the American Chemical Society* 127: 2264–2271.

Hamet, P. and J. Tremblay. (2017). Artificial intelligence in medicine. *Metabolism* 69: S36–S40.

Hassen, W. M., C. Chaix, A. Abdelghani, F. Bessueille, D. Leonard and N. Jaffrezic-Renault. (2008). An impedimetric DNA sensor based on functionalized magnetic nanoparticles for HIV and HBV detection. *Sens Actuators B: Chem* 134: 2755–2760.

Haug, P. J. (1993). Uses of diagnostic expert systems in clinical care. *Proc Annu Symp Comput Appl Med Care* 379–383.

He, J., S. L. Baxter, J. Xu, J. Xu, X. Zhou and K. Zhang. (2019). The practical implementation of artificial intelligence technologies in medicine. *Nature Medicine* 25(1): 30–36.

Hopfield, J. J. (1982). Neural networks and physical systems with emergent collective computational abilities. *Proc Natl Acad Sci USA* 79(8): 2554–2558.

Hosny, A., C. Parmar, J. Quackenbush, L. H. Schwartz and J. W. L. Aerts Hugo. 2018. Artificial intelligence in radiology. *Nature Reviews Cancer* 18(8): 500–510.

Huang, T., P. D. Nallathamby and X. H. N. Xu. (2008). Photostable single-molecule nanoparticle optical biosensors for real-time sensing of single cytokine molecules and their binding reactions. *Journal of the American Chemical Society* 130: 17095–17105.

Huber, F., H. P. Lang, J. Zhang, D. Rimoldi and C. Gerber. (2015). Nanosensors for cancer detection, *Swiss Med Wkly* 145 w14092 8 pp.

Janowczyk, A. and A. Madabhushi. (2016). Deep learning for digital pathology image analysis: A comprehensive tutorial with selected use cases. *J Pathol Inform* 7: 29.

Javaid, M., A. Haleem, R. P. Singh, S. Rab and R. Suman. (2021). Exploring the potential of nanosensors: A brief overview. *Sensors International* 2: 100130.

Kaisti, M. (2017). Detection principles of biological and chemical FET sensors. *Biosens Bioelectron* 98: 437–448.

Kim, T. H. (2010). Emerging approach of natural language processing in opinion mining: A review. pp. 121–128. *In*: Tomar, G. S., W. I. Grosky, T. H. Kim, S. Mohammed and S. K. Saha (eds.). Ubiquitous Computing and Multimedia Applications. Berlin: Springer.

Kitano, H. (2002). Systems biology: A brief overview. *Science* 295: 1662–1664.

Ledley, R. S. and L. B. Lusted. (1959). Reasoning foundations of medical diagnosis; symbolic logic, probability, and value theory aid our understanding of how physicians reason. *Science* 130: 9–21.

Liu, Y., A. Jain, C. Eng, D. H. Way, K. Lee, P. Bui et al. (2020). A deep learning system for differential diagnosis of skin diseases. *Nature Medicine* 26(6): 900–908.

Lodwick, G. S., T. E. Keats and J. P. Dorst. (1963). The coding of Roentgen images for computer analysis as applied to lung cancer. *Radiology* 81: 185–200.

Lu, W., A. K. Singh, S. A. Khan, D. Senapati, H. Yu and P. C. Ray. (2010). Gold nano-popcorn-based targeted diagnosis, nanotherapy treatment, and in situ monitoring of photothermal therapy response of prostate cancer cells using surface-enhanced Raman spectroscopy. *Journal of the American Chemical Society* 132(51): 18103–18114.

Lussier, F., T. Brulé, M. J. Bourque, C. Ducrot, L. É. Trudeau and J. F. Masson. 2017. Dynamic SERS nanosensor for neurotransmitter sensing near neurons. *Faraday Discussions* 205: 387–407.

Moulin, A. M., S. J. O'Shea and M. E. Welland. (2000). Microcantilever based biosensors. *Ultramicroscopy* 82: 23–31.

Mousavi, S. M., S. A. Hashemi, M. Zarei, A. M. Amani and A. Babapoor. (2018). Nanosensors for chemical and biological and medical applications. *Med Chem (Los Angeles)* 8: 205–217.

Natesan, M., S. W. Wu, C. I. Chen, S. M. R. Jensen, N. Karlovac, B. K. Dyas et al. (2019). A smartphone-based rapid telemonitoring system for Ebola and Marburg disease surveillance. *ACS Sens* 4: 61–68.

Mamoshina, P., A. Vieira, E. Putin and A. Zhavoronkov. (2016). Applications of deep learning in biomedicine. *Mol Pharm* 13(5): 1445–1454.

Maruvada, P., W. Wang, P. D. Wagner and S. Srivastava. (2005). Biomarkers in molecular medicine: Cancer detection and diagnosis. *BioTechniques* supplement 9–15.

McDonald, R. J., K. M. Schwartz, L. J. Echel, F. E. Diehn, C. H. Hunt, B. J. Bartholmari et al. (2015). The effects of changes in utilization and technological advancements of cross-sectional imaging on radiologist workload. *Acad Radiol* 22: 1191–1198.

Mousavi, S. M., S. A. Hashemi, M. Zarei, A. M. Amani and A. Babapoor. (2018). Nanosensors for chemical and biological and medical applications. *Med Chem (Los Angeles)* 8: 205–217.

Peng, Y., Y. Zhang and L. Wang. 2010. Artificial intelligence in biomedical engineering and informatics: An introduction and review. *Artif Intell Med* 48: 71–73.

Perfe´zou, M., A. Turner and A. Merkoc. (2012). Cancer detection using nanoparticle-based sensors. *Chem Soc Rev* 41: 2606–2622.

Rong, G., A. Mendez, E. B. Assi, B. Zhao and M. Sawan. (2020). Artificial intelligence in healthcare: Review and prediction case studies. *Engineering* 6: 291–301.

Ronkainen, N. J., H. B. Halsall and W. R. Heinemanb. (2010). Electrochemical biosensors. *Chem Soc Rev* 39: 1747–1763.

Ruckh, T. T. and H. A. Clark. (2014). Implantable nanosensors: toward continuous physiologic monitoring. *Anal Chem* 86(3): 1314–1323.

Russo, L., J. Leva Bueno, J. F. Bergua, M. Costantini, M. Giannetto, V. Puntes et al. (2018). Low-cost strategy for the development of a rapid electrochemical assay for bacteria detection based on Au/Ag nanoshells. *ACS Omega* 3: 18849–18856.

Sang, S., Y. Zhao, W. Zhang, P. Li, J. Hu and G. Li. (2014). Surface stress-based biosensors. *Biosens Bioelectron* 51: 124–135.

Santos, D. H., M. B. G. Garcia and A. C. Garcia. (2002). Metal-nanoparticles based electroanalysis *Electroanalysis* 14(18): 1225–1235.

Saylan, Y., S. Akgonullu, H. Yavuz, S. Unal and A. Denizli. (2019). Molecularly imprinted polymer based-sensors for medical applications. *Sensors* 19: 1279–1298.

Saylan, Y. and A. Denizli. (2020). Virus detection using nanosensors. pp. 501–511. *In:* Han, B., V. K. Tomar, T. A. Nguyen, A. Farmani and P. K. Singh (eds.). Nanosensors for Smart Cities. Elsevier.

Schaal, S. (1999). Is imitation learning the route to humanoid robots? *Trends Cogn Sci* 3(6): 233–242.

Singh, P. and R. D. S. Yadava. (2018). Effect of surface stress on resonance frequency of microcantilever sensors. *IEEE Sensors J* 18(18): 7529–7536.

Singh, P. and R. D. S. Yadava. (2020). Nanosensors for health care. pp. 433–450 *In:* Han, B., V. K. Tomar, T. A. Nguyen, A. Farmani and P. K. Singh (eds.). Nanosensors for Smart Cities. Elsevier.

Stoney, G. G. (1909). The tension of the metallic films deposited by electrolysis. *Proc R Soc Lon Ser A* 82: 172–177.

Tam, P. D., N. Van Hieu, N. D. Chien, A. T. Le and M. A. Tuan. (2009). DNA sensor development based on multi-wall carbon nanotubes for label-free influenza virus (type A) detection. *J Immunol Methods* 350: 118–124.

Tang, L. and J. Li. (2017). Plasmon-based colorimetric nanosensors for ultrasensitive molecular diagnostics. *ACS Sensors* 2(7): 857–875.

The´venot, D. R., K. Toth, R. A. Durst and G. S. Wilson. (2001). Electrochemical biosensors: Recommended definitions and classification. *Anal Lett* 34: 635–659.

Topol, E. J. (2019). High-performance medicine: The convergence of human and artificial intelligence. *Nature Medicine* 25(1): 44–56.

Torres, K. Y. C., Z. Dai, N. Rubinova, Y. Xiang, E. Pretsch, J. Wang et al. (2006). Potentiometric biosensing of proteins with ultrasensitive ionselective microelectrodes and nanoparticle labels. *J Am Chem Soc* 128: 13676–13677.

Tran, B. X., G. T. Vu, G. H. Ha, Q. H. Vuong, M. T. Ho, T. T. Vuong et al. (2019). Global evolution of research in artificial intelligence in health and medicine: a bibliometric study. *J Clin Med* 8(3): 360.

Verma, N. and A. Bhardwaj. (2015). Biosensor technology for pesticides-a review. *Appl Biochem Biotechnol* 175: 3093–3119.

Waggoner, P. S. and H. G. Craighead. (2007). Micro- and nanomechanical sensors for environmental, chemical, and biological detection. *Lab-on-a-Chip* 7: 1238–1255.

Wang, F. and A. Preininger. (2019). AI in health: State of the art, challenges, and future directions. *Yearbook of Medical Informatics* 28(01): 016–026.

Wooldridge, M. and N. R. Jennings. (1995). Intelligent agents: Theory and practice. *Knowl Eng Rev* 10(2): 115–152.

Yadava, R. D. S. (2012). Modeling, simulation, and information processing for development of a polymeric electronic nose system. pp. 411–502. *In:* Korotcenkov, G. (ed.). Chemical Sensors Simulation and Modeling, Momentum Press, New York, vol. 3.

Yang, D., K. Jiang, D. Zhao, C. Yu, Z. Cao, S. Xie et al. (2018). Intelligent and connected vehicles: current status and future perspectives. *Sci China Technol Sci* 61(10): 1446–1471.

Yang, J., P. Carey IV, F. Ren, B. C. Lobo, M. Gebhard, M. E. Leon, J. Lin and S. J. Pearton. (2020). Nanosensor networks for health-care applications. pp. 405–417. *In:* Han, B., V. K. Tomar, T. A. Nguyen, A. Farmani and P. K. Singh (eds.). Nanosensors for Smart Cities. Elsevier.

Yu, K. H., A. L. Beam and I. S. Kohane. (2018a). Artificial intelligence in healthcare. *Nat Biomed Eng* 2(10): 719–731.

Yu, X., F. Chen, R. Wang and Y. Li. (2018b). Whole-bacterium SELEX of DNA aptamers for rapid detection of *E. coli* O157:H7 using a QCM sensor. *J Biotechnol* 266: 39–49.

Zhu, J., H. Gan, J. Wu and H. Ju. (2018). Molecular machine powered surface programmatic chain reaction for highly sensitive electrochemical detection of protein. *Anal Chem* 90: 5503–5508.

Zhou, Y. M. W., H. Yang, Y. Ding and X. Luo. 2011. A label-free biosensor based on silver nanoparticles array for clinical detection of serum p53 in head and neck squamous cell carcinoma. *International Journal of Nanomedicine* 6: 381–386.

Index

For Product Safety Concerns and Information please contact our EU
representative GPSR@taylorandfrancis.com
Taylor & Francis Verlag GmbH, Kaufingerstraße 24, 80331 München, Germany